Graphs, Network
and Algorithm

Graphs, Networks, and Algorithms

M. N. S. SWAMY

Dean of Engineering and Computer Science

Concordia University
Montreal, Canada

K. THULASIRAMAN

Professor of Computer Science

Indian Institute of Technology
Madras, India

A WILEY INTERSCIENCE PUBLICATION

JOHN WILEY & SONS New York · Chichester · Brisbane · Toronto

Library of Congress Cataloging in Publication Data:

Swamy, M. N. S.
Graphs, networks, and algorithms.

Includes bibliographies and indexes.
1. Graph theory. 2. Electric networks.
3. Algorithms. I. Thulasiraman, K., joint author.
II. Title.

QA166.S88 511'.5 80-17782
ISBN 0-471-03503-3

Printed in the United States of America

10 9 8 7 6 5 4 3 2 1

Dedicated to Our Parents
and Teachers

अज्ञानतिमिरान्धस्य ज्ञानाञ्जनशलाकया ।
चक्षुरुन्मीलितं येन तस्मै श्रीगुरवे नमः ॥

from *Vishwasara Tantra*

Salutations to the Guru who with the collyrium stick of knowledge has opened the eyes of one blinded by the disease of ignorance.

Preface

During the last two decades, graph theory, a branch of combinatorial mathematics, has been gaining increasing popularity among workers concerned with various aspects of science and engineering . Having had its origin in the solution of puzzles and games, such as the Königsberg bridge problem and Hamilton's game, it has now become a powerful tool in the understanding and solution of problems arising in the study of many large complex systems. In fact, there are several systems whose study becomes much simpler with the aid of graph theory. This is not surprising, since binary relations among objects of a set can be conveniently represented by graphs, and descriptions of systems involve such relations among their different subsystems. In addition, graph theory has proved useful in the study of problems arising in certain other branches of mathematics such as group theory and matrix theory. Every time a new area of application of graph theory emerged, the need arose for the introduction and study of new concepts or for a further study of certain known concepts. This need, in turn, led to a flurry of research activities on various related concepts. This continuous interaction has significantly contributed to the rapid growth of this branch of mathematics.

Several books have been written, discussing different aspects of graph theory: analysis, design, enumeration, algorithms, and applications. In this volume we attempt to discuss in a unified manner the theory of graphs, its application in the study of electrical networks, and the theory underlying several graph algorithms. Before we get down to the details regarding the scope and contents of the book, we shall discuss briefly how graph theory has proved useful in the study of electrical networks, operations research, and computer science—the three major disciplines which are of interest to us and which have been interacting with graph theory to their mutual benefit and with much success.

It is well known that Euler, with his solution of the Königsberg bridge problem, laid the foundation for the theory of graphs. However, its first application to a problem in physical science did not arise till 1847, when Kirchhoff developed the theory of trees for its application in the study of electrical networks. In the study of these networks, Kirchhoff's laws play a fundamental role. These laws specify the relationships among the voltage variables as well as those among the current variables of a network. For a given electrical network, these relationships do not depend on the nature of the elements used; rather, they depend only on the way the various elements are connected or, in other words, on the graph of the network. In fact, the circuits and cutsets of the graph of a network completely specify the equations describing Kirchhoff's voltage and current laws. The question then arises whether every circuit and every cutset of a network are necessary to specify these equations. The answers to this and other related questions require a thorough study of the properties of circuits, cutsets, and trees of a graph. This explains the role of graph theory as an important analytical tool in the study of electrical networks. Many important discoveries in network theory are mainly graph-theoretic in nature, and the efforts leading to such discoveries have resulted in significant contributions to the theory of graphs.

Transport networks and communication nets can be conveniently represented by graphs, and therefore operations researchers concerned with problems such as flows, shortest paths, or invulnerable network design arising in the study of these networks have found the graph-theoretic approach useful. Their studies have contributed significantly toward the rapid development of graph theory in the last two decades. The theory of network flows developed by Ford and Fulkerson has thrown much light on several combinatorial questions and has yielded new proofs for many important theorems in graph theory. Kirchhoff's law of the conservation of flow (similar to Kirchhoff's current law for an electrical network) again plays a central role in the development of network flow theory. In recent years, there has been a great deal of interest in the design of communication nets having certain specified properties. This problem usually reduces to one of constructing an extremal graph (that is, a graph with a minimum or maximum number of edges) having a specified value for one of the topological parameters. Studies in this area have resulted in some important contributions to what are called the extremal problems in graph theory.

Computer science is the most recent addition to the growing list of application areas of graph theory. To a computer scientist, graph theory is a convenient language for expressing his concepts and many of the results therein have direct relevance to the problems with which he is concerned. Extensive use of graph theory has recently been made in the study of what is referred to as the code optimization problem in compiler construction. Many graph-theoretic concepts not known earlier have been developed in the course of this investigation. Besides its use as a tool in the study of his problems, graph theory has another important appeal for the computer scientist. Among the main concerns of computer science are the design of efficient algorithms and the study of their complexity, and graph theory (in general, combinatorial theory) offers a lot of scope for the design of such algorithms. For several problems, efficient graph-theoretical algorithms have been found recently. Computer scientists have also identified certain problems for which "efficient" algorithms are not likely to exist. This is indeed a significant contribution of computer science to graph theory.

The foregoing discussion highlights the role that the theory of graphs has come to play in engineering and science. This also explains its growing emphasis in university curricula. While a detailed knowledge of the several concepts and results of graph theory is not necessary before one can start studies in any one area of application, we believe that a thorough knowledge of the various concepts and a deep training in the techniques used in the study of these concepts will be very useful to conceive of new applications which are not obvious otherwise. In fact, this is also true of many other branches of mathematics such as complex variable theory, matrix theory, or transform theory which are essential in the study of systems engineering.

This book is addressed to students in mathematics, electrical engineering, and computer science. As its name implies, the book is organized into three parts, dealing with the theory of graphs, electrical networks, and graph algorithms.

In Part I (Chapters 1 to 10) we discuss the theory of graphs. here we aim to give a sound introduction to several basic concepts and results in graph theory. The topics discussed in this part include trees, Hamilton and Euler graphs, directed graphs, matrices of a graph, planarity, connectivity, matching, and coloring. We have also included in this part an introduction to matroid theory. Among the matroid topics presented here are Minty's self-dual axiom system which makes obvious the duality between circuits and cutsets of a graph, the arc coloring lemma, the greedy algorithm, and its intimate relationship with matroids. In recent years electrical network theorists have taken an interest in the theory of matroids since it provides much insight into some of the problems they are concerned with. While for a mathematician there is a lot of scope in matroid theory for generalizing graph-theoretical concepts, for a computer scientist there is scope for designing matroid algorithms.

Part II (Chapters 11 to 13) is concerned with a discussion of those aspects of electrical network theory whose development is essentially graph-theoretic. In Chapter 11 we discuss, among other things, the principal partition of a graph and its application in the mixed-variable method of network analysis and a graph-theoretic proof of the no-gain property of resistance networks. In Chapter 12 we discuss several results in the theory of resistance networks and a method for realizing circuit and cutset matrices. In the concluding chapter of this part we develop topological formulas for network functions. These formulas follow easily from the properties of matrices of a graph presented in Part I. We also discuss in Part II Tellegen's theorem which is essentially graph-theoretic in nature, and its application in the computation of network sensitivities. It is surprising that such an important theorem lay dormant for so many years without receiving much attention.

Part III, which deals with graph algorithms, is organized into two chapters: Chapter 14 on algorithms for the analysis of graphs and Chapter 15 on algorithms that concern optimization problems on graphs. Our main concern here is the theory underlying the design, proof of correctness, and analysis of several graph algorithms. Among other things we include algorithms for flow graph reducibility, dominators, shortest paths, matchings, optimum binary search trees, network flows, and optimum branchings. We also include Hopcroft and Karp's analysis of a bipartite matching algorithm and Edmonds and Karp's analysis of Ford-Fulkerson's labeling algorithm. We see in this part several deep contributions to graph theory made by computer scientists. A major omission from this book is a discussion of $\mathcal{NP}$-complete problems, however, this topic is beyond the scope of this book.

As regards prerequisites, a mathematically inclined graduate student will have little difficulty in following the discussions of Parts I and III. Our development in Part II assumes that the student has already had an exposure to a basic course in electrical network analysis.

Different courses may be organized based on the contents of the book, some possibilities are suggested below.

1. A course on "Graph Theory" for students in mathematics, electrical engineering, and computer science may be organized based on Chapters 1 to 10 and Section 15.7. For computer science students several sections in Chapter 6 may be omitted, and Sections 14.3 and 14.4 on depth-first search algorithms may be included.
2. A course on "Graphs and Electrical Networks" for graduate students in electrical engineering may be organized based on Chapters 1 to 7 and 11 to 13, omitting Sections 3.2 and 5.6 to 5.8.
3. A course on "Algorithmic Graph Theory" for students in mathematics, electrical engineering, and computer science may be organized based on Part III and relevant background material from Part I. For computer science students Sections 14.5 and 14.6 can be elaborated upon by including a discussion of algorithms for set manipulation problems which would help in the discussion of program graph reducibility and dominator algorithms.

Based on this book the authors have taught at Concordia University, Montreal, Canada, courses on "Graph Theory" for students of mathematics and electrical engineering, and graduate courses on "Graphs and Electrical Networks" for students of electrical engineering. The latter course has also been taught by the second author (K. Thulasiraman) at the Indian Institute of Technology, Madras, India.

The second author (K. Thulasiraman) has used portions of Part I in a course on "Combinatorics and Graph Theory" and Part III in a course on "Design and Analysis of Algorithms" for graduate students in computer science at the Indian Institute of Technology, Madras.

We express our sincere thanks to Dr. P. K. Rajan, North Dakota State University, Fargo and Mr. R. Jayakumar, Indian Institute of Technology,

Madras, who read major portions of the manuscript of the book, pointed out several omissions and errors, and offered many suggestions that aided us in our presentation. We also thank Professors V. G. K. Murti and C. R. Muthukrishnan of the Indian Institute of Technology, Madras, Professor V. Ramachandran and Dr. L. Roytman of Concordia University, Montreal, Dr. K. Sankara Rao, North Dakota State University, Fargo, Dr. H. Narayanman, Indian Institute of Technology, Bombay, and Messrs. A. Mohan and P. Narendran, former students of the Indian Institute of Technology, Madras, for reading different parts of the manuscript and offering helpful comments; Dr. S. A. Choudum, Madurai University, Madurai, India, for drawing our attention to certain simpler proofs in graph theory; Dr. V. V. Bapeswara Rao, Indian Institute of Technology, Madras, for permission to use parts of his unpublished work; and Professors V. Chvátal, McGill University, Montreal, G. Kishi, Tokyo Institute of Technology, Tokyo, L. Lovász, Hungary, R. E. Tarjan, Stanford University, Stanford, California, and K. R. Parthasarathy and his graduate students at the Indian Institute of Technology, Madras, for helpful comments and suggestions.

K. Thulasiraman recalls with pleasure his association with Professor V. G. K. Murti who initiated him into research. It is very rarely that one is blessed with such a teacher.

We thank Concordia University, Montreal, for its generous support. K. Thulasiraman wishes to thank the Indian Institute of Technology, Madras, and Concordia University for their support and encouragement, which made his coauthorship of the book possible.

We thank our wives, Leela Swamy and Santha Thulasiraman, and our children for their patience and understanding during the entire period of our efforts.

Finally we would like to thank Gloria Miller and Kamala Ramachandran for their excellent typing of the manuscript.

M. N. S. Swamy
K. Thulasiraman

Montreal, Canada
Madras, India
September 1980

Contents

PART I GRAPH THEORY

Graphs, Networks, and Algorithms

I

Graph Theory

Chapter 1

Basic Concepts

We begin our study with an introduction in this chapter to several basic concepts in the theory of graphs. A few results involving these concepts will be established. These results, while illustrating the concepts, will also serve to introduce the reader to certain techniques commonly used in proving theorems in graph theory.

1.1 SOME BASIC DEFINITIONS

A *graph* $G=(V, E)$ consists of two sets: a finite set V of elements called *vertices* and a finite set E of elements called *edges*. Each edge is identified with a pair of vertices. If the edges of a graph G are identified with ordered pairs of vertices, then G is called a *directed* or an *oriented* graph. Otherwise G is called an *undirected* or a *nonoriented* graph. Our discussions in the first four chapters of this book are concerned with undirected graphs.

We use the symbols $v_1, v_2, v_3, \ldots$ to represent the vertices and the symbols $e_1, e_2, e_3, \ldots$ to represent the edges of a graph. The vertices v_i and v_j associated with an edge e_l are called the *end vertices* of e_l. The edge e_l is then denoted as $e_l=(v_i, v_j)$. Note that while the elements of E are distinct, more than one edge in E may have the same pair of end vertices. All edges having the same pair of end vertices are called *parallel edges*. Further, the end vertices of an edge need not be distinct. If $e_l=(v_i, v_i)$, then the edge e_l is called a *self-loop* at vertex v_i. A graph is called a *simple graph* if it has no parallel edges or self-loops. A graph G is of *order n* if its vertex set has n elements.

A graph with no edges is called an *empty graph*. A graph with no vertices (and hence no edges) is called a *null graph*.

Pictorially a graph can be represented by a diagram in which a vertex is represented by a dot or a circle and an edge is represented by a line segment connecting the dots or the circles which represent the end vertices of the edge. For example, if

$$V=\{v_1, v_2, v_3, v_4, v_5, v_6\}$$

and

$$E=\{e_1, e_2, e_3, e_4, e_5\},$$

such that

$$e_1=(v_1, v_2),$$

$$e_2=(v_1, v_4),$$

$$e_3=(v_5, v_6),$$

$$e_4=(v_1, v_2),$$

$$e_5=(v_5, v_5),$$

then the graph $G=(V, E)$ is represented as in Fig. 1.1. In this graph e_1 and e_4 are parallel edges and e_5 is a self-loop.

An edge is said to be *incident on* its end vertices. Two vertices are *adjacent* if they are the end vertices of some edge. If two edges have a common end vertex, then these edges are said to be *adjacent*.

For example, in the graph of Fig. 1.1, edge e_1 is incident on vertices v_1 and v_2; v_1 and v_4 are two adjacent vertices, while e_1 and e_2 are two adjacent edges.

The number of edges incident on a vertex v_i is called the *degree* of the vertex, and it is denoted by $d(v_i)$. Sometimes the degree of a vertex is also

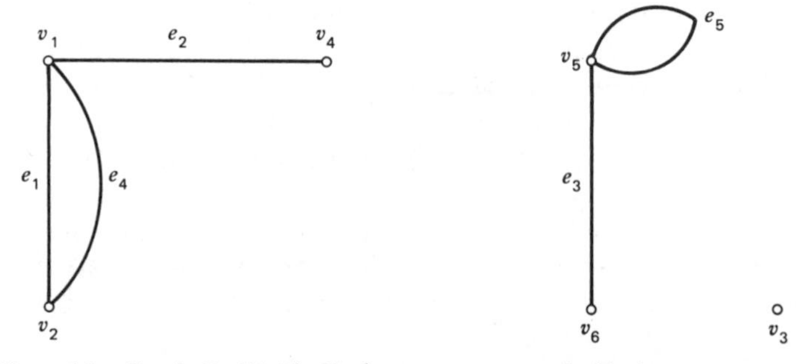

Figure 1.1. Graph $G=(V, E)$. $V=\{v_1, v_2, v_3, v_4, v_5, v_6\}$; $E=\{e_1, e_2, e_3, e_4, e_5\}$.

referred to as its *valency*. A vertex of degree 1 is called a *pendant vertex*. The only edge incident on a pendant vertex is called a *pendant edge*. A vertex of degree 0 is called an *isolated vertex*. By definition, a self-loop at a vertex v_i contributes 2 to the degree of v_i. $\delta(G)$ and $\Delta(G)$ denote, respectively, the minimum and maximum degrees in G.

In the graph G of Fig. 1.1,

$$d(v_1)=3,$$

$$d(v_2)=2,$$

$$d(v_3)=0,$$

$$d(v_4)=1,$$

$$d(v_5)=3,$$

$$d(v_6)=1.$$

Note that v_3 is an isolated vertex, v_4 and v_6 are pendant vertices, and e_2 is a pendant edge. For G it can be verified that the sum of the degrees of the vertices is equal to 10, whereas the number of edges is equal to 5. Thus the sum of the degrees of the vertices of G is equal to twice the number of edges of G and hence an even number. It may be further verified that in G the number of vertices of odd degree is also even. These interesting results are not peculiar to the graph of Fig. 1.1. In fact, they are true for all graphs as the following theorems show.

THEOREM 1.1. The sum of the degrees of the vertices of a graph G is equal to $2m$, where m is the number of edges of G.

Proof

Since each edge is incident on two vertices, it contributes 2 to the sum of the degrees of the graph G. Hence all the edges together contribute $2m$ to the sum of the degrees of G. ■

THEOREM 1.2. The number of vertices of odd degree in any graph is even.

Proof

Let the number of vertices in a graph G be equal to n. Let, without any loss of generality, the degrees of the first r vertices $v_1, v_2, \ldots, v_r$ be even and those of the remaining $(n-r)$ vertices be odd. Then

$$\sum_{i=1}^{n} d(v_i) = \sum_{i=1}^{r} d(v_i) + \sum_{i=r+1}^{n} d(v_i). \tag{1.1}$$

By Theorem 1.1, the sum on the left-hand side of (1.1) is even. The first sum on the right-hand side is also even, because each term in this sum is even. Hence the second sum on the right-hand side should be even. Since each term in this sum is odd, it is necessary that there be an even number of terms in this sum. In other words, $(n-r)$, the number of vertices of odd degree, should be even. ∎

1.2 SUBGRAPHS AND COMPLEMENTS

Consider a graph $G=(V, E)$. $G'=(V', E')$ is a *subgraph* of G if V' and E' are, respectively, subsets of V and E such that an edge (v_i, v_j) is in E' only if v_i and v_j are in V'. G' will be called a *proper subgraph* of G if either E' is a proper subset of E or V' is a proper subset of V. If all the vertices of a graph G are present in a subgraph G' of G, then G' is called a *spanning subgraph* of G.

For example, consider the graph G shown in Fig. 1.2a. The graph G' shown in Fig. 1.2b is a subgraph of G. Its vertex set is $\{v_1, v_2, v_4, v_5\}$. In fact, it is a proper subgraph of G. The graph G'' of Fig. 1.2c is a spanning subgraph of G.

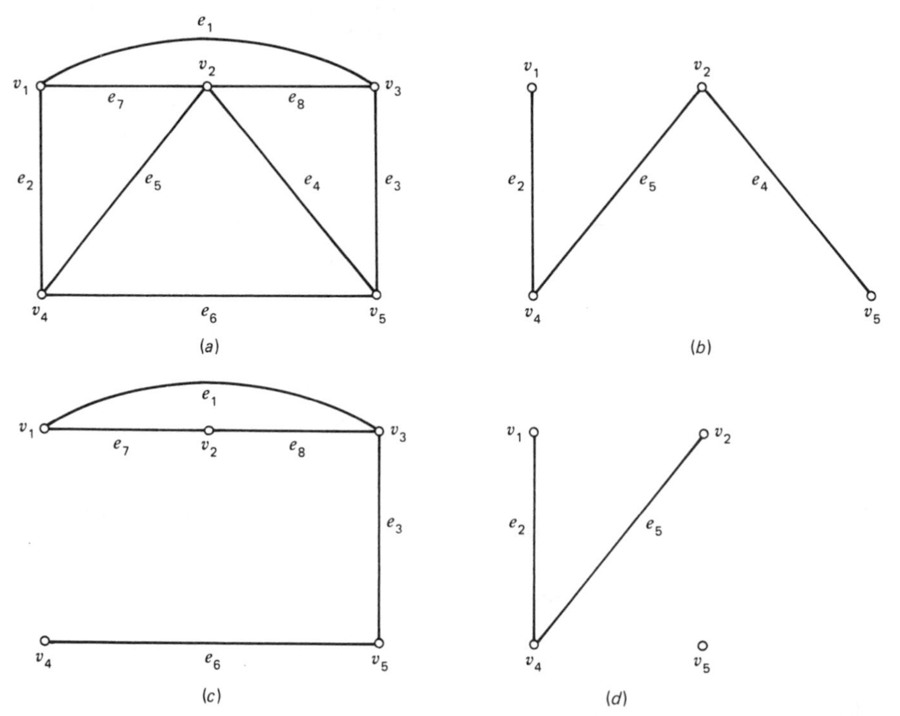

Figure 1.2. A graph and some of its subgraphs. (a) Graph G. (b) Subgraph G'. (c) Subgraph G''. (d) Subgraph G'''.

Some of the vertices in a subgraph may be isolated vertices. For example, the graph G''' shown in Fig. 1.2*d* is a subgraph of G with an isolated vertex.

If a subgraph $G'=(V', E')$ of a graph G has no isolated vertices, then it can be seen from the definition of a subgraph that every vertex in V' is the end vertex of some edge in E'. Thus in such a case, E' uniquely specifies V' and hence the subgraph G'. The subgraph G' is then called the *induced subgraph of G on the edge set E′* (or simply *edge-induced subgraph of G*) and is denoted as $\langle E' \rangle$.

Note that the vertex set V' of $\langle E' \rangle$ is the smallest subset of V containing all the end vertices of the edges in E'. The subgraphs G' and G'' of Fig. 1.2*b* and *c* are edge-induced subgraphs of the graph G of Fig. 1.2*a*, whereas G''' shown in Fig. 1.2*d* is not an edge-induced subgraph.

Next we define a vertex-induced subgraph.

Let V' be a subset of the vertex set V of a graph $G=(V, E)$. Then the subgraph $G'=(V', E')$ is the *induced subgraph of G on the vertex set V′* (or simply *vertex-induced subgraph of G*) if E' is a subset of E such that edge (v_i, v_j) is in E' if and only if v_i and v_j are in V'. In other words, if v_i and v_j are in V', then every edge in E having v_i and v_j as its end vertices should be in E'. Note that, in this case, V' completely specifies E' and thus the subgraph G'. Hence the vertex-induced subgraph $G'=(V', E')$ is denoted simply as $\langle V' \rangle$. As an example, the graph shown in Fig. 1.3 is a vertex-induced subgraph of the graph G of Fig. 1.2*a*.

Note that the edge set E' of the vertex-induced subgraph on the vertex set V' is the largest subset of E such that the end vertices of all of its edges are in V'.

A subgraph G' of a graph G is said to be a *maximal subgraph* of G with respect to some property P if G' has the property P and G' is not a proper subgraph of any other subgraph of G having the property P.

A subgraph G' of a graph G is said to be a *minimal subgraph* of G with respect to some property P if G' has the property P and no subgraph of G having the property P is a proper subgraph of G'.

Maximal and *minimal subsets* of a set with respect to a property are defined in a similar manner.

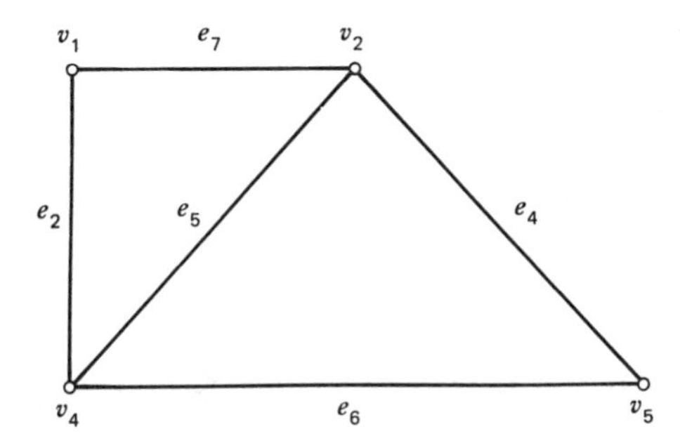

Figure 1.3. A vertex induced subgraph of the graph G of Fig. 1.2*a*.

For example, the vertex set V' of an edge-induced subgraph $\langle E' \rangle$ of a graph $G=(V, E)$ is the minimal subset of V containing the end vertices of all the edges of E'. On the other hand, the edge set E' of a vertex-induced subgraph $\langle V' \rangle$ is the maximal subset of E such that the end vertices of all of its edges are in V'.

Later we see that a "component" (Section 1.4) of a graph G is a maximal "connected" subgraph of G, and a "spanning tree" (Chapter 2) of a connected graph G is a minimal "connected" spanning subgraph of G.

Next we define the complement of a graph.

Graph $\overline{G}=(V, E')$ is called the *complement* of a simple graph $G=(V, E)$ if the edge (v_i, v_j) is in E' if and only if it is not in E. In other words, two vertices v_i and v_j are adjacent in $\overline{G}$ if and only if they are not adjacent in G. A graph and its complement are shown in Fig. 1.4. As another example, consider the graph G shown in Fig. 1.5a. In this graph there is an edge between every pair of vertices. Hence in the complement $\overline{G}$ of G there will be no edge between any pair of vertices; that is, $\overline{G}$ will contain only isolated vertices. This is shown in Fig. 1.5b.

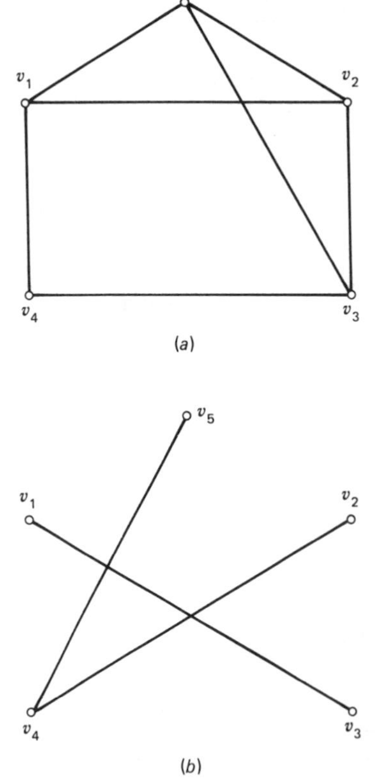

Figure 1.4. A graph and its complement. (a) Graph G. (b) Graph $\overline{G}$, complement of G.

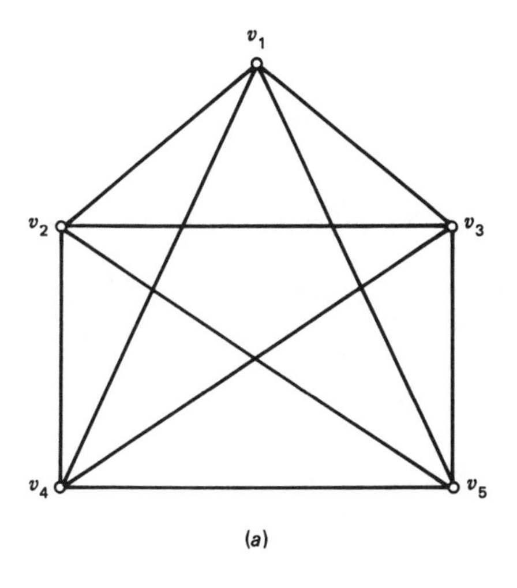

v_1

v_2 v_3

v_4 v_5

(b)

Figure 1.5. A graph and its complement. (*a*) Graph G. (*b*) Graph $\bar{G}$, complement of G.

Let $G'=(V', E')$ be a subgraph of a graph $G=(V, E)$. The subgraph $G''=(V, E-E')$ of G is called the *complement of* G' *in* G. For example, in Fig. 1.2, subgraph G'' is the complement of G' in the graph G.

The following example illustrates some of the ideas presented thus far.

Suppose we want to prove the following:

At any party with six people there are three mutual acquaintances or three mutual nonacquaintances.

Representing people by the vertices of a graph and acquaintance relationship among the people by edges connecting the corresponding vertices, we can see that the above assertion can also be stated as follows:

In any simple graph G with six vertices there are three mutually adjacent vertices or three mutually nonadjacent vertices.

In view of the definition of the complement of a graph, we see that the above statement is equivalent to the following:

For any simple graph G with six vertices, G or $\overline{G}$ contains three mutually adjacent vertices.

To prove this, we may proceed as follows:

Consider any vertex v of a simple graph G with six vertices. Note that if v is not adjacent to three vertices in G, then it will be adjacent to three vertices in $\overline{G}$. So, without any loss of generality, we may assume that, in G, v is adjacent to some three vertices v_1, v_2, and v_3. If any two of these vertices, say v_1 and v_2, are adjacent in G, then the vertices v, v_1, and v_2 are mutually adjacent in G, and the assertion is proved.

If no two of the three vertices v_1, v_2, and v_3 are adjacent in G, then it means that v_1, v_2, and v_3 are mutually nonadjacent in G. Hence, by the definition of a complement, the vertices v_1, v_2, and v_3 are mutually adjacent in $\overline{G}$, and the assertion is again proved.

1.3 WALKS, TRAILS, PATHS, AND CIRCUITS

A *walk* in a graph $G=(V, E)$ is a finite alternating sequence of vertices and edges $v_0, e_1, v_1, e_2, \ldots, v_{k-1}, e_k, v_k$ beginning and ending with vertices such that v_{i-1} and v_i are the end vertices of the edge e_i, $1 \leqslant i \leqslant k$. Alternately, a walk can be considered as a finite sequence of vertices $v_0, v_1, v_2, \ldots, v_k$, such that (v_{i-1}, v_i), $1 \leqslant i \leqslant k$, is an edge in the graph G. This walk is usually called a v_0–v_k walk with v_0 and v_k referred to as the *end* or *terminal vertices* of this walk. All other vertices are *internal vertices* of this walk. Note that in a walk, edges and vertices can appear more than once.

A walk is *open* if its end vertices are distinct; otherwise it is *closed*.

In the graph G of Fig. 1.6, the sequence $v_1, e_1, v_2, e_2, v_3, e_8, v_6, e_9, v_5, e_7, v_3, e_{11}, v_6$ is an open walk, whereas the sequence $v_1, e_1, v_2, e_3, v_5, e_7, v_3, e_2, v_2, e_1, v_1$ is a closed walk.

A walk is a *trail* if all its edges are distinct. A trail is *open* if its end vertices are distinct; otherwise, it is *closed*. In Fig. 1.6, $v_1, e_1, v_2, e_2, v_3, e_8, v_6, e_{11}, v_3$ is an open trail, whereas $v_1, e_1, v_2, e_2, v_3, e_7, v_5, e_3, v_2, e_4, v_4, e_5, v_1$ is a closed trail.

An open trail is a *path* if all its vertices are distinct.

A closed trail is a *circuit* if all its vertices except the end vertices are distinct.

For example, in Fig. 1.6 the sequence v_1, e_1, v_2, e_2, v_3 is a path, whereas the sequence $v_1, e_1, v_2, e_3, v_5, e_6, v_4, e_5, v_1$ is a circuit.

An edge of a graph G is said to be a *circuit edge* of G if there exists a circuit in G containing the edge. Otherwise the edge is called a *noncircuit edge*. In Fig. 1.6, all edges except e_{12} are circuit edges.

The number of edges in a path is called the *length of the path*. Similarly the *length of a circuit* is defined.

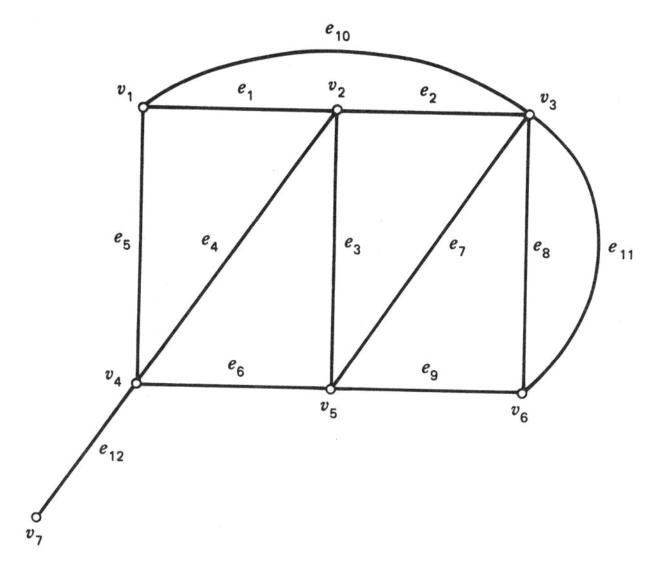

Figure 1.6. Graph G.

The following properties of paths and circuits should be noted:

1. In a path, the degree of each vertex which is not an end vertex is equal to 2; the end vertices have degrees equal to 1.
2. In a circuit, every vertex is of degree 2, and so of even degree. The converse of this statement, namely, the edges of a subgraph in which every vertex is of even degree form a circuit, is not true. A more general question is discussed in Chapter 3.
3. In a path the number of vertices is one more than the number of edges, whereas in a circuit the number of edges is equal to the number of vertices.

1.4 CONNECTEDNESS AND COMPONENTS OF A GRAPH

An important concept in graph theory is that of connectedness.

Two vertices v_i and v_j are said to be *connected* in a graph G if there exists a v_i–v_j path in G. A vertex is connected to itself.

A graph G is *connected* if there exists a path between every pair of vertices in G.

For example, the graph of Fig. 1.6 is connected.

Consider a graph $G=(V, E)$ which is not connected. Then the vertex set V of G can be partitioned* into subsets $V_1, V_2, \ldots, V_p$ such that the vertex-induced

*A set V is said to be *partitioned* into subsets $V_1, V_2, \ldots, V_p$ if $V_1 \cup V_2 \cup \cdots \cup V_p = V$ and $V_i \cap V_j = \varnothing$ for all i and j, $i \neq j$. $\{V_1, V_2, \ldots, V_p\}$ is then called a *partition* of V.

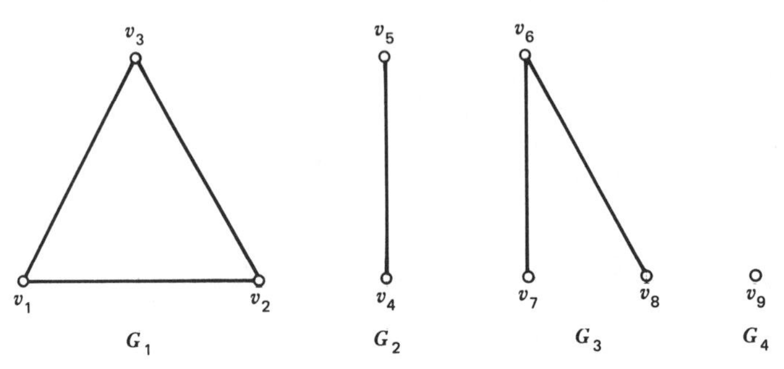

Figure 1.7. Graph G with components G_1, G_2, G_3, and G_4.

subgraphs $\langle V_i \rangle$, $i=1,2,\ldots,p$, are connected and no vertex in subset V_i is connected to any vertex in subset V_j, $j \neq i$. The subgraphs $\langle V_i \rangle, i=1,2,\ldots,p$, are called the *components* of G. It may be seen that a component of a graph G is a maximal connected subgraph of G; that is, a component of G is not a proper subgraph of any other connected subgraph of G.

For example, the graph G of Fig. 1.7 is not connected. Its four components G_1, G_2, G_3, and G_4 have vertex sets $\{v_1, v_2, v_3\}$, $\{v_4, v_5\}$, $\{v_6, v_7, v_8\}$, and $\{v_9\}$, respectively.

Note that an isolated vertex by itself should be treated as a component since, by definition, a vertex is connected to itself. Further, note that if a graph G is connected, it has only one component which is the same as G itself.

We next consider some properties of connected graphs.

THEOREM 1.3. In a connected graph, any two longest paths have a common vertex.

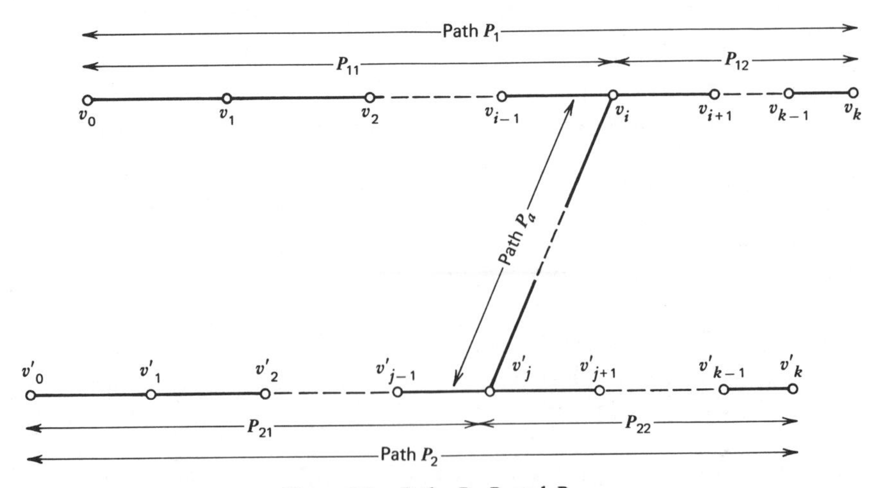

Figure 1.8. Paths P_1, P_2 and P_a.

Proof

Consider any two longest paths P_1 and P_2 in a connected graph G. Let P_1 be denoted by the vertex sequence $v_0, v_1, v_2, \ldots, v_k$ and P_2 by the sequence $v'_0, v'_1, v'_2, \ldots, v'_k$.

Assume that P_1 and P_2 have no common vertex. Since the graph G is connected, then for some $i, 0 \leqslant i \leqslant k$, and some $j, 0 \leqslant j \leqslant k$, there exists a v_i–v'_j path P_a such that all the vertices of P_a other than v_i and v'_j are different from those of P_1 and P_2. The paths P_1, P_2, and P_a may be as shown in Fig. 1.8. Let

$$t_1 = \text{length of } v_0\text{–}v_i \text{ path } P_{11},$$

$$t_2 = \text{length of } v_i\text{–}v_k \text{ path } P_{12},$$

$$t'_1 = \text{length of } v'_0\text{–}v'_j \text{ path } P_{21},$$

$$t'_2 = \text{length of } v'_j\text{–}v'_k \text{ path } P_{22},$$

$$t_a = \text{length of path } P_a.$$

The paths P_{11}, P_{12}, P_{21}, and P_{22} are also shown in Fig. 1.8. Note that

$$t_1 + t_2 = t'_1 + t'_2 = \text{length of a longest path in } G$$

and

$$t_a > 0.$$

Without any loss of generality, let

$$t_1 \geqslant t_2$$

and

$$t'_1 \geqslant t'_2$$

so that

$$t_1 + t'_1 \geqslant t_1 + t_2 = t'_1 + t'_2.$$

Now it may be verified that the paths P_{11}, P_a, and P_{21} together constitute a v_0–v'_0 path with its length equal to $t_1 + t'_1 + t_a > t_1 + t_2$, because $t_a > 0$. This contradicts that $t_1 + t_2$ is the length of a longest path in G. ■

The following theorem is a very useful one; it is used often in the discussions of the next chapter. In this theorem as well as in the rest of the book, we abbreviate $\{x\}$ to x whenever it is clear that we are referring to a set rather than an element.

THEOREM 1.4. If a graph $G=(V, E)$ is connected, then the graph $G'=(V, E-e)$ that results after removing a circuit edge e is also connected. ■

We leave the proof of this theorem as an exercise.

1.5 OPERATIONS ON GRAPHS

In this section, we introduce a few operations involving graphs. The first three operations are binary operations involving two graphs, and the last four are unary operations, that is, operations defined with respect to a single graph.

Consider two graphs, $G_1=(V_1, E_1)$ and $G_2=(V_2, E_2)$. The *union* of G_1 and G_2, denoted as $G_1 \cup G_2$, is the graph $G_3=(V_1 \cup V_2, E_1 \cup E_2)$; that is, the vertex set of G_3 is the union of V_1 and V_2, and the edge set of G_3 is the union of E_1 and E_2.

For example, two graphs G_1 and G_2 and their union are shown in Fig. 1.9*a*, *b* and *c*.

The *intersection* of G_1 and G_2, denoted as $G_1 \cap G_2$, is the graph $G_3=(V_1 \cap V_2, E_1 \cap E_2)$. That is, the vertex set of G_3 consists of only those vertices present in both G_1 and G_2, and the edge set of G_3 consists of only those edges present in both G_1 and G_2.

The intersection of the graphs G_1 and G_2 of Fig. 1.9*a* and *b* is shown in Fig. 1.9*d*.

The *ring sum* of two graphs G_1 and G_2, denoted as $G_1 \oplus G_2$, is the induced graph G_3 on the edge set $E_1 \oplus E_2$. In other words, G_3 has no isolated vertices and consists of only those edges present either in G_1 or in G_2, but not in both. The ring sum of the graphs of Fig. 1.9*a* and *b* is shown in Fig. 1.9*e*.

It may be easily verified that the three operations defined above are commutative, that is,

$$G_1 \cup G_2 = G_2 \cup G_1,$$

$$G_1 \cap G_2 = G_2 \cap G_1,$$

$$G_1 \oplus G_2 = G_2 \oplus G_1.$$

Also note that these operations are binary, that is, they are defined with respect to two graphs. Of course, the definitions of these operations can be extended in an obvious way to include more than two graphs.

Next we discuss four unary operations on a graph.

Vertex removal If v_i is a vertex of a graph $G=(V, E)$, then $G-v_i$ is the induced subgraph of G on the vertex set $V-v_i$; that is, $G-v_i$ is the graph obtained after removing from G the vertex v_i and all the edges incident on v_i.

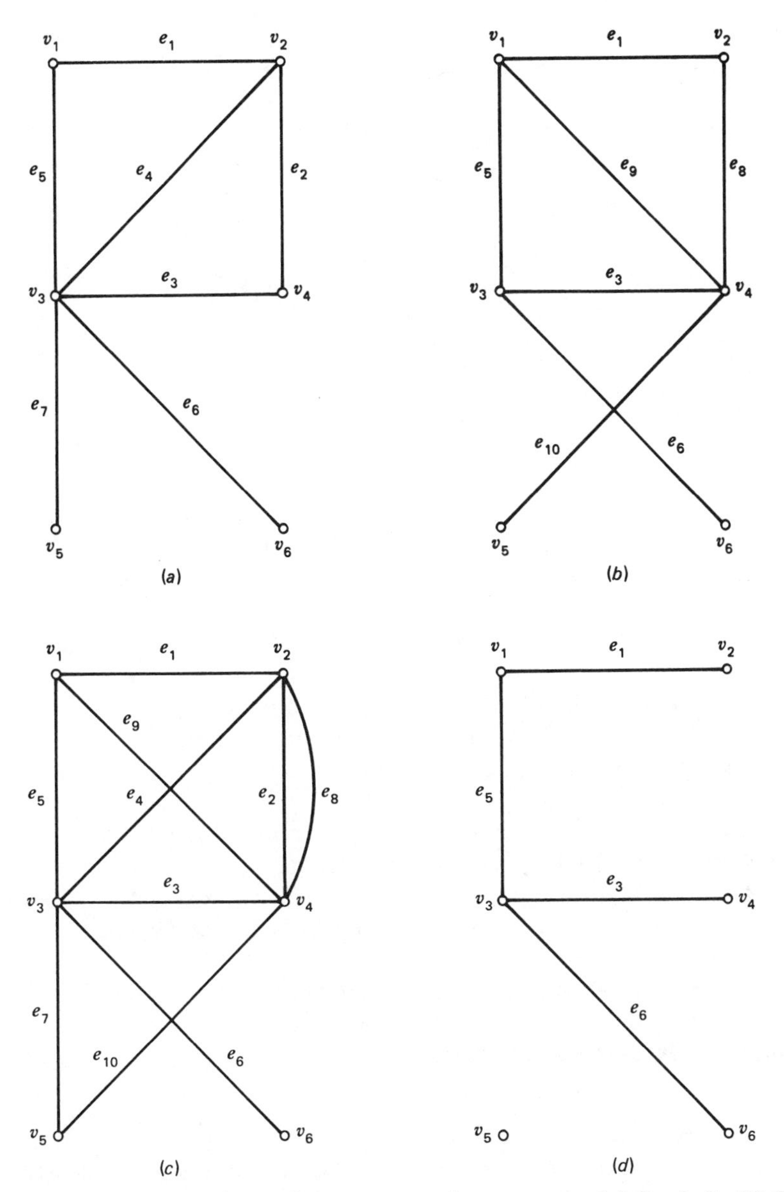

Figure 1.9. Union, intersection, and ring sum operations on graphs. (*a*) Graph G_1. (*b*) Graph G_2. (*c*) $G_1 \cup G_2$. (*d*) $G_1 \cap G_2$. (*e*) $G_1 \oplus G_2$.

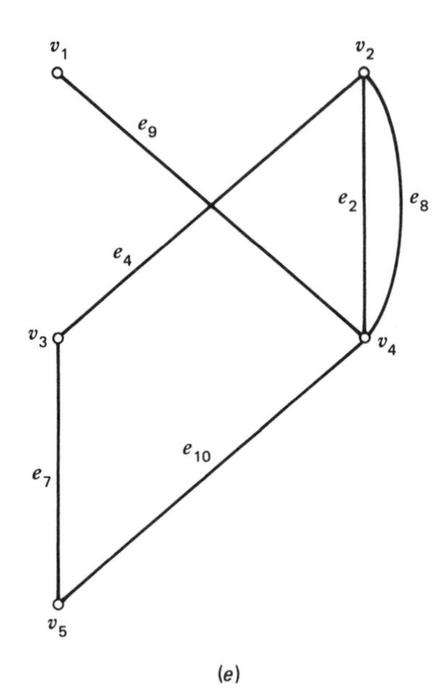

Figure 1.9. *(continued)*

Edge removal If e_i is an edge of a graph $G=(V, E)$, then $G-e_i$ is the subgraph of G that results after removing from G the edge e_i. Note that the end vertices of e_i are not removed from G.

The removal of a set of vertices or edges from a graph is defined as the removal of single vertices or edges in succession.

If $G_1=(V', E')$ is a subgraph of the graph $G=(V, E)$, then by $G-G_1$ we refer to the graph $G'=(V, E-E')$. Thus $G-G_1$ is the complement in G of the subgraph G_1.

The vertex removal and edge removal operations are illustrated in Fig. 1.10.

The short-circuiting or identifying operation to be considered next is a familiar one for electrical engineers.

Short-circuiting or identifying A pair of vertices v_i and v_j in a graph G are said to be *short-circuited* (or *identified*) if the two vertices are replaced by a new vertex such that all the edges in G incident on v_i and v_j are now incident on the new vertex. For example, short-circuiting v_3 and v_4 in the graph of Fig. 1.11*a* results in the graph shown in Fig. 1.11*b*.

Contraction By *contraction* of an edge e we refer to the operation of removing e and identifying its end vertices. A graph G is *contractible* to a graph H if H can be obtained from G by a sequence of contractions. For

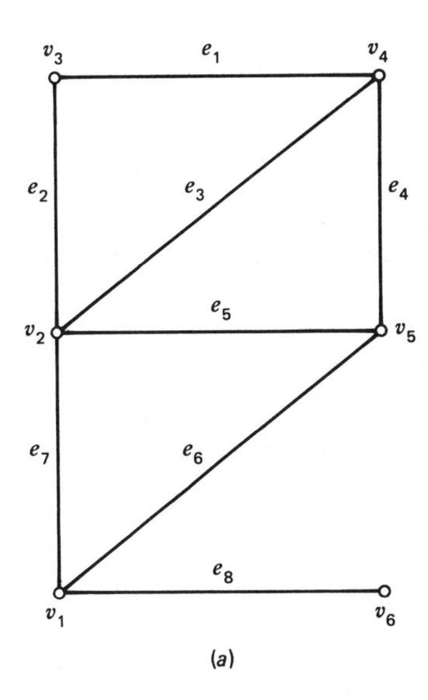

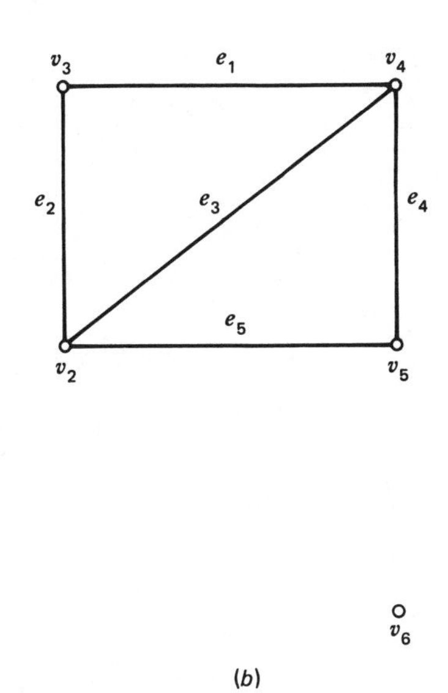

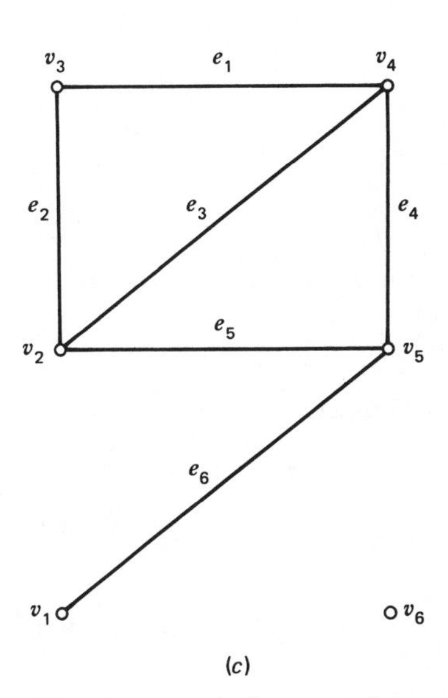

Figure 1.10. Vertex and edge removal operations on a graph. (*a*) G. (*b*) $G-v_1$. (*c*) $G-\{e_7, e_8\}$.

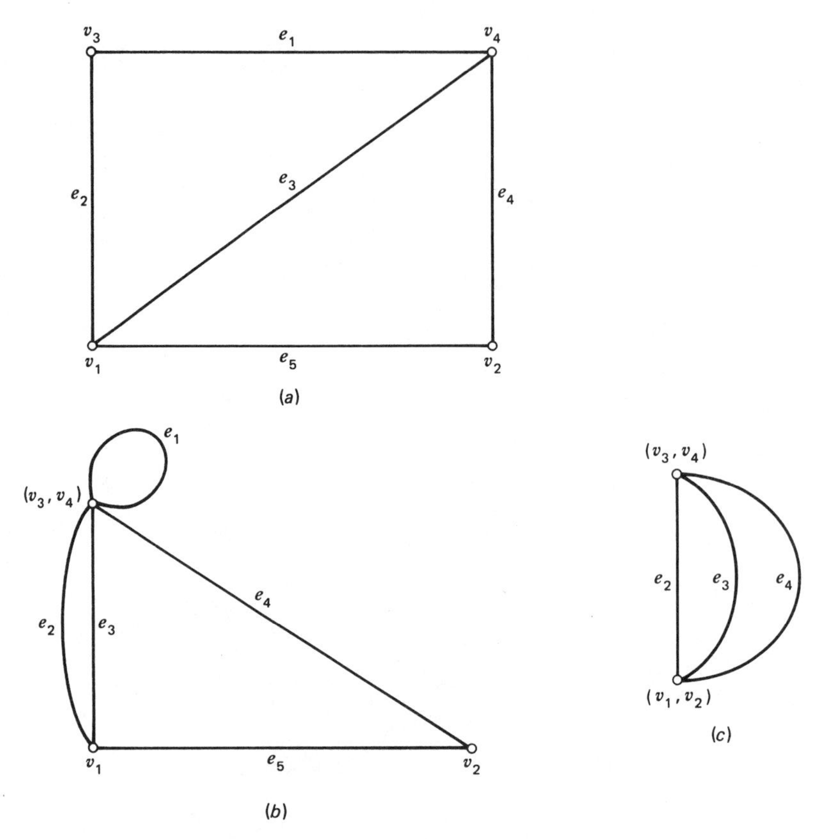

Figure 1.11. Short-circuiting and contracting in a graph. (*a*) Graph G. (*b*) Graph G after short-circuiting v_3 and v_4. (*c*) Graph G after contracting e_1 and e_5.

example, the graph shown in Fig. 1.11*c* is obtained by contracting the edges e_1 and e_5 of the graph G in Fig. 1.11*a*.

1.6 SPECIAL GRAPHS

Certain special classes of graphs which occur frequently in the theory of graphs are introduced next.

A *complete graph* G is a simple graph in which every pair of vertices is adjacent. If a complete graph G has n vertices, then it will be denoted by K_n. It may be seen that K_n has $n(n-1)/2$ edges. As an example, the graph K_5 is shown in Fig. 1.12.

A graph G is *regular* if all the vertices of G are of equal degree. If G is regular with $d(v_i)=r$ for all vertices v_i in G, then G is called *r-regular*. A 4-regular graph is shown in Fig. 1.13. It may be noted that K_n is an $(n-1)$-regular graph.

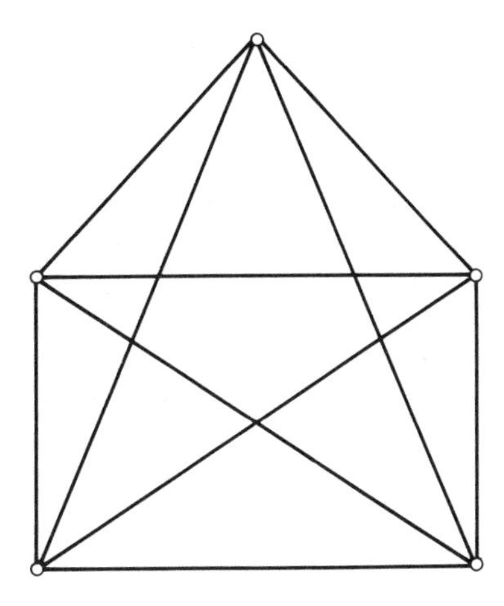

Figure 1.12. Graph K_5.

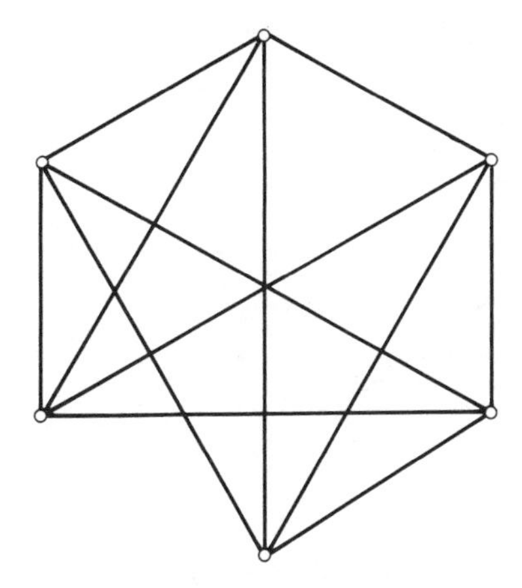

Figure 1.13. A 4-regular graph.

A graph $G=(V, E)$ is *bipartite* if its vertex set V can be partitioned into two subsets V_1 and V_2 such that each edge of E has one end vertex in V_1 and another in V_2; (V_1, V_2) is referred to as a *bipartition* of G. If in a simple bipartite graph G, with bipartition (V_1, V_2), there is an edge (v_i, v_j) for every vertex v_i in V_1 and every vertex v_j in V_2, then G is called a *complete bipartite* graph, and will be denoted by $K_{m,n}$ if V_1 has m vertices and V_2 has n vertices.

A bipartite graph and the complete bipartite graph $K_{3,4}$ are shown in Fig. 1.14.

A graph $G=(V, E)$ is *k-partite* if it is possible to partition V into k subsets $V_1, V_2, \ldots, V_k$ such that each edge of G has one end vertex in some V_i and the other in some V_j, $i \neq j$. A *complete k-partite graph* G is a simple k-partite graph with vertex set partition $\{V_1, V_2, \ldots, V_k\}$ and with the additional property that

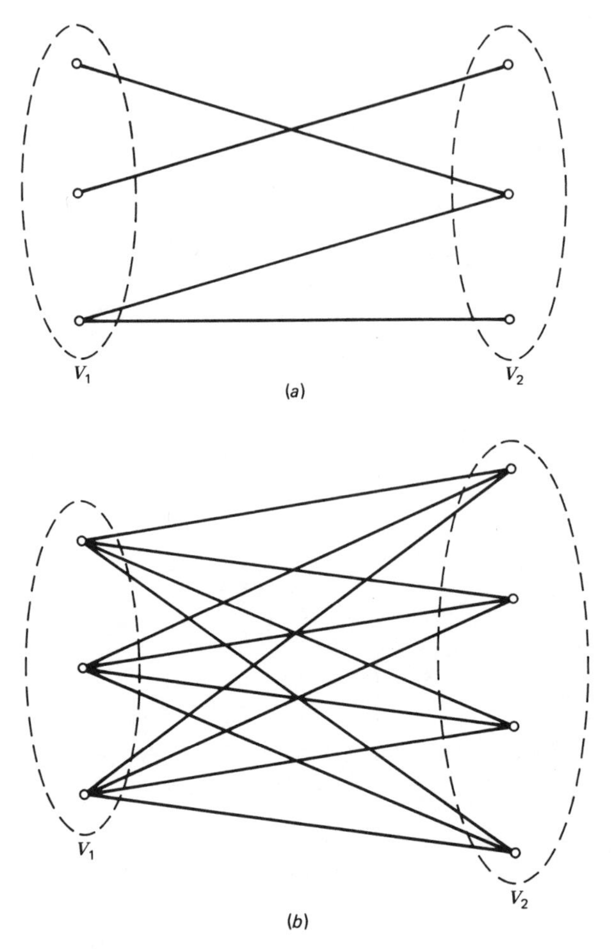

Figure 1.14. (*a*) A bipartite graph. (*b*) The complete bipartite graph $K_{3,4}$.

for every vertex v_i in V_r and every vertex v_j in V_s, $r \neq s$, $1 \leqslant r, s \leqslant k, (v_i, v_j)$ is an edge in G. A complete 3-partite graph is shown in Fig. 1.15.

1.7 CUT-VERTICES AND SEPARABLE GRAPHS

A vertex v_i of a graph G is a *cut-vertex* of G if the graph $G - v_i$ consists of a greater number of components than G. If G is connected, then $G - v_i$ will contain at least two components, that is, $G - v_i$ will not be connected. According to this definition, an isolated vertex is not a cut-vertex.

We call a graph *trivial* if it has only one vertex. Thus a trivial graph has no cut-vertex.

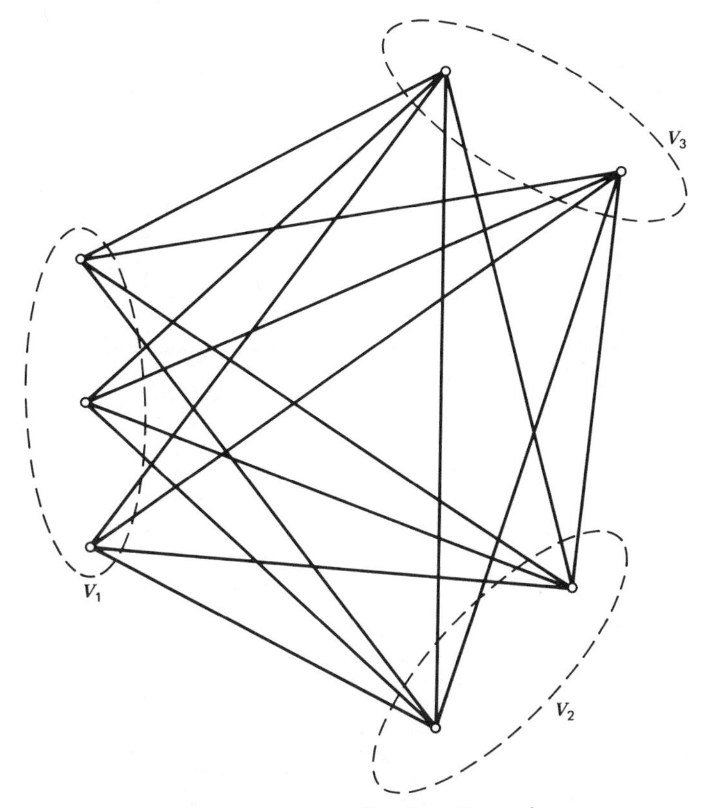

Figure 1.15. A complete 3-partite graph.

A *nonseparable graph* is a connected graph with no cut-vertices. All other graphs are *separable*. (Note that a graph that is not connected is separable.)

The graph G shown in Fig. 1.16a is separable. It has three cut-vertices v_1, v_2, and v_3.

A *block* of a separable graph G is a maximal nonseparable subgraph of G. The blocks of the separable graph G of Fig. 1.16a are shown in Fig. 1.16b.

The following theorem presents an equivalent definition of a cut-vertex.

THEOREM 1.5. A vertex v is a cut-vertex of a connected graph G if and only if there exist two vertices u and w distinct from v such that v is on every u–w path.

Proof

Necessity Since v is a cut-vertex of G, $G-v$ is, by definition, not connected. Let G_1 be one of the components of $G-v$. Let V_1 be the vertex set of G_1, and V_2 the complement of V_1 in $V-v$.

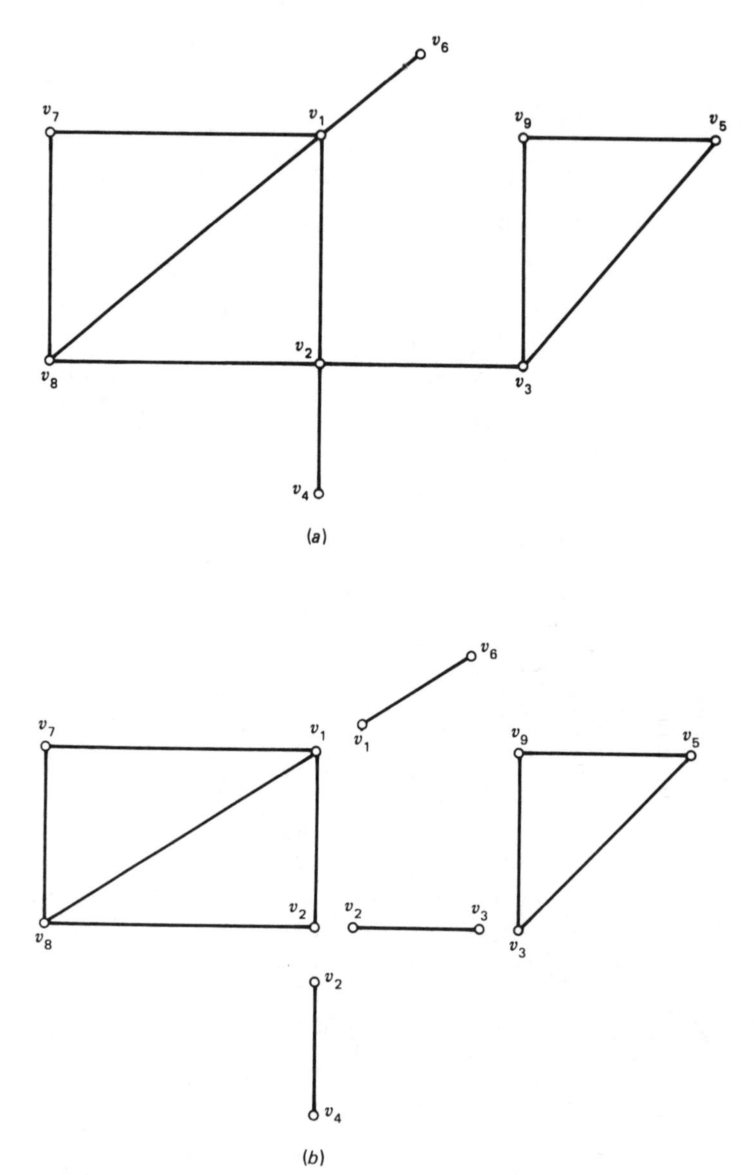

Figure 1.16. A separable graph and its blocks. (a) A separable graph G. (b) Blocks of G.

Let u and w be two vertices such that u is in V_1 and w is in V_2. Consider any u–w path in G. If the cut-vertex v does not lie on this path, then this path is also in $G-v$; that is, the vertices u and w are connected in $G-v$. However, this is a contradiction since u and w are in different components of $G-v$. Hence the vertex v lies on every u–w path.

Sufficiency If v is on every u–w path, then the vertices u and w are not connected in $G-v$. Thus the graph $G-v$ is not connected. Hence, by definition, v is a cut-vertex. ■

An equivalent definition of a separable graph given below follows from Theorem 1.5:

A connected graph G is separable if and only if there exists a vertex v in G such that it is the only vertex common to two proper nontrivial subgraphs G_1 and G_2 whose union is equal to G.

Suppose a graph G is separable. Is it possible that all the vertices of G are cut-vertices? The answer is "no" as we prove in the next theorem.

THEOREM 1.6. Every nontrivial connected graph contains at least two vertices which are not cut-vertices.

Proof

We prove the theorem by induction on the number of vertices of a graph.

The theorem is true for every connected graph with two vertices, for neither of these vertices is a cut-vertex.

Assume that the theorem is true for all nontrivial connected graphs having less than n vertices, where $n>2$. Let G be a connected graph having n vertices. If G has no cut-vertices, then the theorem is proved.

Otherwise, let v be a cut-vertex of G. Further let $G_1, G_2, \ldots, G_k$ be the k components of the graph $G-v$. If any G_i is trivial then its only vertex is not a cut-vertex.

Consider any nontrivial component G_j. By the induction hypothesis, G_j contains vertices v_1 and v_2 which are not cut-vertices of G_j. It is obvious that if one of these vertices is not adjacent to v in G, then this vertex is not a cut-vertex of G. On the other hand, if both v_1 and v_2 are adjacent to v in G, then neither v_1 nor v_2 is a cut-vertex of G.

Thus every component of $G-v$ has at least one vertex which is not a cut-vertex of G. Hence the graph G has at least two vertices which are not cut-vertices. ■

1.8 ISOMORPHISM AND 2-ISOMORPHISM

The two graphs shown in Fig. 1.17 are "seemingly" different. However, one can be redrawn to look exactly like the other. Thus these two graphs are

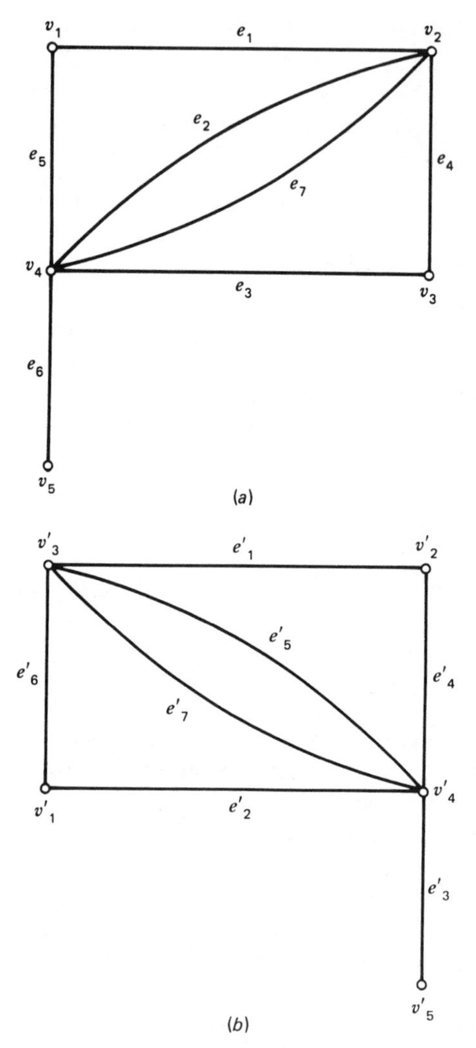

Figure 1.17. Isomorphic graphs. (*a*) G_1. (*b*) G_2.

"equivalent" in some sense. This equivalence is stated more precisely as follows:

Two graphs G_1 and G_2 are said to be *isomorphic* if there exists a one-to-one correspondence between their vertex sets and a one-to-one correspondence between their edge sets so that the corresponding edges of G_1 and G_2 are incident on the corresponding vertices of G_1 and G_2. In other words, if vertices v_1 and v_2 in G_1 correspond respectively to vertices v'_1 and v'_2 in G_2, then an edge in G_1 with end vertices v_1 and v_2 should correspond to an edge in G_2 with v'_1 and v'_2 as end vertices and vice versa.

According to the above definition, the two graphs shown in Fig. 1.17 are isomorphic. The correspondences between their vertex sets and edge sets are as follows:

vertex correspondence

$$v_1 \leftrightarrow v_2', \quad v_2 \leftrightarrow v_3', \quad v_3 \leftrightarrow v_1'$$

$$v_4 \leftrightarrow v_4', \quad v_5 \leftrightarrow v_5'.$$

edge correspondence

$$e_1 \leftrightarrow e_1', \quad e_2 \leftrightarrow e_5', \quad e_3 \leftrightarrow e_2',$$

$$e_4 \leftrightarrow e_6', \quad e_5 \leftrightarrow e_4', \quad e_6 \leftrightarrow e_3',$$

$$e_7 \leftrightarrow e_7'.$$

Consider next the two separable graphs G_1 and G_2 shown in Fig. 1.18*a* and *b*. These two graphs are not isomorphic. Suppose we "split" the cut-vertex v_1 in G_1 into two vertices so as to get the two edge-disjoint graphs* shown in Fig. 1.18*c*. If we perform a similar "splitting" operation on the cut-vertex v_1' in G_2, we get the two edge-disjoint graphs shown in Fig. 1.18*d*. It may be seen that the graphs of Fig. 1.18*c* and *d* are isomorphic. Thus the two graphs G_1 and G_2 become isomorphic after splitting the cut-vertices. Such graphs are said to be 1-*isomorphic*.

2-isomorphism to be defined next is a more general type of isomorphism. Two graphs G_1 and G_2 are 2-*isomorphic* if they become isomorphic after one or more applications of either or both of the following operations:

1. "*Splitting*" a cut-vertex in G_1 and/or G_2 into two vertices to get two edge-disjoint graphs.
2. If one of the graphs, say G_1, has two subgraphs G_1' and G_1'' which have exactly two common vertices v_1 and v_2, the interchange of the names of these vertices in one of the subgraphs. (Geometrically this operation is equivalent to "*turning around*" one of the subgraphs G_1' and G_1'' at the common vertices v_1 and v_2.)

Consider the graphs G_1 and G_2 shown in Fig. 1.19*a* and *b*. After performing in G_1 a "splitting" operation on vertex v_2 and a "turning around" operation at vertices v_1 and v_2, we get the graph G_1' shown in Fig. 1.19*c*. A "splitting"

*Two graphs are *edge-disjoint* if they have no edge in common, and *vertex-disjoint* if they have no vertex in common.

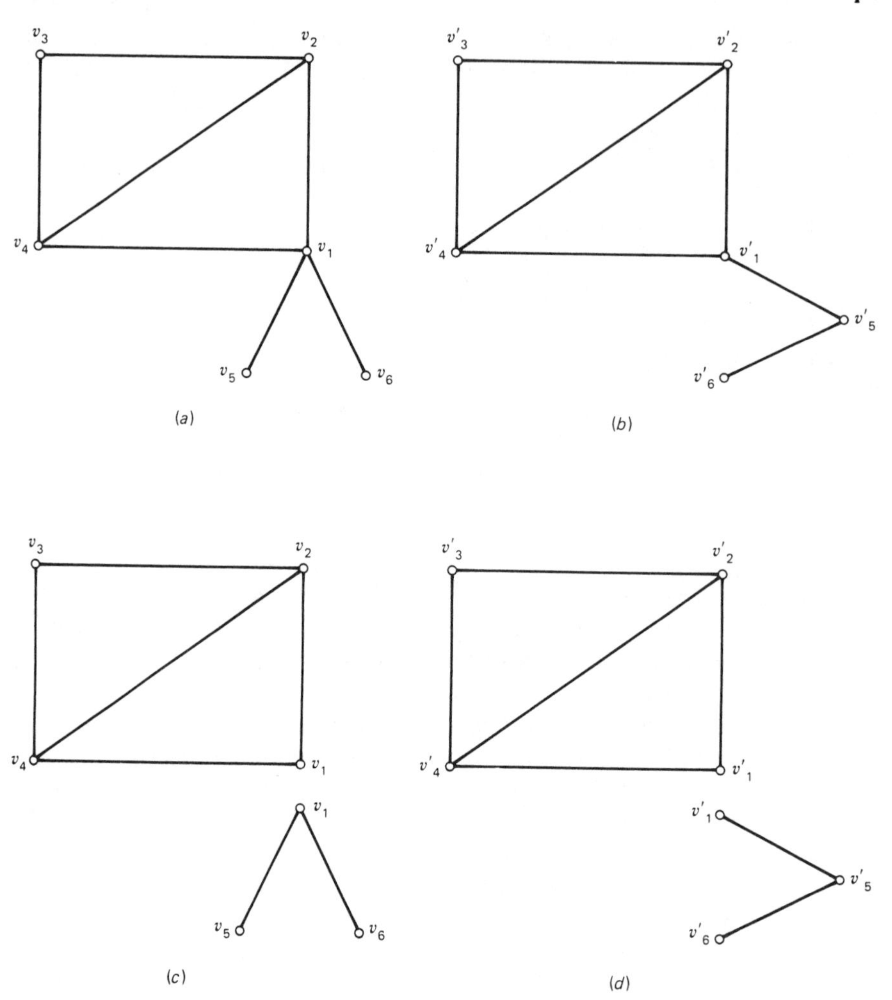

Figure 1.18. 1-isomorphic graphs. (*a*) G_1. (*b*) G_2. (*c*) Graph after splitting v_1 in G_1. (*d*) Graph after splitting v'_1 in G_2.

operation on vertex v'_2 in G_2 results in the graph G'_2 shown in Fig. 1.19*d*. These two graphs G'_1 and G'_2 are isomorphic. Hence G_1 and G_2 are 2-isomorphic.

The next theorem presents an important result on 2-isomorphic graphs.

THEOREM 1.7. Two graphs G_1 and G_2 are 2-isomorphic if and only if there exists a one-to-one correspondence between their edge sets such that the circuits in one graph correspond to the circuits in the other. ■

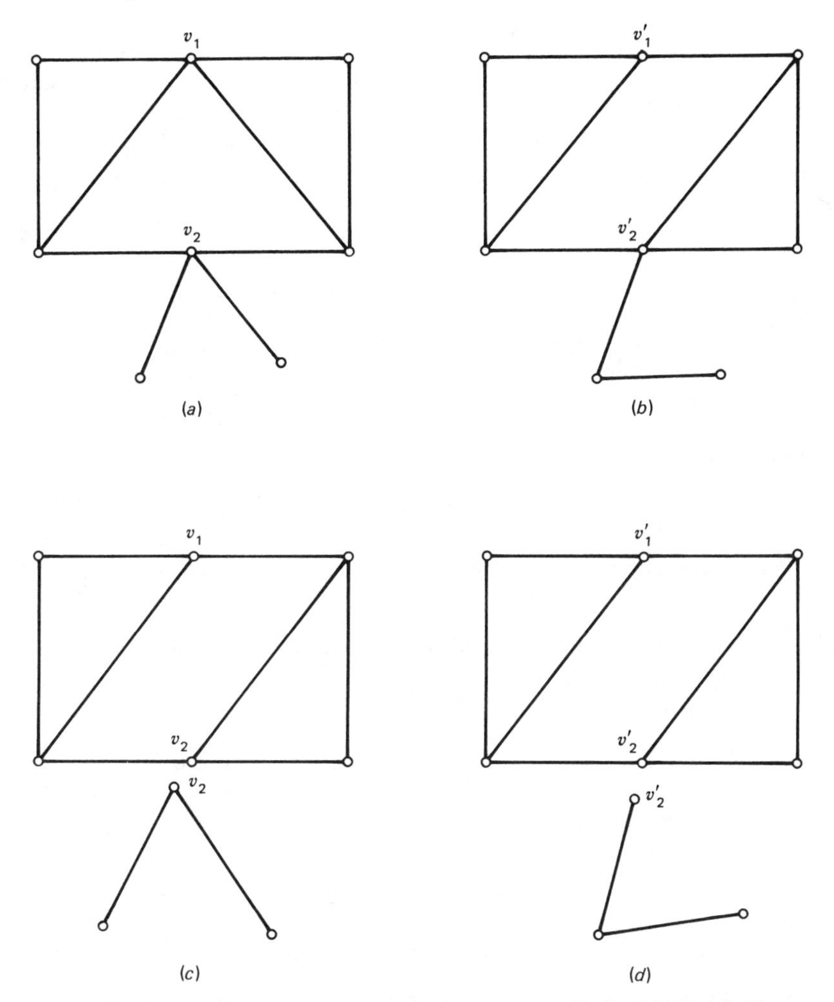

Figure 1.19. Illustration of 2-isomorphism. (*a*) G_1. (*b*) G_2. (*c*) G'_1. (*d*) G'_2.

The fact that circuits in G_1 will correspond to circuits in G_2, when G_1 and G_2 are 2-isomorphic, is fairly obvious. However, the proof of the converse of this result is too lengthy to be discussed here. Whitney's original paper [1.1] on 2-isomorphic graphs discusses this.

1.9 FURTHER READING

Berge [1.2] and Harary [1.3] are two excellent references for several of the topics covered in this part of the book. Berge also discusses hypergraphs and matroids. Bondy and Murty [1.4] is a very well written text which includes a number of unsolved problems in graph theory. Wilson [1.5] gives an elegant introduction to the theory of graphs and has a chapter on matroids. Other text books on graph theory include Liu [1.6], Behzad and Chartrand [1.7], and Deo [1.8]. Several books and monographs dealing with special topics such as enumeration, algorithms, etc., are also available. These are referred to in the appropriate chapters.

1.10 EXERCISES

1.1 Let G be a graph with n vertices and m edges such that the vertices have degree k or $k+1$. Prove that if G has n_k vertices of degree k and n_{k+1} vertices of degree $k+1$, then $n_k=(k+1)n-2m$.

1.2 Prove or disprove:

(a) The union of any two distinct closed walks joining two vertices contains a circuit.

(b) The union of any two distinct paths joining two vertices contains a circuit.

1.3 If in a graph G there is a path between any two vertices a and b, and a path between any two vertices b and c, then prove that there is a path between a and c.

1.4 Let P_1 and P_2 be two distinct paths between any two vertices of a graph. Prove that $P_1 \oplus P_2$ is a circuit or the union of some edge-disjoint circuits of the graph.

1.5 Prove that a closed trail with all its vertices of degree 2 is a circuit.

1.6 Show that if two distinct circuits of a graph G contain an edge e, then in G there exists a circuit that does not contain e.

1.7 Show that in a simple graph G with $\delta(G) \geqslant k$ there is a path of length at least k. Also show that G has a circuit of length at least $k+1$, if $k \geqslant 2$.

1.8 Prove that a graph $G=(V, E)$ is connected if and only if, for every partition (V_1, V_2) of V with V_1 and V_2 nonempty, there is an edge of G joining a vertex in V_1 to a vertex in V_2.

1.9 Prove that a simple graph G with n vertices and k components can have at most $(n-k)(n-k+1)/2$ edges. Deduce from this result that G must be connected if it has more than $(n-1)(n-2)/2$ edges.

1.10 Prove that if a graph G (connected or disconnected) has exactly two vertices of odd degree, then there must be a path joining these two vertices.

1.11 Prove that if a simple graph G is not connected, then its complement $\overline{G}$ is.

1.12 If G is a graph with n vertices and m edges such that $m<n-1$, then prove that G is not connected.

1.13 If, for a graph G with n vertices and m edges, $m \geqslant n$, then prove that G contains a circuit edge.

1.14 Prove that a simple n-vertex graph G is connected if $\delta(G) \geqslant (n-1)/2$.

1.15 Show that a simple graph G with at least two vertices contains two vertices of the same degree.

1.16 Show that if a graph $G=(V, E)$ is simple and connected but not complete, then G has three vertices u, v, and w such that the edges (u, v) and (v, w) are in E and the edge (u, w) is not in E.

1.17 If vertices u and v are connected in a graph G, the *distance* between u and v in G, denoted by $d(u, v)$, is the length of a shortest u–v path in G. If there is no u–v path in G, then we define $d(u, v)$ to be infinite. The *diameter* of G is the maximum distance between two vertices of G. Show that if G has diameter greater than 3, then $\overline{G}$ has diameter less than 3.

1.18 The *girth* of a graph G is the length of a shortest circuit in G. If G has no circuits, we define the girth of G to be infinite. Show that a k-regular graph of girth 4 has at least $2k$ vertices.

1.19 A simple graph G is *self-complementary* if it is isomorphic to its complement $\overline{G}$. Prove that the number of vertices of a self-complementary graph must be of the form $4k$ or $4k+1$ where k is an integer.

1.20 Prove that an n-vertex simple graph is not bipartite if it has more than $n^2/4$ edges.

1.21 Prove that a graph is bipartite if and only if all its circuits are of even length.

1.22 Construct a simple cubic graph with $2n (n \geqslant 3)$ vertices having no triangles.

Note A graph is *cubic* if it is 3-regular. A *triangle* is a circuit of length 3.

1.23 Prove that if v is a cut-vertex of a simple graph G, then v is not a cut-vertex of $\overline{G}$.

1.24 Prove that the following properties of a graph G with $n \geqslant 3$ vertices are equivalent:

(a) G is nonseparable.

(b) Every two vertices of G lie on a common circuit.

(c) For any vertex v and any edge e of G there exists a circuit containing both.

(d) every two edges of G lie on a common circuit.

(e) Given two vertices and one edge of G, there is a path joining the vertices which contains the edge.

(f) For every three distinct vertices of G there is a path joining any two of them which contains the third.

(g) For every three distinct vertices of G there is a path joining any two of them which does not contain the third.

1.25 Show that if a graph G has no circuits of even length, then each block of G is either K_1 or K_2 or a circuit of odd length.

1.26 Show that a connected graph that is not a block has two blocks that have a common vertex. Let $b(v)$ denote the number of blocks of a graph $G=(V, E)$ containing vertex v. Show that the number of blocks of G is equal to $p+\Sigma_{v\in V}(b(v)-1)$, where p is the number of components of G.

1.27 Let $c(B)$ denote the number of cut-vertices of a connected graph G which are vertices of the block B. Then the number $c(G)$ of cut-vertices of G is given by

$$c(G)-1=\sum_{\text{all blocks}}[c(B)-1].$$

1.28 A *bridge* of a graph G is an edge e such that $G-e$ has more components than G. Prove the following:

(a) An edge e of a connected graph G is a bridge if and only if there exist vertices u and w such that e is on every u–w path of G.

(b) An edge of a graph G is a bridge of G if and only if it is on no circuit of G.

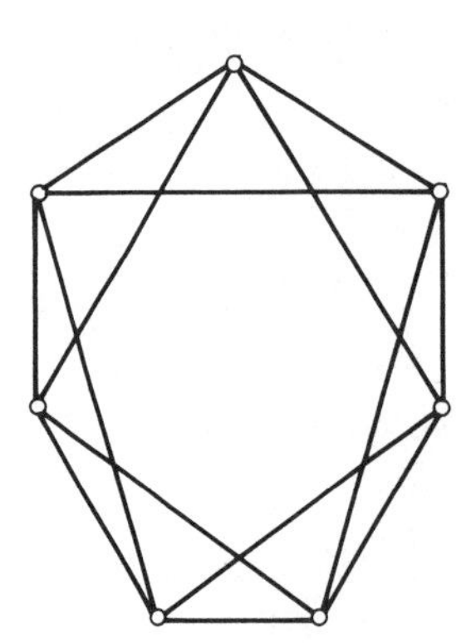

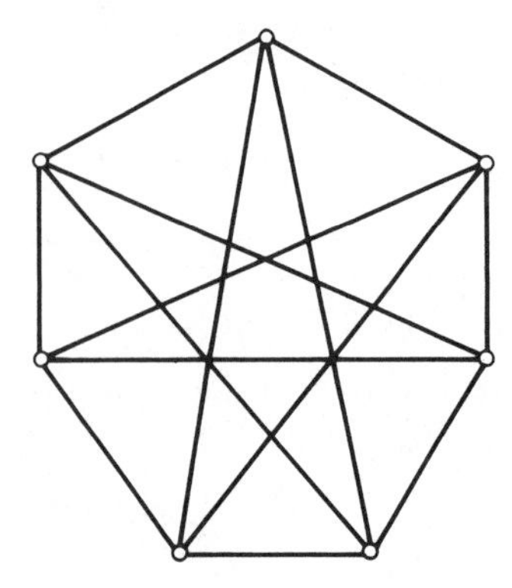

Figure 1.20.

1.29 Are the graphs shown in Fig. 1.20 isomorphic? Why?

1.30 Show that the two graphs of Fig. 1.21 are not isomorphic.

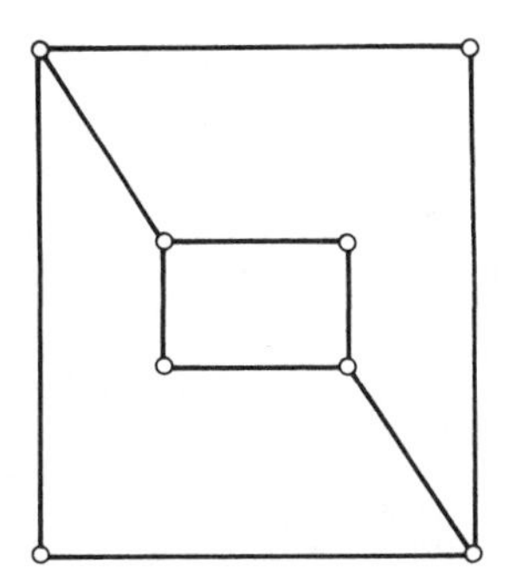
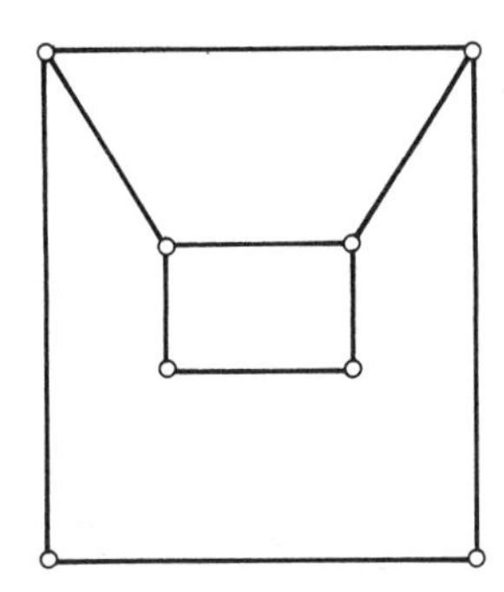

Figure 1.21.

1.31 Determine all nonisomorphic simple graphs of order 3 and those of order 4.

Note There are exactly four nonisomorphic graphs on three vertices, and 11 on four vertices.

1.32 Prove that any two simple connected graphs with n vertices, all of degree 2, are isomorphic.

1.11 REFERENCES

1.1 H. Whitney, "2-Isomorphic Graphs," *Am. J. Math.*, Vol. 55, 245–254 (1933).

1.2 C. Berge, *Graphs and Hypergraphs*, North Holland, Amsterdam, 1973.

1.3 F. Harary, *Graph Theory*, Addison-Wesley, Reading, Mass., 1969.

1.4 J. A. Bondy and U. S. R. Murty, *Graph Theory with Applications*, Macmillan, London, 1976.

1.5 R. J. Wilson, *Introduction to Graph Theory*, Oliver and Boyd, Edinburgh, 1972.

1.6 C. L. Liu, *Introduction to Combinatorial Mathematics*, McGraw-Hill, New York, 1968.

1.7 M. Behzad and G. Chartrand, *Introduction to the Theory of Graphs*, Allyn and Bacon, Boston, 1971.

1.8 N. Deo, *Graph Theory with Applications to Engineering and Computer Science*, Prentice-Hall, Englewood Cliffs, N.J., 1974.

Chapter 2

Trees, Cutsets, and Circuits

The graphs that are encountered in most of the applications are connected. Among connected graphs trees have the simplest structure and are perhaps the most important ones. If connected graphs are important, then the set of edges disconnecting a connected graph should be of equal importance. This leads us to the concept of a cutset. In this chapter, we study trees and cutsets and several results associated with them. We also bring out the relationship between them and circuits.

2.1 TREES, SPANNING TREES, AND COSPANNING TREES

A graph is said to be *acyclic* if it has no circuits. A *tree* is a connected acyclic graph. A *tree* of a graph G is a connected acyclic subgraph of G. A *spanning tree* of a graph G is a tree of G having all the vertices of G. A connected subgraph of a tree T is called a *subtree* of T.

Consider, for example, the graph G shown in Fig. 2.1*a*. The graphs G_1 and G_2 of Fig. 2.1*b* are two trees of G. The graphs G_3 and G_4 of Fig. 2.1*c* are two of the spanning trees of G.

The *cospanning tree* T^* of a spanning tree T of a graph G is the subgraph of G having all the vertices of G and exactly those edges of G which are not in T. Note that a cospanning tree may not be connected. The cospanning trees G_3^* and G_4^* of the spanning trees G_3 and G_4 of Fig. 2.1*c* are shown in Fig. 2.1*d*.

The edges of a spanning tree T are called the *branches* of T, and those of the corresponding cospanning tree T^* are called *links* or *chords*.

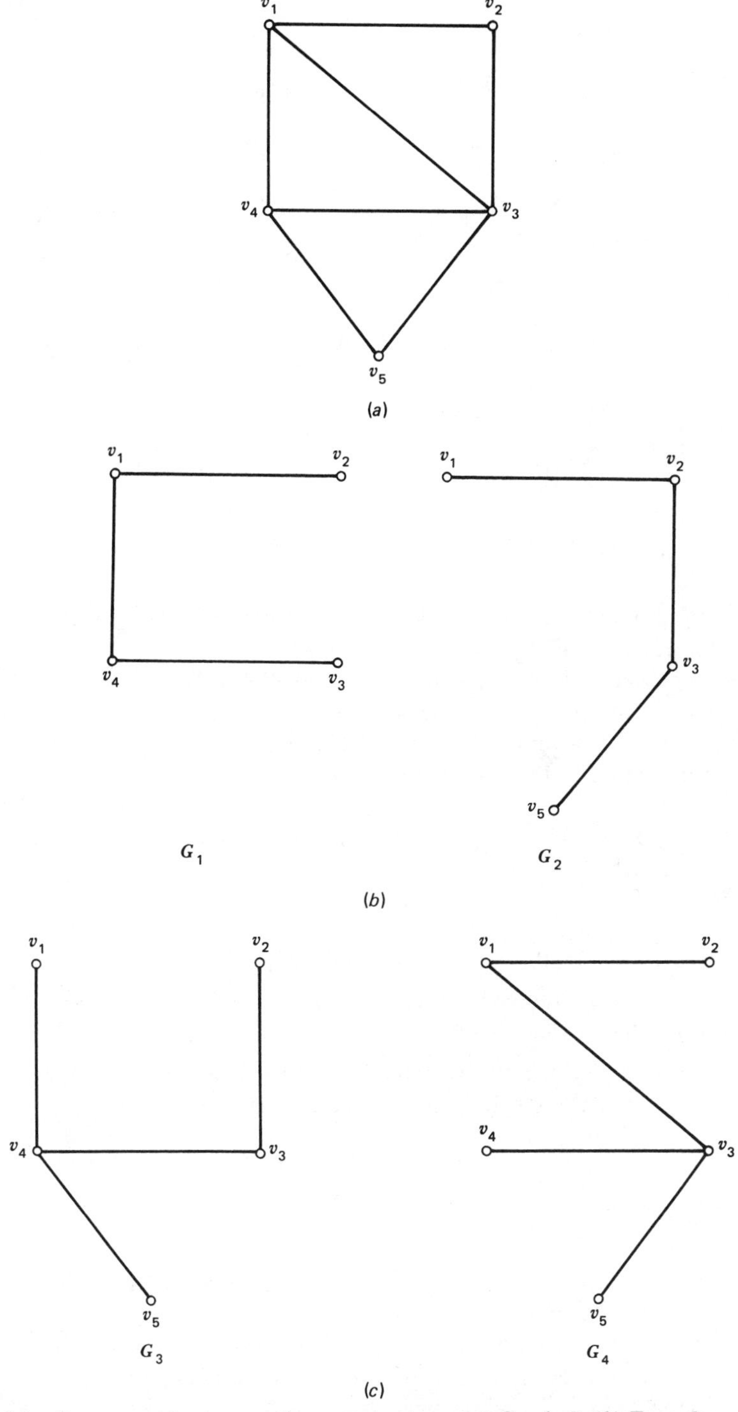

Figure 2.1. Trees, spanning trees, and cospanning trees. (*a*) Graph G. (*b*) Trees G_1 and G_2 of G. (*c*) Spanning trees G_3 and G_4 of G. (*d*) Cospanning trees G_3^* and G_4^* of G.

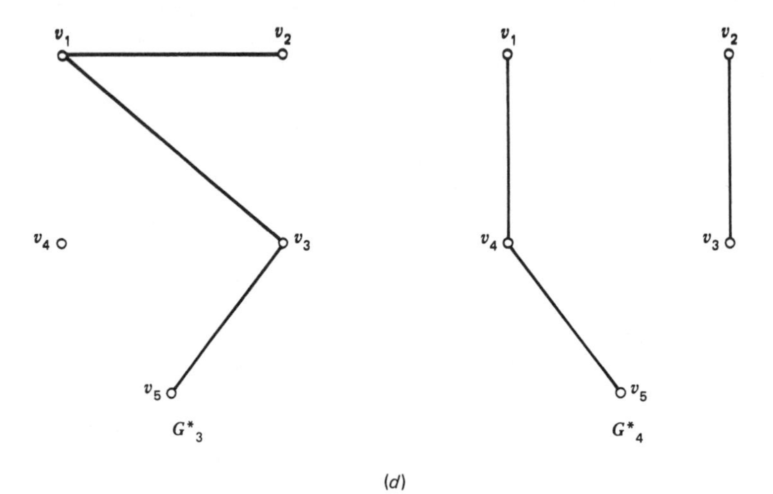

Figure 2.1. (*continued*)

A spanning tree T uniquely determines its cospanning tree T^*. As such, we refer to the edges of T^* as the chords or links of T.

We now proceed to discuss several properties of a tree.

While the definition of a tree as a connected acyclic graph is conceptually simple, there exist several other equivalent ways of characterizing a tree. These are discussed in the following theorem.

THEOREM 2.1. The following statements are equivalent for a graph G with n vertices and m edges:

1. G is a tree.
2. There exists exactly one path between any two vertices of G.
3. G is connected and $m=n-1$.
4. G is acyclic and $m=n-1$.
5. G is acyclic and if any two nonadjacent vertices of G are connected by an edge then the resulting graph has exactly one circuit.

Proof

1⇒2 See Exercise 1.2b.

2⇒3 We first note that G is connected because there exists a path between any two vertices of G.

We prove that $m=n-1$ by induction on the number of vertices of G. This is obvious for connected graphs with one or two vertices. Assume that this is true for connected graphs with fewer than n vertices.

Consider any edge e in G. The edge e constitutes the only path between the end vertices of e. Hence in $G-e$ there is no path between these vertices. Thus $G-e$ is not connected. Further, it should contain exactly two components, for otherwise the graph G will not be connected.

Let G_1 and G_2 be the two components of $G-e$. Let n_1 and m_1 be, respectively, the numbers of vertices and edges in G_1. Similarly, let n_2 and m_2 be defined for G_2. Then we have

$$n=n_1+n_2$$

and

$$m=m_1+m_2+1.$$

Note that G_1 and G_2 satisfy the hypothesis of statement 2, namely, there exists exactly one path between any two vertices in G_1 and G_2. Since $n_1<n$ and $n_2<n$, we have by the induction hypothesis

$$m_1=n_1-1$$

and

$$m_2=n_2-1.$$

Hence

$$m=n_1-1+n_2-1+1=n-1.$$

3⇒4 Let the graph G be renamed as G_0; that is, $G_0=G$.

Suppose that G_0 has some circuits. Then let G_1 be the graph that results after removing from G_0 a circuit edge, say e_1; that is, $G_1=G_0-e_1$. Since G_0 is connected, it follows from Theorem 1.4 that G_1 is also connected and has all the n vertices of G_0. Further, the number of edges in G_1 is equal to $m-1$.

If G_1 is not acyclic, let e_2 be a circuit edge of G_1. Again the graph $G_2=G_1-e_2=G_0-e_1-e_2$ should be connected and should have all the n vertices of G_1. Also, G_2 has $m-2$ edges. If G_2 is not acyclic, repeat this process until we get a connected graph G_p which is acyclic. Note that G_p has n vertices and $m-p$ edges.

Since G_p is connected and acyclic, it must be a tree. Hence it follows from the previous statement of the theorem that

$$m-p=n-1.$$

Since by hypothesis $m=n-1$, we get $p=0$. Thus the graph $G=G_0$ is acyclic.

4⇒5 Let $G_1, G_2, \ldots, G_p$ be the p components of G with n_i and m_i denoting, respectively, the numbers of vertices and edges in the component G_i. Then

$$m = m_1 + m_2 + \cdots + m_p$$

and

$$n = n_1 + n_2 + \cdots + n_p.$$

Each component G_i is connected, and it is also acyclic because G is. Hence G_i is a tree. Then, by statement 3 of the theorem,

$$m_i = n_i - 1 \qquad \text{for all } 1 \leqslant i \leqslant p.$$

Hence we get

$$m = \sum_{i=1}^{p} m_i = \sum_{i=1}^{p} (n_i - 1) = n - p.$$

However, by hypothesis

$$m = n - 1.$$

Hence we get

$$p = 1.$$

Thus G consists of exactly one component. Hence it is connected. Since it is also acyclic, it is a tree. Then, by statement 2 of the theorem, there exists exactly one path between any two distinct vertices of G. Hence if we add an edge $e = (v_1, v_2)$ to G, then this edge together with the unique path between v_1 and v_2 will form exactly one circuit in the resulting graph.

5⇒1 Suppose that G is not connected. Consider any two vertices v_a and v_b which are in different components of G. Then v_a and v_b are not connected in G.

Addition of an edge (v_a, v_b) to G does not produce a circuit since in G there is no path between v_a and v_b. This, however, contradicts the hypothesis. Hence the assumption that G is not connected is false. Thus G is connected.

Since G is also acyclic, it must be a tree, by definition. ■

It should be clear that each of the statements 1 through 5 of the above theorem represents a set of necessary and sufficient conditions for a graph G to be a tree. An immediate consequence of this theorem is the following.

Corollary 2.1.1 Consider a subgraph G' of an n-vertex graph G. Let G' have n vertices and m' edges. Then the following statements are equivalent:

1. G' is a spanning tree of G.
2. There exists exactly one path between any two vertices of G'.
3. G' is connected and $m' = n - 1$.
4. G' is acyclic and $m' = n - 1$.
5. G' is acyclic and if any two nonadjacent vertices of G' are connected by an edge then the resulting graph has exactly one circuit. ■

A condition that is not covered by the above corollary, but can be proved easily, is stated next.

Corollary 2.1.2 A subgraph G' of an n-vertex graph G is a spanning tree of G if and only if G' is acyclic, connected, and has $n-1$ edges. ■

It should now be obvious that a subgraph of an n-vertex graph G having any three of the following properties should be a spanning tree of G:

1. It has n vertices.
2. It is connected.
3. It has $n-1$ edges.
4. It is acyclic.

The question then arises whether any two of these four properties will be sufficient to define a spanning tree.

This question is answered next. (See also Exercise 2.3.)

THEOREM 2.2. A subgraph G' of an n-vertex graph G is a spanning tree of G if and only if G' is acyclic and has $n-1$ edges.

Proof

Necessity follows from Theorem 2.1, statement 4.

To show the sufficiency, we have to prove that G' is connected and has all the n vertices of G.

Let G' consist of p components $G_1, G_2, \ldots, G_p$, with n_i denoting the number of vertices in component G_i. Let n' be the number of vertices in G'. Then

$$n' = \sum_{i=1}^{p} n_i.$$

Each G_i is connected. It is also acyclic because G is. Thus each G_i is a tree and hence has $n_i - 1$ edges. Thus the total number of edges in G' is equal to

$$\sum_{i=1}^{p} (n_i - 1) = n' - p.$$

But by hypothesis

$$n' - p = n - 1.$$

Since $n' \leqslant n$ and $p \geqslant 1$, it is clear that the above equation is true if and only if $n' = n$ and $p = 1$. Thus G' is connected and has n vertices. Since it is also acyclic, it is, by definition, a spanning tree of G. ■

Suppose a graph G has a spanning tree T. Then G should be connected because the subgraph T of G is connected and has all the vertices of G. Next, we would like to prove the converse of this result, namely, that a connected graph has at least one spanning tree.

If a connected graph G is acyclic, then it is its own spanning tree. If not, let e_1 be some circuit edge of G. Then, by Theorem 1.4, the graph $G_1 = G - e_1$ is connected and has all the vertices of G. If G_1 is not acyclic, repeat the process until we get a connected acyclic graph G_p which has all the vertices of G. This graph G_p will then be a spanning tree of G.

The results of the above discussion are summarized in the next theorem.

THEOREM 2.3. A graph G is connected if and only if it has a spanning tree. ■

Since a spanning tree T of a graph G is acyclic, every subgraph of T is an acyclic subgraph of G. Is it then true that every acyclic subgraph of G is a subgraph of some spanning tree of G? The answer is "yes" as proved in the next theorem.

THEOREM 2.4. A subgraph G' of a connected graph G is a subgraph of some spanning tree of G if and only if G' is acyclic.

Proof

Necessity is obvious.

To prove the sufficiency, let T be a spanning tree of a graph G. Consider the graph $G_1 = T \cup G'$. It is obvious that G' is a subgraph of G_1. G_1 is connected and has all the vertices of G because T is a subgraph of G_1. If G_1 is acyclic, then it is a spanning tree of which G' is a subgraph, and the theorem is proved. (Note that if G_1 is acyclic, $G_1 = T$ and G' is then a subgraph of T.)

Suppose G_1 has a circuit C_1. Since G' is acyclic, it follows that not all the edges of C_1 are in G'. Thus C_1 must have at least one edge, say e_1, which is not in G'. Removal of this circuit edge e_1 from G_1 results in the graph $G_2 = G_1 - e_1$ which is also connected and has all the vertices of G_1. Note that G' is a subgraph of G_2. If G_2 is acyclic, then it is a required spanning tree. If not, repeat the process until a spanning tree of which G' is a subgraph is obtained. ■

Next we prove an interesting theorem on the minimum number of pendant vertices, that is, vertices of degree 1, in a tree.

THEOREM 2.5. In a nontrivial tree there are at least two pendant vertices.

Proof

Suppose a tree T has n vertices. Then, by Theorem 2.1, it has $n-1$ edges. We also have, by Theorem 1.1, that

$$\sum_{i=1}^{n} d(v_i) = 2 \times \text{number of edges in } T.$$

Thus

$$d(v_1) + d(v_2) + \cdots + d(v_n) = 2n - 2.$$

This equation will be true only if at least two of the terms on its left-hand side are equal to 1, that is, T has at least two pendant vertices. ■

2.2 *k*-TREES, SPANNING *k*-TREES, AND FORESTS

A *k-tree* is an acyclic graph consisting of k components. Obviously, each component of a k-tree is a tree by itself. Note that a 1-tree is the same as a tree.

If a k-tree is a spanning subgraph of a graph G, then it is called a *spanning k-tree of G*.

The *cospanning k-tree* T^* of a spanning k-tree T of G is the spanning subgraph of G containing exactly those edges of G which are not in T.

For example, the graph of Fig. 2.2*b* is a 2-tree of the graph G shown in Fig. 2.2*a*. A spanning 3-tree T of G and the corresponding cospanning 3-tree T^* are shown in Fig. 2.2*c* and *d*.

Let the k components of a spanning k-tree of an n-vertex graph G be denoted by $T_1, T_2, \ldots, T_k$. If n_i is the number of vertices in T_i, then

$$n = n_1 + n_2 + \cdots + n_k.$$

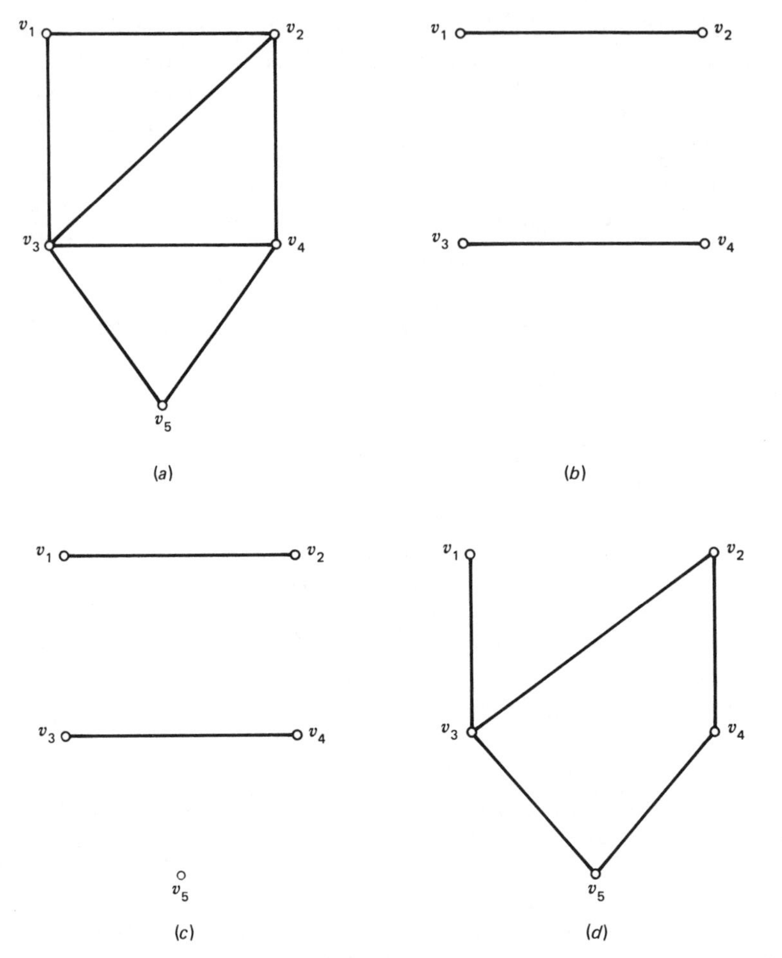

Figure 2.2. Illustrations of the definitions of a k-tree, a spanning k-tree, and a cospanning k-tree. (a) Graph G. (b) A 2-tree of G. (c) A spanning 3-tree T of G. (d) Cospanning 3-tree T^*.

Since each T_i is a tree, we have, by Theorem 2.1,

$$m_i = n_i - 1,$$

where m_i is the number of edges in T_i.

Thus the total number of edges in the spanning k-tree T is equal to

$$\sum_{i=1}^{k} m_i = \sum_{i=1}^{k} (n_i - 1) = n - k.$$

If m is the number of edges in G, then the cospanning k-tree T^* will have $m - n + k$ edges.

A *forest* of a graph G is a spanning k-tree of G, where k is the number of components in G.

If a graph G has p components, then for any spanning k-tree of $G, k \geqslant p$. Since a forest T of G is a spanning k-tree of G with $k = p$, it is necessary that each component of T be a spanning tree of one of the components of G. Thus a forest T of a graph G with p components $G_1, G_2, \ldots, G_p$ consists of p components $T_1, T_2, \ldots, T_p$ such that T_i is a spanning tree of G_i, $1 \leqslant i \leqslant p$.

The *co-forest* T^* of a forest T of a graph G is the spanning subgraph of G containing exactly those edges of G which are not in T.

Note that forest and spanning tree are synonymous in the case of a connected graph.

A forest T and the corresponding co-forest T^* of a graph are shown in Fig. 2.3.

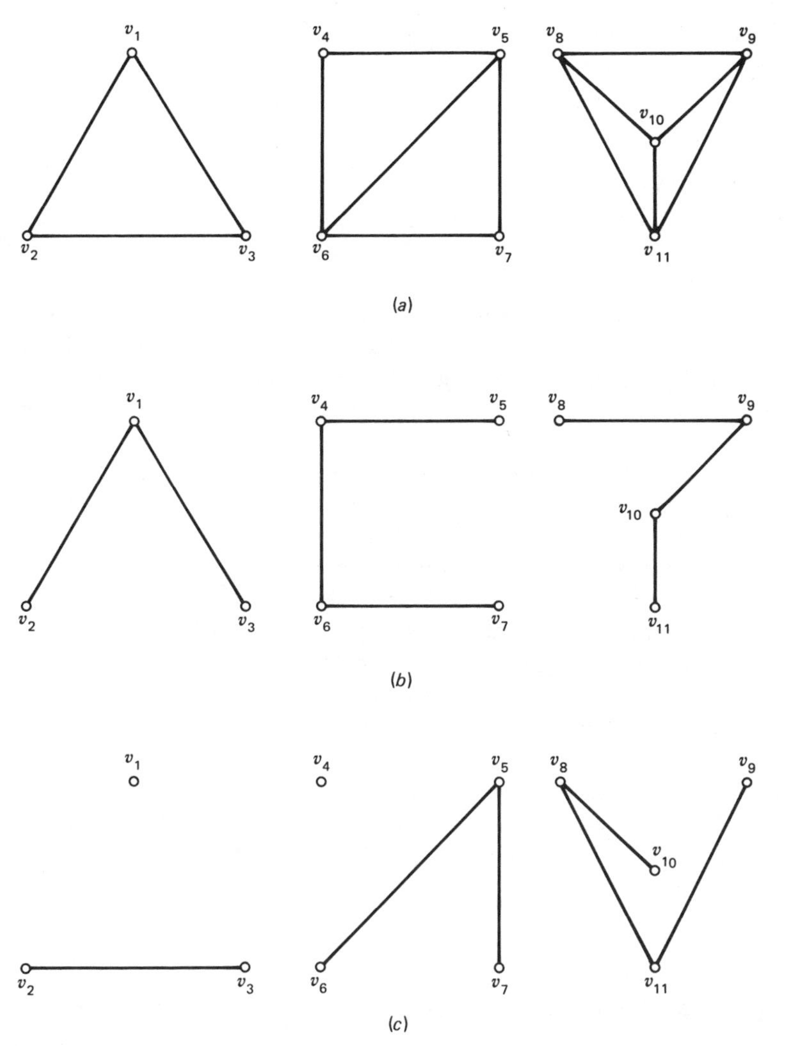

Figure 2.3. Forest and co-forest. (a) Graph G. (b) A forest T of G. (c) Co-forest T^*.

2.3 RANK AND NULLITY

Consider a graph G with m edges, n vertices, and k components. The *rank* of G, denoted by $\rho(G)$, is defined equal to $n-k$, and the *nullity* of G, denoted by $\mu(G)$, is defined equal to $m-n+k$. Note that

$$\rho(G)+\mu(G)=m.$$

It follows from the definition of a forest and a co-forest that the rank $\rho(G)$ of a graph G is equal to the number of edges in a forest of G and the nullity $\mu(G)$ of G is equal to the number of edges in a co-forest of G.

The numbers $\rho(G)$ and $\mu(G)$ are among the most important ones associated with a graph. As we see in Chapter 4, they define the dimensions of the cutset and circuit subspaces of a graph.

2.4 FUNDAMENTAL CIRCUITS

Consider a spanning tree T of a connected graph G. Let the branches of T be denoted by $b_1, b_2, \ldots, b_{n-1}$, and let the chords of T be denoted by $c_1, c_2, \ldots, c_{m-n+1}$, where m is the number of edges in G, and n is the number of vertices in G.

While T is acyclic, by Theorem 2.1 the graph $T \cup c_i$ contains exactly one circuit C_i. This circuit consists of the chord c_i and those branches of T which lie in the unique path in T between the end vertices of c_i. The circuit C_i is called the *fundamental circuit* of G with respect to the chord c_i of the spanning tree T.

The set of all the $m-n+1$ fundamental circuits $C_1, C_2, \ldots, C_{m-n+1}$ of G with respect to the chords of the spanning tree T of G is known as the *fundamental set of circuits* of G with respect to T.

An important feature of the fundamental circuit C_i is that it contains exactly one chord, namely, chord c_i. Further, chord c_i is not present in any other fundamental circuit with respect to T. Because of these properties, the edge set of no fundamental circuit can be expressed as the ring sum of the edge sets of some or all of the remaining fundamental circuits. We also show in Chapter 4 that every circuit of a graph G can be expressed as the ring sum of some fundamental circuits of G with respect to a spanning tree of G. It is for these reasons that the "fundamental" circuits are called so.

A graph G and a set of fundamental circuits of G are shown in Fig. 2.4.

2.5 CUTSETS

A *cutset* S of a connected graph G is a minimal set of edges of G such that its removal from G disconnects G, that is, the graph $G-S$ is disconnected.

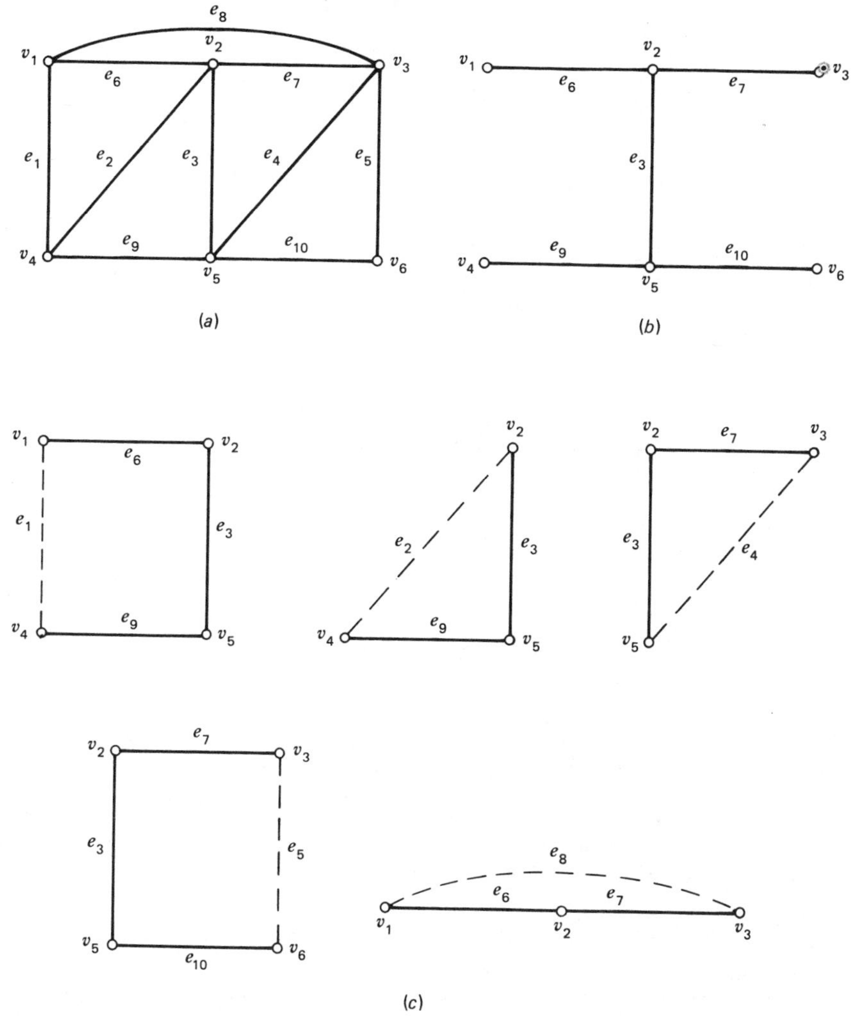

Figure 2.4. A set of fundamental circuits of a graph G. (a) Graph G. (b) A spanning tree T of G. (c) Set of five fundamental circuits of G with respect to T. (Chords are indicated by dashed lines.)

For example, consider the subset $S_1 = \{e_1, e_3, e_7, e_{10}\}$ of edges of the graph G in Fig. 2.5a. The removal of S_1 from G results in the graph $G_1 = G - S_1$ of Fig. 2.5b. G_1 is disconnected. Furthermore, the removal of any proper subset of S_1 cannot disconnect G. Thus S_1 is a cutset of G.

Consider next the set $S_2 = \{e_1, e_2, e_5, e_7, e_9\}$. The graph $G_2 = G - S_2$ shown in Fig. 2.5c is disconnected. However, the set $S_2' = \{e_1, e_2, e_5, e_9\}$ which is a proper subset of S_2 also disconnects G. The graph $G_3 = G - S_2'$ is shown in Fig. 2.5d. Thus S_2 is not a cutset of G.

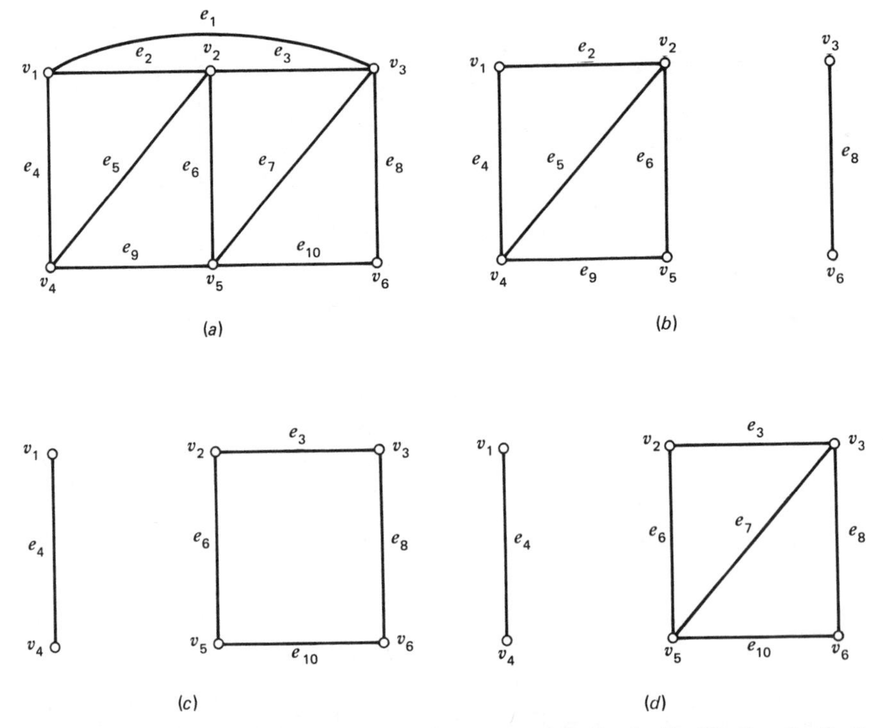

Figure 2.5. Illustration of the definition of a cutset. (*a*) Graph G. (*b*) $G_1 = G - S$, $S = \{e_1, e_3, e_7, e_{10}\}$. (*c*) $G_2 = G - S_2$, $S_2 = \{e_1, e_2, e_5, e_7, e_9\}$. (*d*) $G_3 = G - S_2'$, $S_2' = \{e_1, e_2, e_5, e_9\}$.

Note that, by the definition of a cutset given above, if S is a cutset of a graph G, then the ranks of G and $G - S$ differ by at least 1; that is, $\rho(G) - \rho(G - S) \geq 1$.

Seshu and Reed [2.1] define a cutset as follows:

A cutset S of a connected graph G is a minimal set of edges of G such that removal of S disconnects G into exactly two components; that is, $\rho(G) - \rho(G - S) = 1$.

The question now arises whether these two definitions of a cutset are equivalent. The answer is "yes" and its proof is left as an exercise (Exercise 2.15).

2.6 CUT

We now define the concept of a cut which is closely related to that of a cutset.

Consider a connected graph G with vertex set V. Let V_1 and V_2 be two mutually disjoint subsets of V such that $V = V_1 \cup V_2$; that is, V_1 and V_2 have no common vertices and together contain all the vertices of V. Then the set S of all those edges of G having one end vertex in V_1 and the other in V_2 is

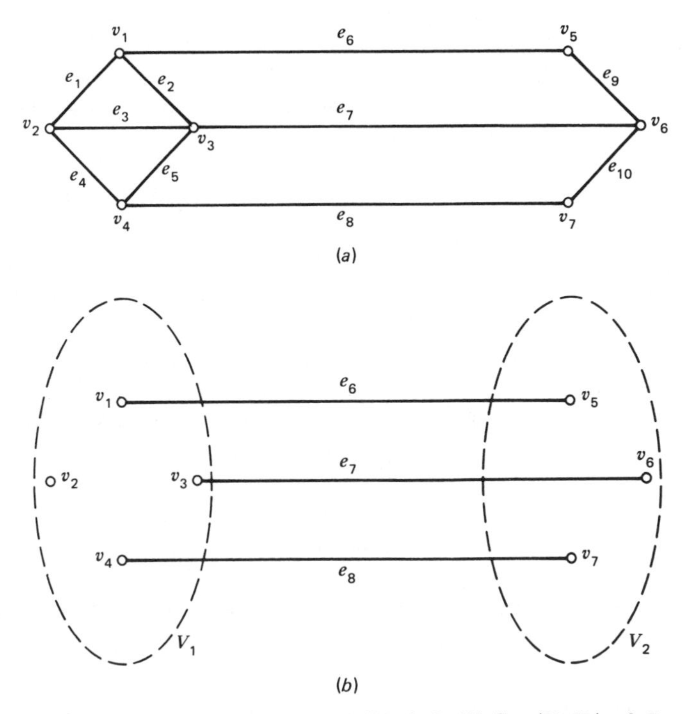

Figure 2.6. Definition of a cut. (*a*) Graph G. (*b*) Cut $\langle V_1, V_2 \rangle$ of G.

called a *cut* of G. This is usually denoted by $\langle V_1, V_2 \rangle$. Reed [2.2] refers to a cut as a *seg* (the set of edges segregating the vertex set V).

For example, for the graph G shown in Fig. 2.6, if $V_1 = \{v_1, v_2, v_3, v_4\}$ and $V_2 = \{v_5, v_6, v_7\}$, then the cut $\langle V_1, V_2 \rangle$ of G is equal to the set $\{e_6, e_7, e_8\}$ of edges.

Note that the cut $\langle V_1, V_2 \rangle$ of G is the minimal set of edges of G whose removal disconnects G into two graphs G_1 and G_2 which are induced subgraphs of G on the vertex sets V_1 and V_2. G_1 and G_2 may not be connected. If both these graphs are connected, then $\langle V_1, V_2 \rangle$ is also the minimal set of edges disconnecting G into exactly two components. Then, by definition, $\langle V_1, V_2 \rangle$ is a cutset of G.

Suppose that for a cutset S of G, V_1 and V_2 are, respectively, the vertex sets of the two components G_1 and G_2 of $G - S$. Then S is the cut $\langle V_1, V_2 \rangle$.

Thus we have the following theorem.

THEOREM 2.6.

1. A cut $\langle V_1, V_2 \rangle$ of a connected graph G is a cutset of G if the induced subgraphs of G on vertex sets V_1 and V_2 are connected.
2. If S is a cutset of a connected graph G, and V_1 and V_2 are the vertex sets of the two components of $G - S$, then $S = \langle V_1, V_2 \rangle$. ■

Any cut $\langle V_1, V_2 \rangle$ in a connected graph G contains a cutset of G, since the removal of $\langle V_1, V_2 \rangle$ from G disconnects G. In fact, we can prove that a cut in a graph G is the union of some edge-disjoint cutsets of G. Formally, we state this in the following theorem.

THEOREM 2.7. A cut in a connected graph G is the union of some edge-disjoint cutsets of G. ■

The proof of this theorem is not difficult, and it is left as an exercise (Exercise 2.19).

Consider next a vertex v_1 in a connected graph. The set of edges incident on v_1 forms the cut $\langle v_1, V-v_1 \rangle$. The removal of these edges disconnects G into two subgraphs. One of these subgraphs containing only the vertex v_1 is, by definition, connected. The other subgraph is the induced subgraph G' of G on the vertex set $V-v_1$. Thus the cut $\langle v_1, V-v_1 \rangle$ is a cutset if and only if G' is connected. However, G' is connected if and only if v_1 is not a cut-vertex (Section 1.7). Thus we have the following theorem.

THEOREM 2.8. The set of edges incident on a vertex v in a connected graph G is a cutset of G if and only if v is not a cut-vertex of G. ■

For example, consider the separable graph G shown in Fig. 2.7*a*. v_1 is a cut-vertex of G. The induced subgraph of G on the vertex set $V-v_1=$

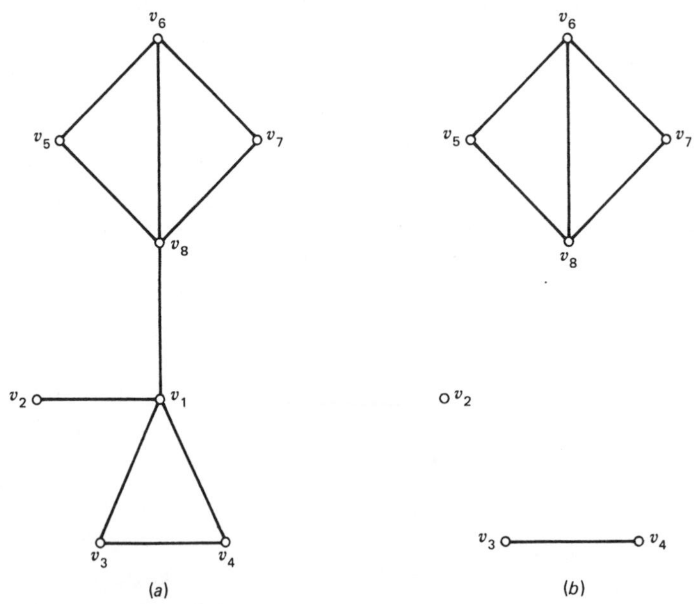

Figure 2.7. Illustration of Theorem 2.8. (*a*) Graph G. (*b*) Induced subgraph of G on the vertex set $\{v_2, v_3, v_4, v_5, v_6, v_7, v_8\}$.

$\{v_2, v_3, v_4, v_5, v_6, v_7, v_8\}$ is shown in Fig. 2.7*b*. This subgraph consists of three components and is not connected. Thus the edges incident on the cut-vertex v_1 do not form a cutset of G.

2.7 FUNDAMENTAL CUTSETS

It was shown in Section 2.4 that a spanning tree of a connected graph can be used to obtain a set of fundamental circuits of the graph. We show in this section how a spanning tree can also be used to define a set of fundamental cutsets.

Consider a spanning tree T of a connected graph G. Let b be a branch of T. Now, removal of the branch b disconnects T into exactly two components T_1 and T_2. Note that T_1 and T_2 are trees of G. Let V_1 and V_2, respectively, denote the vertex sets of T_1 and T_2. V_1 and V_2 together contain all the vertices of G.

Let G_1 and G_2 be, respectively, the induced subgraphs of G on the vertex sets V_1 and V_2. It can be seen that T_1 and T_2 are, respectively, the spanning trees of G_1 and G_2. Hence, by Theorem 2.3, G_1 and G_2 are connected. This, in turn, proves (Theorem 2.6) that the cut $\langle V_1, V_2\rangle$ is a cutset of G. This cutset is known as the *fundamental cutset* of G with respect to the branch b of the spanning tree T of G. The set of all the $n-1$ fundamental cutsets with respect to the $n-1$ branches of a spanning tree T of a connected graph G is known as the *fundamental set of cutsets* of G with respect to the spanning tree T.

Note that the cutset $\langle V_1, V_2\rangle$ contains exactly one branch, namely, the branch b of T. All the other edges of $\langle V_1, V_2\rangle$ are links of T. This follows from the fact that $\langle V_1, V_2\rangle$ does not contain any edge of T_1 or T_2. Further, branch b is not present in any other fundamental cutset with respect to T. Because of these properties, the edge set of no fundamental cutset can be expressed as the ring sum of the edge sets of some or all of the remaining fundamental cutsets. We show in Chapter 4 that every cutset of a graph G can be expressed as the ring sum of some of the fundamental cutsets of G with respect to a spanning tree T of G.

A graph G and a set of fundamental cutsets of G are shown in Fig. 2.8.

2.8 SPANNING TREES, CIRCUITS, AND CUTSETS

In this section, we discuss some interesting results which relate cutsets and circuits to spanning trees and cospanning trees, respectively. These results will bring out the "dual" nature of circuits and cutsets. They will also lead to alternate characterizations of cutsets and circuits in terms of spanning trees and cospanning trees, respectively.

It is obvious that removal of a cutset S from a connected graph G destroys all the spanning trees of G. A little thought will indicate that a cutset is a

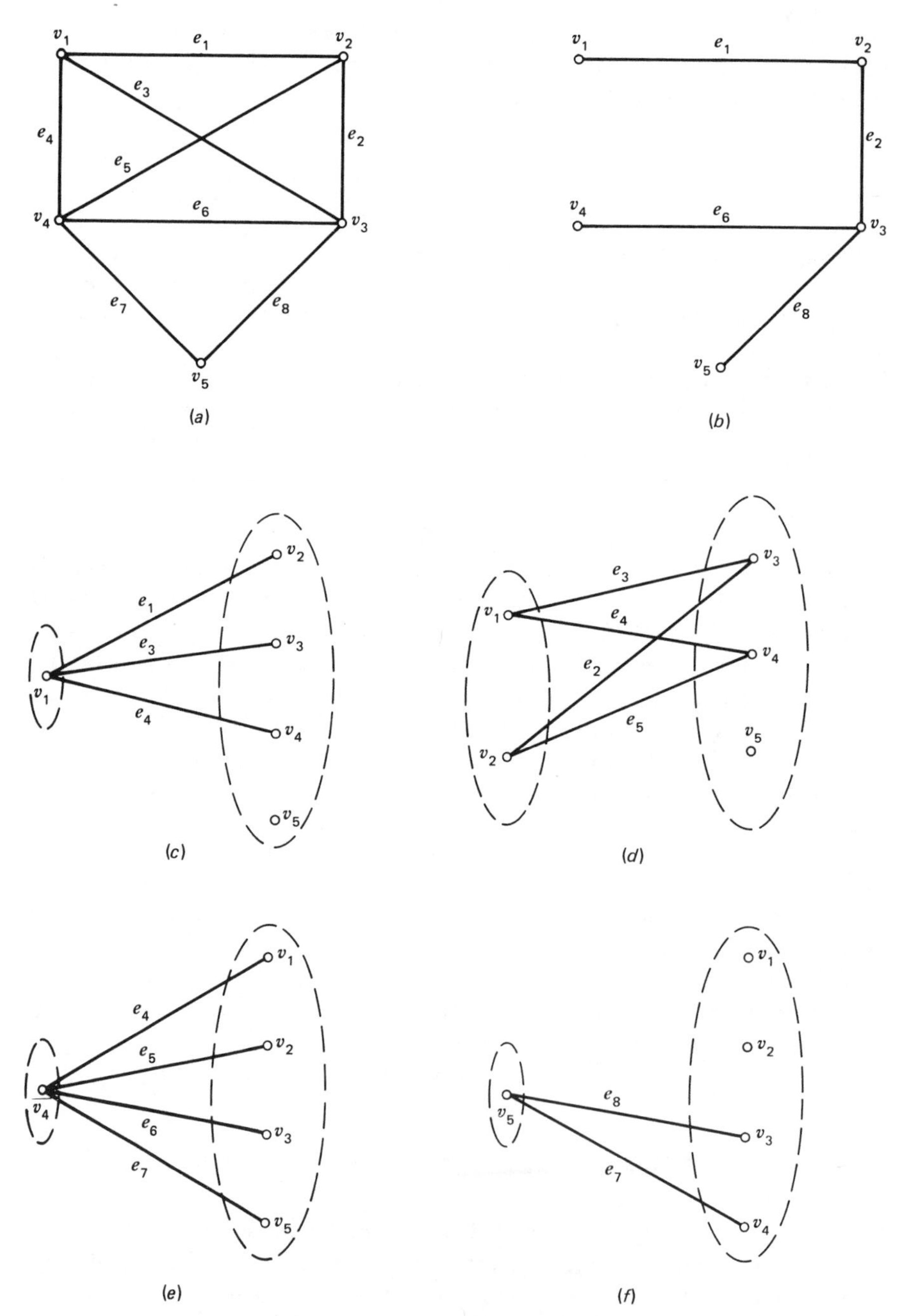

Figure 2.8. A set of fundamental cutsets of a graph. (*a*) Graph *G*. (*b*) Spanning tree *T* of *G*. (*c*) Fundamental cutset with respect to branch e_1. (*d*) Fundamental cutset with respect to branch e_2. (*e*) Fundamental cutset with respect to branch e_6. (*f*) Fundamental cutset with respect to branch e_8.

minimal set of edges whose removal from G destroys all spanning trees of G. However, the converse of this result is not so obvious. The first few theorems of this section discuss these questions and similar ones relating to circuits.

THEOREM 2.9. A cutset of a connected graph G contains at least one branch of every spanning tree of G.

Proof

Suppose that a cutset S of G contains no branch of a spanning tree T of G. Then the graph $G-S$ will contain the spanning tree T and hence, by Theorem 2.3, $G-S$ is connected. This, however, contradicts that S is a cutset of G. ■

THEOREM 2.10. A circuit of a connected graph G contains at least one edge of every cospanning tree of G.

Proof

Suppose that a circuit C of G contains no edge of the cospanning tree T^* of a spanning tree T of G. Then the graph $G-T^*$ will contain the circuit C. Since $G-T^*$ is the same as the spanning tree T, this means that the spanning tree T contains a circuit. However, this is contrary to the definition of a spanning tree. ■

THEOREM 2.11. A set S of edges of a connected graph G is a cutset of G if and only if S is a minimal set of edges containing at least one branch of every spanning tree of G.

Proof

Necessity If the set S of edges of G is a cutset of G, then, by Theorem 2.9, it contains at least one branch of every spanning tree of G. If it is not a minimal such set, then a proper subset S' of S will contain at least one branch of every spanning tree of G. Then $G-S'$ will contain no spanning tree of G and it will be disconnected. Thus removal of a proper subset S' of the cutset S of G will disconnect G. This, however, would contradict the definition of a cutset. Hence the necessity.

Sufficiency If S is a minimal set of edges containing at least one branch of every spanning tree of G, then the graph $G-S$ will contain no spanning tree, and hence it will be disconnected. Suppose S is not a cutset. Then a proper subset S' of S will be a cutset. Then, by the necessity part of the theorem, S' will be a minimal set of edges containing at least one branch of every spanning tree of G. This, however, will contradict that S is a minimal such set. Hence the sufficiency. ■

The above theorem gives a characterization of a cutset in terms of spanning trees. We would like to establish next a similar characterization for a circuit in terms of cospanning trees.

Consider a set C of edges constituting a circuit in a graph G. By Theorem 2.10, C contains at least one edge of every cospanning tree of G. We now show that no proper subset C' of C has this property.

It is obvious that C' does not contain a circuit. Hence, by Theorem 2.4, we can construct a spanning tree T containing C'. The cospanning tree T^* corresponding to T has no common edge with C'. Hence for every proper subset C' of C, there exists at least one cospanning tree T^* which has no common edge with C'. In fact, this statement is true for every acyclic subgraph of a graph. Thus we have the following theorem.

THEOREM 2.12. A circuit of a connected graph G is a minimal set of edges of G containing at least one edge of every cospanning tree of G. ■

The converse of the above theorem follows next.

THEOREM 2.13. The set C of edges of a connected graph G is a circuit of G if it is a minimal set containing at least one edge of every cospanning tree of G.

Proof

As shown earlier, the set C cannot be acyclic since there exists, for every acyclic subgraph G' of G, a cospanning tree not having in common any edge with G'. Thus C has at least one circuit C'. Suppose that C' is a proper subset of C. Then, by Theorem 2.12, C' is a minimal set of edges containing at least one edge of every cospanning tree of G. This, however, contradicts the hypothesis that C is a minimal such set. Hence no proper subset of C is a circuit. Since C is not acyclic, C must be a circuit. ■

Theorems 2.12 and 2.13 establish that a set C of edges of a connected graph G is a circuit if and only if it is a minimal set of edges containing at least one edge of every cospanning tree of G.

The new characterizations of a cutset and a circuit as given by Theorems 2.11, 2.12, and 2.13 clearly bring out the dual nature of the concepts of circuits and cutsets. This duality is explored further in Chapter 10, where we discuss the theory of matroids.

The next theorem relates circuits and cutsets without involving trees.

THEOREM 2.14. A circuit and a cutset of a connected graph have an even number of common edges.

Proof

Let C be a circuit and S a cutset of a connected graph G. Let V_1 and V_2 be the vertex sets of the two connected subgraphs G_1 and G_2 of $G-S$.

If C is a subgraph of G_1 or of G_2, then obviously the number of edges common to C and S is equal to zero, an even number.

Suppose that C and S have some common edges. Let us traverse the circuit C starting from a vertex, say v_1, in the set V_1. Since the traversing should end at v_1, it is necessary that every time we meet with an edge of S leading us from a vertex in V_1 to a vertex in V_2, there must be an edge of S leading us from a vertex in V_2 back to a vertex in V_1. This is possible only if C and S have an even number of common edges. ■

The above theorem is a very important one. It forms the basis of the orthogonality relationship between cutsets and circuits. This relationship is discussed in Chapter 4.

We would like to point out that the converse of Theorem 2.14 is not quite true. However, we show in Chapter 4 that a set S of edges of a graph G is a cutset (circuit) or the union of some edge-disjoint cutsets (circuits) if and only if S has an even number of edges in common with every circuit (cutset).

Fundamental circuits and fundamental cutsets of a connected graph have been defined with respect to a spanning tree of a graph. It is, therefore, not surprising that fundamental circuits and cutsets are themselves related as proved next.

THEOREM 2.15.

1. The fundamental circuit with respect to a chord of a spanning tree T of a connected graph consists of exactly those branches of T whose fundamental cutsets contain the chord.
2. The fundamental cutset with respect to a branch of a spanning tree T of a connected graph consists of exactly those chords of T whose fundamental circuits contain the branch.

Proof

1. Let C be the fundamental circuit of a connected graph G with respect to a chord c_1 of a spanning tree T of G. Let C contain, in addition to the chord c_1, the branches $b_1, b_2, \ldots, b_k$ of T.

   Suppose S_i, $1 \leqslant i \leqslant k$, is the fundamental cutset of G with respect to the branch b_i, $1 \leqslant i \leqslant k$, of T. The branch b_i is the only branch common to both C and S_i. The chord c_1 is the only chord in C. Since C and S_i must have an even number of common edges, it is necessary that the fundamental cutset S_i contain c_1. Next we show that no other fundamental cutset of T contains c_1.

   Suppose the fundamental cutset S_{k+1} with respect to some branch b_{k+1}

of T contains c_1. Then c_1 will be the only common edge between C and S_{k+1}. This will contradict Theorem 2.14. Thus the chord c_1 is present only in those cutsets defined by the branches $b_1, b_2, \ldots, b_k$.

2. Proof of this part is similar to that of part 1. ■

2.9 FURTHER READING

The concept of a tree is central in the development of several results relating circuits and cutsets of a graph. It so happens that the spanning trees, circuits, and cutsets of a connected graph are, respectively, the bases, circuits, and cocircuits of a matroid that can be defined on the edge set of a graph. So the results of this chapter will be of help in understanding our development of matroid theory in Chapter 10.

Electrical network theory is one of the earliest areas of applications of graph theory. The results of this chapter and those of Chapters 4 and 6 form the foundation of the graph-theoretic study of electrical networks. The pioneering work of Seshu and Reed [2.1] and textbooks by Kim and Chien [2.3], Chen [2.4], and Mayeda [2.5] are highly recommended for further reading on this topic.

Several questions relating to trees have been extensively studied in the literature. Some of these are discussed in the subsequent chapters of the book.

2.10 EXERCISES

2.1 Show that there are exactly six nonisomorphic trees on six vertices, and 11 on seven vertices. Draw all of them.

2.2 Show that a tree is a bipartite graph.

2.3 Consider a subgraph G_s of a connected graph G having n vertices. Show that except for the pair (b) and (d), no other pair of the following conditions implies that G_s is a spanning tree of G:

(a) G_s contains n vertices.
(b) G_s contains $n-1$ edges.
(c) G_s is connected.
(d) G_s contains no circuits.

2.4 Prove that each pendant edge (the edge incident on a pendant vertex) in a connected graph G is contained in every spanning tree of G.

2.5 Prove that each edge of a connected graph G is a branch of some spanning tree of G.

2.6 Prove that in a tree every vertex of degree greater than 1 is a cut-vertex.

2.7 Prove that each edge of a nonseparable graph G can be made a chord of some cospanning tree of G.

2.8 Prove or disprove: Any two edges of a nonseparable graph can be contained in some fundamental circuit.

2.9 Under what conditions can any two edges of a graph G be made chords of some cospanning tree of G?

2.10 Prove that a nonseparable graph has nullity equal to 1 if and only if it is a circuit.

2.11 Show that the nullity of any graph is nonnegative. Give an example of a graph with nullity equal to 0.

2.12 Consider the following two operations on a graph G:

(a) If only two edges $e_1=(v_1,v_a)$ and $e_2=(v_2,v_a)$ are incident on vertex v_a, then replace e_1 and e_2 by a single edge connecting v_1 and v_2.

(b) Replace any edge (v_1,v_2) by two edges (v_1,v_a) and (v_2,v_a) where v_a is a new vertex not in G.

Prove that the nullity of G is invariant under the above operations.

2.13 A connected graph G is *minimally connected* if for every edge e of G the graph $G-e$ is not connected. Prove that a connected graph is a tree if and only if it is minimally connected.

2.14 Prove that a subgraph G_s of a connected graph G is a spanning tree of G if and only if it is a maximal subgraph of G containing no circuits.

2.15 Show that a cutset of a connected graph G is a minimal set S of edges of G such that removal of S disconnects G into exactly two components; that is, $\rho(G)-\rho(G-S)=1$.

2.16 Prove that every connected graph contains a cutset.

2.17 Prove that a subgraph of a connected graph G can be included in a cospanning tree of G if and only if it contains no cutsets of G.

2.18 Prove that a subset S of edges of a connected graph G forms a cospanning tree of G if and only if it is a maximal subset of edges containing no cutsets of G.

2.19 Prove that a cut in a connected graph G is a cutset or the union of some edge-disjoint cutsets of G.

2.20 Prove that every cutset of a nonseparable graph with more than two vertices contains at least two edges.

2.21 Prove that a graph G is nonseparable if and only if every two edges of G lie on a common cutset.

2.22 Let C be a circuit in a graph G. Let a and b be any two edges in C. Prove that there exists a cutset S such that $S \cap C = \{a, b\}$.

2.23 Let T_1 and T_2 be spanning trees of a connected graph G. Show that if e is any edge of T_1, then there exists an edge f of T_2 with the property that $(T_1 - e) \cup f$ (the graph obtained from T_1 by replacing e by f) is also a spanning tree of G. Show also that T_1 can be "transformed" into T_2 by replacing the edges of T_1 one at a time by the edges of T_2 in such a way that at each stage we obtain a spanning tree of G.

2.24 (a) Let C_1 and C_2 be two circuits in a graph G. Let e_1 be an edge that is in both C_1 and C_2, and let e_2 be an edge that is in C_1 but not in C_2. Prove that there exists a circuit C_3 which is such that $C_3 \subseteq (C_1 \cup C_2) - e_1$, and e_2 is in C_3.

(b) Repeat part (a) when the term "circuit" is replaced by a "cutset."

Note This result is one of the postulates used by Whitney [2.6] to define "circuits" of a matroid (Chapter 10).

2.25 Let T be an arbitrary tree on $k+1$ vertices. Show that if G is simple and $\delta(G) \geqslant k$, then G has a subgraph isomorphic to T.

2.26 Show that a graph G contains k edge-disjoint spanning trees if and only if for each partition $(V_1, V_2, \ldots, V_l)$ of V the number of edges which have their end vertices in different parts of the partition is at least $k(l-1)$ (Nash-Williams [2.7] and Tutte [2.8]).

2.27 A *center* of a graph $G = (V, E)$ is a vertex u such that $\max_{v \in V}\{d(u, v)\}$ is as small as possible, where $d(u, v)$ is the distance between u and v. Show that a tree has exactly one center or two adjacent centers.

2.28 The *tree graph* of an n-vertex connected graph G is the graph whose vertices are the spanning trees $T_1, T_2, \ldots, T_r$ of G, with T_i and T_j adjacent if and only if they have exactly $n-2$ edges in common. Show that the tree graph of any connected graph is connected (Cummins [2.9]).

Hint See Exercise 2.23.

2.11 REFERENCES

2.1 S. Seshu and M. B. Reed, *Linear Graphs and Electrical Networks*, Addison-Wesley, Reading, Mass., 1961.

2.2 M. B. Reed, "The Seg: A New Class of Subgraphs," *IRE Trans. Circuit Theory*, Vol. CT-8, 17–22 (1961).

2.3 W. H. Kim and R. T. Chien, *Topological Analysis and Synthesis of Communication Networks*, Columbia University Press, New York, 1962.

2.4 W. K. Chen, *Applied Graph Theory*: *Graphs and Electrical Networks*, North-Holland, Amsterdam, 1971.

2.5 W. Mayeda, *Graph Theory*, Wiley-Interscience, New York, 1972.

2.6 H. Whitney, "On the Abstract Properties of Linear Dependence," *Am. J. Math.*, Vol. 57, 509–533 (1935).

2.7 C. St. J. A. Nash-Williams, "Edge-Disjoint Spanning Trees of Finite Graphs," *J. London Math. Soc.*, Vol. 36, 445–450 (1961).

2.8 W. T. Tutte, "On the Problem of Decomposing a Graph into *n* Connected Factors," *J. London Math. Soc.*, Vol. 36, 221–230 (1961).

2.9 R. L. Cummins, "Hamilton Circuits in Tree Graphs," *IEEE Trans. Circuit Theory*, Vol. CT-13, 82–90 (1966).

Chapter 3

Eulerian and Hamiltonian Graphs

Many discoveries in graph theory can be traced to attempts to solve "practical" problems—puzzles, games, and so on. One of these problems was the celebrated Königsberg bridge problem. This problem may be stated as follows:

There were two islands in the Pregel river in Königsberg, Germany. These islands were connected to each other and to the banks of the river by seven bridges as shown in Fig. 3.1*a*. The problem was to start at any one of the four land areas (marked as A, B, C, and D in Fig. 3.1*a*), walk through each bridge exactly once, and then return to the starting point, that is, to establish a closed walk across all the seven bridges without recrossing any of them.

Many were convinced that there was no solution to this problem. However, Euler, the great Swiss mathematician, proved so in 1736 and laid the foundation of the theory of graphs. Euler first showed that the problem was equivalent to establishing a closed trail along the edges of the graph of Fig. 3.1*b*, where the vertices A, B, C, and D represent the land areas, and the edges represent the bridges connecting the land areas. He then generalized the problem and established a characterization of graphs in which such a closed trail exists. These graphs have come to be known as Eulerian graphs. The discussions of Section 3.1 concern these graphs.

In 1859 Sir William Hamilton, another great mathematician invented a game. The game challenges the player to establish a closed route along the edges of a dodecahedron which passes through each vertex exactly once. Hamilton's game, in graphical terms, is equivalent to determining whether in

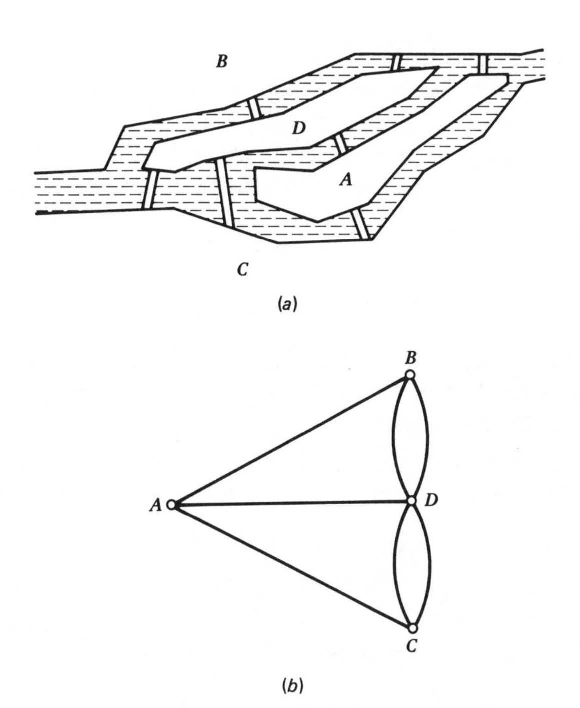

Figure 3.1. (*a*) Königsberg bridge problem. (*b*) Graph of the Königsberg bridge problem.

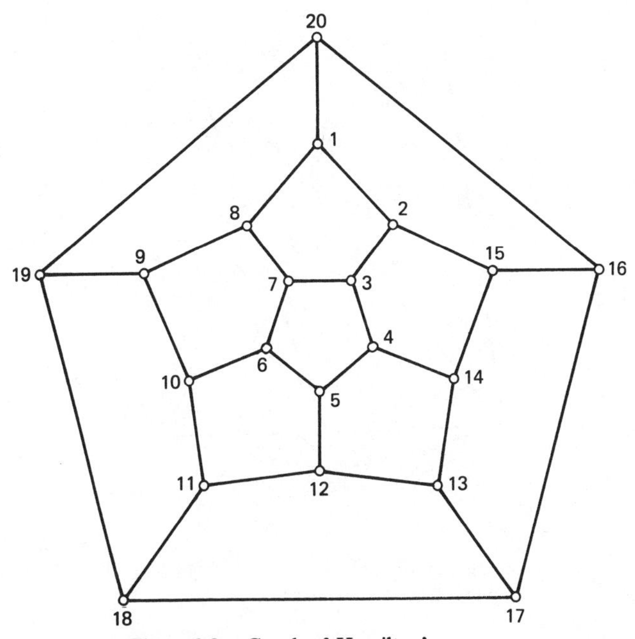

Figure 3.2. Graph of Hamilton's game.

the graph of the dodecahedron shown in Fig. 3.2 there exists a spanning circuit, that is, a circuit containing all the 20 vertices. It may be verified that the sequence of vertices 1,2,...,20, 1 constitutes one such circuit in the graph. All graphs in which a spanning circuit exists have come to be known as Hamiltonian graphs. The discussions of Section 3.2 concern these graphs.

3.1 EULERIAN GRAPHS

An *Euler trail* in a graph G is a closed trail containing all the edges of G. An *open Euler trail* is an open trail containing all the edges of G. A graph possessing an Euler trail is an *Eulerian graph*.

Consider the graph G_1 shown in Fig. 3.3*a*. The sequence of edges e_1, e_2, e_3, e_4, e_5, e_6, e_7, e_8, e_9, e_{10}, e_{11}, and e_{12} constitutes an Euler trail in G_1. Hence G_1 is Eulerian.

In the graph G_2 of Fig. 3.3*b*, the sequence of edges e_1, e_2, e_3, e_4, e_5, e_6, e_7, e_8, e_9, e_{10}, e_{11}, e_{12}, and e_{13} constitutes an open Euler trail. However, there is no Euler trail in G_2. Hence G_2 is not Eulerian.

A non-Eulerian graph G_3 with no open Euler trail is shown in Fig. 3.3*c*.

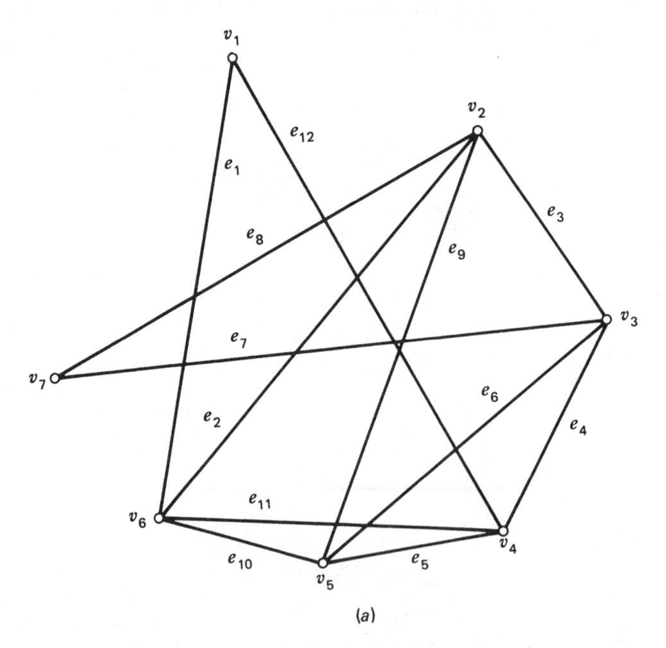

Figure 3.3. (*a*) G_1, an Eulerian graph. (*b*) G_2, a non-Eulerian graph having an open Euler trail. (*c*) G_3, a non-Eulerian graph with no open Euler trail.

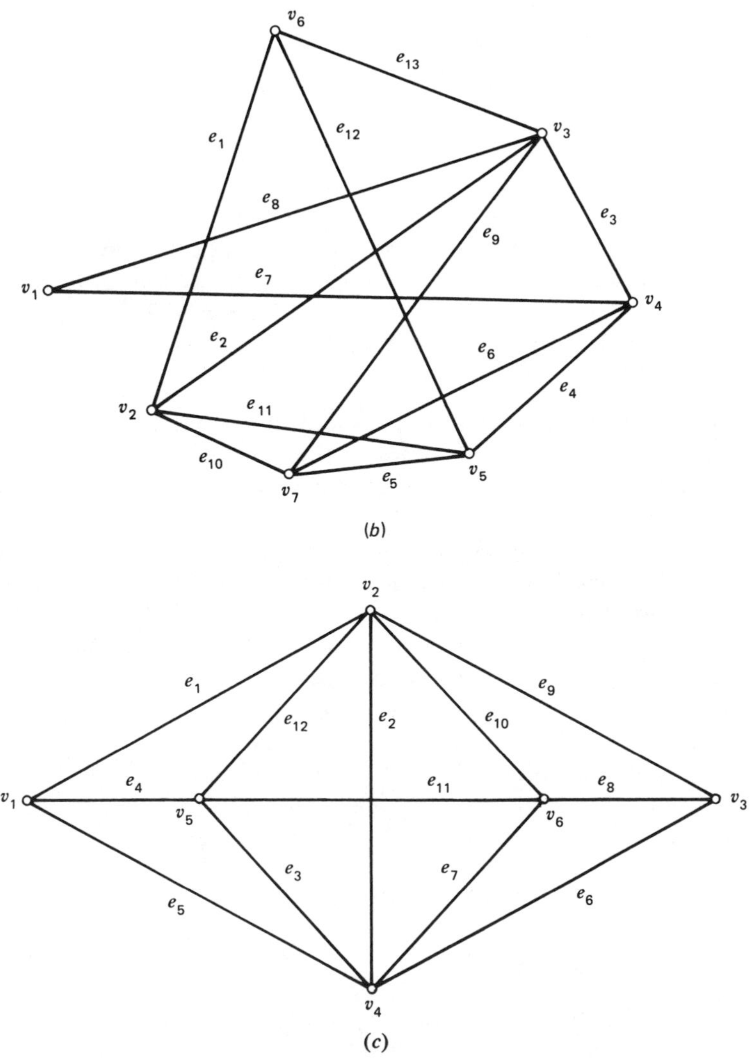

Figure 3.3. *(continued)*

The following theorem gives simple and useful characterizations of Eulerian graphs.

THEOREM 3.1. The following statements are equivalent for a connected graph G:

1. G is Eulerian.
2. The degree of every vertex in G is even.
3. G is the union of some edge-disjoint circuits.

Proof

1⇒2 Let T be an Euler trail in G. Suppose we traverse T starting from any vertex, say v_1, in G. Let T be

$$v_1 = x_1, e_1, x_2, e_2, x_3, \ldots, e_{r-1}, x_r, e_r, x_{r+1} = v_1$$

where, of course, all the edges are distinct; the vertices $x_2, \ldots, x_r$ may not all be distinct and some of these vertices may be v_1. Then it is clear that the pair of successive edges e_i and e_{i+1}, $1 \leqslant i \leqslant r-1$, contributes 2 to the degree of the vertex x_{i+1}. In addition, vertex v_1 gets a contribution of 2 to its degree from the initial and the final edges e_1 and e_r. Thus all the vertices are of even degree.

2⇒3 Since G is connected and every vertex in G has even degree, it follows that the degree of each vertex in G is greater than 1. Thus G has no pendant vertices. Hence G is not a tree by Theorem 2.5. This means that G has at least one circuit, say C_1.

Consider the graph $G_1 = G - C_1$. Since every vertex in C_1 is also of even degree, it follows that every vertex in G_1 must have even degree. However, G_1 may be disconnected.

If G_1 is totally disconnected, that is, G_1 contains only isolated vertices, then $G = C_1$ and statement 3 is proved. Otherwise G_1 has at least one circuit C_2.

Consider next the graph $G_2 = G_1 - C_2 = G - C_1 - C_2$. Again every vertex in G_2 has even degree. If G_2 is totally disconnected, then $G_2 = C_1 \cup C_2$. Otherwise repeat the procedure until we obtain a totally disconnected graph $G_n = G - C_1 - C_2 - \cdots - C_n$, where $C_1, C_2, \ldots, C_n$ are circuits of G, no two of which have common edges. Then

$$G = C_1 \cup C_2 \cup \cdots \cup C_n$$

and statement 3 is proved.

3⇒1 Let G be the union of the edge-disjoint circuits $C_1, C_2, \ldots, C_n$. Consider any of these circuits, say C_1. Since G is connected, there must be at least one circuit, say C_2, which has a common vertex v_1 with C_1. Let T_{12} be a closed trail beginning at v_1 and traversing C_1 and C_2 in succession. This trail obviously contains all the edges of C_1 and C_2.

Again, since G is connected, T_{12} must have a common vertex v_2 with at least one circuit, say C_3, different from C_1 and C_2. The closed trail T_{123} beginning at v_2 and traversing T_{12} and C_3 in succession will contain all the edges of C_1, C_2, and C_3.

Repeat this procedure until the closed trail $T_{123\cdots n}$ containing all the edges of $C_1, C_2, \ldots, C_n$ is obtained. This closed trail is an Euler trail in G. Hence G is Eulerian. ■

By this theorem, the graph G_1 of Fig. 3.3*a* is Eulerian because every vertex in G_1 is of even degree, whereas the graphs G_2 and G_3 of Fig. 3.3*b* and *c* are not Eulerian since they contain some vertices of odd degree. It may also be verified that the Eulerian graph G_1 is the union of the edge-disjoint circuits whose edge sets are given below:

$$\{e_4, e_5, e_6\}$$
$$\{e_3, e_7, e_8\}$$
$$\{e_2, e_9, e_{10}\}$$
$$\{e_1, e_{11}, e_{12}\}.$$

The following result is a consequence of statement 3 of Theorem 3.1.

Corollary 3.1.1 Every vertex of an Eulerian graph is contained in some circuit. ■

Though an Euler trail does not exist in a graph which contains some vertices of odd degree, it is possible to construct in such a graph a set of edge-disjoint open trails which together contain all the edges of the graph. This is proved in the next theorem.

THEOREM 3.2. Let $G=(V, E)$ be a connected graph with $2k$ odd degree vertices, $k \geqslant 1$. Then E can be partitioned into subsets $E_1, E_2, \ldots, E_k$ such that each E_i constitutes an open trail in G.

Proof

Let r_i and s_i, $1 \leqslant i \leqslant k$, be the $2k$ odd degree vertices of G. Now to G add k new vertices $w_1, w_2, \ldots, w_k$ together with $2k$ edges (r_i, w_i) and (s_i, w_i), $1 \leqslant i \leqslant k$. In the resulting graph G' every vertex is of even degree, and hence G' is Eulerian.

It may be noted that in any Euler trail of G', the edges (r_i, w_i) and (s_i, w_i) for any $1 \leqslant i \leqslant k$ appear consecutively. Removal of these $2k$ edges will then result in k edge-disjoint open trails of G such that each edge of G is present in precisely one of these trails. These open trails give the required partition of E. ■

Corollary 3.2.1 Let G be a connected graph with exactly two odd degree vertices. Then G has an open trail (which begins at one of the odd degree vertices and ends at the other) containing all the edges of G. ■

For example, the graph G_2 of Fig. 3.3*b* has exactly two odd degree vertices v_6 and v_3, and the open trail $\{e_1, e_2, e_3, e_4, e_5, e_6, e_7, e_8, e_9, e_{10}, e_{11}, e_{12}, e_{13}\}$ contains all the edges of G_2. This trail begins at v_6 and ends at v_3.

The graph G_3 of Fig. 3.3*c* has four odd degree vertices. This graph has two edge-disjoint open trails constituted by the following sets of edges:

$$\{e_1, e_2, e_3, e_4, e_5\}$$
$$\{e_6, e_7, e_8, e_9, e_{10}, e_{11}, e_{12}\}.$$

A graph G is said to be *randomly Eulerian from a vertex* v if, whenever we start from v and traverse along the edges of G in an arbitrary way, we eventually obtain an Euler trail.

It should be noted that if a graph G is randomly Eulerian from a vertex v, then it should be possible to extend every closed v–v trail not containing all the edges to an Euler trail of G. In other words, if an Eulerian graph G is not randomly Eulerian from a vertex v, then there must be a closed v–v trail containing all the edges incident on v but not containing all the edges of G.

For example, consider the Eulerian graph of Fig. 3.4. This graph is randomly Eulerian from vertices v_1 and v_2. It is not randomly Eulerian from the other vertices. It may be verified that for each vertex v_i different from v_1 and v_2 there exists a closed v_i–v_i trail containing all the edges incident on v_i but not containing all the edges of G. For example, the closed v_3–v_3 trail consisting of the edges e_4, e_1, e_2, and e_3 has this property.

The next theorem gives a characterization of a graph which is randomly Eulerian from a vertex v.

THEOREM 3.3. An Eulerian graph G is randomly Eulerian from a vertex v if and only if every circuit of G contains v.

Proof

Necessity Suppose graph G is randomly Eulerian from a vertex v. Assume that there exists a circuit C in G which does not contain v. Consider the graph

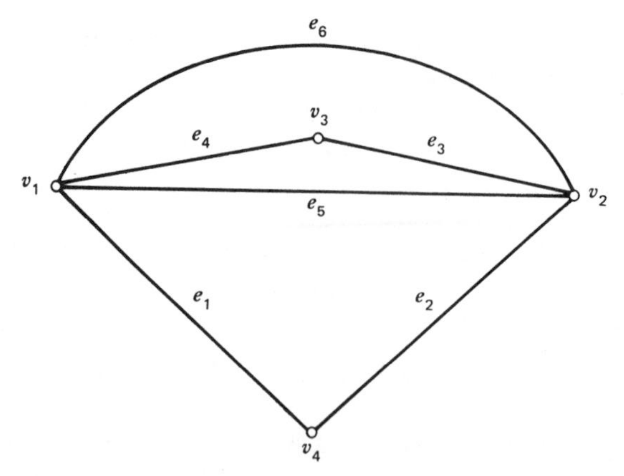

Figure 3.4. A graph randomly Eulerian from two vertices.

$G' = G - C$. Every vertex in G' is of even degree. G' may not be connected. However, G'', the component of G' containing v, is Eulerian, and it contains all the edges incident on v. Thus in G'', there exists an Euler trail T starting and ending at vertex v. This trail necessarily contains all the edges incident on v. Therefore it cannot be extended to include the edges of C. This contradicts the fact that G is randomly Eulerian from v.

Sufficiency Let vertex v in an Eulerian graph G be present in every circuit of G. Assume that G is not randomly Eulerian from v. Then there exists a closed v–v trail T containing all the edges incident on v, but not containing all the edges of G. Further, there exists a vertex $u \neq v$ such that it is the end vertex of an edge not in T.

On removing from G the edges of T, a graph G' in which v is an isolated vertex results. In G' every vertex is of even degree. So the component of G' containing u is an Eulerian graph. By Corollary 3.1.1 there is a circuit containing u. This circuit obviously does not contain vertex v. This contradicts the hypothesis that v is in every circuit of G. ■

It may be verified that in the graph G of Fig. 3.4, vertices v_1 and v_2 are present in every circuit of G. Thus G is randomly Eulerian from both these vertices. On the other hand, for each one of the other vertices there exists a circuit not containing it.

A graph is *randomly Eulerian* if it is randomly Eulerian from each of its vertices. It then follows from Theorem 3.3 that all the vertices of a randomly Eulerian graph G are in exactly one circuit C of G and there is no other circuit in G. In other words, G is randomly Eulerian if and only if it is a circuit.

3.2 HAMILTONIAN GRAPHS

A *Hamilton circuit* in a graph G is a circuit containing all the vertices of G. A *Hamilton path* in G is a path containing all the vertices of G.

A graph G is defined to be *Hamiltonian* if it has a Hamilton circuit.

The graph G_1 shown in Fig. 3.5*a* is Hamiltonian because the sequence of edges $e_1, e_2, e_3, e_4, e_5, e_6$ constitutes a Hamilton circuit in G_1. The graph of Fig. 3.5*b* has a Hamilton path formed by the edges e_1, e_2, e_3, e_4, but it has no Hamilton circuit.

Whereas an Euler trail is a closed walk passing through each edge exactly once, a Hamilton circuit is a closed walk passing through each vertex exactly once. Thus there is a striking similarity between an Eulerian graph and a Hamiltonian graph. This may lead one to expect that there exists a simple, useful, and elegant characterization of a Hamiltonian graph, as in the case of Eulerian graphs. Such is not the case; in fact, development of such a characterization is a major unsolved problem in graph theory. However,

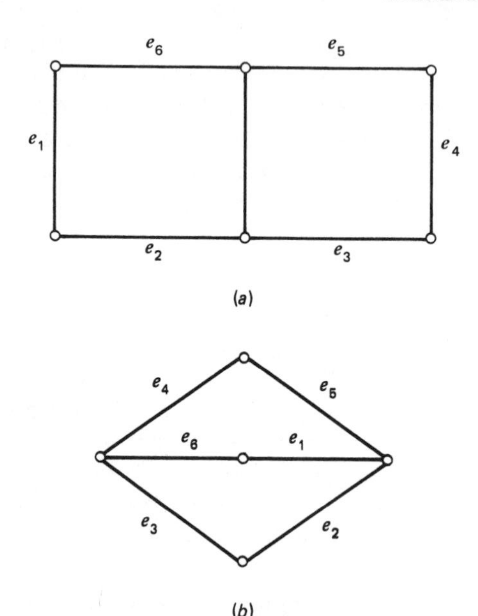

Figure 3.5. (a) A Hamiltonian Graph. (b) A non-Hamiltonian graph having a Hamilton path.

several sufficient conditions have been established for a simple graph to be Hamiltonian. (Recall that a graph is simple if it has neither parallel edges nor self-loops.) We consider some of these conditions in this section.

A sequence $d_1 \leqslant d_2 \leqslant \cdots \leqslant d_n$ is said to be *graphic* if there is a graph G with n vertices $v_1, v_2, \ldots, v_n$ such that the degree $d(v_i)$ of v_i equals d_i for each i. $(d_1, d_2, \ldots, d_n)$ is then called the *degree sequence* of G.

If

$$S: \quad d_1 \leqslant d_2 \leqslant \cdots \leqslant d_n$$

and

$$S^*: \quad d_1^* \leqslant d_2^* \leqslant \cdots \leqslant d_n^*$$

are graphic sequences such that $d_i^* \geqslant d_i$ for $1 \leqslant i \leqslant n$, then S^* is said to *majorize* S.

The following result is due to Chvátal [3.1].

THEOREM 3.4. A simple graph $G=(V, E)$ of order n, with degree sequence $d_1 \leqslant d_2 \leqslant \cdots \leqslant d_n$ is Hamiltonian if

$$d_k \leqslant k < \tfrac{1}{2}n \Rightarrow d_{n-k} \geqslant n-k. \tag{3.1}$$

Proof

First note that if $d_k \leqslant k$, then the number of vertices with degrees not exceeding k is at least k. Similarly if $d_{n-k} \geqslant n-k$, then the number of vertices whose degrees are not exceeded by $n-k$ is at least $k+1$. Further, if a graphic sequence satisfies (3.1), then so does every graphic sequence which majorizes it.

We now prove the theorem by contradiction.

Let there be a simple non-Hamiltonian graph whose degree sequence satisfies (3.1). Then this graph is a spanning subgraph of a simple maximal non-Hamiltonian graph $G=(V, E)$ whose degree sequence $d_1 \leqslant d_2 \leqslant \cdots \leqslant d_n$ also satisfies (3.1).

Let u and v be two nonadjacent vertices in G such that $d(u)+d(v)$ is as large as possible and $d(u) \leqslant d(v)$. Since G is maximal non-Hamiltonian, it follows that addition of an edge joining u and v will result in a Hamiltonian graph. Thus in G, there is a Hamilton path $u=u_1, u_2, u_3, \ldots, u_n=v$ with u and v as the end vertices (Fig. 3.6). Let

$$S=\{i|(u_1, u_{i+1}) \in E\},$$

$$T=\{i|(u_i, u_n) \in E\}.$$

Now there is no $j \in S \cap T$. For if $j \in S \cap T$, then the edges (u_1, u_{j+1}) and (u_j, u_n) would be in G, and so the cyclic sequence of vertices $u_j, u_{j-1}, \ldots,$ $u_1, u_{j+1}, u_{j+2}, \ldots, u_n, u_j$ would form a Hamilton circuit in G.

Since the vertex $u_n=v$ is neither in S nor in T, it follows that $S \cup T \subseteq$ $\{1, 2, \ldots, n-1\}$. Therefore,

$$d(u)+d(v)=|S|+|T|<n$$

and

$$d(u)<\tfrac{1}{2}n,$$

where $|X|$ denotes the number of elements in set X.

Since $S \cap T=\varnothing$, no u_j with $j \in S$ is adjacent to v. The choice of $d(u)$ and $d(v)$ then implies that for $j \in S$, $d(u_j) \leqslant d(u)$. Thus there are at least $|S|=d(u)$ vertices whose degrees do not exceed $d(u)$. So if we set $k=d(u)$, then we get $d_k \leqslant k<\frac{1}{2}n$, and therefore by (3.1) $d_{n-k} \geqslant n-k$. This means that there are at

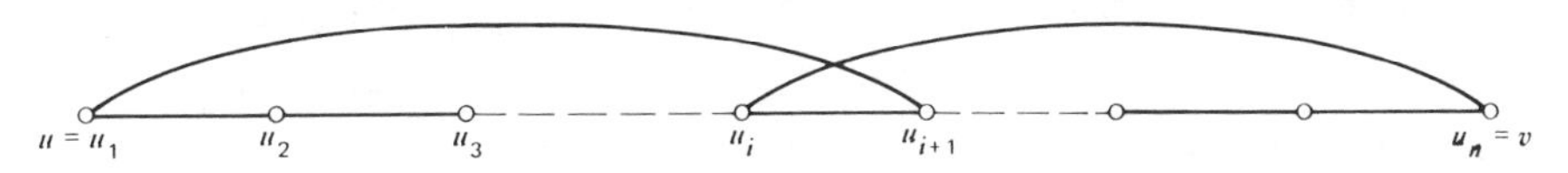

Figure 3.6.

least $k+1$ vertices whose degrees are not exceeded by $n-k$. The vertex u can be adjacent to at most k of these $k+1$ vertices because $d(u)=k$. Thus there is a vertex w with $d(w)\geqslant n-k$, which is not adjacent to u. But then $d(u)+d(w) \geqslant n>d(u)+d(v)$ contradicting the choice of $d(u)$ and $d(v)$. ■

We establish, in the following corollary, the sufficient conditions developed by Dirac [3.2], Ore [3.3], Pósa [3.4], and Bondy [3.5] for a graph to be Hamiltonian.

Corollary 3.4.1 A simple graph $G=(V,E)$ of order $n\geqslant 3$ with degree sequence $d_1\leqslant d_2\leqslant \cdots \leqslant d_n$ is Hamiltonian if one of the following conditions is satisfied:

1. (Dirac) $1\leqslant k\leqslant n \Rightarrow d_k\geqslant \frac{1}{2}n$.
2. (Ore) $(u,v)\notin E\Rightarrow d(u)+d(v)\geqslant n$.
3. (Pósa) $1\leqslant k<\frac{1}{2}n\Rightarrow d_k>k$.
4. (Bondy) $j<k, d_j\leqslant j, d_k\leqslant k-1\Rightarrow d_j+d_k\geqslant n$.

Proof

We prove by showing that all these conditions imply (3.1).

1⇒2 Clearly any degree sequence which satisfies condition 1 also satisfies condition 2.

2⇒3 If this is not true, then there exists a t such that $1\leqslant t<\frac{1}{2}n$ and $d_t\leqslant t$. Now suppose there exists an l with $l<t$ and $(v_l,v_t)\notin E$. Then

$$d_l+d_t\leqslant 2d_t<2\cdot\tfrac{1}{2}n=n,$$

contradicting condition 2. Therefore the induced subgraph of G on the vertices $v_1, v_2,\ldots, v_t$ is a complete graph.

Since $d_t\leqslant t$, every vertex v_i, $1\leqslant i\leqslant t$, is adjacent to at most one vertex $v_j, t+1\leqslant j\leqslant n$. Further, $t<\frac{1}{2}n$ implies that $n-t>t$. So there exists a vertex $v_j, t+1\leqslant j\leqslant n$, which is not adjacent to any v_i, $1\leqslant i\leqslant t$. Thus $d_j\leqslant n-t-1$. But then

$$\begin{aligned} d_t+d_j &\leqslant t+n-t-1 \\ &<n. \end{aligned}$$

Thus we have $(v_t,v_j)\notin E$ and $d_t+d_j<n$ contradicting condition 2.

3⇒4 If this is not true, then there exists a $j<k$ such that $d_j\leqslant j$, $d_k\leqslant k-1$ and $d_j+d_k<n$. Then $d_j<\frac{1}{2}n$. If we now set $t=d_j<\frac{1}{2}n$, we get

$$d_t=d_{d_j}\leqslant d_j=t.$$

Therefore we have $t < \frac{1}{2}n$ and $d_t \leqslant t$ contradicting condition 3.

4⇒(3.1) If this is not true, then there exists a t such that $d_t \leqslant t < \frac{1}{2}n$ and $d_{n-t} \leqslant n-t-1$. But then

$$\begin{aligned} d_t + d_{n-t} &\leqslant t + n - t - 1 \\ &= n-1, \end{aligned}$$

contradicting condition 4. ■

It is easy to see that if a graphic sequence satisfies any one of the conditions stated in Theorem 3.4 and Corollary 3.4.1, so does every graphic sequence which majorizes it. Chvátal's condition, the strongest of these five conditions, has the interesting property that it is the best condition of this kind. That is, if a graphic sequence fails to satisfy Chvátal's condition, then it is majorized by the degree sequence of a non-Hamiltonian graph [3.1].

Though, in general, it is difficult to establish the non-Hamiltonian nature of a graph, in certain cases it may be possible to do so by the use of some shrewd arguments. This is illustrated in Liu [3.6] through the following example.

Consider the graph G shown in Fig. 3.7. We show that there is no Hamilton path in G.

Among all the edges incident on any vertex at most two can be included in any Hamilton path. In the graph G, vertex v_8 is of degree 5, and hence at least three of the edges incident on v_8 cannot be included in any Hamilton path. The same is true of the vertices v_{10} and v_{12}. Since the degrees of v_2, v_4, v_6, and v_{16} are equal to 3, at least one of the three edges incident on each of these vertices cannot be included in a Hamilton path. Thus at least 13 of the 27 edges of G cannot be included in any Hamilton path. Hence there are not enough edges to form a Hamilton path on the 16 vertices of G. Thus G has no Hamilton path.

A graph is *randomly Hamiltonian from a vertex v* if any path starting from v can be extended to a Hamilton v–v circuit. A graph is *randomly Hamiltonian* if it is randomly Hamiltonian from each of its vertices.

The following theorem completely characterizes randomly Hamiltonian graphs. Proof of this theorem may be found in Behzad and Chartrand [3.7].

THEOREM 3.5. A simple graph of order n is randomly Hamiltonian if and only if it is a circuit or a complete graph or the complete bipartite graph $K_{n/2, n/2}$, the last one being possible only when n is even. ■

We conclude this section with a reference to the traveling salesman problem. The problem is as follows:

A salesman is required to visit a number of cities. What is the route he should take, if he has to start at his home city, visit each city exactly once, and then return home traveling the shortest distance?

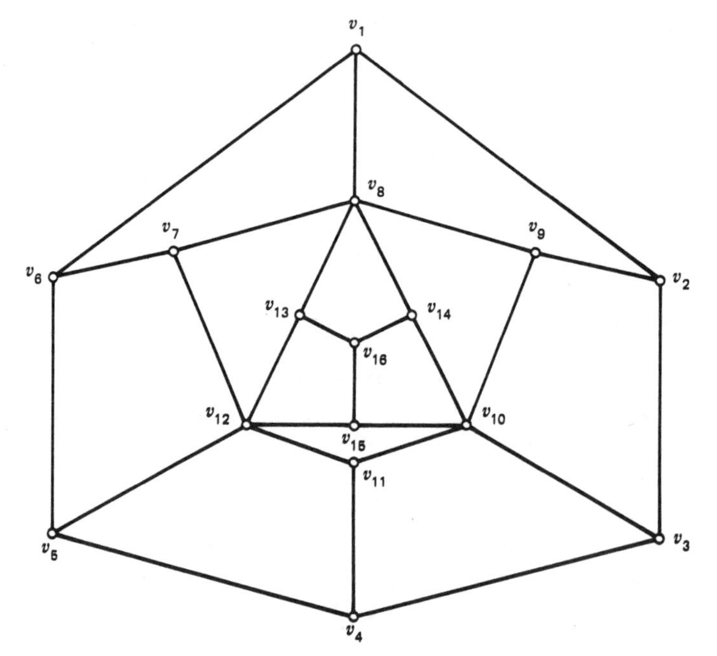

Figure 3.7. A non-Hamiltonian graph.

Suppose we represent cities by the vertices of a graph and roads by edges connecting the vertices. The length of a road may be represented as a weight associated with the corresponding edge. If between every pair of vertices there is a road connecting them, then it may be seen that the traveling salesman problem is equivalent to finding a shortest Hamilton circuit in a complete graph in which each edge is associated with a weight.*

In a complete graph of order n there exist $(n-1)!/2$ Hamilton circuits. One "brute-force" approach to solving the traveling salesman problem is to generate all the $(n-1)!/2$ Hamilton circuits and then pick the shortest one. The labor in this approach is too great (even for a computer), even for values of n as small as 50. For arbitrary values of n, no efficient algorithm for solving this problem exists. For more discussions, Lin [3.8], Belmore and Nemhauser [3.9], and Held and Karp [3.10], [3.11] may be consulted.

3.3 FURTHER READING

Berge [3.12] proves, in a unified manner, several sufficient conditions for a graph to be Hamiltonian. This book also discusses the question of partitioning the edge set of a graph into paths, circuits, and so on. See also Bondy and Murty [3.13].

* The *weight* of a circuit is the sum of the weights of the edges in the circuit.

We mentioned in Section 3.2 that if the degree sequence of a graph does not satisfy Chvátal's condition (3.1), then it is majorized by the degree sequence of a non-Hamiltonian graph. Further there are graphic sequences which do not satisfy (3.1), but which are necessarily degree sequences of Hamiltonian graphs. See Nash-Williams [3.14]. Bondy and Chvátal [3.15] have generalized Theorem 3.4.

For a detailed review of results on Hamilton circuits, Nash-Williams [3.16] and Lesniak-Foster [3.17] may be referred.

3.4 EXERCISES

3.1 Let G be a connected graph with $2k$ odd degree vertices. Prove that the edge set E of G cannot be partitioned into subsets $E_1, E_2, \ldots, E_l, l<k$, such that each E_i forms an open trail in G.

3.2 Prove that if a graph G is randomly Eulerian from a vertex v, then v is the only cut-vertex or G has no cut-vertices.

3.3 Let G be an Eulerian graph with $n \geqslant 3$ vertices. Prove that G is randomly Eulerian from none, one, two, or all of its vertices.

3.4 If a graph G is randomly Eulerian from a vertex v, then prove that $\Delta(G)=d(v)$, where $\Delta(G)$ is the maximum degree in G.

3.5 Let G be a randomly Eulerian graph from a vertex v. If $d(u)=d(v)$, $u \neq v$, then prove that G is randomly Eulerian from u.

3.6 Do there exist graphs in which an Euler trail is also a Hamilton circuit? Characterize such graphs.

3.7 Show that the graph of Fig. 3.8 does not have a Hamilton path.

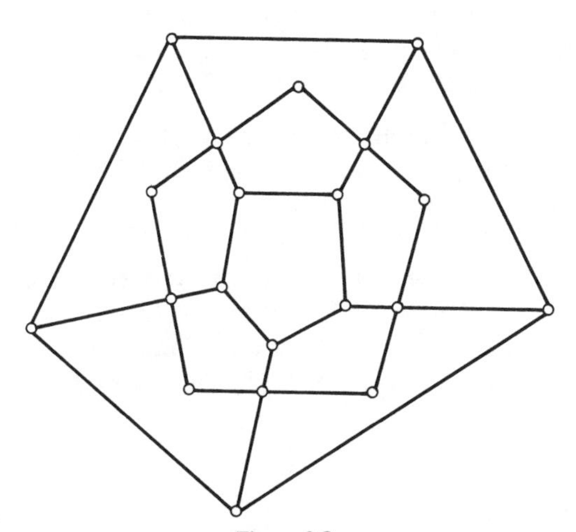

Figure 3.8.

3.8 Let G be a connected simple graph with $n>2\delta(G)$ vertices. Prove that the length of a longest path of G should be greater than or equal to $2\delta(G)$, where $\delta(G)$ is the minimum degree in G (Dirac [3.2]).

3.9 Let G be a simple graph with n vertices. Prove that G has a Hamilton path if for every pair of vertices u and v in G, $d(u)+d(v)\geqslant n-1$.

3.10 Let $d_1 \leqslant d_2 \leqslant \cdots \leqslant d_n$ be the degrees of a simple graph with $n \geqslant 2$ vertices. If

$$d_k \leqslant k-1 \leqslant \tfrac{1}{2}n-1 \Rightarrow d_{n+1-k} \geqslant n-k,$$

then prove that G has a Hamilton path (Chvátal [3.1]).

3.11 Let G be a simple graph with n vertices and m edges such that $n \geqslant 3$ and $m \geqslant (n^2-3n+6)/2$. Prove that G is Hamiltonian (Ore [3.3]).

3.12 Let G be a simple graph with n vertices, and u and v be nonadjacent vertices in G such that

$$d(u)+d(v) \geqslant n.$$

Denote by G' the graph obtained by adding edge (u,v) to G. Show that G is Hamiltonian if and only if G' is Hamiltonian.

3.13 The *closure* of a simple graph G with n vertices is the graph obtained by recursively joining pairs of nonadjacent vertices whose degree sum is at least n until no such pair remains. Show that the closure of a graph is well defined.

Note An easy consequence of Exercise 3.12 is that a graph is Hamiltonian if and only if its closure is Hamiltonian. So G is Hamiltonian if its closure is complete. This result can be used to prove several sufficient conditions for a graph to be Hamiltonian. See Bondy and Chvátal [3.15] and Bondy and Murty [3.13].

3.14 Prove that the maximum number of pairwise edge-disjoint Hamilton circuits in the complete graph K_n is $\lfloor (n-1)/2 \rfloor$.

Note $\lfloor x \rfloor$ means the greatest integer less than or equal to x.

3.15 A Graph g is *Hamilton-connected* if for every pair of distinct vertices u and v of G, there exists a Hamilton u–v path. Prove that a simple graph G of order $n \geqslant 3$ is Hamilton-connected if for every pair of nonadjacent vertices u and v of G, $d(u)+d(v) \geqslant n+1$ (Erdös and Gallai [3.18]).

3.16 Show that the tree graph (defined in Exercise 2.28) of a connected graph is Hamiltonian (Cummins [3.19], Shank [3.20]).

3.5 REFERENCES

3.1 V. Chvátal, "On Hamilton's Ideals," *J. Combinatorial Theory B*, Vol. 12, 163–168 (1972).

3.2 G. A. Dirac, "Some Theorems on Abstract Graphs," *Proc. London Math. Soc.*, Vol. 2, 69–81 (1952).

3.3 O. Ore, "Arc Coverings of Graphs," *Ann. Mat. Pura Appl.*, Vol. 55, 315–321 (1961).

3.4 L. Pósa,"A Theorem Concerning Hamilton Lines," *Magyar Tud. Akad. Mat. Kutató Int. Közl.*, Vol. 7, 225–226 (1962).

3.5 J. A. Bondy, "Properties of Graphs with Constraints on Degrees," *Studia Sci. Math. Hungar.*, Vol. 4, 473–475 (1969).

3.6 C. L. Liu, *Introduction to Combinatorial Mathematics*, McGraw-Hill, New York, 1968.

3.7 M. Behzad and G. Chartrand, *Introduction to the Theory of Graphs*, Allyn and Bacon, Boston, 1971.

3.8 S. Lin, "Computer Solutions of the Traveling Salesman Problem," *Bell Syst. Tech. J.*, Vol. 44, 2245–2269 (1965).

3.9 M. Belmore and G. L. Nemhauser, "The Traveling Salesman Problem: A Survey," *Operations Res.*, Vol. 16, 538–558 (1968).

3.10 M. Held and R. M. Karp, "The Traveling Salesman Problem and Minimum Spanning Trees," *Operations Res.*, Vol. 18, 1138–1162 (1970).

3.11 M. Held and R. M. Karp, "The Traveling Salesman Problem and Minimum Spanning Trees: Part II," *Math. Programming*, Vol. 1, 6–25 (1971).

3.12 C. Berge, *Graphs and Hypergraphs*, North Holland, Amsterdam, 1973.

3.13 J. A. Bondy and U. S. R. Murty, *Graph Theory with Applications*, Macmillan, London, 1976, chap. 4.

3.14 C. St. J. A. Nash-Williams, "Hamilton Arcs and Circuits," in *Recent Trends in Graph Theory*, Springer, Berlin, 1971, pp. 197–210.

3.15 J. A. Bondy and V. Chvátal, "A Method in Graph Theory," *Discrete Math.*, Vol. 15, 111–135 (1976).

3.16 C. St. J. A. Nash-Williams, "Hamiltonian Circuits," in *Studies in Graph Theory*, Part II, MAA Press, 1975, pp. 301–360.

3.17 L. Lesniak-Foster, "Some Recent Results in Hamiltonian Graphs," *J. Graph Theory*, Vol. 1, 27–36 (1977).

3.18 P. Erdös and T. Gallai, "On Maximal Paths and Circuits of Graphs," *Acta Math. Acad. Sci. Hung.*, Vol. 10, 337–356 (1959).

3.19 R. L. Cummins, "Hamilton Circuits in Tree Graphs," *IEEE Trans. Circuit Theory*, Vol. CT-13, 82–90 (1966).

3.20 H. Shank, "A Note on Hamilton Circuits in Tree Graphs," *IEEE Trans. Circuit Theory*, Vol. CT-15, 86 (1968).

Chapter 4

Graphs and Vector Spaces

Modern algebra has proved to be a very valuable tool for engineers and scientists in the study of their problems. Identifying the algebraic structure associated with a set of objects has been found to be very useful since the powerful and elegant results relating to the algebraic structure can then be brought to bear upon the study of such a set. Systems theory, electrical network theory, coding theory, switching circuits (sequential and combinational), and computer science are some examples of areas which have benefited by taking such an approach.

In this chapter, we show that a vector space can be associated with a graph, and we study in detail the properties of two important subspaces of this vector space, namely, the cutset and circuit spaces. We see in Part II of this book that these subspaces indeed define the voltage and current spaces of an electrical network.

In the first two sections of this chapter, we provide an introduction to some elementary algebraic concepts and results which will be used in the discussions of the subsequent sections. For more detailed discussions of these concepts and related results in linear algebra MacLane and Birkhoff [4.1], Halmos [4.2], and Hohn [4.3] may be consulted.

4.1 GROUPS AND FIELDS

Consider a finite set $S=\{a, b, c, \dots\}$. Let $+$ denote a binary operation defined on S. This operation assigns to every pair of elements a and b of S a unique element denoted by $a+b$. The set S is said to be *closed* under $+$ if the element $a+b$ belongs to S whenever a and b are in S.

The operation + is said to be *associative* if

$$a+(b+c)=(a+b)+c, \qquad \text{for all } a, b, \text{ and } c \text{ in } S.$$

Further, it is said to be *commutative* if

$$a+b=b+a, \qquad \text{for all } a \text{ and } b \text{ in } S.$$

We are now in a position to define a group.

A set S with a binary operation +, called addition, is a *group* if the following postulates hold:

1. S is closed under +.
2. + is associative.
3. There exists a unique element e in S such that $a+e=e+a=a$ for all a in S. The element e is called the *identity* of the group.
4. For each element a in S there exists a unique element b such that $b+a=a+b=e$. The element b is called the *inverse* of a and vice versa. Clearly the identity element e is its own inverse.

A group is said to *abelian* if the operation + is commutative.

A common example of a group is the set $S=\{\ldots,-2,-1,0,1,2,\ldots\}$ of all integers, with + defined as the usual addition operation. In this group, 0 is the identity element and $-a$ is the inverse of a for all a in S. Note that this group is abelian.

Is the set S of all integers with the multiplication operation a group? (No. Why?)

An important example of a group is the set $Z_p=\{0,1,2,\ldots,p-1\}$ of integers with modulo* p addition operation. In this group, 0 is the identity element. Further the integer $p-a$ is the inverse of the integer a, for all a not equal to 0. Of course, 0 is its own inverse. This group is also abelian. As an example, the addition table for Z_5 is given below:

+	0	1	2	3	4
0	0	1	2	3	4
1	1	2	3	4	0
2	2	3	4	0	1
3	3	4	0	1	2
4	4	0	1	2	3

Next we define a field.

A set F with two operations + and $\cdot$, called, respectively, addition and multiplication, is a *field* if the following postulates are satisfied:

*If $a=mp+q$, $0<q<p-1$, then in modulo arithmetic $a=q(\text{modulo } p)$.

1. F is an abelian group under $+$, with the identity element denoted as e.
2. The set $F-\{e\}$ is an abelian group under $\cdot$, the multiplication operation.
3. The multiplication operation is distributive with respect to addition; that is,

$$a\cdot(b+c)=(a\cdot b)+(a\cdot c) \qquad \text{for all } a, b, \text{and } c \text{ in } F.$$

As an example, consider again the set $Z_p=\{0,1,2,\ldots,p-1\}$ with addition (modulo p) and multiplication (modulo p) as the two operations. As shown earlier, Z_p is an abelian group under modulo p addition, with 0 as the identity element. It can be shown that the set $Z_p-\{0\}=\{1,2,\ldots,p-1\}$ is a group under modulo p multiplication if and only if p is prime. This group is also abelian. The fact that modulo p multiplication is distributive with respect to modulo p addition may be easily verified. Thus the set Z_p is a field if and only if p is prime.

The field Z_p is usually denoted as $GF(p)$ and is called a *Galois field*. The multiplication table for $GF(5)$ is given below as an illustration:

$\cdot$	0	1	2	3	4
0	0	0	0	0	0
1	0	1	2	3	4
2	0	2	4	1	3
3	0	3	1	4	2
4	0	4	3	2	1

A field that is of special interest to us is $GF(2)$, the set of integers modulo 2. In this field

$$\begin{array}{ll} 0+0=0 & 0\cdot 0=0 \\ 1+0=0+1=1 & 1\cdot 0=0\cdot 1=0 \\ 1+1=0 & 1\cdot 1=1. \end{array}$$

4.2 VECTOR SPACES

Consider a set S with a binary operation $\boxplus$. Let F be a field with $+$ and $\cdot$ denoting, respectively, the addition and multiplication operations. A multiplication operation, denoted by $*$, is also defined between elements in F and those in S. This operation assigns to each ordered pair (α, s), where α is in F and s in S, a unique element denoted by $\alpha * s$ of S. The set S is a *vector space* over F if the following postulates hold:

1. S is an abelian group under $\boxplus$.
2. For any elements α and β in F, and any elements s_1 and s_2 in S,

$$\alpha*(s_1 \boxplus s_2)=(\alpha*s_1) \boxplus (\alpha*s_2)$$

and

$$(\alpha+\beta)*s_1=(\alpha*s_1)\boxed{+}(\beta*s_1).$$

3. For any elements α and β in F and any element s in S,

$$(\alpha\cdot\beta)*s=\alpha*(\beta*s).$$

4. For any element s in S, $1*s=s$, where 1 is the multiplicative identity in F.

Next we give an important example of a vector space.

Consider the set W of all n-vectors† over a field F. (Note that the elements of the n-vectors are chosen from F.) The symbols $+$ and $\cdot$ will, respectively, denote the addition and multiplication operations in F, and the symbols 0 and 1 will denote the additive and multiplicative identities in F. Let $\boxed{+}$, an addition operation on W, and $*$, a multiplication operation between the elements of F and those of W, be defined as follows:

1. If $\omega_1=(\alpha_1,\alpha_2,\ldots,\alpha_n)$ and $\omega_2=(\beta_1,\beta_2,\ldots,\beta_n)$ are elements of W, then

$$\omega_1\boxed{+}\omega_2=(\alpha_1+\beta_1,\alpha_2+\beta_2,\ldots,\alpha_n+\beta_n).$$

2. If α is in F, then

$$\alpha*\omega_1=(\alpha\cdot\alpha_1,\alpha\cdot\alpha_2,\ldots,\alpha\cdot\alpha_n).$$

With $\boxed{+}$ defined as above, we can easily establish that W is an abelian group under $\boxed{+}$, with the n-vector $(0,0,0,\ldots,0)$ as the identity element. Thus W satisfies the first postulate in the definition of a vector space. We can easily show that the elements of W and F also satisfy the other three requirements of a vector space.

For example, the set W of the eight 3-vectors given below is a vector space over $GF(2)$:

$$\omega_0=(0\ 0\ 0),\qquad \omega_1=(0\ 0\ 1),\qquad \omega_2=(0\ 1\ 0),\qquad \omega_3=(0\ 1\ 1),$$

$$\omega_4=(1\ 0\ 0),\qquad \omega_5=(1\ 0\ 1),\qquad \omega_6=(1\ 1\ 0),\qquad \omega_7=(1\ 1\ 1).$$

This vector space is used in all illustrations in this section.

A few important definitions and results (without proof) relating to a vector space are stated next.

Consider a vector space S over a field F.

Vectors and scalars Elements of S are called *vectors* and those of F are called *scalars*.

†An *n-vector* over F is a row vector with n elements from the field F.

Linear combination If an element s in S is expressible as

$$s = (\alpha_1 * s_1) \boxplus (\alpha_2 * s_2) \boxplus \cdots \boxplus (\alpha_j * s_j),$$

where s_i's are vectors and α_i's are scalars, then s is said to be a *linear combination* of $s_1, s_2, \ldots, s_j$.

Linear independence Vectors $s_1, s_2, \ldots, s_j$ are said to be *linearly independent* if no vector in this set is expressible as a linear combination of the remaining vectors in the set.

Basis vectors Vectors $s_1, s_2, \ldots, s_n$ form a *basis* in the vector space S if they are linearly independent and every vector in S is expressible as a linear combination of these vectors. The vectors $s_1, s_2, \ldots, s_n$ are called *basis vectors*.

It can be shown that the representation of a vector as a linear combination of basis vectors is unique for a given basis. A vector space may have more than one basis. However, it can be proved that all the bases have the same number of vectors.

Dimension The dimension of the vector space S, denoted as $\dim(S)$, is the number of vectors in a basis of S.

Subspace If S' is a subset of the vector space S over F, then S' is a subspace of S if S' is also a vector space over F.

Direct sum The *direct sum* $S_1 \boxplus S_2$ of two subspaces S_1 and S_2 of S is the set of all vectors of the form $s_i \boxplus s_j$, where s_i is in S_1 and s_j is in S_2.

It can be shown that $S_1 \boxplus S_2$ is also a subspace, and that its dimension is given by:

$$\dim(S_1 \boxplus S_2) = \dim(S_1) + \dim(S_2) - \dim(S_1 \cap S_2).$$

Note that $S_1 \cap S_2$ is also a subspace whenever S_1 and S_2 are subspaces.

Let us illustrate the above definitions using the vector space W of 3-vectors over $GF(2)$. The vectors $\omega_0, \omega_1, \ldots, \omega_7$ of W are as defined earlier.

1. ω_1 is a linear combination of ω_6 and ω_7, since

$$\omega_1 = (1 * \omega_6) \boxplus (1 * \omega_7).$$

2. Vectors ω_3 and ω_4 are linearly independent, since neither of these two can be expressed in terms of the other. Note that ω_0 and ω_i are not linearly independent, for any i.

3. Vectors ω_1, ω_2, and ω_4 form a basis in W, since they are linearly independent and the remaining vectors can be expressed as linear combinations of these as shown below:

$$\omega_0 = (0*\omega_1) \boxplus (0*\omega_2) \boxplus (0*\omega_4),$$

$$\omega_3 = (1*\omega_1) \boxplus (1*\omega_2),$$

$$\omega_5 = (1*\omega_1) \boxplus (1*\omega_4),$$

$$\omega_6 = (1*\omega_2) \boxplus (1*\omega_4),$$

$$\omega_7 = (1*\omega_1) \boxplus (1*\omega_2) \boxplus (1*\omega_4).$$

It may be verified that the vectors ω_1, ω_3, and ω_7 also form a basis.

4. Dimension of W is equal to 3, since there are three vectors in a basis of W.

5. The sets

$$W' = \{\omega_0, \omega_1, \omega_2, \omega_3\}$$

and

$$W'' = \{\omega_0, \omega_1, \omega_6, \omega_7\}$$

are subspaces of W. It may be verified that $\{\omega_1, \omega_2\}$ is a basis for W', and $\{\omega_1, \omega_6\}$ is a basis for W''. Thus

$$\dim(W') = 2$$

and

$$\dim(W'') = 2.$$

6. If W' and W'' are as defined above, then

$$W' \boxplus W'' = \{\omega_0, \omega_1, \omega_2, \omega_3, \omega_4, \omega_5, \omega_6, \omega_7\}.$$

7. The dimension of $W' \boxplus W''$ is given by:

$$\begin{aligned}\dim(W' \boxplus W'') &= \dim(W') + \dim(W'') - \dim(W' \cap W'') \\ &= 4 - \dim(W' \cap W'').\end{aligned}$$

Since

$$W' \cap W'' = \{\omega_0, \omega_1\},$$

we have

$$\dim(W' \cap W'') = 1.$$

Thus

$$\dim(W' \boxplus W'') = 4 - 1 = 3.$$

This can also be obtained by noting that in this case $W' \boxplus W'' = W$.

Next we define isomorphism between two vector spaces defined over the same field.

Let S and S' be two n-dimensional vector spaces over a field F. S and S' are said to *isomorphic* if there exists a one-to-one correspondence between S and S' such that the following hold true:

1. If vectors s_1 and s_2 of S correspond to the vectors s_1' and s_2' of S, respectively, then the vector $s_1 \boxplus s_2$ corresponds to the vector $s_1' \triangle s_2'$, where $\boxplus$ and $\triangle$ are corresponding operations in S and S'.
2. For any α in F, the vector $\alpha * s$ corresponds to the vector $\alpha \triangle s'$ if s corresponds to s', where $*$ and $\triangle$ are corresponding operations in S and S'.

Consider an n-dimensional vector space S over a field F and the n-dimensional vector space W of n-vectors over F. Let the vectors $s_1, s_2, \ldots, s_n$ form a basis in S. Suppose we define that a vector s in S corresponds to the vector $\omega = (\alpha_1, \alpha_2, \ldots, \alpha_n)$† of W if and only if $s = (\alpha_1 * s_1) \boxplus (\alpha_2 * s_2) \boxplus \cdots \boxplus (\alpha_n * s_n)$. Then it is not difficult to show that this one-to-one correspondence defines an isomorphism between S and W. Thus we have the following important result.

THEOREM 4.1. Every n-dimensional vector space over a field F is isomorphic to the vector space W of n-vectors over F. ■

The above theorem provides the main link connecting vector spaces and matrices. It implies that an n-dimensional vector space over a field F can be studied in terms of the n-dimensional vector space W of all n-vectors over F.

We conclude this section with the definition of two more important concepts—*dot product* (or *inner product*) and *orthogonality*.

Let

$$\omega_1 = (\alpha_1 \quad \alpha_2 \quad \cdots \quad \alpha_n)$$

†$\alpha_1, \ldots, \alpha_n$ are called the *coordinates* of s relative to the basis $\{s_1, s_2, \ldots, s_n\}$.

and

$$\omega_2=(\beta_1 \quad \beta_2 \quad \cdots \quad \beta_n)$$

be two vectors in the vector space W of n-vectors over F. The *dot product* of ω_1 and ω_2, denoted by $\langle\omega_1,\omega_2\rangle$, is a scalar defined as

$$\langle\omega_1,\omega_2\rangle=\alpha_1\cdot\beta_1+\alpha_2\cdot\beta_2+\cdots+\alpha_n\cdot\beta_n.$$

For example, if

$$\omega_1=(0 \quad 1 \quad 0 \quad 0 \quad 1)$$

and

$$\omega_2=(1 \quad 0 \quad 1 \quad 1 \quad 1),$$

then

$$\begin{aligned}\langle\omega_1,\omega_2\rangle&=0\cdot1+1\cdot0+0\cdot1+0\cdot1+1\cdot1\\&=0+0+0+0+1\\&=1.\end{aligned}$$

Vectors ω_i and ω_j are *orthogonal* to each other if $\langle\omega_i,\omega_j\rangle=0$, where 0 is the additive identity in F. For example, the vectors

$$\omega_1=(1 \quad 1 \quad 0 \quad 1 \quad 1)$$

and

$$\omega_2=(1 \quad 1 \quad 1 \quad 0 \quad 0)$$

are orthogonal over $GF(2)$ since

$$\begin{aligned}\langle\omega_1,\omega_2\rangle&=1\cdot1+1\cdot1+0\cdot1+1\cdot0+1\cdot0\\&=1+1+0+0+0\\&=0.\end{aligned}$$

Two subspaces W' and W'' of W are *orthogonal subspaces* of W if each vector in one subspace is orthogonal to every vector in the other subspace.

Two subspaces W' and W'' of W are *orthogonal complements* if they are orthogonal to each other and their direct sum $W' \boxplus W''$ is equal to the vector space W.

For example, consider again the 3-dimensional vector space W of 3-vectors over $GF(2)$. In this vector space, the subspaces $W'=\{\omega_0,\omega_1,\omega_2,\omega_3\}$ and $W''=\{\omega_0,\omega_4\}$ are orthogonal to each other. It may be verified that the direct sum of W' and W'' is equal to W. Hence they are orthogonal complements.

4.3 VECTOR SPACE OF A GRAPH

In this section, we show how we can associate a vector space with a graph and we also identify two important subspaces of this vector space.

Consider a graph $G=(V, E)$. Let W_G denote the collection of all subsets of E, including the empty set $\varnothing$. First we show that W_G is an abelian group under $\oplus$, the ring sum operation between sets. After suitably defining multiplication between the elements of the field $GF(2)$ and those of W_G, we show that W_G is a vector space over $GF(2)$.

The following are easy to verify:

1. W_G is closed under $\oplus$.
2. $\oplus$ is associative.
3. $\oplus$ is commutative.

Further, for any element E_i in W_G,

$$E_i \oplus \varnothing = E_i$$

and

$$E_i \oplus E_i = \varnothing.$$

Thus for the operation $\oplus$, $\varnothing$ is the identity element, and each E_i is its own inverse. Hence W_G is an abelian group under $\oplus$, thereby satisfying the first requirement in the definition of a vector space.

Let $*$, a multiplication operation between the elements of $GF(2)$ and those of W_G, be defined as follows:

For any E_i in W_G,

$$1 * E_i = E_i$$

and

$$0 * E_i = \varnothing.$$

With this definition of $*$, we can easily verify that the elements of W_G satisfy the following other requirements for a vector space:

For any elements α and β in $GF(2)=\{0,1\}$, and any elements E_i and E_j in W_G,

1. $(\alpha+\beta)*E_i=(\alpha*E_i)\oplus(\beta*E_i)$.
2. $\alpha*(E_i\oplus E_j)=(\alpha*E_i)\oplus(\alpha*E_j)$.
3. $(\alpha\cdot\beta)*E_i=\alpha*(\beta*E_i)$.
4. $1*E_i=E_i$.

(Note that 1 is the multiplicative identity in $GF(2)$.)

Thus W_G is a vector space over $GF(2)$.

If $E=\{e_1, e_2, \ldots, e_m\}$, then it is easy to see that the subsets $\{e_1\}$, $\{e_2\}, \ldots, \{e_m\}$ will constitute a basis for W_G. Hence the dimension of W_G is equal to m, the number of edges in G.

Since each edge-induced subgraph of G corresponds to a unique subset of E, and by definition (see Chapter 1), the ring sum of any two edge-induced subgraphs corresponds to the ring sum of their corresponding edge sets, it is clear that the set of all edge-induced subgraphs of G is also a vector space over $GF(2)$ if we define the multiplication operation $*$ as follows:

For any edge-induced subgraph G_i of G,

$$1 * G_i = G_i$$

and

$$0 * G_i = \varnothing,$$

the null graph having no vertices and no edges.

This vector space will also be referred to by the symbol W_G. Note that W_G will include $\varnothing$, the null graph.

We summarize the results of the above discussion in the following theorem.

THEOREM 4.2. For a graph G having m edges W_G is an m-dimensional vector space over $GF(2)$. ■

Since, in this Chapter, we are concerned only with edge-induced subgraphs, we refer to them simply as subgraphs without the adjective "edge-induced." However, we may still use this adjective in some places to emphasize the edge-induced nature of the concerned subgraph.

Next we show that the following subsets of W_G are subspaces:

1. W_C, the set of all circuits (including the null graph $\varnothing$) and unions of edge-disjoint circuits of G.
2. W_S, the set of all cutsets (including the null graph $\varnothing$) and unions of edge-disjoint cutsets of G.

This result will follow once we show that W_C and W_S are closed under $\oplus$, the ring sum operation.

THEOREM 4.3. W_C, the set of all circuits and unions of edge-disjoint circuits of a graph G, is a subspace of the vector space W_G of G.

Proof

By Theorem 3.1, a graph can be expressed as the union of edge-disjoint circuits if and only if every vertex in the graph is of even degree. Hence we may regard W_C as the set of all edge-induced subgraphs of G, in which all vertices are of even degree.

Consider any two distinct members C_1 and C_2 of W_C. C_1 and C_2 are edge-induced subgraphs with the degrees of all their vertices even. Let C_3 denote the ring sum of C_1 and C_2. To prove the theorem, we need only to show that C_3 belongs to W_C. In other words, we should show that in C_3 every vertex is of even degree.

Consider any vertex v in C_3. Obviously, this vertex should be present in at least one of the subgraphs C_1 and C_2. Let X_i, $i=1,2,3$, denote the set of edges incident on v in C_i. Let $|X_i|$ denote the number of edges in X_i. Thus $|X_i|$ is the degree of the vertex v in C_i. Note that $|X_1|$ and $|X_2|$ are even and one of them may be zero. Further $|X_3|$ is nonzero.

Since $C_3=C_1\oplus C_2$, we get

$$X_3=X_1\oplus X_2.$$

Hence

$$|X_3|=|X_1|+|X_2|-2|X_1\cap X_2|.$$

It is now clear from the above equation that $|X_3|$ is even, because $|X_1|$ and $|X_2|$ are both even. In other words, the degree of vertex v in C_3 is even. Since this should be true for all vertices in C_3, it follows that C_3 belongs to W_C, and the theorem is proved. ■

W_C will be referred to as the *circuit subspace* of the graph G.

As an example to illustrate the result of the above theorem, consider the graph G shown in Fig. 4.1*a*. The subgraphs C_1 and C_2 shown in Fig. 4.1*b* and *c* are unions of edge-disjoint circuits of G, since the degree of each vertex in these subgraphs is even. Thus they belong to the subspace W_C of W_G. The ring sum C_3 of C_1 and C_2 is shown in Fig. 4.1*d*. It may be seen that all the vertices in C_3 are of even degree. Hence C_3 also belongs to W_C.

Next we show that W_S, the set of all cutsets and unions of edge-disjoint cutsets of G, is a subspace of W_G.

By Theorem 2.7, a cut is a cutset or union of some edge-disjoint cutsets. Thus every cut of G belongs to W_S. We now prove that every element of W_S is a cut. While doing so we also prove that W_S is a subspace of W_G.

THEOREM 4.4. The ring sum of any two cuts in a graph G is also a cut in G.

Proof

Consider any two cuts $S_1=\langle V_1,V_2\rangle$ and $S_2=\langle V_3,V_4\rangle$ in a graph $G=(V,E)$. Note that

$$V_1\cup V_2=V_3\cup V_4=V$$

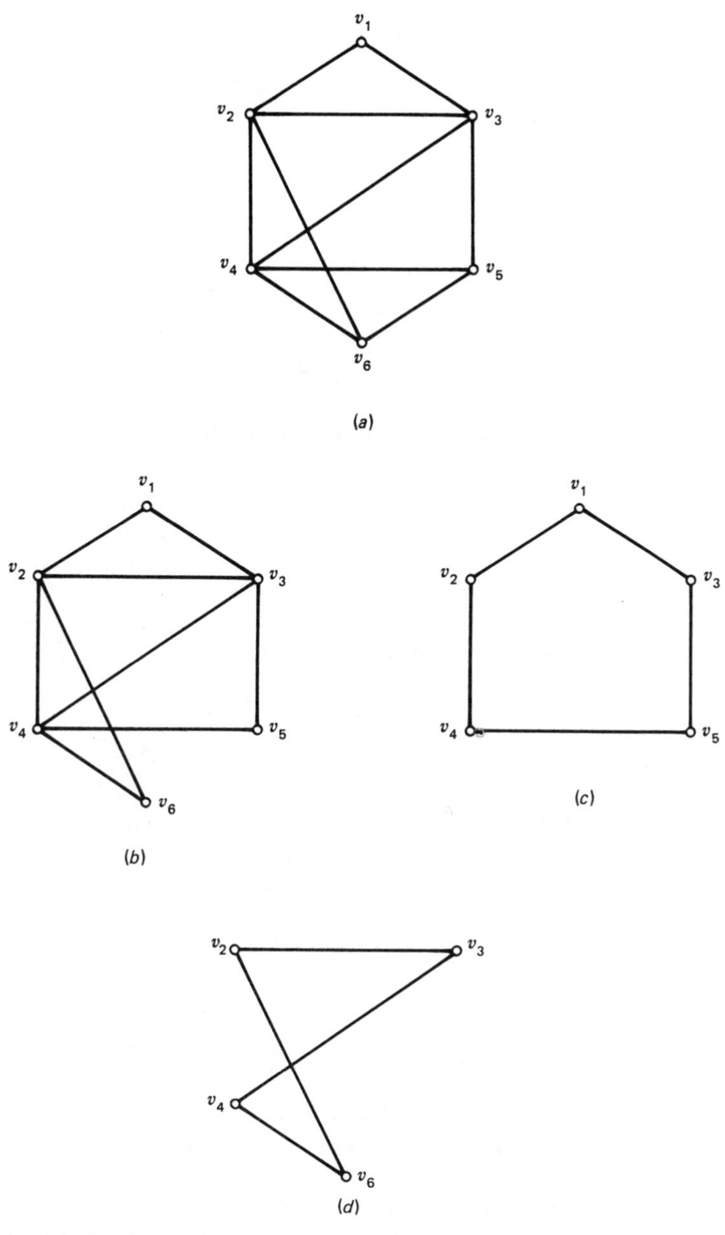

Figure 4.1. (*a*) Graph G. (*b*) Subgraph C_1 of G. (*c*) Subgraph C_2 of G. (*d*) Subgraph $C_1 \oplus C_2$ of G.

and

$$V_1 \cap V_2 = V_3 \cap V_4 = \varnothing.$$

Let

$$A = V_1 \cap V_3,$$
$$B = V_1 \cap V_4,$$
$$C = V_2 \cap V_3,$$
$$D = V_2 \cap V_4.$$

It is easy to see that the sets A, B, C, and D are mutually disjoint. Then

$$S_1 = \langle A \cup B, C \cup D \rangle$$
$$= \langle A, C \rangle \cup \langle A, D \rangle \cup \langle B, C \rangle \cup \langle B, D \rangle$$

and

$$S_2 = \langle A \cup C, B \cup D \rangle$$
$$= \langle A, B \rangle \cup \langle A, D \rangle \cup \langle C, B \rangle \cup \langle C, D \rangle.$$

Hence, we get

$$S_1 \oplus S_2 = \langle A, C \rangle \cup \langle B, D \rangle \cup \langle A, B \rangle \cup \langle C, D \rangle.$$

Since

$$\langle A \cup D, B \cup C \rangle = \langle A, C \rangle \cup \langle B, D \rangle \cup \langle A, B \rangle \cup \langle C, D \rangle,$$

we can write

$$S_1 \oplus S_2 = \langle A \cup D, B \cup C \rangle.$$

Because $A \cup D$ and $B \cup C$ are mutually disjoint and together include all the vertices in V, $S_1 \oplus S_2$ is a cut in G. Hence the theorem. ■

Since the ring sum of two disjoint sets is the same as their union, we get the following corollary of the above theorem.

Corollary 4.4.1 The union of any two edge-disjoint cuts in a graph G is also a cut in G. ■

Since a cutset is also a cut, it is now clear from Corollary 4.4.1 that W_S is the set of all cuts in G.

Further, by Theorem 4.4, W_S is closed under the ring sum operation. Thus we get the following theorem.

THEOREM 4.5. W_S, the set of all cutsets and unions of edge-disjoint cutsets in a graph G, is a subspace of the vector space W_G of G. ■

W_S will be referred to as the *cutset subspace* of the graph G.

As an example illustrating the result of Theorem 4.5, consider the following cuts S_1 and S_2 in the graph G of Fig. 4.2:

$$S_1 = \{e_1, e_3, e_4, e_5, e_6, e_7\}$$
$$= \langle V_1, V_2 \rangle,$$
$$S_2 = \{e_1, e_2, e_4, e_5, e_8\}$$
$$= \langle V_3, V_4 \rangle,$$

where

$$V_1 = \{v_1, v_2, v_4\},$$
$$V_2 = \{v_3, v_5, v_6\},$$
$$V_3 = \{v_1, v_4, v_5\},$$
$$V_4 = \{v_2, v_3, v_6\}.$$

Then

$$S_1 \oplus S_2 = \{e_2, e_3, e_6, e_7, e_8\}.$$

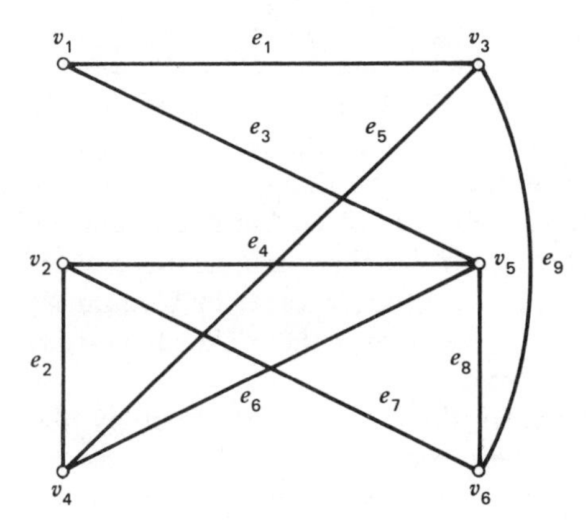

Figure 4.2.

If

$$A=V_1\cap V_3=\{v_1,v_4\},$$

$$B=V_1\cap V_4=\{v_2\},$$

$$C=V_2\cap V_3=\{v_5\},$$

$$D=V_2\cap V_4=\{v_3,v_6\},$$

then it is verified that the set $S_1\oplus S_2=\{e_2,e_3,e_6,e_7,e_8\}$ can be written (as in the proof of Theorem 4.4) as

$$S_1\oplus S_2=\langle A\cup D, B\cup C\rangle=\langle\{v_1,v_4,v_3,v_6\},\{v_2,v_5\}\rangle.$$

4.4 DIMENSIONS OF CIRCUIT AND CUTSET SUBSPACES

In this section we show that the dimensions of the circuit and cutset subspaces of a graph are equal to the nullity and the rank of the graph, respectively. We do this by proving that the set of fundamental circuits and the set of fundamental cutsets with respect to some spanning tree of a connected graph are bases for the circuit and cutset subspaces of the graph, respectively.

Let T be a spanning tree of a connected graph G with n vertices and m edges. The branches of T will be denoted by $b_1, b_2,\dots, b_{n-1}$ and the chords by $c_1, c_2,\dots, c_{m-n+1}$. Let C_i and S_i refer to the fundamental circuit and the fundamental cutset with respect to c_i and b_i, respectively.

By definition, each fundamental circuit contains exactly one chord, and this chord is not present in any other fundamental circuit. Thus no fundamental circuit can be expressed as the ring sum of the other fundamental circuits. Hence the fundamental circuits $C_1, C_2,\dots, C_{m-n+1}$ are independent. Similarly, the fundamental cutsets $S_1, S_2,\dots, S_{n-1}$ are also independent, since each of these contains exactly one branch which is not present in the others.

To prove that $C_1, C_2,\dots, C_{m-n+1}$ ($S_1, S_2,\dots, S_{n-1}$) constitute a basis for the circuit (cutset) subspace of G, we need only to prove that every subgraph in the circuit (cutset) subspace of G can be expressed as a ring sum of C_i's (S_i's).

Consider any subgraph C in the circuit subspace of G. Let C contain the chords $c_{i_1}, c_{i_2},\dots, c_{i_r}$. Let C' denote the ring sum of the fundamental circuits $C_{i_1}, C_{i_2},\dots, C_{i_r}$. Obviously, the chords $c_{i_1}, c_{i_2},\dots, c_{i_r}$ are present in C', and C' contains no other chords of T. Since C also contains these chords and no others, $C'\oplus C$ contains no chords.

We now claim that $C'\oplus C$ is empty. If this is not true, then by the preceding arguments, $C'\oplus C$ contains only branches and hence has no circuits. On the other hand, being a ring sum of circuits, $C'\oplus C$ is, by

Theorem 4.3, a circuit or the union of some edge-disjoint circuits. Thus the assumption that $C' \oplus C$ is not empty leads to a contradiction. Hence $C' \oplus C$ is empty. This implies that $C = C' = C_{i_1} \oplus C_{i_2} \oplus \cdots \oplus C_{i_r}$. In other words, every subgraph in the circuit subspace of G can be expressed as a ring sum of C_i's.

In an exactly similar manner we can prove that every subgraph in the cutset subspace of G can be expressed as a ring sum of S_i's.

Thus we have the following theorem.

THEOREM 4.6. Let a connected graph G have m edges and n vertices. Then

1. The fundamental circuits with respect to a spanning tree of G constitute a basis for the circuit subspace of G, and hence the dimension of the circuit subspace of G is equal to $m-n+1$, the nullity of G.
2. The fundamental cutsets with respect to a spanning tree of G constitute a basis for the cutset subspace of G, and hence the dimension of the cutset subspace of G is equal to $n-1$, the rank of G. ■

It is now easy to see that in the case of a graph G which is not connected, the set of all the fundamental circuits with respect to the chords of a forest of G, and the set of all the fundamental cutsets with respect to the branches of a forest of G are, respectively, bases for the circuit and cutset subspaces of G. Thus we get the following corollary of the previous theorem.

Corollary 4.6.1 If a graph G has m edges, n vertices, and p components, then

1. The dimension of the circuit subspace of G is equal to $m-n+p$, the nullity of G.
2. The dimension of the cutset subspace of G is equal to $n-p$, the rank of G. ■

As an example, consider the graph G shown in Fig. 4.3. The edges marked as b_1, b_2, b_3, b_4 constitute a spanning tree of G. The chords of T are marked as c_1, c_2, c_3, c_4, and c_5. The fundamental circuits C_1, C_2, C_3, C_4, and C_5 with respect to the chords c_1, c_2, c_3, c_4, and c_5, and the fundamental cutsets S_1, S_2, S_3, and S_4 with respect to the branches b_1, b_2, b_3, and b_4 are obtained as:

$$C_1 = \{c_1, b_1, b_2\}, \qquad S_1 = \{b_1, c_1, c_2\},$$

$$C_2 = \{c_2, b_1, b_2, b_3\}, \qquad S_2 = \{b_2, c_1, c_2, c_4, c_5\},$$

$$C_3 = \{c_3, b_3, b_4\}, \qquad S_3 = \{b_3, c_2, c_3, c_5\},$$

$$C_4 = \{c_4, b_2, b_4\}, \qquad S_4 = \{b_4, c_3, c_4\}.$$

$$C_5 = \{c_5, b_2, b_3\}.$$

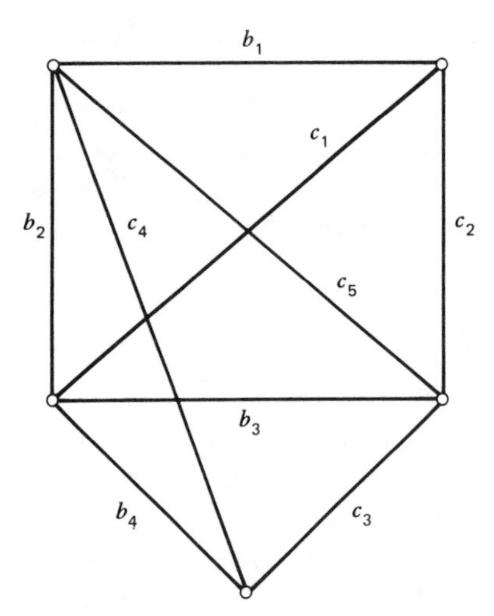

Figure 4.3.

Consider first the subgraph C consisting of the edges b_1, b_2, b_3, c_2, c_3, c_4, and c_5. It may be verified that in C every vertex is of even degree, and hence C belongs to the circuit subspace of G. Since C contains the chords c_2, c_3, c_4, and c_5, it follows from the arguments preceding Theorem 4.6 that C should be equal to the ring sum of the fundamental circuits C_2, C_3, C_4, and C_5. This can be verified to be true as follows:

$$\begin{aligned} C_2 \oplus C_3 \oplus C_4 \oplus C_5 &= \{c_2, b_1, b_2, b_3\} \oplus \{c_3, b_3, b_4\} \oplus \{c_4, b_2, b_4\} \oplus \{c_5, b_2, b_3\} \\ &= \{b_1, b_2, b_3, c_2, c_3, c_4, c_5\} \\ &= C. \end{aligned}$$

Consider next the cut S consisting of the edges b_1, b_3, c_1, c_3, c_5. Again, since S contains the branches b_1 and b_3, it should be equal to the ring sum of the cutsets S_1 and S_3. This also can be verified to be true as follows:

$$\begin{aligned} S_1 \oplus S_3 &= \{b_1, c_1, c_2\} \oplus \{b_3, c_2, c_3, c_5\} \\ &= S. \end{aligned}$$

In the preceding discussions we have shown that, starting from a spanning tree, we can construct bases for the circuit and cutset subspaces of a graph. The bases so constructed are the ones which are commonly used in the study of electrical networks.

4.5 RELATIONSHIP BETWEEN CIRCUIT AND CUTSET SUBSPACES

We establish in this section a characterization for the subgraphs in the circuit subspace of a graph G in terms of those in the cutset subspace of G.

We proved in Section 2.8 (Theorem 2.14) that a circuit and a cutset have an even number of common edges. Since every subgraph in the circuit subspace of a graph is a circuit or the union of edge-disjoint circuits, and every subgraph in the cutset subspace is a cutset or the union of edge-disjoint cutsets, we get the following as an immediate consequence of Theorem 2.14.

THEOREM 4.7. Every subgraph in the circuit subspace of a graph G has an even number of common edges with every subgraph in the cutset subspace of G. ■

In the next theorem, we prove the converse of the above.

THEOREM 4.8.

1. A subgraph of a graph G belongs to the circuit subspace of G if it has an even number of common edges with every subgraph in the cutset subspace of G.
2. A subgraph of a graph G belongs to the cutset subspace of G if it has an even number of common edges with every subgraph in the circuit subspace of G.

Proof

1. We may assume, without any loss of generality, that G is connected. The proof when G is not connected will follow in an exactly similar manner.

   Let T be a spanning tree of G. Let $b_1, b_2, \ldots$ denote the branches of T and $c_1, c_2, \ldots$ denote its chords. Consider any subgraph C of G which has an even number of common edges with every subgraph in the cutset subspace of G. Without any loss of generality, assume that C contains the chords $c_1, c_2, \ldots, c_r$. Let C' denote the ring sum of the fundamental circuits $C_1, C_2, \ldots, C_r$ with respect to the chords $c_1, c_2, \ldots, c_r$.

   Obviously, C' consists of the chords $c_1, c_2, \ldots, c_r$ and no other chords. Hence $C' \oplus C$ consists of no chords.

   C', being the ring sum of some circuits of G, has an even number of common edges with every subgraph in the cutset subspace of G. Since C also has this property, so does $C' \oplus C$.

   We now claim that $C' \oplus C$ is empty. If not, $C' \oplus C$ contains only branches. Let b_i be any branch in $C' \oplus C$. Then b_i is the only edge common between $C' \oplus C$ and the fundamental cutset with respect to b_i. This is not possible since $C' \oplus C$ must have an even number of common edges with every cutset. Thus $C' \oplus C$ should be empty. In other words, $C = C' = C_1 \oplus C_2 \oplus \cdots \oplus C_r$, and hence C belongs to the circuit subspace of G.
2. The proof of this part follows in an exactly similar manner. ■

4.6 ORTHOGONALITY OF CIRCUIT AND CUTSET SUBSPACES

By Theorem 4.1, every n-dimensional vector space over a field F is isomorphic to the vector space of all n-vectors over the same field. Hence W_G, the vector space of a graph G, is isomorphic to the vector space of all m-vectors over $GF(2)$, where m is the number of edges in G.

Let $e_1, e_2, \ldots, e_m$ denote the m edges of G. Suppose we associate each edge-induced subgraph G_i of G with an m-vector w_i such that the jth entry of w_i is equal to 1 if and only if the edge e_j is in G_i. Then the ring sum $G_i \oplus G_j$ of two subgraphs G_i and G_j will correspond to the m-vector $w_i + w_j$, the modulo 2 sum of w_i and w_j. It can now be seen that the association just described indeed defines an isomorphism between W_G and the vector space of all m-vectors over $GF(2)$. In fact, if we choose $\{e_1\}, \{e_2\}, \ldots, \{e_m\}$ as the basis vectors for W_G, then the entries of w_i are the coordinates of G_i relative to this basis.

In view of this isomorphism, we again use the symbol W_G to denote the vector space of all the m-vectors associated with the subgraphs of the graph G. Also, W_C will denote the subspace of m-vectors representing the subgraphs in the circuit subspace of G and similarly W_S will denote the subspace of those representing the subgraphs in the cutset subspace of G.

Consider any two vectors w_i and w_j such that w_i is in W_C and w_j is in W_S. Because every subgraph in W_C has an even number of common edges with those in W_S, it follows that the dot product $\langle w_i, w_j \rangle$ of w_i and w_j is equal to the modulo 2 sum of an even number of 1's. This means $\langle w_i, w_j \rangle = 0$. In other words, the m-vectors in W_C are orthogonal to those in W_S. Thus we have the following theorem.

THEOREM 4.9. The cutset and circuit subspaces of a graph are orthogonal to each other. ■

Consider next the direct sum $W_C \boxplus W_S$. We know that

$$\dim(W_C \boxplus W_S) = \dim(W_C) + \dim(W_S) - \dim(W_C \cap W_S).$$

Since $\dim(W_C) + \dim(W_S) = m$, we get

$$\dim(W_C \boxplus W_S) = m - \dim(W_C \cap W_S).$$

Now the orthogonal subspaces W_C and W_S will also be orthogonal complements of W_G if and only if $\dim(W_C \boxplus W_S) = m$. In other words W_C and W_S will be orthogonal complements if and only if $\dim(W_C \cap W_S) = 0$, that is, $W_C \cap W_S$ is the zero vector whose elements are all equal to zero. Thus we get the following theorem.

THEOREM 4.10. W_C and W_S, the circuit and cutset subspaces of a graph, are orthogonal complements if and only if $W_C \cap W_S$ is the zero vector. ■

Suppose W_C and W_S are orthogonal complements. Then it means that every vector in W_G can be expressed as $w_i + w_j$, where w_i is in W_C and w_j is in W_S. In other words, every subgraph of G can be expressed as the ring sum of two subgraphs, one belonging to the circuit subspace and the other belonging to the cutset subspace. In particular, the graph G itself can be expressed as above.

Suppose W_C and W_S are not orthogonal complements. Then, clearly, there exists a subgraph which cannot be expressed as the ring sum of subgraphs in W_C and W_S. The question then arises whether, in this case too, it is possible to express G as the ring sum of subgraphs from W_C and W_S. The answer is in the affirmative as stated in the next theorem.

THEOREM 4.11. Every graph G can be expressed as the ring sum of two subgraphs one of which is in the circuit subspace and the other is in the cutset subspace of G. ■

See Chen [4.4] and Williams and Maxwell [4.5] for a proof of this theorem.

We conclude this section with an example.

Consider the graph G_a shown in Fig. 4.4a. It may be verified that no nonempty subgraph of this graph belongs to the intersection of the cutset and circuit subspaces. Hence these subspaces of G_a are orthogonal complements. Then the set of fundamental circuits and fundamental cutsets with respect to some spanning tree of G_a constitutes a basis for the vector space of G_a. One such set with respect to the spanning tree formed by the edges e_1, e_2, e_3, and

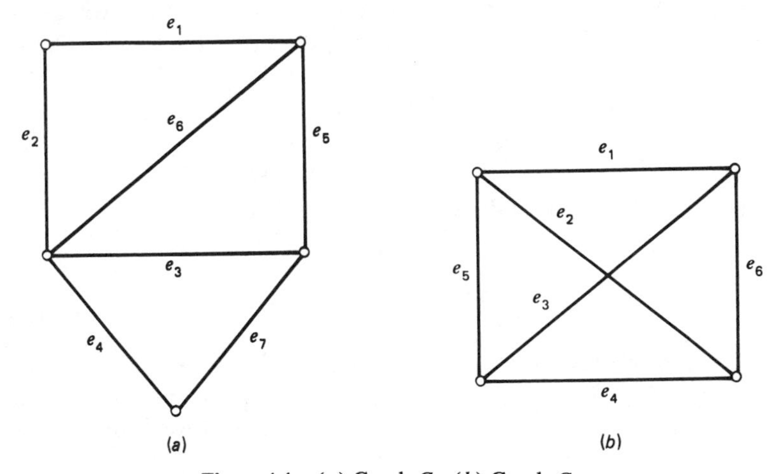

Figure 4.4. (a) Graph G_a. (b) Graph G_b.

e_4 is given below:

fundamental cutset vectors

$$\begin{aligned} S_1 &= (1\ 0\ 0\ 0\ 1\ 1\ 0) \\ S_2 &= (0\ 1\ 0\ 0\ 1\ 1\ 0) \\ S_3 &= (0\ 0\ 1\ 0\ 1\ 0\ 1) \\ S_4 &= (0\ 0\ 0\ 1\ 0\ 0\ 1) \end{aligned}$$

fundamental circuit vectors

$$\begin{aligned} C_1 &= (1\ 1\ 1\ 0\ 1\ 0\ 0) \\ C_2 &= (1\ 1\ 0\ 0\ 0\ 1\ 0) \\ C_3 &= (0\ 0\ 1\ 1\ 0\ 0\ 1) \end{aligned}$$

We can now verify that every vector can be expressed as the ring sum of a circuit vector and a cutset vector. In particular, the vector (1 1 1 1 1 1 1) representing G_a itself can be expressed as:

$$\begin{aligned} (1\,1\,1\,1\,1\,1\,1) &= S_1 \oplus S_2 \oplus S_4 \oplus C_1 \oplus C_2 \\ &= (1\,1\,0\,1\,0\,0\,1) \oplus (0\,0\,1\,0\,1\,1\,0) \end{aligned}$$

where (1 1 0 1 0 0 1) represents a cut and (0 0 1 0 1 1 0) represents a circuit in G_a.

Consider next the graph G_b in Fig. 4.4*b*. In this graph, the edges e_1, e_2, e_3, and e_4 constitute a cut as well as a circuit. Hence the circuit and cutset subspaces of this graph are not orthogonal complements. This means that there exists some subgraph in G_b which cannot be expressed as the ring sum of subgraphs from the cutset and circuit subspaces of G_b. However, according to Theorem 4.11, such a decomposition should be possible for G_b. This is true since

$$G_b = \{e_1, e_2, e_5\} \oplus \{e_3, e_4, e_6\},$$

where $\{e_1, e_2, e_5\}$ is a cut and $\{e_3, e_4, e_6\}$ is a circuit in G_b.

4.7 FURTHER READING

An early paper on vector spaces associated with a graph is by Gould [4.6], where the question of constructing a graph having a specified set of circuits is also discussed. Chen [4.4] and Williams and Maxwell [4.5] are also recommended for further reading on this topic.

4.8 EXERCISES

4.1 Show that the circuits formed by the following sets of edges constitute a basis for the circuit subspace of the graph shown in Fig. 4.5:

$$\{e_1, e_3, e_4, e_6\}; \qquad \{e_2, e_3, e_5, e_7\}; \qquad \{e_1, e_2, e_8\}.$$

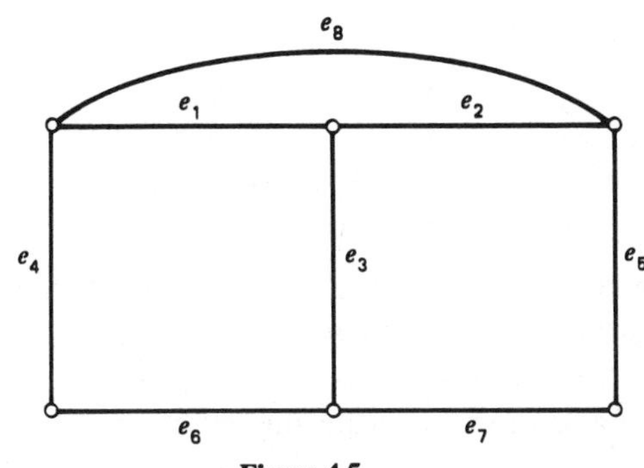

Figure 4.5.

4.2 An *incidence set* at a vertex is the set of edges incident on the vertex. Show that any $n-1$ incidence sets in an n-vertex connected graph form a basis of the cutset subspace of the graph.

4.3 If ρ is the rank and μ is the nullity of a graph G, show that

(a) The number of distinct bases possible for the cutset subspace of G is equal to

$$\frac{1}{\rho!}(2^\rho - 2^0)(2^\rho - 2^1)(2^\rho - 2^2)\cdots(2^\rho - 2^{\rho-1})$$

(b) The number of distinct bases possible for the circuit subspace of G is equal to

$$\frac{1}{\mu!}(2^\mu - 2^0)(2^\mu - 2^1)(2^\mu - 2^2)\cdots(2^\mu - 2^{\mu-1}).$$

4.4 For the graph G shown in Fig. 4.6, obtain the following:

(a) A set of basis vectors for the circuit subspace, which are not fundamental circuit vectors with respect to any spanning tree of G.

(b) A set of basis vectors for the cutset subspace, which are neither incidence sets nor fundamental cutsets with respect to any spanning tree of G.

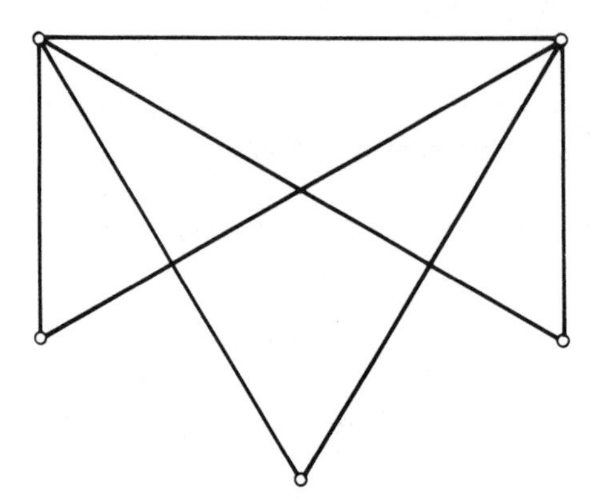

Figure 4.6.

4.5 (a) Test whether the cutset and circuit subspaces of the graph G of Fig. 4.6 are orthogonal complements of the vector space of G.

(b) Express G as the ring sum of two subgraphs, one from the circuit subspace and the other from the cutset subspace.

4.6 Show that every subset of the edge set of a tree is a cut of the tree.

4.7 Show that a subgraph of a graph has an even number of edges if it belongs to both the cutset and the circuits subspaces of the graph.

4.8 A subset E' of edges of a graph is said to be independent if E' contains no circuits. Prove the following:

(a) Every subset of an independent set is independent.

(b) If I and J are independent sets containing k and $k+1$ edges, respectively, then there exists an edge e which is in J but not in I, such that $I \cup \{e\}$ is an independent set.

4.9 Repeat Exercise 4.8 replacing "circuit" by "cutset."

4.9 REFERENCES

4.1 S. MacLane and G. Birkhoff, *Algebra*, Macmillan, New York, 1967.

4.2 P. R. Halmos, *Finite-Dimensional Vector Spaces*, Van Nostrand Reinhold, New York, 1958.

4.3 F. E. Hohn, *Elementary Matrix Algebra*, Macmillan, New York, 1958.

4.4 W. K. Chen, "On Vector Spaces Associated with a Graph," *SIAM J. Appl. Math.*, Vol. 20, 526–529 (1971).

4.5 T. W. Williams and L. M. Maxwell, "The Decomposition of a Graph and the Introduction of a New Class of Subgraphs," *SIAM J. Appl. Math.*, Vol. 20, 385–389 (1971).

4.6 R. Gould, "Graphs and Vectors Spaces," *J. Math. Phys.*, Vol. 37, 193–214 (1958).

Chapter 5

Directed Graphs

In the last four chapters we developed several basic results in the theory of undirected graphs. Undirected graphs are not adequate for representing several situations. For example, in the graph representation of a traffic network where an edge may represent a street, we have to assign directions to the edges to indicate the permissible direction of traffic flow. As another example, a computer program can be modeled by a graph in which an edge represents the flow of control from one set of instructions to another. In such a representation of a program, directions have to be assigned to the edges to indicate the directions of the flow of control. An electrical network is yet another example of a physical system whose representation requires a directed graph. In Chapters 11 through 15, applications involving directed graphs and related algorithms are discussed.

In this chapter we develop several basic results in the theory of directed graphs. We discuss questions relating to the existence of directed Euler circuits and Hamilton trails. We also discuss directed trees and their relationship with directed Euler trails.

5.1 BASIC DEFINITIONS AND CONCEPTS

We begin with the introduction of a few basic definitions and concepts relevant to directed graphs.

A *directed graph* $G=(V,E)$ consists of two sets: a finite set V of elements called *vertices* and a finite set E of elements called *edges*. Each edge is associated with an ordered pair of vertices.

We use the symbols $v_1, v_2, \ldots$ to represent the vertices and the symbols $e_1, e_2, \ldots$ to represent the edges of a directed graph. If $e_l = (v_i, v_j)$, then v_i and v_j are called the *end vertices* of the edge e_l, v_i being called the *initial vertex* and v_j the *terminal vertex* of e_l. All edges having the same pair of initial and terminal vertices are called *parallel edges*. An edge is called a *self-loop* at vertex v_i, if v_i is the initial as well as the terminal vertex of the edge.

In the pictorial representation of a directed graph, a vertex is represented by a dot or a circle and an edge is represented by a line segment connecting the dots or the circles that represent the end vertices of the edge. In addition, each edge is assigned an orientation indicated by an arrow which is drawn from the initial to the terminal vertex.

For example, if

$$V = \{v_1, v_2, v_3, v_4, v_5, v_6, v_7\}$$

and

$$E = \{e_1, e_2, e_3, e_4, e_5, e_6, e_7, e_8\}$$

such that

$$e_1 = (v_1, v_2),$$
$$e_2 = (v_1, v_2),$$
$$e_3 = (v_1, v_3),$$
$$e_4 = (v_3, v_1),$$
$$e_5 = (v_2, v_4),$$
$$e_6 = (v_3, v_4),$$
$$e_7 = (v_4, v_4),$$
$$e_8 = (v_5, v_6),$$

then the directed graph $G = (V, E)$ will be represented as in Fig. 5.1. In this graph e_1 and e_2 are parallel edges, and e_7 is a self-loop.

An edge is said to be *incident on* its end vertices. Two vertices are *adjacent* if they are the end vertices of some edge. If two edges have a common end vertex, then these edges are said to be *adjacent*.

An edge is said to be *incident out of* its initial vertex and *incident into* its terminal vertex. A vertex is called an *isolated vertex* if no edge is incident on it.

The *degree* $d(v_j)$ of a vertex v_j is the number of edges incident on v_j. The *in-degree* $d^-(v_j)$ of v_j is the number of edges incident into v_j and the *out-degree* $d^+(v_j)$ is the number of edges incident out of v_j. δ^+ and δ^- will

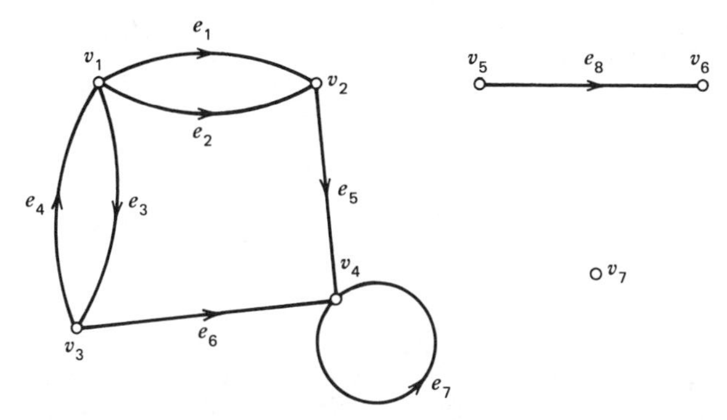

Figure 5.1. A directed graph.

denote the minimum out-degree and the minimum in-degree in a directed graph. Similarly Δ^+ and Δ^- will denote the maximum out-degree and the maximum in-degree, respectively.

For any vertex v, the sets $\Gamma^+(v)$ and $\Gamma^-(v)$ are defined as follows:

$$\Gamma^+(v)=\{w|(v,w)\in E\},$$

$$\Gamma^-(v)=\{w|(w,v)\in E\}.$$

For example, in the graph of Fig. 5.1 $\Gamma^+(v_1)=\{v_2,v_3\}$ and $\Gamma^-(v_4)=\{v_2,v_3,v_4\}$.

Note that a self-loop at a vertex contributes to the in-degree as well as to the out-degree of the vertex. The following result is a consequence of the fact that every edge contributes 1 to the sum of the in-degrees as well as to the sum of the out-degrees of a directed graph.

THEOREM 5.1. In a directed graph, sum of the in-degrees = sum of the out-degrees = m, where m is the number of edges of the graph. ■

Subgraphs and *induced subgraphs* of a directed graph are defined as in the case of undirected graphs (Section 1.2).

The graph that results after ignoring the orientations of a directed graph G is called the *underlying undirected graph* of G. This undirected graph will be denoted by G_u.

A *directed walk* in a directed graph $G=(V,E)$ is a finite sequence of vertices $v_0, v_1, \ldots, v_k$, such that (v_{i-1}, v_i), $1 \leqslant i \leqslant k$, is an edge in G. This directed walk is usually called a v_0–v_k directed walk, v_0 is called the *initial vertex*, v_k is called the *terminal vertex* of this walk, and all other vertices are called *internal vertices*. The initial and terminal vertices of a directed walk are called its *end vertices*. Note that in a directed walk, edges and hence vertices can appear more than once.

A directed walk is *open* if its end vertices are distinct; otherwise it is *closed*.

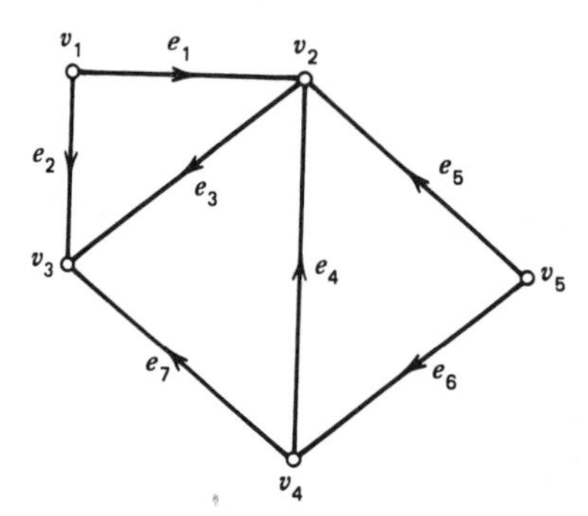

Figure 5.2. An acyclic directed graph.

A directed walk is a *directed trail* if all its edges are distinct. A directed trail is *open* if its end vertices are distinct; otherwise it is *closed.*

An open directed trail is a *directed path* if all its vertices are distinct.

A closed directed trail is a *directed circuit* if all its vertices except the end vertices are distinct.

A directed graph is said to be *acyclic* if it has no directed circuits. For example, the directed graph in Fig. 5.2 is acyclic.

A sequence of vertices in a directed graph G is a *walk* in G if it is a walk in the underlying undirected graph G_u. For example, the sequence $v_1, v_2, v_3, v_4, v_2, v_3$ in the graph of Fig. 5.2 is a walk; but it is not a directed walk.

Trail, *path*, and *circuit* in a directed graph are defined in a similar manner.

A directed graph is *connected* if the underlying undirected graph is connected.

A subgraph of a directed graph G is a *component* of G if it is a component of G_u.

In a directed graph G, a vertex v_i is *strongly connected* to a vertex v_j if in G there exists a directed path from v_i to v_j and a directed path from v_j to v_i. If v_i is strongly connected to v_j, then, obviously, v_j is strongly connected to v_i. Every vertex is strongly connected to itself.

If v_i is strongly connected to v_j, and v_j is strongly connected to v_k, then it is easy to see that v_i is strongly connected to v_k. Therefore, in such a case, we simply state that the vertices v_i, v_j, and v_k are strongly connected.

A directed graph is *strongly connected* if all its vertices are strongly connected. For example, the graph in Fig. 5.3 is strongly connected.

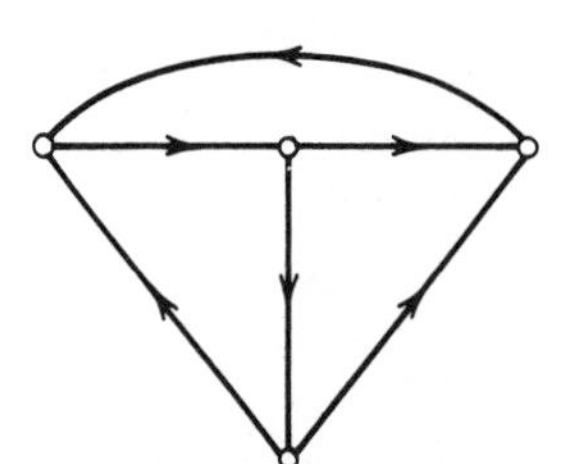

Figure 5.3. A strongly connected directed graph.

A maximal strongly connected subgraph of a directed graph G is called a *strongly connected component* of G. If a directed graph is strongly connected, then it has only one strongly connected component, namely, itself.

Consider a directed graph $G=(V, E)$. It is easy to see that each vertex of G is present in exactly one strongly connected component of G. Thus the vertex sets of the strongly connected components constitute a partition of the vertex set V of G.

For example, the directed graph of Fig. 5.4a has three strongly connected components with $\{v_2, v_3, v_4, v_5\}$, $\{v_1\}$, and $\{v_6\}$ as their vertex sets which form a partition of the vertex set $\{v_1, v_2, v_3, v_4, v_5, v_6\}$ of the directed graph.

It is interesting to note that there may be some edges in a directed graph which do not belong to any strongly connected component of the graph. For example, in the graph of Fig. 5.4a, the edges e_1, e_6, e_7, e_9, and e_{10} are not present in any strongly connected component.

Thus while the "strongly connected" property induces a partition of the vertex set of a graph, it may not induce a partition of the edge set.

Union, intersection, ring sum, and other operations involving directed graphs are defined in exactly the same way as in the case of undirected graphs (see Section 1.5).

The graph that results after contracting all the edges in each strongly connected component of a directed graph G is called the *condensed graph* G_c

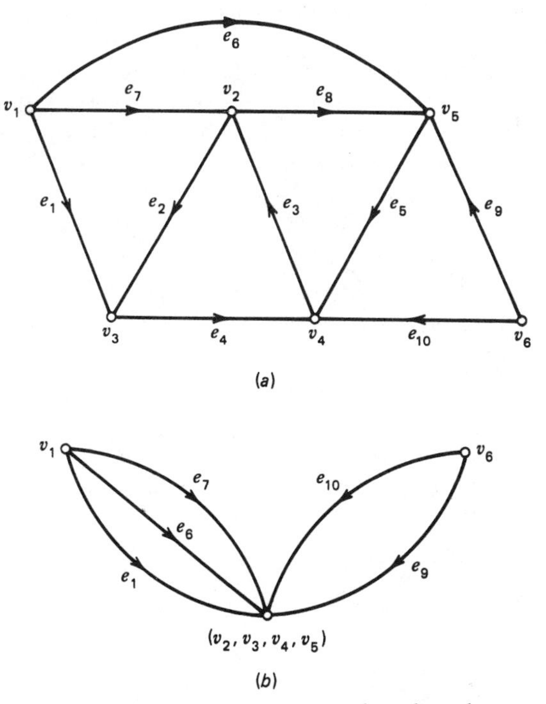

Figure 5.4. A graph and its condensed graph.

of G. For example, the condensed graph of the graph of Fig. 5.4a is shown in Fig. 5.4b.

The vertex in G_c which corresponds to a strongly connected component is called the *condensed image* of the component.

The *rank* and *nullity* of a directed graph are the same as those of the corresponding undirected graph. Thus if a directed graph G has m edges, n vertices, and p components, then the rank ρ and the nullity μ of G are given by

$$\rho = n - p$$

and

$$\mu = m - n + p.$$

Next we define a minimally connected directed graph and study some properties of such graphs.

A directed graph G is *minimally connected* if G is strongly connected and the removal of any edge from G destroys its "strongly connected" property. For example, the graph of Fig. 5.5 is minimally connected.

Obviously, a minimally connected directed graph can have neither parallel edges nor self-loops.

We know that an undirected graph is minimally connected if and only if it is a tree (Exercise 2.13). By Theorem 2.5, a tree has at least two vertices of degree 1. Thus in a minimally connected undirected graph there are at least two vertices of degree 1.

We now establish an analogous result for the case of directed graphs. In a strongly connected directed graph the degree of every vertex should be at least 2, since at each vertex, there should be one edge incident out of the vertex and one edge incident into the vertex. In the following theorem we prove that a minimally connected directed graph has at least two vertices of degree 2.

THEOREM 5.2. If a minimally connected directed graph G has more than one vertex, then G has at least two vertices of degree 2.

Proof

Since G is strongly connected and has more than one vertex, it must have at least one directed circuit. Thus for G, the nullity $\mu \geqslant 1$.

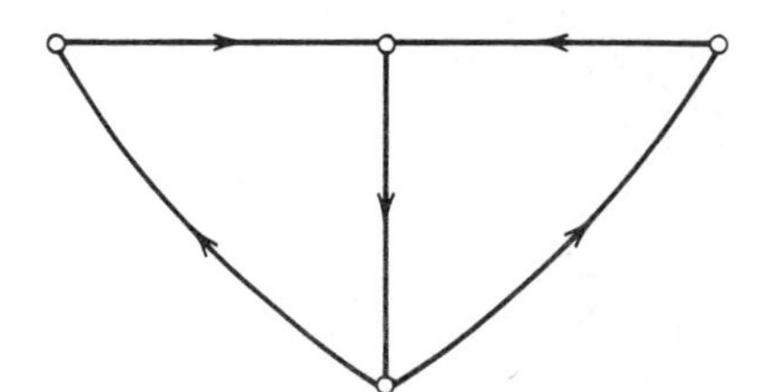

Figure 5.5. A minimally connected directed graph.

We shall prove the theorem by induction on μ.

If $\mu=1$, then G must be a directed circuit, and hence the theorem is true for this case.

Let the theorem be true for all minimally connected directed graphs for which the nullity $\mu<k$, with $k\geqslant 2$. Now consider a minimally connected graph G with $\mu=k$. We shall show that the theorem is true for G.

Two cases now arise.

Case 1 Every directed circuit in G is of length 2.

In this case any two adjacent vertices of G are joined by exactly two edges oriented in different directions. Let G' be a simple undirected graph which has the same vertex set as G and in which two vertices are adjacent if and only if they are adjacent in G. Since G is connected, G' is also connected. Further, G' has no circuits because G has no directed circuit of length greater than 2. Thus G' is a tree. Hence, by Theorem 2.5, it has at least two pendant vertices. These two vertices have degree 2 in G, and the theorem is therefore true for this case.

Case 2 G has a directed circuit C of length $l\geqslant 3$.

Since G is minimally connected, there is only one edge in G between any two adjacent vertices of C, and further, there is no edge in G between any two vertices that are not adjacent in C.

Suppose G' is the graph that results after contracting the edges of C. G' has $(m-l)$ edges and $(n-l+1)$ vertices, where m and n are, respectively, the number of edges and the number of vertices in G. Therefore the nullity of G' is equal to

$$(m-l)-(n-l+1)+1=k-1.$$

Since G' is also minimally connected, it follows from the induction hypothesis that G' has at least two vertices v_1 and v_2 of degree 2. If one of these vertices, say v_1, is the condensed image of C, then C has at least one vertex, say v_3, which is of degree 2. In such a case v_2 and v_3 are two vertices of degree 2 in G.

On the other hand, if neither v_1 nor v_2 is the condensed image of C, then v_1 and v_2 are two vertices of degree 2 in G.

Hence the theorem. ■

We conclude this section with the definition of the "quasi-strongly connected" property.

A graph is said to be *quasi-strongly connected* if for every pair of vertices v_1 and v_2 there is a vertex v_3 from which there is a directed path to v_1 and a directed path to v_2. Note that v_3 need not be distinct from v_1 or v_2.

For example, the graph of Fig. 5.6 is quasi-strongly connected.

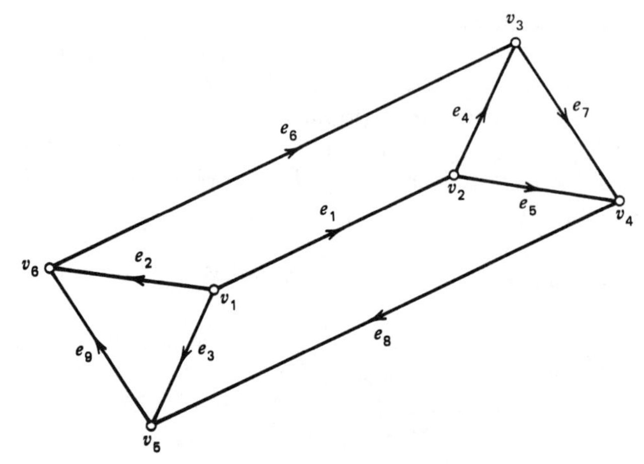

Figure 5.6. A quasi-strongly connected graph.

5.2 GRAPHS AND RELATIONS

A *binary relation* on a set $X=\{x_1, x_2, \dots\}$ is a collection of ordered pairs of elements of X. For example, if the set X is composed of men and R is the relation "is son of," then the ordered pair (x_i, x_j) means that x_i is the son of x_j. This is also denoted as $x_i R x_j$.

A most convenient way of representing a binary relation R on a set X is by a directed graph, the vertices of which stand for the elements of X and the edges stand for the ordered pairs of elements of X defining the relation R.

For example, the relation "is a factor of" on the set $X=\{2,3,4,6,9\}$ is shown in Fig. 5.7.

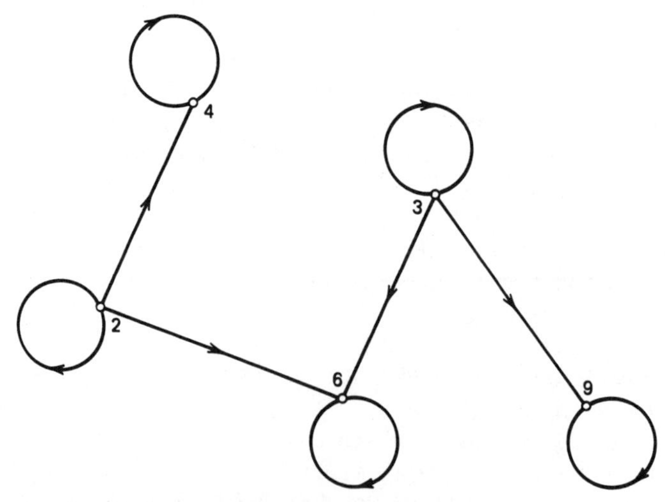

Figure 5.7. A directed graph representing the relation "is a factor of".

Consider a set $X=\{x_1, x_2,\ldots\}$ and a relation R on X.

1. R is *reflexive* if every element x_i is in relation R to itself; that is, for every x_i, $x_i R x_i$.
2. R is *symmetric* if $x_i R x_j$ implies $x_j R x_i$.
3. R is *transitive* if $x_i R x_j$ and $x_j R x_k$ imply $x_i R x_k$.
4. R is an *equivalence relation* if it is reflexive, symmetric, and transitive. If R is an equivalence relation defined on a set S, then we can uniquely partition S into subsets $S_1, S_2,\ldots, S_k$ such that two elements x and y of S belong to S_i if and only if $x R y$. The subsets $S_1, S_2,\ldots, S_k$ are all called the *equivalence classes* induced by the relation R on the set S.

Suppose the set X is composed of positive integers. Then

1. The relation "is a factor of" is reflexive and transitive.
2. The relation "is equal to" is reflexive, symmetric, and transitive, and is therefore an equivalence relation.

The directed graph representing a reflexive relation is called a *reflexive directed graph*. In a similar way, *symmetric* and *transitive directed graphs* are defined. We can now make the following observations about these graphs:

1. In a reflexive directed graph, there is a self-loop at each vertex.
2. In a symmetric directed graph, there are two oppositely oriented edges between any two adjacent vertices. Therefore an undirected graph can be considered as representing a symmetric relation if we associate with each edge two oppositely oriented edges.
3. The edge (v_1, v_2) is present in a transitive graph G if there is a directed path in G from v_1 to v_2.

5.3 DIRECTED TREES OR ARBORESCENCES

A vertex v in a directed graph G is a *root* of G if there are directed paths from v to all the remaining vertices of G.

For example, in the graph G of Fig. 5.6, the vertex v_1 is a root. It is clear that if a graph has a root, it is quasi-strongly connected. In the following theorem we prove that the "quasi-strongly connected" property implies the existence of a root.

THEOREM 5.3. A directed graph G has a root if and only if it is quasi-strongly connected.

Proof

Necessity Obvious.

Sufficiency Consider the vertices $x_1, x_2, \ldots, x_n$ of G. Since G is quasi-strongly connected, there exists a vertex y_2 from which there is a directed path to x_1 and a directed path to x_2. For the same reason, there is a vertex y_3 from which there is a directed path to y_2 and a directed path to x_3. Clearly, y_3 is also connected to x_1 and x_2 by directed paths through y_2. Proceeding in this manner, we see that there is a vertex y_n from which there is a directed path to y_{n-1} and x_n. Obviously y_n is a root since it is also connected to $x_1, \ldots, x_{n-1}$ through y_{n-1}. ■

A directed graph G is a *tree* if the underlying undirected graph is a tree.

A directed graph G is a *directed tree* or *arborescence* if G is a tree and has a root. A vertex v in G is called a *leaf* if $d^+(v)=0$.

For example, the graph of Fig. 5.8 is a directed tree. In this graph the vertex v_1 is a root and it is the only root.

We present in the next theorem a number of equivalent characterizations of a directed tree.

THEOREM 5.4. Let G be a directed graph with $n>1$ vertices. Then the following statements are equivalent:

1. G is a directed tree.
2. There exists a vertex r in G such that there is exactly one directed path from r to every other vertex of G.
3. G is quasi-strongly connected and loses this property if any edge is removed from it.
4. G is quasi-strongly connected and has a vertex r such that

$$d^-(r)=0$$

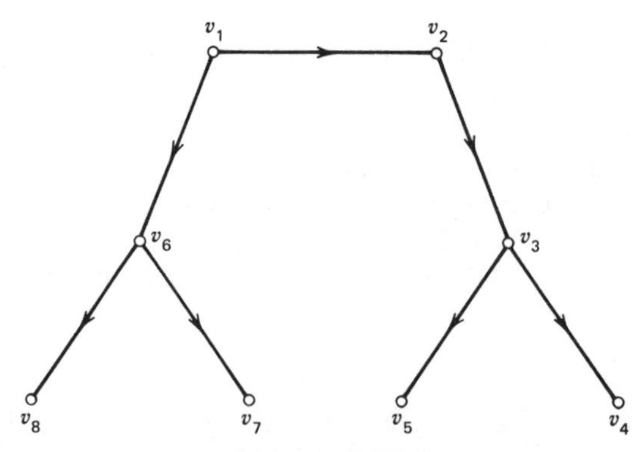

Figure 5.8. A directed tree.

and

$$d^{-}(v)=1, \qquad v \neq r.$$

5. G has no circuits (not necessarily directed circuits) and has a vertex r such that

$$d^{-}(r)=0$$

and

$$d^{-}(v)=1, \qquad v \neq r.$$

6. G is quasi-strongly connected without circuits.

Proof

1⇒2 The root r in G has the desired property.

2⇒3 Obviously G is quasi-strongly connected. Suppose this property is not destroyed when an edge (u, v) is removed from G. Then there is a vertex z such that there are two directed paths from z, one to u and the other to v, which do not use the edge (u, v). Thus in G, there are two directed paths from z to v and hence two directed paths from r to v. This contradicts statement 2.

3⇒4 Since G is quasi-strongly connected, it has a root r. So, for every vertex $v \neq r$,

$$d^{-}(v) \geqslant 1.$$

Suppose, for some vertex v, that

$$d^{-}(v)>1.$$

Then there are two edges, say (x, v) and (y, v), incident into v. Thus there are two distinct paths from r to v, one through (x, v) and the other through (y, v). So even if we remove one of these two edges, the resulting graph still has r as a root and hence remains quasi-strongly connected. This contradicts statement 3 and so,

$$d^{-}(v)=1, \qquad v \neq r.$$

Finally, no edge can be incident into r, for otherwise the graph that results after removing this edge from G will still have r as its root and hence will be quasi-strongly connected. This will again contradict statement 3. Therefore

$$d^{-}(r)=0.$$

4⇒5 The sum of in-degrees in G is equal to $n-1$. Therefore by Theorem 5.1, G has $n-1$ edges. Since G is also connected, by Theorem 2.1, statement 3, it is a tree and hence is circuitless.

5⇒6 G has $n-1$ edges because the sum of in-degrees in G is equal to $n-1$. Since it is also circuitless, it follows from Theorem 2.2 that it is a tree. Therefore, there exists a unique path in G from vertex r to every other vertex. Such a path should be a directed path, for otherwise at least one of the vertices in this path will have its in-degree greater than 1, contradicting statement 5. Thus vertex r is a root of G. Therefore, by Theorem 5.3, G is quasi-stringly connected.

6⇒1 Since G is quasi-strongly connected, it has a root, by Theorem 5.3. Further, it is a tree, because it is connected and has no circuits. ■

A subgraph of a directed graph G is a *directed spanning tree* of G if the subgraph is a directed tree and contains all the vertices of G. For example, the subgraph consisting of the edges e_1, e_2, e_3, e_4, and e_5 is a directed spanning tree of the graph of Fig. 5.6.

We know that a graph G has a spanning tree if and only if G is connected. The corresponding theorem in the case of directed graphs is the following one.

THEOREM 5.5. A directed graph G has a directed spanning tree if and only if G is quasi-strongly connected.

Proof

Necessity If G has a directed spanning tree, then obviously the root of the directed tree is also a root of G. Therefore, by Theorem 5.3, G is quasi-strongly connected.

Sufficiency Suppose G is quasi-strongly connected and it is not a directed tree. Then, by Theorem 5.4, statement 3, there are edges whose removal from G will not destroy the quasi-strongly connected property of G. Therefore, if we remove successively all these edges from G, then the resulting graph is a directed spanning tree of G. ■

The usefulness of directed trees in the topological study of active electrical networks is discussed in Chapter 13.

A *binary tree* is a directed tree in which the out-degree of every vertex is at most 2. For example, the graph of Fig. 5.9 is a binary tree. In the study of certain aspects of computer science, for example, analysis of algorithms, search techniques, and so on, binary trees are found to be very useful. One problem in this context is the following:

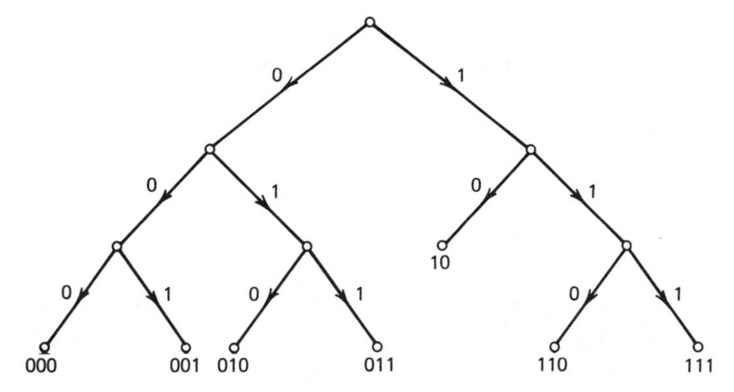

Figure 5.9. A binary tree.

Given n weights $w_1, w_2, \ldots, w_n$ and n lengths $l_1, l_2, \ldots, l_n$, construct a binary tree with n leaves $v_1, v_2, \ldots, v_n$ such that

1. w_i, $1 \leqslant i \leqslant n$, is the weight of vertex v_i.
2. l_i, $1 \leqslant i \leqslant n$, is the length of the path from the root to v_i.
3. $\Sigma_{i=1}^{n} w_i l_i$, called the *weighted path length*, is minimum.

A situation where a problem more general than the above arises is described below.

The set of words such that no word is the beginning of another is called a *prefix code*. Clearly, if a sequence of letters is formed by concatenating the words of a prefix code, the sequence can be decomposed into individual words of the prefix code by reading the sequence from left to right. For example, $\{000, 001, 01, 10, 11\}$ is a prefix code and the sequence 1011001000 formed by concatenating the words of this code is easily decomposed into the words 10, 11, 001, and 000.

Suppose $S = \{0, 1, 2, \ldots, m-1\}$ is an alphabet of m letters. Let T be a directed tree such that

1. The out-degree of each vertex is at most m; and
2. Each out-going edge at a vertex is associated with a letter of the alphabet S such that no two such edges are associated with the same letter.

Then each leaf v can be associated with a word which is formed by concatenating all the letters in the order in which they appear on the edges along the path from the root to v. It may be verified that the words so associated with the leaves form a prefix code. For example, the prefix code corresponding to the directed tree of Fig. 5.9 is the following:

$$\{000, 001, 010, 011, 10, 110, 111\}.$$

Now the problem is: Given m and the lengths $l_1, l_2, \ldots, l_n$, construct a prefix code using the letters of the alphabet $S=\{0,1,2,\ldots,m-1\}$ with $l_1, l_2, \ldots, l_n$ as the lengths of the code words.

This question and certain related ones are considered in Sections 15.2 and 15.3.

5.4 DIRECTED EULERIAN GRAPHS

A *directed Euler trail* in a directed graph G is a closed directed trail which contains all the edges of G. An *open directed Euler trail* is an open directed trail containing all the edges of G. A directed graph possessing a directed Euler trail is called a *directed Eulerian graph*.

The graph G of Fig. 5.10 is a directed Eulerian graph, since the sequence of edges $e_1, e_2, e_3, e_4, e_5, e_6$ constitutes a directed Euler trail in G.

The following theorem gives simple and useful characterizations of directed Eulerian graphs. The proof of this theorem follows along the same lines as that for Theorem 3.1.

THEOREM 5.6. The following statements are equivalent for a connected directed graph G:

1. G is a directed Eulerian graph.
2. For every vertex v of G, $d^-(v)=d^+(v)$.
3. G is the union of some edge-disjoint directed circuits. ■

Consider, for example, the directed Eulerian graph G of Fig. 5.10. It is easy to verify that it satisfies property 2 of the above theorem. Further it is the union of the edge-disjoint circuits $\{e_2, e_3\}$ and $\{e_1, e_4, e_5, e_6\}$.

Another theorem which can be easily proved is the following one.

THEOREM 5.7. A directed connected graph possesses an open directed Euler trail if and only if the following conditions are satisfied:

1. In G there are two vertices v_1 and v_2 such that

$$d^+(v_1)=d^-(v_1)+1$$

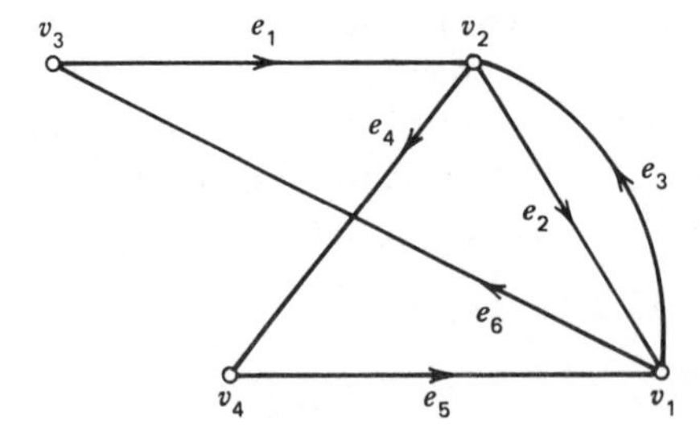

Figure 5.10. A directed Eulerian graph.

and

$$d^-(v_2)=d^+(v_2)+1.$$

2. For every vertex v, different from v_1 and v_2,

$$d^-(v)=d^+(v). \quad ■$$

For example, the graph G of Fig. 5.11 satisfies the conditions of the above theorem. The sequence of edges $e_1, e_2, e_3, e_4, e_5, e_6$ is an open directed Euler trail in G.

An interesting application of Theorem 5.6 is discussed next.

Let $S=\{0,1,\ldots,s-1\}$ be an alphabet of s letters. Obviously we can construct s^n different words of length n using the letters of the alphabet S. A *de Bruijn sequence* is a circular sequence $a_0, a_1, a_2,\ldots, a_{L-1}$ of length $L=s^n$ such that for every word ω of length n, there is a unique i such that

$$\omega=a_i a_{i+1}\cdots a_{i+n-1},$$

where the computation of the indices is modulo L. de Bruijn sequences find application in the study of coding theory and communication circuits. References [5.1] and [5.2] may be consulted for more information on these sequences. The problem which we now wish to consider is the following:

For every $s\geqslant 2$ and every integer n, does there exist a de Bruijn sequence over the alphabet $S=\{0,1,2,\ldots,s-1\}$?

The answer to the above question is in the affirmative, as we shall see now.

We show that for every $s\geqslant 2$ and every n there exists a directed Eulerian graph $G_{s,n}$ such that each directed Euler trail in this graph corresponds to a de Bruijn sequence of length s^n over the alphabet $S=\{0,1,2,\ldots,s-1\}$. The graph $G_{s,n}=(V,E)$ is constructed as follows:

1. V consists of all the s^{n-1} words of length $n-1$ over the alphabet S.
2. E is the set of all the s^n words of length n over S.
3. The edge $b_1b_2\cdots b_n$ has $b_1b_2\cdots b_{n-1}$ as its initial vertex and $b_2b_3\cdots b_n$ as its terminal vertex. Note that at each vertex there are s incoming edges and s outgoing edges. Thus for each vertex v, $d^-(v)=d^+(v)$.

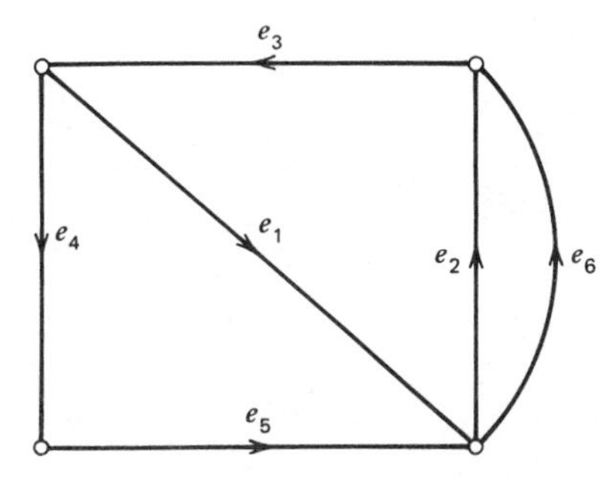

Figure 5.11. A graph with an open directed Euler trail.

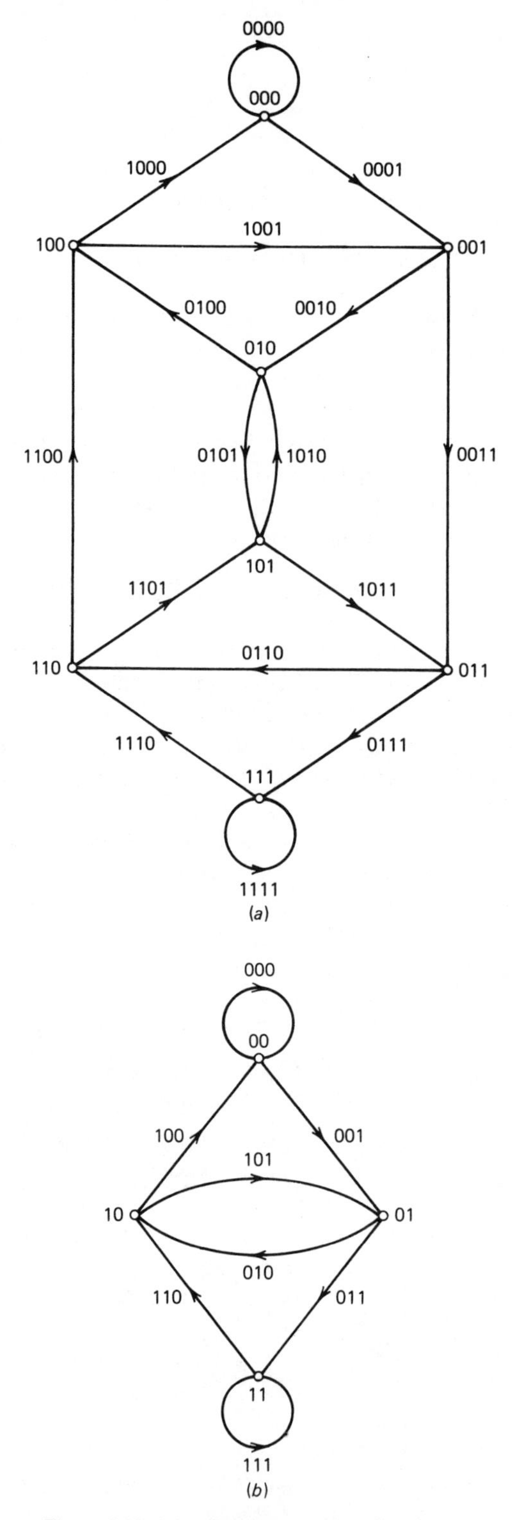

Figure 5.12. (*a*) Graph $G_{2,4}$. (*b*) Graph $G_{2,3}$.

For example, the graphs $G_{2,4}$ and $G_{2,3}$ are as shown in Fig. 5.12*a* and *b*.

Suppose there exists a directed Euler trail C in $G_{s,n}$. If we concatenate the first letters of the words, represented by the edges of $G_{s,n}$, in the order in which they appear in C, then we get a sequence of length s^n such that every subsequence of n consecutive letters in this sequence will correspond to a unique word and no two different subsequences will correspond to the same word. Thus the sequence constructed as above will be a de Bruijn sequence. For example, the sequence of edges 000, 001, 011, 111, 110, 101, 010, 100 is a directed Euler trail in $G_{2,3}$. Concatenating the first letters of these words we get the de Bruijn sequence 00011101.

Thus to show that there exists a de Bruijn sequence for every $s \geqslant 2$ and every n, we have to prove that $G_{s,n}$ is a directed Eulerian graph. This we do in the following theorem.

THEOREM 5.8. $G_{s,n}$ is a directed Eulerian graph.

Proof

$G_{s,n}$ is connected since for any two vertices $v_1 = b_1 b_2 \cdots b_{n-1}$ and $v_2 = c_1 c_2 \cdots c_{n-1}$, there is a directed path from v_1 to v_2 consisting of the following edges:

$$b_1 b_2 \cdots b_{n-1} c_1,\ b_2 b_3 \cdots b_{n-1} c_1 c_2, \ldots, b_{n-1} c_1 c_2 \cdots c_{n-1}.$$

Since, as we have seen before, the in-degree of every vertex in $G_{s,n}$ is equal to its out-degree, it follows from Theorem 5.6 that $G_{s,n}$ is a directed Eulerian graph. ■

Corollary 5.8.1 For every $s \geqslant 2$ and every n there exists a de Bruijn sequence. ■

5.5 DIRECTED SPANNING TREES AND DIRECTED EULER TRAILS

Let G be a directed Eulerian graph without self-loops. In this section we relate the number of directed Euler trails in G to the number of directed spanning trees of G.

Let $v_1, v_2, \ldots, v_n$ denote the vertices of G. Consider a directed Euler trail C in G. Let e_{j_1} be any edge of G incident into v_1. For every $p = 2, 3, \ldots, n$, let e_{j_p} denote the first edge on C to enter vertex v_p after traversing e_{j_1}. For example, in the directed Eulerian graph shown in Fig. 5.10, the sequence of edges $e_1, e_2, e_3, e_4, e_5, e_6$ is a directed Euler trail. If we choose $e_{j_1} = e_5$, then $e_{j_2} = e_1$, $e_{j_3} = e_6$, and $e_{j_4} = e_4$.

Let H denote the subgraph of G on the edge set $\{e_{j_2}, e_{j_3}, \ldots, e_{j_n}\}$.

LEMMA 5.1. Let C be a directed Euler trail of a directed Eulerian graph G. The subgraph H defined as above is a directed spanning tree of G with root v_1.

Proof

Clearly in H, $d^-(v_1)=0$, and $d^-(v_p)=1$ for all $p=2,3,\ldots,n$. Suppose H has a circuit C'. Then v_1 is not in C', for otherwise either $d^-(v_1)>0$ or $d^-(v)>1$ for some other vertex v on C'. For the same reason C' is a directed circuit. Since $d^-(v)=1$ for every vertex v on C', no edge not in C' enters any vertex in C'. This means that the edge e which is the first edge of C to enter a vertex of C' after traversing e_{j_1} does not belong to H, contradicting the definition of H. Thus H has no circuits.

Now it follows from Theorem 5.4, statement 5, that H is a directed spanning tree of G. ■

Given a directed spanning tree H of an n-vertex directed Eulerian graph G without self-loops, let v_1 be the root of H and e_{j_1} an edge incident into v_1 in G. Let e_{j_p} for $p=2,3,\ldots,n$ be the edge entering v_p in H. We now describe a method for constructing a directed Euler trail in G.

1. Start from vertex v_1 and traverse backward on any edge entering v_1 other than e_{j_1} if such an edge exists, or on e_{j_1} if there is no other alternative.
2. In general on arrival at a vertex v_p, leave it by traversing backward on an edge entering v_p which has not yet been traversed and, if possible, other than e_{j_p}. Stop if no untraversed edges entering v_p exist.

In the above procedure every time we reach a vertex $v_p \neq v_1$, there will be an untraversed edge entering v_p because the in-degree of every vertex in G is equal to its out-degree. Thus this procedure terminates only at the vertex v_1 after traversing all the edges incident into v_1.

Suppose there exists in G an untraversed edge (u,v) when the above procedure terminates at v_1. Since the in-degree of u is equal to its out-degree, there exists at least one untraversed edge incident into u. If there is more than one such untraversed edge, then one of these will be the edge y entering u in H. This follows from step 2 of the procedure. This untraversed edge y will lead to another untraversed edge which is also in H. Finally we shall arrive at v_1 and shall find an untraversed edge incident into v_1. This is not possible since all the edges incident into v_1 will have been traversed when the procedure terminates at v_1.

Thus all the edges of G will be traversed during the procedure we described above, and indeed a directed Euler trail is constructed.

Since at each vertex v_p there are $(d^-(v_p)-1)!$ different orders for picking the incoming edges (with e_{j_p} at the end), it follows that the number of distinct directed Euler trails which we can construct from a given directed spanning tree H and e_{j_1} is

$$\prod_{p=1}^{n} (d^-(v_p)-1)!.$$

Further, each different choice of H will yield a different e_{j_p} for some $p=2,3,\ldots,n$, which will in turn result in a different entry to v_p after traversing e_{j_1} in the resulting directed Euler trail.

Finally, since the procedure of constructing a directed Euler trail is the reversal of the procedure for constructing a directed spanning tree, it follows that every directed Euler trail can be constructed from some directed spanning tree.

Thus we have proved the following theorem due to Van Aardenne-Ehrenfest and de Bruijn [5.3].

THEOREM 5.9. The number of directed Euler trails of a directed Eulerian graph G without self-loops is

$$\tau_d(G) \prod_{p=1}^{n} (d^-(v_p)-1)!$$

where $\tau_d(G)$ is the number of directed spanning trees of G with v_1 as root. ■

Since the number of directed Euler trails is independent of the choice of the root, we get the following.

Corollary 5.9.1 The number of directed spanning trees of a directed Eulerian graph is the same for every choice of root. ■

We establish in Section 6.9 a formula for evaluating $\tau_d(G)$.

5.6 DIRECTED HAMILTONIAN GRAPHS

A directed circuit in a directed graph G is a *directed Hamilton circuit* of G if it contains all the vertices of G.

A directed path in G is a *directed Hamilton path* of G if it contains all the vertices of G.

A graph is a *directed Hamiltonian graph* if it has a directed Hamilton circuit.

For example, the sequence of edges $e_1, e_2, e_3, e_4, e_5, e_6$ is a directed Hamilton circuit in the graph of Fig. 5.13a. In Fig. 5.13b, the sequence of edges

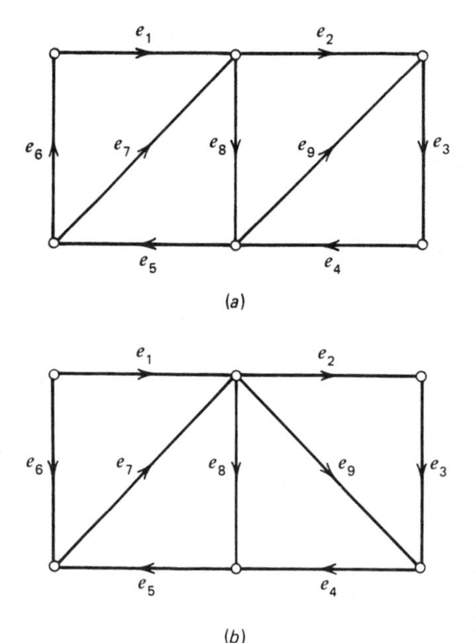

Figure 5.13. (*a*) A directed Hamiltonian graph. (*b*) A directed graph with a Hamilton path but with no Hamilton circuits.

e_1, e_2, e_3, e_4, e_5 is a directed Hamilton path. Note that this latter graph has no directed Hamilton circuit.

Characterizing a directed Hamiltonian graph is as difficult as characterizing an undirected Hamiltonian graph. However, there are several sufficient conditions which guarantee the existence of directed Hamilton circuits and paths. We now discuss a few of these conditions.

A directed graph G is *complete* if its underlying undirected graph is complete.

The following theorem is due to Moon [5.4].

THEOREM 5.10. Let u be any vertex of a strongly connected complete directed graph with $n \geqslant 3$ vertices. For each k, $3 \leqslant k \leqslant n$, there is a directed circuit of length k containing u.

Proof

Let $G=(V, E)$ be a strongly connected complete directed graph with $n \geqslant 3$ vertices. Consider any vertex u of G. Let $S=\Gamma^+(u)$ and $T=\Gamma^-(u)$. Since G is strongly connected, neither S nor T is empty. For the same reason, there is a directed edge from a vertex $v \in S$ to a vertex $w \in T$. Thus u is in a directed circuit of length 3 (see Fig. 5.14).

We prove the theorem by induction on k. Assume that u is in directed circuits of all lengths between 3 and p, where $p < n$. We now show that u is in a directed circuit of length $p+1$.

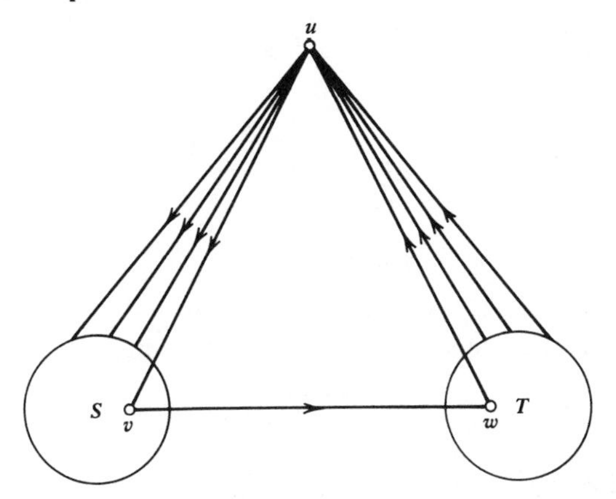

Figure 5.14.

Let C: $u=v_0, v_1, v_2, \ldots, v_p=u$ be a directed circuit of length p. Two cases need to be considered.

Suppose for some vertex v not in C there exist two directed edges (x, v) and (v, y) with x and y on C. Then there exist two adjacent vertices v_i and v_{i+1} on C such that the directed edges (v_i, v) and (v, v_{i+1}) are in G. Thus in this case u is in the directed circuit $u=v_0, v_1, v_2, \ldots, v_i, v, v_{i+1}, \ldots, v_p=u$ of length $p+1$.

Otherwise, let S be the set of all those vertices not in C which are the terminal vertices of edges directed away from the vertices of C, and let T be the set of all those vertices not in C which are the initial vertices of edges directed toward vertices in C. Again, since G is strongly connected, neither S nor T is empty. Further, there exists an edge directed from a vertex $v \in S$ to a vertex $w \in T$. Thus u is in the directed circuit $u=v_0, v, w, v_2, \ldots, v_p=u$ of length $p+1$ (see Fig. 5.15). ■

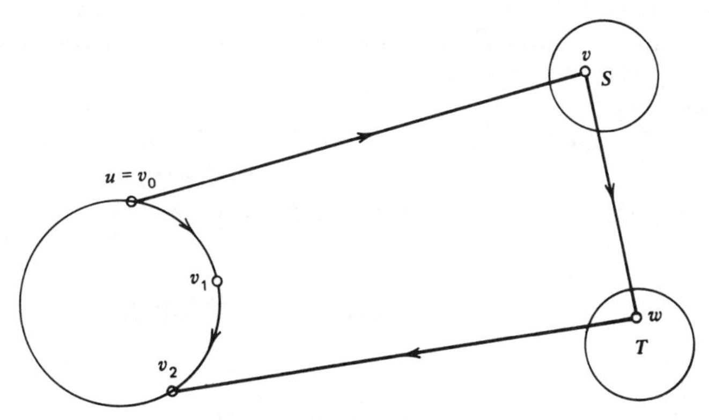

Figure 5.15.

Corollary 5.10.1 A strongly connected complete directed graph is Hamiltonian. ■

Next we state without proof a very powerful theorem due to Ghouila-Houri. The proof of this theorem is quite involved and may be found in [5.5] (pp. 196–199).

THEOREM 5.11. Let G be a strongly connected n-vertex graph without parallel edges and self-loops. If for every vertex v in G

$$d^-(v)+d^+(v)\geqslant n,$$

then G has a directed Hamilton circuit. ■

The following result generalizes Dirac's result (Corollary 3.4.1) for undirected graphs, and it can be proved using Theorem 5.11 and Exercise 5.6.

Corollary 5.11.1 Let G be a directed n-vertex graph without parallel edges or self-loops. If $\min(\delta^-,\delta^+)\geqslant \frac{1}{2}n>1$, then G contains a directed Hamilton circuit. ■

See Bondy and Murty [5.6] (pp. 178–179) for a direct proof of this result.

We conclude this section with a result on the existence of a directed Hamilton path.

THEOREM 5.12. If a directed graph $G=(V, E)$ is complete, then it has a directed Hamilton path.

Proof

Consider a directed path P: $a_1, a_2,\cdots, a_p$ of maximum length in G. Suppose b is a vertex not lying on P. Then the edge $(b, a_1)\notin E$, for otherwise the path $b, a_1, a_2,\cdots, a_p$ will be longer than P. Therefore, $(a_1, b)\in E$.

Now $(b, a_2)\notin E$, for otherwise $a_1, b, a_2, a_3,\cdots, a_p$ will be a path longer than P. Thus $(a_2, b)\in E$.

Repeating the above argument we find that $(a_p, b)\in E$. But this is a contradiction, for $a_1, a_2,\cdots, a_p, b$ will be a path longer than P. Thus there is no vertex outside P, and hence P is a directed Hamilton path in G. ■

5.7 ACYCLIC DIRECTED GRAPHS

In this section we study some properties of an important class of directed graphs, namely, the acyclic directed graphs. As we know, a directed graph is

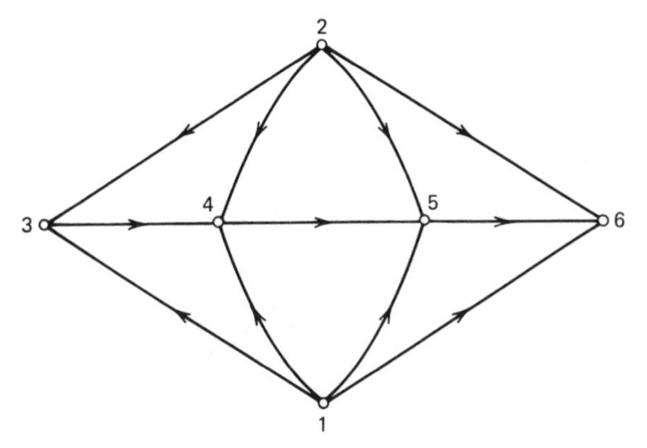

Figure 5.16. An acyclic directed graph.

acyclic if it has no directed circuits. Obviously the simplest example of an acyclic directed graph is a directed tree.

The main result of this section is that we can label the vertices of an n-vertex acyclic directed graph G with integers from the set $\{1,2,\ldots,n\}$ such that the presence of the edge (i, j) in G implies that $i<j$. Note that the edge (i, j) is directed from vertex i to vertex j. Ordering the vertices as above is called *topological sorting*. For example, the vertices of the acyclic directed graph of Fig. 5.16 are topologically sorted. The proof of the main result of this section depends on the following theorem.

THEOREM 5.13. In an acyclic directed graph G there exists at least one vertex with zero in-degree and at least one vertex with zero out-degree.

Proof

Let P: $v_1, v_2, \ldots, v_p$ be a maximal directed path in G. We claim that the in-degree of v_1 and the out-degree of v_p are both equal to zero.

If the in-degree of v_1 is not equal to zero, then there exists a vertex w such that the edge (w, v_1) is in G. We now examine two cases.

Case 1 Suppose $w \neq v_i$, for any i, $1 \leqslant i \leqslant p$. Then there is a directed path P': $w, v_1, v_2, \ldots, v_p$ which contains all the edges of P. This contradicts that P is a maximal directed path.

Case 2 Suppose $w = v_i$ for some i. Then in G, there is a directed circuit C: $v_1, v_2, \ldots, v_i, v_1$. This is again a contradiction since G has no directed circuits.

Thus there is no vertex w such that the edge (w, v_1) is in G. In other words, the in-degree of v_1 is zero.

Following a similar reasoning we can show that the out-degree of v_p is zero. ∎

To topologically sort the vertices of an n-vertex acyclic directed graph G, we proceed as follows.

Select any vertex with zero out-degree. Since G is acyclic, by Theorem 5.13, there is at least one such vertex in G. Label this vertex with the integer n. Now remove from G this vertex and the edges incident on it. Let G' be the resulting graph. Since G' is also acyclic, we can now select a vertex whose out-degree in G' is zero. Label this with the integer $n-1$. Repeat the above procedure until all the vertices are labeled. It is now easy to verify that this procedure results in a topological sorting of the vertices of G.

For example, the labeling of the vertices of the graph of Fig. 5.16 has been done according to the above procedure.

5.8 TOURNAMENTS

A *tournament* is a complete directed graph. It derives its name from its application in the representation of structures of round-robin tournaments. In a round-robin tournament several teams play a game that cannot end in a tie, and each team plays every other team exactly once. In the directed graph representation of the round-robin tournament, vertices represent teams and an edge (v_1, v_2) is present in the graph if the team represented by the vertex v_1 defeats the team represented by the vertex v_2. Clearly, such a directed graph has no parallel edges and self-loops, and there is exactly one edge between any two vertices. Thus it is a complete directed graph and hence a tournament. A tournament is shown in Fig. 5.17.

The teams participating in a tournament can be ranked according to their scores. The *score* of a team is the number of teams it has defeated. This motivates the definition of the score sequence of a tournament.

The *score sequence* of an n-vertex tournament is the sequence $(s_1, s_2, \ldots, s_n)$ such that each s_i is the out-degree of a vertex of the tournament. An interesting characterization of a tournament in terms of the score sequence is given in the following theorem.

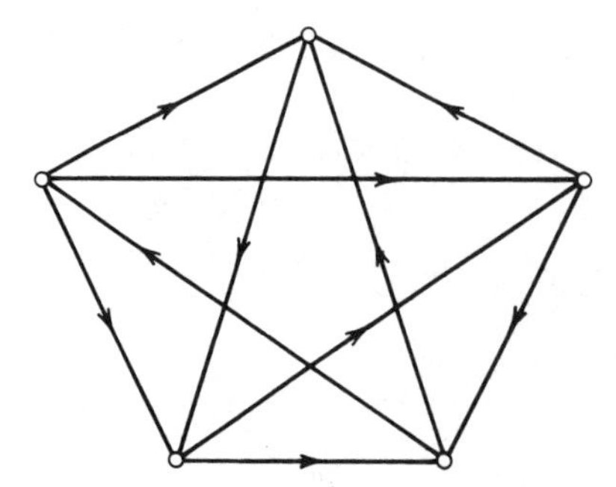

Figure 5.17. A tournament.

THEOREM 5.14. A sequence of nonnegative integers $s_1, s_2, \ldots, s_n$ is the score sequence of a tournament G if and only if

1. $\sum_{i=1}^{n} s_i = \frac{n(n-1)}{2}$.
2. $\sum_{i=1}^{k} s_i \geqslant \frac{k(k-1)}{2}$, for all $k<n$.

Proof

Necessity Note that the sum $\sum_{i=1}^{n} s_i$ is equal to the number of edges in the tournament G (Theorem 5.1). Since a tournament is a complete graph, it has $n(n-1)/2$ edges, where n is the number of vertices. Thus

$$\sum_{i=1}^{n} s_i = \frac{n(n-1)}{2}.$$

To prove condition 2, consider the subtournament on any k vertices $v_1, v_2, \ldots, v_k$. This subtournament has $k(k-1)/2$ edges. Therefore in the entire tournament

$$\sum_{i=1}^{k} s_i \geqslant \frac{k(k-1)}{2},$$

since there may be edges directed from the vertices in the subtournament to those outside the subtournament.

Sufficiency See [5.5] (pp. 107–109) ∎

Suppose we can order the teams in a round-robin tournament such that each team precedes the one it has defeated. Then we can assign the integers $1, 2, \ldots, n$ to the teams to indicate their ranks in this order. Such a ranking is always possible since in a tournament there exists a directed Hamilton path (Theorem 5.12), and it is called *ranking* by a Hamilton path.

Note that ranking by a Hamilton path may not be the same as ranking by the score. Further, a tournament may have more than one directed Hamilton path. In such a case there will be more than one Hamilton path ranking. However, there exists exactly one directed Hamilton path in a transitive tournament. This is stated in the following theorem, which is easy to prove.

THEOREM 5.15. In a transitive tournament there exists exactly one directed Hamilton path. ∎

For other ranking procedures see Bondy and Murty [5.6], Kendall [5.7], Moon and Pullman [5.8].

5.9 FURTHER READING

Berge [5.5] and Harary [5.9] are good references for a number of results on directed graphs. Moon [5.10] is a monograph exclusively devoted to the study of tournaments. See also Nash-Williams [5.11].

Directed trees are used in the computer representation of combinatorial objects. For an extensive discussion of this topic, Knuth [5.12], [5.13], Aho, Hopcroft, and Ullman [5.14], and Reingold, Nievergelt, and Deo [5.15] may be referred.

Directed trees also arise in several other applications such as the topological study of electrical networks. In the topological study of such networks one is interested in computing the number of directed trees of a directed graph. This question is discussed in Chapter 6. Network-theoretic applications of directed trees are discussed in Chapter 13.

Harary, Norman, and Cartwright [5.16] discuss several applications of directed graphs.

5.10 EXERCISES

5.1 Let G be a directed graph without self-loops and parallel edges. Let $\max\{\delta^-, \delta^+\}=k$. Prove the following:

(a) G has a directed path of length at least k.

(b) G has a directed circuit of length at least $k+1$, if $k>0$.

5.2 Show that the edges of an undirected graph $G=(V, E)$ can be oriented so that in the resulting directed graph

$$|d^+(v)-d^-(v)|\leqslant 1 \qquad \text{for all } v\in V.$$

5.3 A *directed cut* in a directed graph is a cut $\langle S, \bar{S}\rangle$ whose edges are all oriented away from S or all oriented toward S. Show that an edge of a directed graph belongs either to a directed circuit or to a directed cutset, but no edge belongs to both. (This is a special case of a more general result known as the "arc coloring lemma" which we prove in Chapter 10. See Theorem 10.31.)

5.4 For a directed graph G with at least one edge, prove that the following are equivalent:

(a) G has no directed circuits.

(b) Each edge of G is in a directed cutset.

5.5 For a connected directed graph G with at least one edge, prove that the following are equivalent:

(a) G is strongly connected.

(b) Every edge of G lies on a directed circuit.
(c) G has no directed cutsets.

5.6 Show that an n-vertex directed graph without parallel edges or self-loops is strongly connected if

$$\min\{\delta^-, \delta^+\} \geqslant \frac{n-1}{2}.$$

5.7 Show that a strongly connected graph which contains a circuit of odd length also contains a directed circuit of odd length.

5.8 Let $G=(V, E)$ be a directed graph such that

(a) $d^+(x)-d^-(x)=l=d^-(y)-d^+(y)$, and
(b) $d^+(v)=d^-(v)$, for $v \in V-\{x, y\}$.

Show that there are l edge-disjoint directed x–y paths in G.

5.9 Show that a strongly connected graph has a spanning directed walk.

5.10 What is the longest circular sequence formed out of three symbols x, y, and z such that no subsequence of four symbols is repeated. Give one such sequence.

5.11 Find a circular sequence of seven 0's and seven 1's such that all 4-digit binary sequences except 0000 and 1111 appear as subsequences of the sequence.

5.12 Prove that a directed Hamilton circuit of $G_{s,n}$ corresponds to a directed Euler trail of $G_{s,n-1}$. Is it true that $G_{s,n}$ always has a directed Hamilton circuit?

5.13 Show that the number of directed Euler trails in a directed graph, having n vertices and $m > 2n$ edges, is even.

5.14 Show that a directed graph $G=(V, E)$ without parallel edges and self-loops has a directed Hamilton path if

$$d^+(v)+d^-(v) \geqslant n-1, \qquad \text{for all } v \in V.$$

5.15 Show that every tournament is strongly connected or can be transformed into a strongly connected tournament by reversing the orientation of just one edge.

5.16 A subset $S \subseteq V$ is an *independent set* of a graph $G=(V, E)$ if no two of the vertices in S are adjacent. Prove that a directed graph G without self-loops has an independent set S such that each vertex v of G not in S can be reached from a vertex in S by a directed path of length at most 2 (Chvátal and Lovász [5.17]).

An interesting corollary of this result is the following: A tournament contains a vertex from which every vertex can be reached by a directed path of length at most 2.

5.17 Show that the scores s_i of an n-vertex tournament satisfy

$$\sum_{i=1}^{n} s_i^2 = \sum_{i=1}^{n} (n-1-s_i)^2.$$

5.11 REFERENCES

5.1 S. W. Golomb, *Shift Register Sequences*, Holden-Day, San Francisco, 1967.

5.2 M. Hall, Jr., *Combinatorial Theory*, Blaisdell, Waltham, Mass., 1967.

5.3 T. Van Aardenne-Ehrenfest and N. G. de Bruijn, "Circuits and Trees in Oriented Linear Graphs," *Simon Stevin*, Vol. 28, 203–217 (1951).

5.4 J. W. Moon, "On Subtournaments of a Tournament," *Canad. Math. Bull.*, Vol. 9, 297–301 (1966).

5.5 C. Berge, *Graphs and Hypergraphs*, North Holland, Amsterdam, 1973.

5.6 J. A. Bondy and U. S. R. Murty, *Graph Theory with Applications*, Macmillan, London, 1976.

5.7 M. G. Kendall, "Further Contributions to the Theory of Paired Comparisons," *Biometrics*, Vol. 11, 43–62 (1955).

5.8 J. W. Moon and N. J. Pullman, "On Generalized Tournament Matrices," *SIAM Rev.*, Vol. 12, 384–399 (1970).

5.9 F. Harary, *Graph Theory*, Addison-Wesley, Reading, Mass., 1969.

5.10 J. W. Moon, *Topics on Tournaments*, Holt, Rinehart and Winston, New York, 1968.

5.11 C. St. J. A. Nash-Williams, "Hamiltonian Circuits," in *Studies in Graph Theory*, Part II, MAA Press, 1975, pp. 301–360.

5.12 D. E. Knuth, *The Art of Computer Programming*, Vol. 1: *Fundamental Algorithms*, Addison-Wesley, Reading, Mass., 1968.

5.13 D. E. Knuth, *The Art of Computer Programming*, Vol. 3: *Sorting and Searching*, Addison-Wesley, Reading, Mass., 1973.

5.14 A. V. Aho, J. E. Hopcroft, and J. D. Ullman, *The Design and Analysis of Computer Algorithms*, Addison-Wesley, Reading, Mass., 1974.

5.15 E. M. Reingold, J. Nievergelt, and N. Deo, *Combinatorial Algorithms: Theory and Practice*, Prentice-Hall, Englewood Cliffs, N.J., 1977.

5.16 F. Harary, R. Z. Norman, and D. Cartwright, *Structural Models: An Introduction to the Theory of Directed Graphs*, Wiley, New York, 1965.

5.17 V. Chvátal and L. Lovász, "Every Directed Graph Has a Semi-Kernel," in *Hypergraph Seminar* (Eds. C. Berge and D. K. Ray-Chaudhuri), Springer, New York, 1974, p. 175.

Chapter 6

Matrices of a Graph

In this chapter we introduce the incidence, circuit, cut, and adjacency matrices of a graph and establish several properties of these matrices which help to reveal the structure of a graph. The incidence, circuit, and cut matrices arise in the study of electrical networks because these matrices are the coefficient matrices of Kirchhoff's equations which describe a network. Thus the properties of these matrices and other related results to be established in this chapter will be used extensively in our discussions in Part II of the book. The properties of the adjacency matrix to be discussed here form the basis of the signal flow graph approach, which is a very powerful tool in the study of linear systems. Signal flow graph theory is developed in Section 6.11.

Our discussions of incidence, circuit, and cut matrices are mainly with respect to directed graphs. However, these discussions become valid for undirected graphs too if addition and multiplication are in $GF(2)$, the field of integers modulo 2.

6.1 INCIDENCE MATRIX

Consider a graph G with n vertices and m edges and having no self-loops. The *all-vertex incidence matrix* $A_c = [a_{ij}]$ of G has n rows, one for each vertex, and m columns, one for each edge. The element a_{ij} of A_c is defined as follows:

G is directed

$$a_{ij} = \begin{cases} 1, & \text{if the } j\text{th edge is incident on the } i\text{th vertex and oriented away from it;} \\ -1, & \text{if the } j\text{th edge is incident on the } i\text{th vertex and oriented toward it;} \\ 0, & \text{if the } j\text{th edge is not incident on the } i\text{th vertex.} \end{cases}$$

G is undirected

$$a_{ij} = \begin{cases} 1, & \text{if the } j\text{th edge is incident on the } i\text{th vertex;} \\ 0, & \text{otherwise.} \end{cases}$$

A row of A_c will be referred to as an *incidence vector* of G. Two graphs and their all-vertex incidence matrices are shown in Fig. 6.1*a* and *b*.

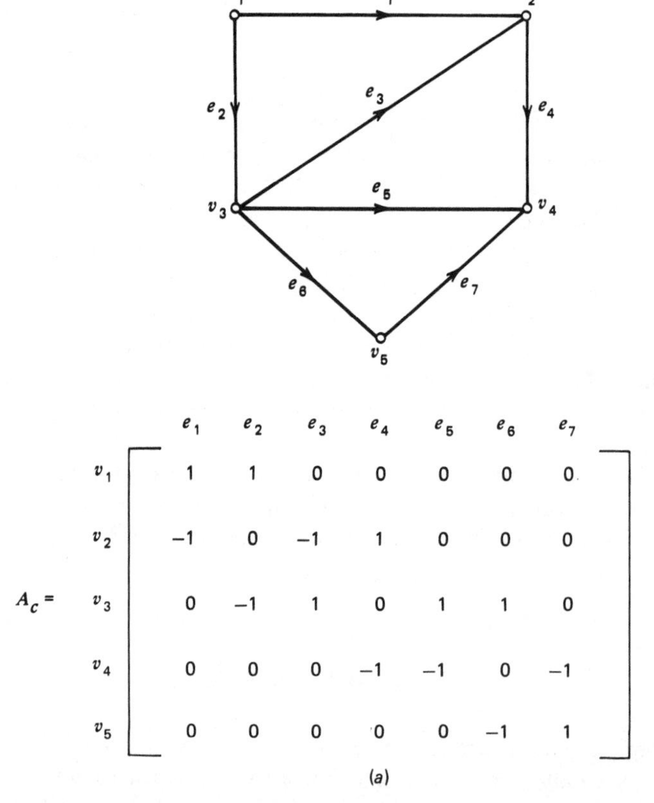

$A_c =$

	e_1	e_2	e_3	e_4	e_5	e_6	e_7
v_1	1	1	0	0	0	0	0
v_2	−1	0	−1	1	0	0	0
v_3	0	−1	1	0	1	1	0
v_4	0	0	0	−1	−1	0	−1
v_5	0	0	0	0	0	−1	1

(*a*)

Figure 6.1. (*a*) A directed graph G and its all-vertex incidence matrix. (*b*) An undirected graph G and its all-vertex incidence matrix.

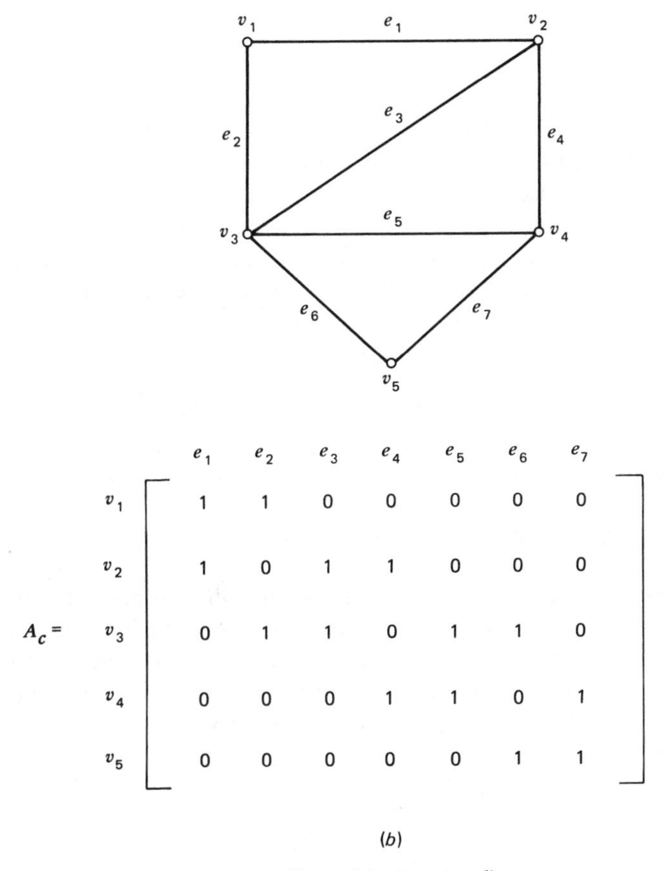

$$A_c = \begin{array}{c} \\ v_1 \\ v_2 \\ v_3 \\ v_4 \\ v_5 \end{array} \begin{array}{c} \begin{array}{ccccccc} e_1 & e_2 & e_3 & e_4 & e_5 & e_6 & e_7 \end{array} \\ \left[\begin{array}{ccccccc} 1 & 1 & 0 & 0 & 0 & 0 & 0 \\ 1 & 0 & 1 & 1 & 0 & 0 & 0 \\ 0 & 1 & 1 & 0 & 1 & 1 & 0 \\ 0 & 0 & 0 & 1 & 1 & 0 & 1 \\ 0 & 0 & 0 & 0 & 0 & 1 & 1 \end{array} \right] \end{array}$$

(*b*)

Figure 6.1. (*continued*)

It should be clear from the preceding definition that each column of A_c contains exactly two nonzero entries, one $+1$ and one -1. Therefore we can obtain any row of A_c from the remaining $n-1$ rows. Thus any $n-1$ rows of A_c contain all the information about A_c. In other words, the rows of A_c are linearly dependent.

An $(n-1)$-rowed submatrix A of A_c will be referred to as an *incidence matrix* of G. The vertex which corresponds to the row of A_c which is not in A will be called the *reference vertex* of A. Note that

$$\operatorname{rank}(A) = \operatorname{rank}(A_c) \leqslant n-1. \tag{6.1}$$

Now we show that in the case of a connected graph, rank of A_c is in fact equal to $n-1$. This result is based on the following theorem.

THEOREM 6.1. The determinant of any incidence matrix of a tree is equal to ± 1.

Proof

Proof is by induction on the number n of vertices in a tree.

Any incidence matrix of a tree on two vertices is just a 1×1 matrix with its only entry being equal to ± 1. Thus the theorem is true for $n=2$. Note that the theorem does not arise for $n=1$.

Let the theorem be true for $2\leqslant n\leqslant k$. Consider a tree T with $k+1$ vertices. Let A denote an incidence matrix of T. By Theorem 2.5, T has at least two pendant vertices. Let the ith vertex of T be a pendant vertex, and let this not be the reference vertex of A. If the only edge incident on this vertex is the lth one, then in A

$$a_{il}=\pm 1 \qquad \text{and} \qquad a_{ij}=0, \qquad j\neq l.$$

If we now expand the determinant of A by the ith row, then

$$\det(A)=\pm(-1)^{i+l}\det(A') \tag{6.2}$$

where A' is obtained by removing the ith row and the lth column from A.

Suppose T' is the graph that results after removing the ith vertex and the lth edge from T. Clearly T' is a tree because the ith vertex is a pendant vertex and the lth edge is a pendant edge in T. Further it is easy to verify that A' is an incidence matrix of T'. Since T' is a tree on $n-1$ vertices, we have by the induction hypothesis that

$$\det A'=\pm 1. \tag{6.3}$$

This result in conjunction with (6.2) proves the theorem for $n=k+1$. ■

Since a connected graph has at least one spanning tree, it follows from the above theorem that in any incidence matrix A of a connected graph with n vertices there exists a nonsingular submatrix of order $n-1$. Thus for a connected graph,

$$\operatorname{rank}(A)=n-1. \tag{6.4}$$

Since $\operatorname{rank}(A_c)=\operatorname{rank}(A)$, we get the following theorem.

THEOREM 6.2. The rank of the all-vertex incidence matrix of an n-vertex connected graph G is equal to $n-1$, the rank of G. ■

An immediate consequence of the above theorem is the following.

Corollary 6.2.1 If an n-vertex graph has p components, then the rank of its all-vertex incidence matrix is equal to $n-p$, the rank of G. ■

6.2 CUT MATRIX

To define the cut matrix of a directed graph we need to assign an orientation to each cut of the graph.

Consider a directed graph $G=(V, E)$. If V_a is a nonempty subset of V, then we may recall (Chapter 2) that the set of edges connecting the vertices in V_a to those in $\bar{V}_a$ is a cut, and this cut is denoted as $\langle V_a, \bar{V}_a\rangle$. The orientation of $\langle V_a, \bar{V}_a\rangle$ may be assumed to be either from V_a to $\bar{V}_a$ or from $\bar{V}_a$ to V_a. Suppose we assume that the orientation is from V_a to $\bar{V}_a$. Then the orientation of an edge in $\langle V_a, \bar{V}_a\rangle$ is said to agree with the orientation of the cut $\langle V_a, \bar{V}_a\rangle$ if the edge is oriented from a vertex in V_a to a vertex in $\bar{V}_a$.

The *cut matrix* $Q_c=[q_{ij}]$ of a graph G with m edges has m columns and as many rows as the number of cuts in G. The entry q_{ij} is defined as follows:

G is directed

$$q_{ij}=\begin{cases} 1, & \text{if the } j\text{th edge is in the } i\text{th cut and its orientation agrees with the cut orientation;} \\ -1, & \text{if the } j\text{th edge is in the } i\text{th cut and its orientation does not agree with the cut orientation;} \\ 0, & \text{if the } j\text{th edge is not in the } i\text{th cut.} \end{cases}$$

G is undirected

$$q_{ij}=\begin{cases} 1, & \text{if the } j\text{th edge is in the } i\text{th cut;} \\ 0, & \text{otherwise.} \end{cases}$$

A row of Q_c will be referred to as a *cut vector*.

Three cuts of the directed graph of Fig. 6.1*a* are shown in Fig. 6.2. In each case the cut orientation is shown in dashed lines. The submatrix of Q_c corresponding to these three cuts is as given below:

$$\begin{array}{c} \\ \text{cut 1} \\ \text{cut 2} \\ \text{cut 3} \end{array} \begin{array}{c} \begin{array}{ccccccc} e_1 & e_2 & e_3 & e_4 & e_5 & e_6 & e_7 \end{array} \\ \begin{bmatrix} 1 & 0 & 1 & 0 & 1 & 1 & 0 \\ 1 & 1 & 0 & 0 & 0 & -1 & 1 \\ 0 & -1 & 1 & 0 & 1 & 1 & 0 \end{bmatrix} \end{array}$$

The corresponding submatrix in the undirected case can be obtained by replacing the -1's in the above matrix by $+1$'s.

Consider next any vertex v. The nonzero entries in the corresponding incidence vector represent the edges incident on v. These edges form the cut $\langle v, V-v\rangle$. If we assume that the orientation of this cut is from v to $V-v$, then we can see from the definitions of cut and incidence matrices that the row in Q_c corresponding to the cut $\langle v, V-v\rangle$ is the same as the row in A_c corresponding to the vertex v. Thus A_c is a submatrix of Q_c.

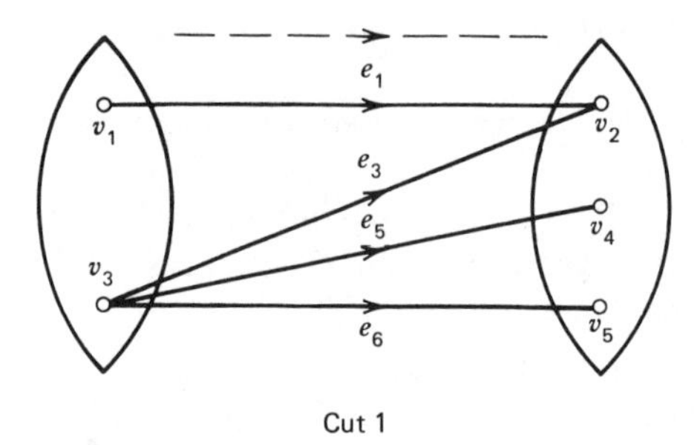

Cut 1

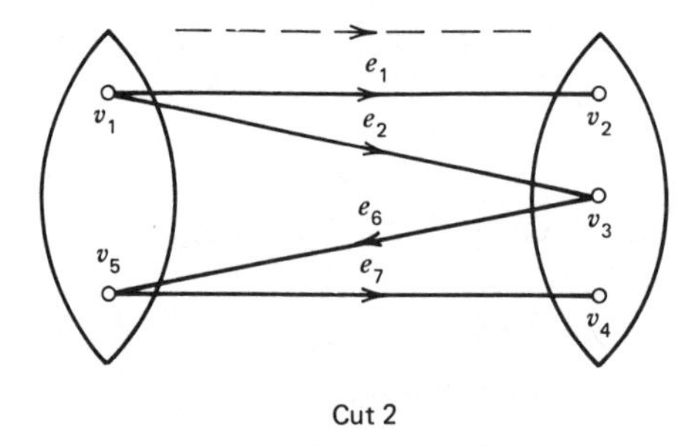

Cut 2

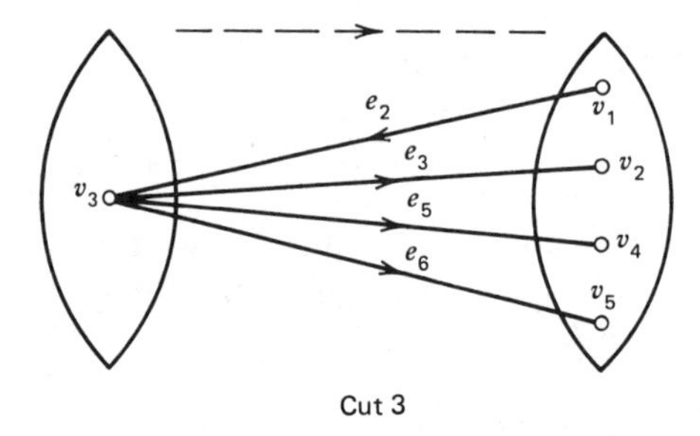

Cut 3

Figure 6.2. Some cuts of the graph of Fig. 6.1a

Now we proceed to show that the rank of Q_c is equal to that of A_c. To do so we need the following theorem.

THEOREM 6.3. Each row in the cut matrix Q_c can be expressed, in two ways, as a linear combination of the rows of the matrix A_c. In each case, the nonzero coefficients in the linear combination are all $+1$ or all -1.

Proof

Let $\langle V_a, \overline{V}_a \rangle$ be the ith cut in a graph G with n vertices and m edges, and let q_i be the corresponding cut vector. Let $V_a = \{v_1, v_2, \ldots, v_r\}$ and $\overline{V}_a =$

$\{v_{r+1}, v_{r+2}, \ldots, v_n\}$. For $1 \leqslant i \leqslant n$, let a_i denote the incidence vector corresponding to the vertex v_i.

We assume, without any loss of generality, that the orientation of $\langle V_a, \overline{V}_a \rangle$ is from V_a to $\overline{V}_a$, and prove the theorem by establishing that

$$q_i = a_1 + a_2 + \cdots + a_r = -(a_{r+1} + a_{r+2} + \cdots + a_n). \tag{6.5}$$

Let v_p and v_q be the end vertices of the kth edge, $1 \leqslant k \leqslant m$. Let this edge be oriented from v_p to v_q so that

$$\begin{aligned} a_{pk} &= 1, \\ a_{qk} &= -1, \\ a_{jk} &= 0, \qquad j \neq p, q. \end{aligned} \tag{6.6}$$

Now four cases arise.

Case 1 $v_p \in V_a$ and $v_q \in \overline{V}_a$, that is, $p \leqslant r$ and $q \geqslant r+1$, so that $q_{ik} = 1$.
Case 2 $v_p \in \overline{V}_a$ and $v_q \in V_a$, that is, $p \geqslant r+1$ and $q \leqslant r$, so that $q_{ik} = -1$.
Case 3 $v_p, v_q \in V_a$, that is, $p, q \leqslant r$, so that $q_{ik} = 0$.
Case 4 $v_p, v_q \in \overline{V}_a$, that is, $p, q \geqslant r+1$, so that $q_{ik} = 0$.

It is easy to verify, using (6.6), that the following is true in each of these four cases.

$$\begin{aligned} q_{ik} &= (a_{1k} + a_{2k} + \cdots + a_{rk}) \\ &= -(a_{r+1,k} + a_{r+2,k} + \cdots + a_{nk}). \end{aligned} \tag{6.7}$$

Now (6.5) follows since the above equation is valid for all $1 \leqslant k \leqslant m$. Hence the theorem. ■

To illustrate the above theorem, consider the cut 1 in Fig. 6.2. This cut separates the vertices in $V_a = \{v_1, v_3\}$ from those in $\overline{V}_a$. The cut orientation is from V_a to $\overline{V}_a$. So the corresponding cut vector can be expressed as follows:

$$\begin{aligned} [1 \quad 0 \quad 1 \quad 0 \quad 1 \quad 1 \quad 0] &= a_1 + a_3 \\ &= -a_2 - a_4 - a_5 \end{aligned}$$

where $a_1, a_2, \ldots, a_5$ are the rows of the matrix A in Fig. 6.1a.

An important consequence of Theorem 6.3 is that rank $(Q_c) \leqslant$ rank(A_c). However, rank$(Q_c) \geqslant$ rank(A_c), because A_c is a submatrix of Q_c. Therefore, we get

$$\text{rank}(Q_c) = \text{rank}(A_c).$$

Then Theorem 6.2 and Corollary 6.2.1, respectively, would lead to the following theorem.

THEOREM 6.4. The rank of the cut matrix Q_c of an n-vertex connected graph G is equal to $n-1$, the rank of G. ■

Corollary 6.4.1 The rank of the cut matrix Q_c of an n-vertex graph G with p components is equal to $n-p$, the rank of G. ■

As the above discussions show, the all-vertex incidence matrix A_c is an important submatrix of the cut matrix Q_c. Next we identify another important submatrix of Q_c.

We know that a spanning tree T of an n-vertex connected graph G defines a set of $n-1$ fundamental cutsets—one fundamental cutset for each branch of T. The submatrix of Q_c corresponding to these $n-1$ fundamental cutsets is known as the *fundamental cutset matrix* Q_f of G with respect to T.

Let $b_1, b_2, \ldots, b_{n-1}$ denote the branches of T. Suppose we arrange the columns and the rows of Q_f so that

1. For $1 \leqslant i \leqslant n-1$, the ith column corresponds to the branch b_i; and
2. The ith row corresponds to the fundamental cutset defined by b_i.

If, in addition, we assume that the orientation of a fundamental cutset is so chosen as to agree with that of the defining branch, then the matrix Q_f can be displayed in a convenient form as follows:

$$Q_f = [U | Q_{fc}] \tag{6.8}$$

where U is the unit matrix of order $n-1$ and its columns correspond to the branches of T.

For example, the fundamental cutset matrix Q_f of the connected graph of Fig. 6.1a with respect to the spanning tree $T = \{e_1, e_2, e_6, e_7\}$ is as given below:

$$Q_f = \begin{array}{c} \\ e_1 \\ e_2 \\ e_6 \\ e_7 \end{array} \begin{array}{c} \begin{array}{cccccccc} e_1 & e_2 & e_6 & e_7 & & e_3 & e_4 & e_5 \end{array} \\ \left[\begin{array}{cccc|ccc} 1 & 0 & 0 & 0 & 1 & -1 & 0 \\ 0 & 1 & 0 & 0 & -1 & 1 & 0 \\ 0 & 0 & 1 & 0 & 0 & 1 & 1 \\ 0 & 0 & 0 & 1 & 0 & 1 & 1 \end{array}\right] \end{array}. \tag{6.9}$$

It is clear from (6.8) that the rank of Q_f is equal to $n-1$, the rank of Q_c. Thus every cut vector (which may be a cutset vector) can be expressed as a linear combination of the fundamental cutset vectors.

6.3 CIRCUIT MATRIX

A circuit can be traversed in one of two directions, clockwise or anticlockwise. The direction we choose for traversing a circuit defines its orientation. The orientation of a circuit can be pictorially shown by an arrow as in Fig. 6.3.

Consider an edge e which has v_i and v_j as its end vertices. Suppose that this edge is oriented from v_i to v_j and that it is present in circuit C. Then we say that the orientation of e agrees with the orientation of the circuit if v_i appears before v_j when we traverse C in the direction specified by its orientation. For example, in Fig. 6.3, the orientation of e_1 agrees with the circuit orientation whereas that of e_4 does not.

The *circuit matrix* $B_c = [b_{ij}]$ of a graph G with m edges has m columns and as many rows as the number of circuits in G. The entry b_{ij} is defined as follows:

G is directed

$$b_{ij} = \begin{cases} 1, & \text{if the } j\text{th edge is in the } i\text{th circuit and its orientation agrees with the circuit orientation;} \\ -1, & \text{if the } j\text{th edge is in the } i\text{th circuit and its orientation does not agree with the circuit orientation;} \\ 0, & \text{if the } j\text{th edge is not in the } i\text{th circuit.} \end{cases}$$

G is undirected

$$b_{ij} = \begin{cases} 1, & \text{if the } j\text{th edge is in the } i\text{th circuit;} \\ 0, & \text{otherwise.} \end{cases}$$

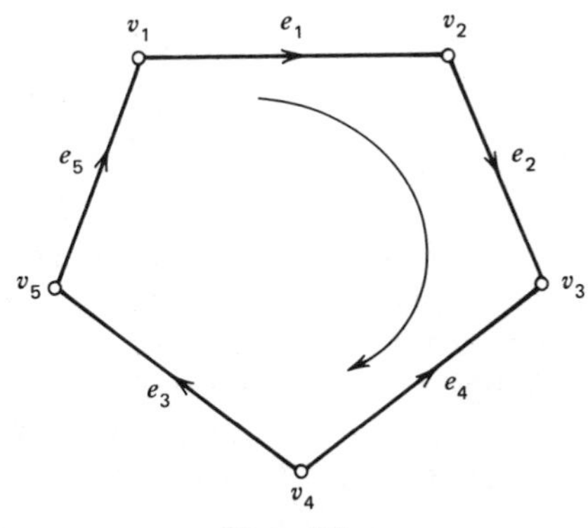

Figure 6.3.

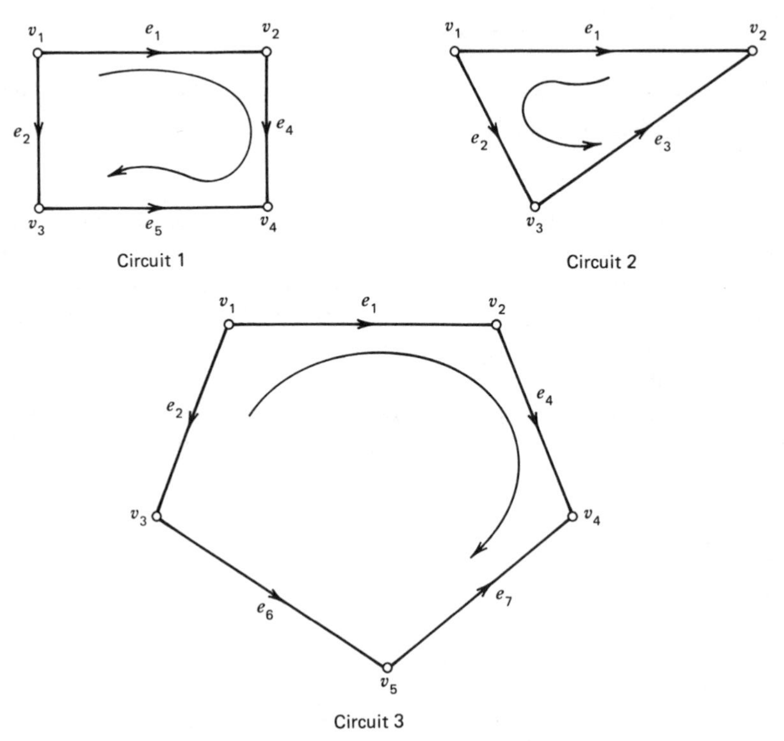

Figure 6.4. Some circuits of the graph of Fig. 6.1*a*.

A row of B_c will be referred to as a *circuit vector* of G.

As an example, consider again the graph of Fig. 6.1*a*. Three circuits of this graph with their orientations are shown in Fig. 6.4. The submatrix of B_c corresponding to these three circuits is given below.

$$\begin{array}{l} \\ \text{circuit 1} \\ \text{circuit 2} \\ \text{circuit 3} \end{array} \begin{array}{c} \begin{array}{ccccccc} e_1 & e_2 & e_3 & e_4 & e_5 & e_6 & e_7 \end{array} \\ \begin{bmatrix} 1 & -1 & 0 & 1 & -1 & 0 & 0 \\ -1 & 1 & 1 & 0 & 0 & 0 & 0 \\ 1 & -1 & 0 & 1 & 0 & -1 & -1 \end{bmatrix} \end{array}$$

The corresponding submatrix for the undirected graph of Fig. 6.1*b* can be obtained by replacing the -1's in the above matrix by $+1$'s.

Next we identify an important submatrix of B_c.

Consider any spanning tree T of a connected graph G having n vertices and m edges. Let $c_1, c_2, \ldots, c_{m-n+1}$ be the chords of T. We know that these $m-n+1$ chords define a set of $m-n+1$ fundamental circuits. The submatrix of B_c corresponding to these fundamental circuits is known as the *fundamental circuit matrix* B_f of G with respect to the spanning tree T.

Suppose we arrange the columns and rows of B_f so that

1. For $1 \leqslant i \leqslant m-n+1$, the ith column corresponds to the chord c_i; and
2. The ith row corresponds to the fundamental circuit defined by c_i.

If, in addition, we choose the orientation of a fundamental circuit to agree with that of the defining chord, then the matrix B_f can be written as

$$B_f = \begin{bmatrix} U & | & B_{ft} \end{bmatrix} \tag{6.10}$$

where U is the unit matrix of order $m-n+1$ and its columns correspond to the chords of T.

For example, the fundamental circuit matrix of the graph of Fig. 6.1a with respect to the spanning tree $T=\{e_1, e_2, e_6, e_7\}$ is as given below:

$$B_f = \begin{array}{c} \\ e_3 \\ e_4 \\ e_5 \end{array} \begin{array}{c} \begin{array}{ccccccc} e_3 & e_4 & e_5 & e_1 & e_2 & e_6 & e_7 \end{array} \\ \left[\begin{array}{ccc|cccc} 1 & 0 & 0 & -1 & 1 & 0 & 0 \\ 0 & 1 & 0 & 1 & -1 & -1 & -1 \\ 0 & 0 & 1 & 0 & 0 & -1 & -1 \end{array}\right] \end{array} \tag{6.11}$$

It is obvious from (6.10) that the rank of B_f is equal to $m-n+1$. Since B_f is a submatrix of B_c, we get

$$\text{rank}(B_c) \geqslant m-n+1. \tag{6.12}$$

We show in the next section that the rank of B_c in the case of a connected graph is equal to $m-n+1$.

Now note that the approach used in the previous section to establish the rank of Q_c cannot be used in the case of B_c. (Why?) Does this mean that the "duality" which we observed (Chapter 4) between circuits and cutsets is merely accidental? No, we shall see that the arguments of the next section would further confirm this "duality."

6.4 ORTHOGONALITY RELATION

We showed in Section 4.6 that in the case of an undirected graph every circuit vector is orthogonal to every cut vector. Now we prove that this result is true in the case of directed graphs too. Our proof is based on the following theorem.

THEOREM 6.5. If a cut and a circuit in a directed graph have $2k$ edges in common, then k of these edges have the same relative orientations in the cut and in the circuit, and the remaining k edges have one orientation in the cut and the opposite orientation in the circuit.

Proof

Consider a cut $\langle V_a, \overline{V}_a \rangle$ and a circuit C in a directed graph. Suppose we traverse C starting from a vertex in V_a. Then, for every edge e_1 which leads us from a vertex in V_a to a vertex in $\overline{V}_a$, there is an edge e_2 which leads us from a vertex in $\overline{V}_a$ to a vertex in V_a. The proof of the theorem will follow if we note that if $e_1(e_2)$ has the same relative orientation in the cut and in the circuit, then $e_2(e_1)$ has one orientation in the cut and the opposite orientation in the circuit (see Fig. 6.5). ∎

Now we prove the main result of this section.

THEOREM 6.6 (ORTHOGONALITY RELATION). If the columns of the circuit matrix B_c and the cut matrix Q_c are arranged in the same edge order, then

$$B_c Q_c^t = 0.$$

Proof

Consider a circuit and a cut which have $2k$ edges in common. The inner product of the corresponding circuit and cut vectors is equal to zero, since, by Theorem 6.5, it is the sum of k 1's and k -1's. The proof of the theorem now follows because each entry of the matrix $B_c Q_c^t$ is the inner product of a circuit vector and a cut vector. ∎

The orthogonality relation is a very profound result with interesting applications in graph theory and (as we discuss in Part II of the book) in network theory. Now we use this relation to establish the rank of the circuit matrix B_c.

Consider a connected graph G with n vertices and m edges. Let B_f and Q_f be the fundamental circuit and cutset matrices of G with respect to a

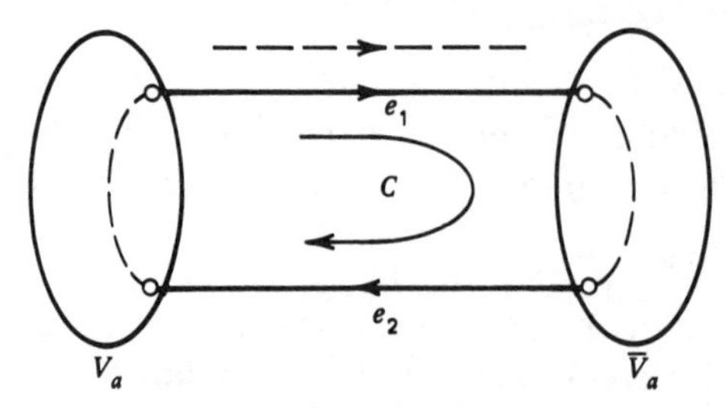

Figure 6.5.

spanning tree T. If the columns of B_f and Q_f are arranged in the same edge order, then we can write B_f and Q_f as

$$B_f = [\,B_{ft} \quad U\,]$$

and

$$Q_f = [\,U \quad Q_{fc}\,].$$

By the orthogonality relation

$$B_f Q_f^t = 0,$$

that is,

$$[\,B_{ft} \quad U\,]\begin{bmatrix} U \\ Q_{fc}^t \end{bmatrix} = 0,$$

that is,

$$B_{ft} = -Q_{fc}^t. \tag{6.13}$$

Let $\beta = [\beta_1, \beta_2, \ldots, \beta_\rho \,|\, \beta_{\rho+1}, \ldots, \beta_m]$, where ρ is the rank of G, be a circuit vector with its columns arranged in the same edge order as B_f and Q_f. Then, again by the orthogonality relation,

$$\beta Q_f^t = [\,\beta_1, \beta_2, \ldots, \beta_\rho \,|\, \beta_{\rho+1}, \ldots, \beta_m\,]\begin{bmatrix} U \\ Q_{fc}^t \end{bmatrix} = 0.$$

Therefore

$$[\,\beta_1, \beta_2, \ldots, \beta_\rho\,] = -[\,\beta_{\rho+1}, \beta_{\rho+2}, \ldots, \beta_m\,]Q_{fc}^t = [\,\beta_{\rho+1}, \beta_{\rho+2}, \ldots, \beta_m\,]B_{ft}.$$

Using the above equation we can write $[\beta_1, \beta_2, \ldots, \beta_m]$ as

$$[\,\beta_1, \beta_2, \ldots, \beta_m\,] = [\,\beta_{\rho+1}, \beta_{\rho+2}, \ldots, \beta_m\,][\,B_{ft} \quad U\,] = [\,\beta_{\rho+1}, \beta_{\rho+2}, \ldots, \beta_m\,]B_f. \tag{6.14}$$

Thus any circuit vector can be expressed as a linear combination of the fundamental circuit vectors. So

$$\text{rank}\,(B_c) \leqslant \text{rank}\,(B_f) = m - n + 1.$$

Combining the above inequality with (6.12) establishes the following theorem and its corollary.

THEOREM 6.7. The rank of the circuit matrix B_c of a connected graph G with n vertices and m edges is equal to $m-n+1$, the nullity of G. ∎

Corollary 6.7.1 The rank of the circuit matrix B_c of a graph G with n vertices, m edges, and p components is equal to $m-n+p$, the nullity of G. ■

Suppose $\alpha=[\alpha_1,\alpha_2,\ldots,\alpha_\rho,\ \alpha_{\rho+1},\ldots,\alpha_m]$ is a cut vector such that its columns are arranged in the same edge order as B_f and Q_f, then we can start from the relation

$$\alpha B_f^t=0$$

and prove that

$$\alpha=\left[\ \alpha_1,\alpha_2,\ldots,\alpha_\rho\right]Q_f, \tag{6.15}$$

by following a procedure similar to that used in establishing (6.14). Thus every cut vector can be expressed as a linear combination of the fundamental cutset vectors. Since rank $(Q_f)=n-1$, we get

$$\text{rank}(Q_c)=\text{rank}(Q_f)=n-1.$$

The above is thus an alternate proof of Theorem 6.4.

Note that the ring sum of two subgraphs corresponds to modulo 2 addition of the corresponding vectors. Therefore in the case of undirected graphs, (6.14) and (6.15) are merely restatements of what we have already proved in the course of our arguments leading to Theorem 4.6, namely, a circuit (cut) can be expressed as a ring sum of fundamental circuits (fundamental cutsets).

6.5 SUBMATRICES OF CUT, INCIDENCE, AND CIRCUIT MATRICES

In this section we characterize those submatrices of Q_c, A_c, and B_c which correspond to circuits, cutsets, spanning trees, and cospanning trees and discuss some properties of these submatrices.

THEOREM 6.8.

1. There exists a linear relationship among the columns of the cut matrix Q_c which correspond to the edges of a circuit.
2. There exists a linear relationship among the columns of the circuit matrix B_c which correspond to the edges of a cutset.

Proof

1. Let us partition Q_c into columns so that

$$Q_c=\left[Q^{(1)},Q^{(2)},\ldots,Q^{(m)}\right].$$

Let $\beta=[\beta_1,\beta_2,\ldots,\beta_m]$ be a circuit vector. Then by the orthogonality relation, we have

$$Q_c\beta'=0$$

or

$$\beta_1 Q^{(1)}+\beta_2 Q^{(2)}+\cdots+\beta_m Q^{(m)}=0. \tag{6.16}$$

If we assume, without loss of generality, that the first r elements of β are nonzero and the remaining ones are zero, then we have from (6.16)

$$\beta_1 Q^{(1)}+\beta_2 Q^{(2)}+\cdots+\beta_r Q^{(r)}=0.$$

Thus there exists a linear relationship among the columns $Q^{(1)}, Q^{(2)},\ldots, Q^{(r)}$ of Q_c which correspond to the edges of a circuit.

2. The proof in this case follows along the same lines as that for part 1. ■

Corollary 6.8.1 There exists a linear relationship among the columns of the incidence matrix which correspond to the edges of a circuit.

Proof

The result follows from Theorem 6.8, part 1, because the incidence matrix is a submatrix of Q_c. ■

THEOREM 6.9. A square submatrix of order $n-1$ of any incidence matrix A of an n-vertex connected graph G is nonsingular if and only if the edges which correspond to the columns of the submatrix form a spanning tree of G.

Proof

Necessity Consider the $n-1$ columns of a nonsingular submatrix of A. Since these columns are linearly independent, by Corollary 6.8.1, there is no circuit in the corresponding subgraph of G. Since this circuitless subgraph has $n-1$ edges, it follows from Theorem 2.2 that it is a spanning tree of G.

Sufficiency This follows from Theorem 6.1 ■

Thus the spanning trees of a connected graph are in one-to-one correspondence with the nonsingular submatrices of the matrix A. This result is in fact the basis of a formula for counting spanning trees which we discuss in Section 6.7.

THEOREM 6.10. Consider a connected graph G with n vertices and m edges. Let Q be a submatrix of Q_c with $n-1$ rows and of rank $n-1$. A square submatrix of Q of order $n-1$ is nonsingular if and only if the edges corresponding to the columns of this submatrix form a spanning tree of G.

Proof

Necessity Let the columns of the matrix Q be rearranged so that

$$Q=[Q_{11} \quad Q_{12}]$$

with Q_{11} nonsingular. Since the columns of Q_{11} are linearly independent, by Theorem 6.8, part 1, there is no circuit in the corresponding subgraph of G. This circuitless subgraph has $n-1$ edges and is therefore, by Theorem 2.2, a spanning tree of G.

Sufficiency Suppose we rearrange the columns of Q so that

$$Q=[Q_{11} \quad Q_{12}]$$

and the columns of Q_{11} correspond to the edges of a spanning tree T. Then the fundamental cutset matrix Q_f with respect to T is

$$Q_f=[U \quad Q_{fc}].$$

Since the row of Q can be expressed as linear combinations of the rows of Q_f, we can write Q as

$$\begin{aligned} Q=[Q_{11} \quad Q_{12}] &= DQ_f \\ &= D[U \quad Q_{fc}]. \end{aligned}$$

Thus

$$Q_{11}=DU=D.$$

Now D is nonsingular, because both Q and Q_f are of maximum rank $n-1$. So Q_{11} is nonsingular, and the sufficiency of the theorem follows. ■

A dual theorem is presented next.

THEOREM 6.11. Consider a connected graph G with n vertices and m edges. Let B be a submatrix of the circuit matrix B_c of G with $m-n+1$ rows and of rank $m-n+1$. A square submatrix of B of order $m-n+1$ is nonsingular if and only if the columns of this submatrix correspond to the edges of a cospanning tree.

Proof

Necessity Let the columns of the matrix B be rearranged so that

$$B=[B_{11} \quad B_{12}]$$

with B_{11} nonsingular. Since the columns of B_{11} are linearly independent, by Theorem 6.8, part 2, the corresponding subgraph contains no cutsets of G. Further this subgraph has $m-n+1$ edges. Thus it is a maximal (why?) subgraph of G containing no cutsets and is therefore a cospanning tree of G (see Exercise 2.18).

Sufficiency Suppose we rearrange the columns of B so that

$$B=[B_{11} \quad B_{12}]$$

and the columns of B_{11} correspond to the edges of a cospanning tree of G. Then the fundamental circuit matrix B_f with respect to this cospanning tree is

$$B_f=[U \quad B_{ft}].$$

Since the rows of B can be expressed as linear combinations of the rows of B_f, we can write B as

$$\begin{aligned} B=[B_{11} \quad B_{12}]&=DB_f \\ &=D[U \quad B_{ft}] \end{aligned}$$

so that $B_{11}=DU=D$.

Now D is nonsingular because both B and B_f are of maximum rank $m-n+1$. Thus B_{11} is nonsingular and the sufficiency of the theorem follows. ■

To illustrate Theorems 6.10 and 6.11, consider the graph G of Fig. 6.1a and the matrices Q_f and B_f in (6.9) and (6.11). The 4th-order square submatrix of Q_f corresponding to the spanning tree $\{e_1, e_3, e_5, e_7\}$ and the 3rd-order submatrix of B_f corresponding to the cospanning tree $\{e_2, e_4, e_6\}$ are, respectively,

$$\begin{bmatrix} 1 & 1 & 0 & 0 \\ 0 & -1 & 0 & 0 \\ 0 & 0 & 1 & 0 \\ 0 & 0 & 1 & 1 \end{bmatrix}$$

and

$$\begin{bmatrix} 1 & 0 & 0 \\ -1 & 1 & -1 \\ 0 & 0 & -1 \end{bmatrix}.$$

It may be verified that the determinants of these matrices are nonzero, and thus these matrices are nonsingular.

We conclude this section with the study of an interesting property of the inverse of a nonsingular submatrix of the incidence matrix.

THEOREM 6.12. Let A_{11} be a nonsingular submatrix of order $n-1$ of an incidence matrix A of an n-vertex connected graph G. Then the nonzero elements in each row of A_{11}^{-1} are either all 1 or all -1.

Proof

Let A be the incidence matrix with v_r as the reference vertex. Assume that

$$A=[A_{11} \quad A_{12}],$$

where A_{11} is nonsingular. We know from Theorem 6.9 that the edges corresponding to the columns of A_{11} constitute a spanning tree T of G. Then Q_f, the fundamental cutset matrix with respect to T, will be

$$Q_f=[U \quad Q_{fc}].$$

By Theorem 6.3, each cut vector can be expressed as a linear combination of the rows of the incidence matrix. So we can write Q_f as

$$Q_f=[U \quad Q_{fc}]=D[A_{11} \quad A_{12}].$$

Thus

$$D=A_{11}^{-1}.$$

Consider now the ith row q_i of Q_f. Let the corresponding cutset be $\langle V_a,\bar{V}_a\rangle$.

Let

$$V_a=\{v_1,v_2,\ldots,v_k\}$$

and

$$\bar{V}_a=\{v_{k+1},v_{k+2},\ldots,v_n\}.$$

Suppose that the orientation of the cutset $\langle V_a,\bar{V}_a\rangle$ is from V_a to $\bar{V}_a$. Then we get from (6.5) that

$$q_i=a_1+a_2+\cdots+a_k \tag{6.17}$$

$$=-(a_{k+1}+a_{k+2}+\cdots+a_n), \tag{6.18}$$

where a_i is the ith row of A_c.

Note that row a_r corresponding to v_r will not be present in A. So if $v_r \in V_a$, then to represent q_i as a linear combination of the rows of A we have to write q_i as in (6.18). If $v_r \in \bar{V}_a$, then we have to write q_i as (6.17). In both cases the nonzero coefficients in the linear combination are either all 1 or all -1.

Thus the nonzero elements in each row of $D = A_{11}^{-1}$ are either all 1 or all -1. ■

The proof used in the above theorem suggests the following simple procedure for evaluating A_{11}^{-1}.

1. From A_{11}, construct the tree T for which A_{11} is the incidence matrix with vertex v_n as the reference vertex.
2. Label the edges of T as $e_1, e_2, \ldots, e_{n-1}$ so that e_i corresponds to the ith column of A_{11}. Similarly label the vertices of T as $v_1, v_2, \ldots, v_{n-1}$ so that v_i corresponds to the ith row of A_{11}.
3. Let the ith column of A_{11}^{-1} correspond to v_i. For each i, $1 \leqslant i \leqslant n-1$, do the following: Remove e_i from T and let T_1 and T_2 be the two components of the resulting disconnected graph. Let V_a be the vertex set of T_1 and $\bar{V}_a$ the vertex set of T_2, and let e_i be oriented from a vertex in V_a to a vertex in $\bar{V}_a$. If $v_n \in V_a$, then the ith row of A_{11}^{-1} has -1 in all columns that correspond to the vertices in $\bar{V}_a$; all the other elements of the ith row will be zero. If $v_n \in \bar{V}_a$, then the ith row has 1 in all the columns that correspond to the vertices in V_a; all the other elements in the ith row will be zero.

As an illustration, let

$$A_{11} = \begin{array}{c} \\ v_1 \\ v_2 \\ v_3 \\ v_4 \end{array} \begin{array}{c} \begin{array}{cccc} e_1 & e_2 & e_3 & e_4 \end{array} \\ \left[\begin{array}{cccc} 0 & -1 & 0 & 0 \\ 1 & 1 & 1 & 0 \\ -1 & 0 & 0 & 1 \\ 0 & 0 & 0 & -1 \end{array} \right] \end{array}.$$

The tree T for which A_{11} is an incidence matrix is shown in Fig. 6.6a. The reference vertex is v_5.

To obtain the first row of A_{11}^{-1}, we first remove e_1 from T and get the two components T_1 and T_2 shown in Fig. 6.6b. Thus

$$V_a = \{v_1, v_2, v_5\}$$

$$\bar{V}_a = \{v_3, v_4\}.$$

Since the reference vertex is in V_a, the first row of A_{11}^{-1} will have -1 in the columns corresponding to the vertices v_3 and v_4 of $\bar{V}_a$, and it is given below:

$$[0 \quad 0 \quad -1 \quad -1].$$

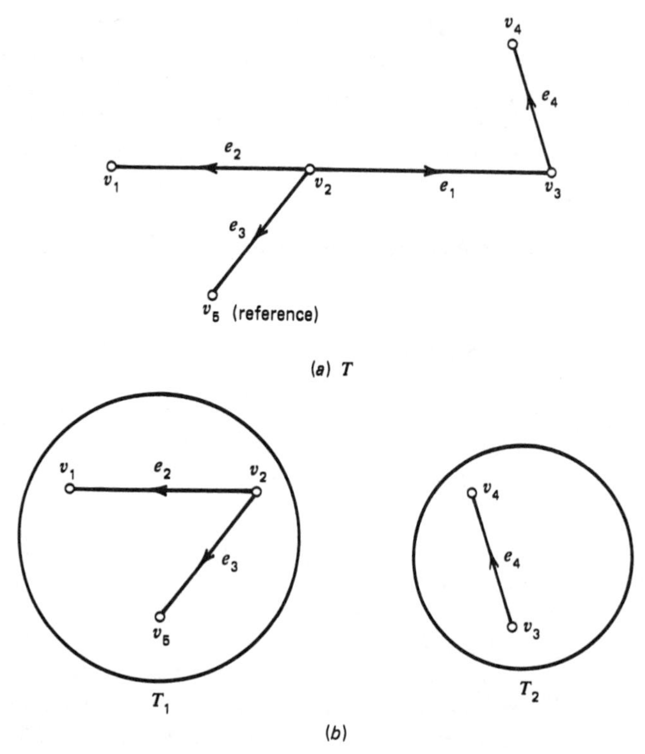

Figure 6.6. (*a*) Tree T. (*b*) Subtrees T_1 and T_2.

In a similar way we get all the other rows of A_{11}^{-1}, and A_{11}^{-1} is as given below:

$$A_{11}^{-1} = \begin{bmatrix} 0 & 0 & -1 & -1 \\ -1 & 0 & 0 & 0 \\ 1 & 1 & 1 & 1 \\ 0 & 0 & 0 & -1 \end{bmatrix}.$$

6.6 UNIMODULAR MATRICES

A matrix is *unimodular* if the determinant of each of its square submatrices is 1, -1, or 0.

We show in this section that the matrices A_c, Q_f, and B_f are all unimodular.

THEOREM 6.13. The incidence matrix A_c of a directed graph is unimodular.

Proof

We prove the theorem by induction on the order of a square submatrix of A_c.

Obviously the determinant of every square submatrix of A_c of order 1 is 1, -1, or 0. Assume, as the induction hypothesis, that the determinant of every square submatrix of order less than k is equal to 1, -1, or 0.

Consider any nonsingular square submatrix of A_c of order k. Every column in this matrix contains at most two nonzero entries, one $+1$ and/or one -1. Since the submatrix is nonsingular, not every column can have both $+1$ and -1. For the same reason, in this submatrix there can be no column consisting of only zeros. Thus there is at least one column that contains exactly one nonzero entry. Expanding the determinant of the submatrix by this column and using the induction hypothesis, we find that the desired determinant is ± 1. ■

Let Q_f be the fundamental cutset matrix of an n-vertex connected graph G with respect to some spanning tree T. Let the branches of T be $b_1, b_2, \ldots, b_{n-1}$.

Let G' be the graph which is obtained from G by identifying or short-circuiting the end vertices of one of the branches, say, the branch b_1. Then $T-\{b_1\}$ is a spanning tree of G'. Let us now delete from Q_f the row corresponding to branch b_1 and denote the resulting matrix as Q'_f. Then it is not difficult to show that Q'_f is the fundamental cutset matrix of G' with respect to the spanning tree $T-\{b_1\}$. Thus the matrix that results after deleting any row from Q_f is a fundamental cutset matrix of some connected graph. Generalizing this, we can state that each matrix formed by some rows of Q_f is a fundamental cutset matrix of some connected graph.

For example, consider the graph G of Fig. 6.1a and the fundamental cutset matrix Q_f given in (6.9). The submatrix consisting of the two rows of Q_f corresponding to the branches e_1 and e_6 is given below:

$$\begin{array}{c} \\ e_1 \\ e_6 \end{array} \begin{array}{c} \begin{array}{ccccccc} e_1 & e_2 & e_6 & e_7 & e_3 & e_4 & e_5 \end{array} \\ \begin{bmatrix} 1 & 0 & 0 & 0 & 1 & -1 & 0 \\ 0 & 0 & 1 & 0 & 0 & 1 & 1 \end{bmatrix} \end{array}.$$

It can be verified that this matrix is the fundamental cutset matrix of the graph shown in Fig. 6.7 with respect to the spanning tree $\{e_1, e_6\}$. This graph is obtained from the graph of Fig. 6.1a by identifying the end vertices of e_2 as well as the end vertices of e_7.

THEOREM 6.14. Any fundamental cutset matrix Q_f of a connected graph G is unimodular.

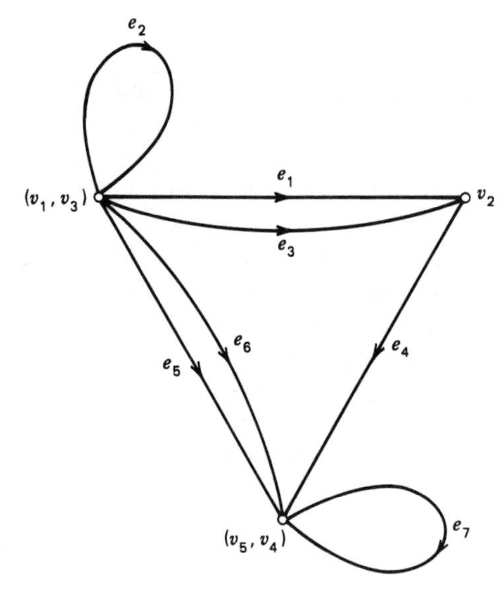

Figure 6.7.

Proof

Let Q_f be the fundamental cutset matrix of G with respect to a spanning tree T. Then

$$Q_f = \begin{bmatrix} U & Q_{fc} \end{bmatrix}.$$

Let an incidence matrix A of G be partitioned as $A = [A_{11} \quad A_{12}]$ where the columns of A_{11} correspond to the branches of T. We know from Theorem 6.9 that A_{11} is nonsingular. Now we can write Q_f as

$$Q_f = \begin{bmatrix} U & Q_{fc} \end{bmatrix} = A_{11}^{-1}\begin{bmatrix} A_{11} & A_{12} \end{bmatrix}.$$

If C is any square submatrix of Q_f of order $n-1$, where n is the number of vertices of G, and D is the corresponding submatrix of A, then $C = A_{11}^{-1}D$. Since $\det D = \pm 1$ or 0 and $\det A_{11}^{-1} = \pm 1$, we get

$$\det C = \pm 1 \text{ or } 0. \tag{6.19}$$

Consider next any square submatrix H of Q_f of order less than $n-1$. From the arguments preceding this theorem we know that H is a submatrix of a fundamental cutset matrix of some connected graph. Therefore, $\det H = \pm 1$ or 0, the proof of which follows along the same lines as that used to prove (6.19).

Thus the determinant of every square submatrix of Q_f is ± 1 or 0 and hence Q_f is unimodular. ∎

Next we show that B_f is unimodular.

THEOREM 6.15. Any fundamental circuit matrix B_f of a connected graph G is unimodular.

Proof

Let B_f and Q_f be the fundamental circuit and cutset matrices of G with respect to a spanning tree T. If $Q_f=[U \quad Q_{fc}]$, then we know from (6.13) that

$$B_f=\left[-Q_{fc}^t \quad U\right].$$

Since Q_{fc} is unimodular, Q_{fc}^t is also unimodular. It is now a simple exercise to show that $[-Q_{fc}^t \quad U]$ is also unimodular. ■

6.7 THE NUMBER OF SPANNING TREES

We derive in this section a formula for counting the number of spanning trees in a connected graph. This formula is based on Theorem 6.9 and a result in matrix theory, known as the Binet-Cauchy theorem.

A major determinant or briefly a *major* of a matrix is a determinant of maximum order in the matrix. Let P be a matrix of order $p\times q$ and Q be a matrix of order $q\times p$ with $p\leqslant q$. The majors of P and Q are of order p. If a major of P consists of the columns $i_1, i_2,\ldots, i_p$ of P, then the corresponding major of Q is formed by the rows $i_1, i_2,\ldots, i_p$ of Q. For example, if

$$P=\begin{bmatrix} 1 & -1 & 3 & 3 \\ 2 & 2 & -1 & 2 \end{bmatrix} \quad \text{and} \quad Q=\begin{bmatrix} 1 & 2 \\ 2 & -1 \\ -3 & 1 \\ 1 & 2 \end{bmatrix},$$

then for the major

$$\begin{vmatrix} -1 & 3 \\ 2 & -1 \end{vmatrix}$$

of P,

$$\begin{vmatrix} 2 & -1 \\ -3 & 1 \end{vmatrix}$$

is the corresponding major of Q.

THEOREM 6.16 (BINET-CAUCHY). If P is a $p\times q$ matrix and Q is a $q\times p$ matrix, with $p\leqslant q$, then

$$\det(PQ)=\sum(\text{product of the corresponding majors of } P \text{ and } Q). \quad ■$$

Proof of this theorem may be found in Hohn [6.1].

As an illustration, if the matrices P and Q are given as earlier, then applying the Binet-Cauchy theorem we get

$$\det(PQ)=\begin{vmatrix}1 & -1\\ 2 & 2\end{vmatrix}\begin{vmatrix}1 & 2\\ 2 & -1\end{vmatrix}+\begin{vmatrix}1 & 3\\ 2 & -1\end{vmatrix}\begin{vmatrix}1 & 2\\ -3 & 1\end{vmatrix}+\begin{vmatrix}1 & 3\\ 2 & 2\end{vmatrix}\begin{vmatrix}1 & 2\\ 1 & 2\end{vmatrix}+$$

$$\begin{vmatrix}-1 & 3\\ 2 & -1\end{vmatrix}\begin{vmatrix}2 & -1\\ -3 & 1\end{vmatrix}+\begin{vmatrix}-1 & 3\\ 2 & 2\end{vmatrix}\begin{vmatrix}2 & -1\\ 1 & 2\end{vmatrix}+\begin{vmatrix}3 & 3\\ -1 & 2\end{vmatrix}\begin{vmatrix}-3 & 1\\ 1 & 2\end{vmatrix}$$

$$=-167.$$

THEOREM 6.17 Let G be a connected undirected graph and A an incidence matrix of a directed graph which is obtained by assigning arbitrary orientations to the edges of G. Then

$$\tau(G)=\det(AA^t),$$

where $\tau(G)$ is the number of spanning trees of G.

Proof

By the Binet-Cauchy theorem

$$\det(AA^t)=\sum\left(\begin{matrix}\text{product of the}\\ \text{corresponding majors of } A \text{ and } A^t\end{matrix}\right). \tag{6.20}$$

Note that the corresponding majors of A and A^t both have the same value equal to 1, -1, or 0 (Theorem 6.13). Therefore each nonzero term in the sum on the right-hand side of (6.20) has the value 1. Furthermore, a major of A is nonzero if and only if the edges corresponding to the columns of the major form a spanning tree.

Thus there is a one-to-one correspondence between the nonzero terms in the sum on the right-hand side of (6.20) and the spanning trees of G. Hence the theorem. ■

As an example, consider the graph G shown in Fig. 6.8a. A directed graph obtained by assigning arbitrary orientations to the edges of G is shown in Fig. 6.8b. The incidence matrix of this directed graph with vertex v_4 as the reference is given by

$$A=\begin{matrix} & e_1 & e_2 & e_3 & e_4 & e_5\\ v_1 & 1 & 0 & 0 & 1 & 0\\ v_2 & 0 & -1 & 1 & -1 & 0\\ v_3 & 0 & 0 & -1 & 0 & -1\end{matrix}.$$

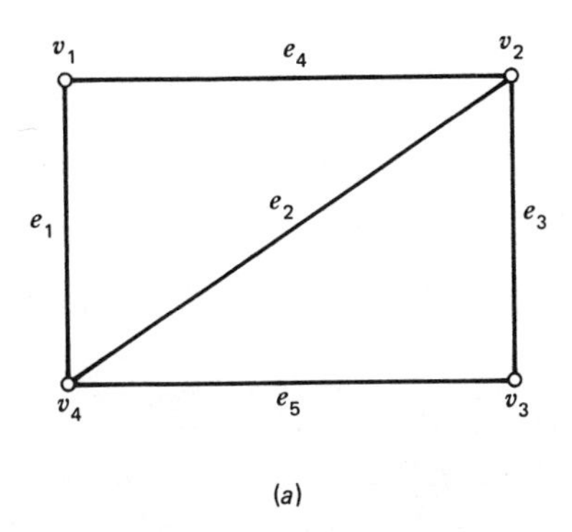

(a)

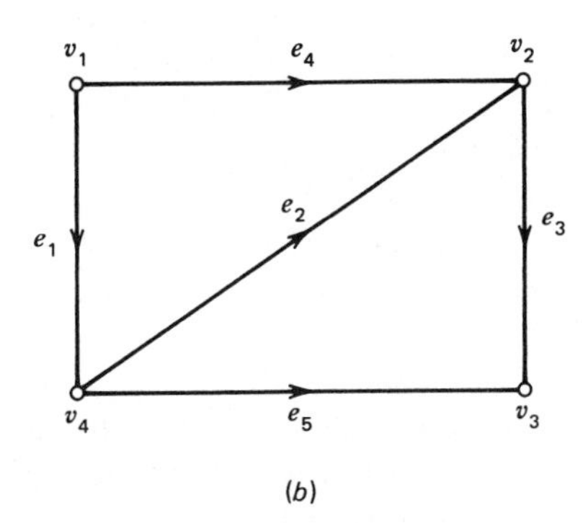

(b)

Figure 6.8.

So

$$AA^t = \begin{bmatrix} 2 & -1 & 0 \\ -1 & 3 & -1 \\ 0 & -1 & 2 \end{bmatrix}$$

and

$$\det(AA^t) = 8.$$

The eight spanning trees of G are

$$\{e_1, e_4, e_5\}, \quad \{e_1, e_3, e_4\}, \quad \{e_1, e_3, e_5\}, \quad \{e_1, e_2, e_3\},$$

$$\{e_1, e_2, e_5\}, \quad \{e_3, e_4, e_5\}, \quad \{e_2, e_3, e_4\}, \quad \{e_2, e_4, e_5\}.$$

Let $v_1, v_2, \ldots, v_n$ denote the vertices of an undirected graph G without self-loops. The *degree matrix* $K=[k_{ij}]$ of G is an $n \times n$ matrix defined as follows:

$$k_{ij}=\begin{cases} -p, & \text{if } i\neq j \text{ and there are } p \text{ parallel edges} \\ & \text{connecting } v_i \text{ and } v_j. \\ d(v_i), & \text{if } i=j. \end{cases}$$

We can easily see that $K=A_cA_c^t$ and is independent of our choice of edge orientations for arriving at the all-vertex incidence matrix A_c. If v_r is the reference vertex of A, then AA^t is obtained by removing the rth row and the rth column of K. Thus the matrix AA^t used in Theorem 6.17 can be obtained by an inspection of the graph G.

It is clear from the definition of the degree matrix that the sum of all the elements in each row of K equals zero. Similarly the sum of all the elements in each column of K equals zero. A square matrix with these properties is called an *equi-cofactor matrix*. As its name implies, all the cofactors of an equi-cofactor matrix are equal [6.2]. Thus from Theorem 6.17 we get the following result, originally due to Kirchhoff [6.3].

THEOREM 6.18. All the cofactors of the degree matrix of a connected undirected graph have the same value equal to the number of spanning trees of G. ∎

Next we derive a formula for counting the number of distinct spanning trees which can be constructed on n labeled vertices. Clearly this number is the same as the number of spanning trees of K_n, the complete graph on n labeled vertices.

THEOREM 6.19 (CAYLEY). There are n^{n-2} labeled trees on $n\geqslant 2$ vertices.

Proof

In the case of K_n, the matrix AA^t is of the form

$$\begin{bmatrix} n-1 & -1 & \dots & -1 \\ -1 & n-1 & \dots & -1 \\ \cdot & \cdot & \dots & \cdot \\ \cdot & \cdot & \dots & \cdot \\ \cdot & \cdot & \dots & \cdot \\ -1 & -1 & \dots & n-1 \end{bmatrix}.$$

By Theorem 6.17, the determinant of this matrix gives the number of spanning trees of K_n, which is the same as the number of labeled trees on n vertices.

To compute $\det(AA^t)$, subtract the first column of AA^t from all the other columns of AA^t. Then we get

$$\begin{bmatrix} n-1 & -n & -n & \dots & -n \\ -1 & n & 0 & \dots & 0 \\ -1 & 0 & n & \dots & 0 \\ \cdot & \cdot & \cdot & \cdots & \cdot \\ \cdot & \cdot & \cdot & \cdots & \cdot \\ \cdot & \cdot & \cdot & \cdots & \cdot \\ -1 & 0 & 0 & \dots & n \end{bmatrix}.$$

Now adding to the first row of the above matrix every one of the other rows, we get

$$\begin{bmatrix} 1 & 0 & 0 & \dots & 0 \\ -1 & n & 0 & \dots & 0 \\ -1 & 0 & n & \dots & 0 \\ \cdot & \cdot & \cdot & \cdots & \cdot \\ \cdot & \cdot & \cdot & \cdots & \cdot \\ -1 & 0 & 0 & \dots & n \end{bmatrix}.$$

The determinant of this matrix is n^{n-2}. The theorem now follows since addition of any two rows or any two columns of a matrix does not change the value of the determinant of the matrix. ■

Several proofs of Cayley's theorem [6.4] are available in the literature. See Moon [6.5] and Prüfer [6.6].

6.8 THE NUMBER OF SPANNING 2-TREES

In this section we relate the cofactors of the matrix AA^t of a graph G to the number of spanning 2-trees of the appropriate type. For this purpose, we need symbols to denote 2-trees in which certain specified vertices are required to be in different components. We use the symbol $T_{ijk\dots,rst\dots}$ to denote spanning 2-trees in which the vertices $v_i, v_j, v_k, \dots$ are required to be in one component and the vertices $v_r, v_s, v_t\dots$ are required to be in the other component of the 2-tree. The number of these spanning 2-trees in the graph G will be denoted by $\tau_{ijk\dots,rst\dots}$. For example, in the graph of Fig. 6.8a, the edges e_1 and e_3 form a spanning 2-tree of the type $T_{14,23}$, and $\tau_{14,23}=1$.

In the following, we denote by A an incidence matrix of the directed graph which is obtained by assigning arbitrary orientations to the edges of the graph G. However, we shall refer to A as an incidence matrix of G. We shall assume, without any loss of generality, that v_n is the reference vertex for A, and the ith row of A corresponds to vertex v_i. Δ_{ij} will denote the (i, j) cofactor of AA^t.

Let A_{-i} denote the matrix obtained by removing from A its ith row. If G' is the graph obtained by short-circuiting the vertices v_i and v_n in G, then we can easily verify the following:

1. A_{-i} is the incidence matrix of G' with v_n as the reference vertex.
2. A set of edges forms a spanning tree of G' if and only if these edges form a spanning 2-tree $T_{i,n}$ of G.

Thus there exists a one-to-one correspondence between the nonzero majors of A_{-i} and the spanning 2-trees of the type $T_{i,n}$.

THEOREM 6.20. For a connected graph G,

$$\Delta_{ii} = \tau_{i,n}.$$

Proof

Clearly $\Delta_{ii} = \det(A_{-i}A^t_{-i})$. Proof is immediate since the nonzero majors of A_{-i} correspond to the spanning 2-trees $T_{i,n}$ of G and vice versa. ∎

Consider next the (i, j) cofactor Δ_{ij} of AA^t which is given by

$$\Delta_{ij} = (-1)^{i+j} \det(A_{-i}A^t_{-j}). \tag{6.21}$$

By the Binet-Cauchy theorem,

$$\det(A_{-i}A^t_{-j}) = \sum \left(\begin{matrix} \text{product of the} \\ \text{corresponding majors of } A_{-i} \text{ and } A^t_{-j} \end{matrix} \right). \tag{6.22}$$

Each nonzero major of A_{-i} corresponds to a spanning 2-tree of the type $T_{i,n}$, and each nonzero major of A_{-j} corresponds to a spanning 2-tree of the type $T_{j,n}$. Therefore the nonzero terms in the sum on the right-hand side of (6.22) correspond to the spanning 2-trees of the type $T_{ij,n}$. Each one of these nonzero terms is equal to a determinant of the type $\det(F_{-i}F^t_{-j})$, where F is the incidence matrix of a 2-tree of the type $T_{ij,n}$.

THEOREM 6.21. Let F denote the incidence matrix of a 2-tree $T_{ij,n}$ with v_n as the reference vertex. If the ith row of F corresponds to vertex v_i, then

$$\det(F_{-i}F^t_{-j}) = (-1)^{i+j}.$$

Proof

Let T_1 and T_2 denote the two components of $T_{ij,n}$. Assume that v_n is in T_2. By interchanging some of its rows and the corresponding columns, we can write the matrix FF^t as

$$S=\left[\begin{array}{c|c} C & 0 \\ \hline 0 & D \end{array}\right],$$

where

1. C is the degree matrix of T_1; and
2. D can be obtained by removing from the degree matrix of T_2 the row and the column corresponding to v_n.

Let row k' of S correspond to vertex v_k.

Interchanging some rows and the corresponding columns of a matrix does not alter the values of the cofactors of the matrix. So

$$\begin{aligned}(i', j') \text{ cofactor of } S &= (i, j) \text{ cofactor of } (FF^t) \\ &= (-1)^{i+j} \det(F_{-i} F_{-j}^t). \qquad (6.23)\end{aligned}$$

By Theorem 6.18 all the cofactors of C have the same value equal to the number of spanning trees of T_1. So we have

$$(i', j') \text{ cofactor of } C = 1.$$

Furthermore,

$$\det D = 1.$$

So

$$\begin{aligned}(i', j') \text{ cofactor of } S &= [(i', j') \text{ cofactor of } C][\det D] \\ &= 1. \qquad (6.24)\end{aligned}$$

Now, from (6.23) and (6.24) we get

$$\det(F_{-i} F_{-j}^t) = (-1)^{i+j}. \quad \blacksquare$$

Proof of the above theorem is by Sankara Rao, Bapeswara Rao, and Murti [6.7].

The following result is due to Mayeda.

THEOREM 6.22. For a connected graph G,

$$\Delta_{ij} = \tau_{ij,n}.$$

Proof

Since each nonzero term in the sum on the right-hand side of (6.22) is equal to a determinant of the type given in Theorem 6.21, we get

$$\det(A_{-i}A'_{-j}) = (-1)^{i+j}\tau_{ij,n}.$$

So

$$\Delta_{ij} = (-1)^{i+j}\det(A_{-i}A'_{-j})$$
$$= \tau_{ij,n}. \quad \blacksquare$$

To illustrate Theorems 6.20 and 6.22 consider again the graph of Fig. 6.8*a*. If vertex v_4 is the reference vertex for A, then

$$A_{-2} = \begin{bmatrix} 1 & 0 & 0 & 1 & 0 \\ 0 & 0 & -1 & 0 & -1 \end{bmatrix}$$

and

$$A_{-3} = \begin{bmatrix} 1 & 0 & 0 & 1 & 0 \\ 0 & -1 & 1 & -1 & 0 \end{bmatrix}.$$

So

$$\tau_{2,4} = \det(A_{-2}A'_{-2})$$
$$= \det\left(\begin{bmatrix} 2 & 0 \\ 0 & 2 \end{bmatrix}\right)$$
$$= 4$$

and

$$\tau_{23,4} = (-1)^{2+3}\det(A_{-2}A'_{-3})$$
$$= -\det\left(\begin{bmatrix} 2 & -1 \\ 0 & -1 \end{bmatrix}\right)$$
$$= 2.$$

In this graph the spanning 2-trees of the type $T_{2,4}$ are

$$\{e_4, e_5\}, \qquad \{e_3, e_4\}, \qquad \{e_1, e_3\}, \qquad \{e_1, e_5\},$$

and the spanning 2-trees of the type $T_{23,4}$ are

$$\{e_1, e_3\}, \qquad \{e_3, e_4\}.$$

6.9 THE NUMBER OF DIRECTED SPANNING TREES IN A DIRECTED GRAPH

In this section we discuss a method due to Tutte [6.8] for computing the number of directed spanning trees in a given directed graph having a specified vertex as root. This method is in fact a generalization of the method given in Theorem 6.17 to compute the number of spanning trees of a graph, and it is given in terms of the in-degree matrix defined below.

The *in-degree matrix* $K=[k_{pq}]$ of a directed graph $G=(V, E)$ without self-loops and with $V=\{v_1, v_2, \ldots, v_n\}$ is an $n\times n$ matrix defined as follows:

$$k_{pq}=\begin{cases} -\omega, & \text{if } p\neq q \text{ and there are } \omega \text{ parallel edges directed from } v_p \text{ to } v_q. \\ d^-(v_p), & \text{if } p=q. \end{cases}$$

Let K_{ij} denote the matrix obtained by removing row i and column j from K.

Tutte's method is based on the following theorem.

THEOREM 6.23. A directed graph $G=(V, E)$ with no self-loops and with $V=\{v_1, v_2, \ldots, v_n\}$ is a directed tree with v_r as the root if and only if its in-degree matrix K has the following properties:

1. $k_{pp}=\begin{cases} 0, & \text{if } p=r, \\ 1, & \text{if } p\neq r. \end{cases}$
2. $\det(K_{rr})=1$.

Proof

Necessity Suppose the given directed graph G is a directed tree with v_r as the root. Clearly G is acyclic. So (see Section 5.7) we can label the vertices of G with the numbers $1,2,\ldots, n$ in such a way that (i, j) is a directed edge of G only if $i<j$. Then in such a numbering, the root vertex would receive the number 1. If the ith row and the ith column of the new in-degree matrix K' of G correspond to the vertex assigned the number i, then we can easily see that K' has the following properties:

$$k'_{11}=0,$$

$$k'_{pp}=1, \qquad \text{for } p\neq 1,$$

$$k'_{pq}=0, \qquad \text{if } p>q.$$

Therefore

$$\det(K'_{11})=1.$$

The matrix K' can be obtained by interchanging some rows and the corresponding columns of K. Such an interchange does not change the value of the determinant of any submatrix of K. So

$$\det(K_{rr})=\det(K'_{11})=1.$$

Sufficiency Suppose the in-degree matrix K of the graph G satisfies the two properties given in the theorem. If G is not a directed tree, then by property 1 and Theorem 5.4, statement 5, it contains a circuit C. The root vertex v_r cannot be in C, for this would imply that $d^-(v_r)>0$ or that $d^-(v)>1$ for some other vertex v in C, contradicting property 1. In a similar way, we can show that

1. C must be a directed circuit; and
2. No edge not in C is incident into any vertex in C.

Consider now the submatrix K_s of K consisting of the columns corresponding to the vertices in C. Because of the above properties, each row of K_s corresponding to a vertex in C has exactly one $+1$ and one -1. All the other rows contain only zero elements. Thus the sum of the columns of K_s is zero. In other words, the sum of the columns of K which correspond to the vertices in C is zero. Since v_r is not in C, this is true in the case of the matrix K_{rr} too, contradicting property 2. Hence the sufficiency. ∎

We now develop Tutte's method for computing the number τ_d of the directed spanning trees of a directed graph G having vertex v_r as the root. Assume that G has no self-loops.

For any graph g, let $K(g)$ denote its in-degree matrix, and let K' be the matrix obtained from K by replacing its rth column by a column of zeros. Denote by S the collection of all the subgraphs of G in each of which $d^-(v_r)=0$ and $d^-(v_p)=1$ for $p\neq r$. Clearly

$$|S|=\prod_{p=1}^{n} d^-(v_p).$$

Further, for any subgraph $g\in S$, the corresponding in-degree matrix satisfies property 1 given in Theorem 6.23.

It is well known in matrix theory that the determinant of a square matrix is a linear function of its columns. For example, if

$$P=[p_1,p_2,\ldots,p'_i+p''_i,\ldots,p_n]$$

is a square matrix with the columns $p_1, p_2, \ldots, p_i' + p_i'', \ldots, p_n$, then

$$\det P = \det[\, p_1, p_2, \ldots, p_i', \ldots, p_n] + \det[\, p_1, p_2, \ldots, p_i'', \ldots, p_n].$$

Using the linearity of the determinant function and the fact that the sum of all the entries in each column of the matrix $K'(G)$ is equal to zero, we can write $\det K'(G)$ as the sum of $|S|$ determinants each of which satisfies the property 1 given in Theorem 6.23. It can be seen that there is a one-to-one correspondence between these determinants and the in-degree matrices of the subgraphs in S. Thus

$$\det K'(G) = \sum_{g \in S} \det K'(g).$$

So

$$\det K'_{rr}(G) = \sum_{g \in S} \det K'_{rr}(g).$$

Since

$$\det K'_{rr}(G) = \det K_{rr}(G),$$

and

$$\det K'_{rr}(g) = \det K_{rr}(g), \qquad \text{for all } g \in S,$$

we get

$$\det K_{rr}(G) = \sum_{g \in S} \det K_{rr}(g).$$

From Theorem 6.23 it follows that each determinant in the sum on the right-hand side of the above equation is nonzero and equal to 1 if and only if the corresponding subgraph in S is a directed spanning tree. Thus we have proved the following theorem.

THEOREM 6.24. Let K be the in-degree matrix of a directed graph G without self-loops. Let the ith row of K correspond to vertex v_i of G. Then the number τ_d of the directed spanning trees of G having v_r as its root is given by

$$\tau_d = \det K_{rr},$$

where K_{rr} is the matrix obtained by removing from K its rth row and its rth column. ■

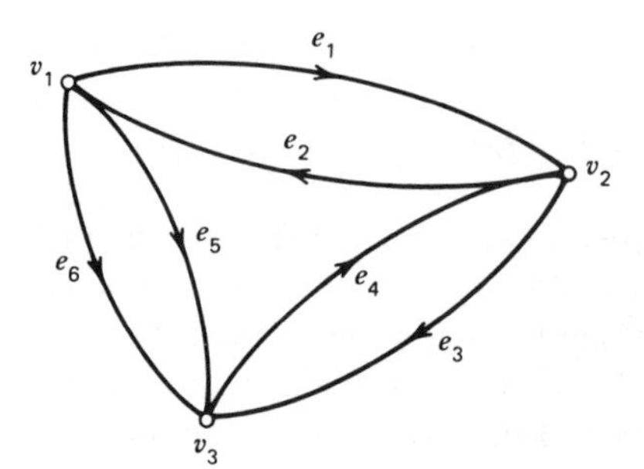

Figure 6.9.

We now illustrate the above theorem and the arguments leading to its proof.

Consider the directed graph G shown in Fig. 6.9. Let us compute the number of directed spanning trees with vertex v_1 as the root.

The in-degree matrix K of G is

$$K=\begin{bmatrix} 1 & -1 & -2 \\ -1 & 2 & -1 \\ 0 & -1 & 3 \end{bmatrix}$$

and

$$K'=\begin{bmatrix} 0 & -1 & -2 \\ 0 & 2 & -1 \\ 0 & -1 & 3 \end{bmatrix}.$$

We can write det K' as

$$\det K'=\begin{vmatrix} 0 & -1 & -2 \\ 0 & 2 & -1 \\ 0 & -1 & 3 \end{vmatrix}=\begin{vmatrix} 0 & -1 & -2 \\ 0 & 1 & -1 \\ 0 & 0 & 3 \end{vmatrix}+\begin{vmatrix} 0 & 0 & -2 \\ 0 & 1 & -1 \\ 0 & -1 & 3 \end{vmatrix}$$

$$=\begin{vmatrix} 0 & -1 & -1 \\ 0 & 1 & 0 \\ 0 & 0 & 1 \end{vmatrix}+\begin{vmatrix} 0 & -1 & -1 \\ 0 & 1 & 0 \\ 0 & 0 & 1 \end{vmatrix}+\begin{vmatrix} 0 & -1 & 0 \\ 0 & 1 & -1 \\ 0 & 0 & 1 \end{vmatrix}$$

$$+\begin{vmatrix} 0 & 0 & -1 \\ 0 & 1 & 0 \\ 0 & -1 & 1 \end{vmatrix}+\begin{vmatrix} 0 & 0 & -1 \\ 0 & 1 & 0 \\ 0 & -1 & 1 \end{vmatrix}+\begin{vmatrix} 0 & 0 & 0 \\ 0 & 1 & -1 \\ 0 & -1 & 1 \end{vmatrix}.$$

The six determinants on the right-hand side in the above equation correspond to the subgraphs on the following sets of edges:

$$\{e_1, e_5\}, \quad \{e_1, e_6\}, \quad \{e_1, e_3\},$$

$$\{e_4, e_5\}, \quad \{e_4, e_6\}, \quad \{e_4, e_3\}.$$

Removing the first row and the first column from the above determinants, we get

$$\det K'_{11}=\begin{vmatrix} 2 & -1 \\ -1 & 3 \end{vmatrix}=\begin{vmatrix} 1 & 0 \\ 0 & 1 \end{vmatrix}+\begin{vmatrix} 1 & 0 \\ 0 & 1 \end{vmatrix}+\begin{vmatrix} 1 & -1 \\ 0 & 1 \end{vmatrix}$$

$$+\begin{vmatrix} 1 & 0 \\ -1 & 1 \end{vmatrix}+\begin{vmatrix} 1 & 0 \\ -1 & 1 \end{vmatrix}+\begin{vmatrix} 1 & -1 \\ -1 & 1 \end{vmatrix}$$

$$=5.$$

The five directed spanning trees with v_1 as root are

$$\{e_1, e_5\}, \quad \{e_1, e_6\}, \quad \{e_1, e_3\}, \quad \{e_4, e_5\}, \quad \{e_4, e_6\}.$$

6.10 ADJACENCY MATRIX

Let $G=(V, E)$ be a directed graph with no parallel edges. Let $V=\{v_1, v_2, \ldots, v_n\}$. The *adjacency matrix* $M=[m_{ij}]$ of G is an $n \times n$ matrix with m_{ij} defined as follows:

$$m_{ij}=\begin{cases} 1, & \text{if } (v_i, v_j) \in E. \\ 0, & \text{otherwise.} \end{cases}$$

For example, the graph of Fig. 6.10 has the following adjacency matrix:

$$M=\begin{array}{c} \\ v_1 \\ v_2 \\ v_3 \\ v_4 \end{array}\begin{array}{c} \begin{array}{cccc} v_1 & v_2 & v_3 & v_4 \end{array} \\ \begin{bmatrix} 1 & 0 & 1 & 0 \\ 1 & 0 & 0 & 0 \\ 0 & 1 & 0 & 1 \\ 0 & 1 & 0 & 1 \end{bmatrix} \end{array}.$$

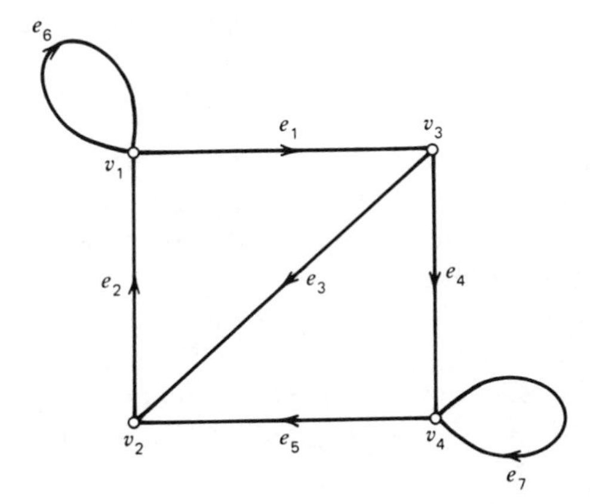

Figure 6.10.

In the case of an undirected graph, $m_{ij}=1$ if and only if there is an edge connecting v_i and v_j.

We now study some results involving the adjacency matrix.

THEOREM 6.25. The (i, j) entry $m_{ij}^{(r)}$ of M^r is equal to the number of directed walks of length r from v_i to v_j.

Proof

Proof is by induction on r. Obviously, the result is true for $r=1$.

As the inductive hypothesis we shall assume that the theorem is true for M^{r-1}. Then

$$m_{ij}^{(r)}=\sum_{k=1}^{n}\left(m_{ik}^{(r-1)}\cdot m_{kj}\right)$$

$$=\sum_{k=1}^{n}\begin{pmatrix}\text{number of directed walks of}\\ \text{length } r-1 \text{ from } v_i \text{ to } v_k\end{pmatrix}\cdot m_{kj}.$$

For each k, the corresponding term in the sum in the above equation gives the number of directed walks of length r from v_i to v_j whose last edge is from v_k to v_j. The theorem follows for M^r, if we note that the number of directed walks of length r from v_i to $v_j=\Sigma_{k=1}^{n}$ (number of directed walks of length r from v_i to v_j whose last edge is from v_k to v_j). ∎

For example, consider the 3rd power of the matrix M of the graph of Fig. 6.10:

$$M^3=\begin{bmatrix}2 & 2 & 1 & 2\\ 1 & 1 & 1 & 1\\ 2 & 1 & 1 & 1\\ 2 & 1 & 1 & 1\end{bmatrix}.$$

The entry (1,4) in this matrix gives the number of directed walks of length 3 from v_1 to v_4. These walks are

$$(e_6, e_1, e_4) \quad \text{and} \quad (e_1, e_4, e_7).$$

We next study an important theorem due to Harary [6.9] which forms the foundation of the signal flow approach for the solution of linear algebraic equations. To introduce the theorem, we need some terminology.

A *1-factor* of a directed graph G is a spanning subgraph of G in which the in-degree and the out-degree of every vertex are both equal to 1. Obviously, such a subgraph is a collection of vertex-disjoint directed circuits including self-loops of G. For example, the two 1-factors of the graph of Fig. 6.10 are shown in Fig. 6.11*a* and *b*.

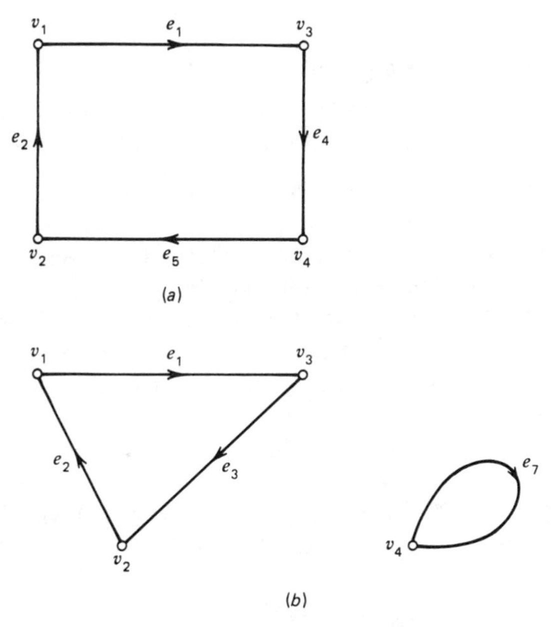

Figure 6.11. The two 1-factors of the graph of Fig. 6.10.

Consider now a permutation $j_1, j_2, \ldots, j_n$ of the integers $1,2,\ldots,n$. This permutation is even if an even number of interchanges are required to rearrange it as $1,2,3,\ldots,n$. An odd permutation is defined in a similar way. For example, consider the permutation 1 3 4 2. The following sequence of two interchanges will rearrange this permutation as 1 2 3 4:

1. Interchange 3 and 2.
2. Interchange 3 and 4.

Thus this permutation is an even permutation.

THEOREM 6.26. Let H_i, $i=1,2,\ldots,p$, be the 1-factors of an n-vertex directed graph G. Let L_i denote the number of directed circuits in H_i and let M be the adjacency matrix of G. Then

$$\det M = (-1)^n \sum_{i=1}^{p} (-1)^{L_i}.$$

Proof

From the definition of a determinant

$$\det M = \sum_{(j)} \varepsilon_{j_1, j_2, \ldots, j_n} m_{1j_1} \cdot m_{2j_2} \cdot \cdots \cdot m_{nj_n}, \tag{6.25}$$

where

1. $(j)=(j_1, j_2,\ldots, j_n)$ is a permutation of $1,2,\ldots, n$.
2. $\varepsilon_{j_1, j_2,\ldots, j_n}=1$ if (j) is an even permutation; otherwise it is equal to -1.
3. The sum $\Sigma_{(j)}$ is taken over all permutations of $1,2,\ldots, n$.

A nonzero term $m_{1j_1}\cdot m_{2j_2}\cdot\cdots\cdot m_{nj_n}$ corresponds to the set of edges $(v_1, v_{j_1}),(v_2, v_{j_2}),\ldots,(v_n, v_{j_n})$. Each v_k appears exactly twice in this set, once as a first element and once as a second element of some ordered pairs. This means that in the subgraph consisting of the edges (v_1, v_{j_1}), $(v_2, v_{j_2}),\ldots,(v_n, v_{j_n})$ the in-degree and the out-degree of each vertex are both equal to 1. Thus each nonzero term in the sum in (6.25) corresponds to a 1-factor of G. Conversely, each 1-factor corresponds to a nonzero term $m_{1j_1}\cdot m_{2j_2}\cdot\cdots\cdot m_{nj_n}$.

For example, the 1-factor in Fig. 6.11*b* corresponds to the term $m_{13}\cdot m_{32}\cdot m_{21}\cdot m_{44}$.

We now have to fix the sign of $\varepsilon_{j_1, j_2,\ldots, j_n}$. Let C be a directed circuit in the 1-factor corresponding to $j_1, j_2,\ldots, j_n$. Let C consist of the ω edges

$$(v_{i_1}, v_{i_2}),(v_{i_2}, v_{i_3}),\ldots,(v_{i_\omega}, v_{i_1}).$$

The initial vertices of these edges form the array

$$i_1, i_2,\ldots, i_\omega,$$

and their terminal vertices form the array

$$i_2, i_3,\ldots, i_1.$$

It is easy to show that $\omega-1$ interchanges are sufficient to rearrange the array $i_2, i_3,\ldots, i_1$ as $i_1, i_2,\ldots, i_\omega$.

Let there be L directed circuits in the 1-factor corresponding to $j_1, j_2,\ldots, j_n$. Let the lengths of these directed circuits be $\omega_1, \omega_2,\ldots, \omega_L$. Then we need

$$(\omega_1-1)+(\omega_2-1)+\cdots+(\omega_L-1)=W$$

interchanges to rearrange $j_1, j_2,\ldots, j_n$ as $1,2,\ldots, n$. So,

$$\varepsilon_{j_1, j_2,\ldots, j_n}=(-1)^W=(-1)^{n+L}.$$

Thus summarizing,

1. Each nonzero term $m_{1j_1}\cdot\cdots\cdot m_{nj_n}$ corresponds to a 1-factor H_i of G (note that the value of this term is 1), and
2. $\varepsilon_{j_1, j_2,\ldots, j_n}=(-1)^{n+L_i}$, where L_i is the number of directed circuits in H_i.

Thus (6.25) reduces to

$$\det M = (-1)^n \sum_{i=1}^{p} (-1)^{L_i}. \quad \blacksquare$$

For example, consider the 1-factors of Fig. 6.11. The corresponding L_i's are

$$L_1 = 1,$$

$$L_2 = 2.$$

Thus the determinant of the adjacency matrix of the graph of Fig. 6.10 is

$$(-1)^1 + (-1)^2 = 0.$$

This may be verified by direct expansion of the determinant.

Suppose in a directed graph G, each edge (i, j) is assigned a weight w_{ij}. Then we may define the *adjacency matrix* $M = [m_{ij}]$ of G as follows:

$$m_{ij} = \begin{cases} w_{ij}, & \text{if there is an edge directed from vertex } i \text{ to vertex } j. \\ 0, & \text{otherwise.} \end{cases}$$

Let us define the *weight product* $w(H)$ of a subgraph H of G as the product of the weights of all the edges of H. If H has no edges, that is, H is empty, we define $w(H) = 1$. Then we can obtain the following as an easy extension of Theorem 6.26.

THEOREM 6.27. The determinant of the adjacency matrix M of an n-vertex weighted directed graph G is given by

$$\det M = (-1)^n \sum_H (-1)^{L_H} w(H)$$

where H is a 1-factor of G, and L_H is the number of directed circuits in H. $\blacksquare$

6.11 THE COATES AND MASON GRAPHS

In this section we develop a graph-theoretic approach for the solution of linear algebraic equations. Two closely related methods due to Coates [6.10] and Mason [6.11], [6.12] are discussed.

6.11.1 Coates' Method

Consider a linear system described by the set of equations

$$AX = Bx_{n+1}, \tag{6.26}$$

where $A = [a_{ij}]$ is an $n \times n$ nonsingular matrix, X is a column vector of unknown variables $x_1, x_2, \ldots, x_n$, B is a column vector of elements $b_1, b_2, \ldots, b_n$, and x_{n+1} is the input variable. We can solve for x_k as

$$\frac{x_k}{x_{n+1}} = \frac{\sum_{i=1}^{n} b_i \Delta_{ik}}{\det A}, \tag{6.27}$$

where Δ_{ik} is the (i, k) cofactor of A.

Let A' denote the matrix obtained from A by adding $-B$ to the right of A and then adding a row of zeros at the bottom of the resulting matrix. Let us now associate with A' a weighted directed graph $G_c(A')$ as follows.

$G_c(A')$ has $n+1$ vertices $x_1, x_2, \ldots, x_{n+1}$. If $a_{ji} \neq 0$, then in $G_c(A')$ there is an edge directed from x_i to x_j with weight a_{ji}. Clearly A' is the transpose of the adjacency matrix of $G_c(A')$. $G_c(A')$ is called the *Coates flow graph* or simply the *Coates graph* associated with A'. We shall also refer to it as the Coates graph associated with the system of equations (6.26).

As an example consider the set of equations:

$$\begin{bmatrix} 2 & 1 & 1 \\ 1 & -1 & -2 \\ 2 & 1 & 2 \end{bmatrix} \begin{bmatrix} x_1 \\ x_2 \\ x_3 \end{bmatrix} = \begin{bmatrix} 1 \\ 0 \\ -1 \end{bmatrix} x_4. \tag{6.28}$$

The matrix A' in this case is

$$A' = \begin{bmatrix} 2 & 1 & 1 & -1 \\ 1 & -1 & -2 & 0 \\ 2 & 1 & 2 & 1 \\ 0 & 0 & 0 & 0 \end{bmatrix}.$$

The Coates graph $G_c(A')$ associated with the set of equations (6.28) is shown in Fig. 6.12a.

Note that we may regard each vertex x_i, $1 \leqslant i \leqslant n$, of $G_c(A')$ as representing an equation in (6.26). For example, the ith equation in (6.26) can be obtained by equating to zero the sum of the products of weights of the edges directed into x_i and the variables corresponding to the vertices from which these variables originate. Furthermore, the Coates graph $G_c(A)$ associated with A can be obtained by removing from $G_c(A')$ the vertex x_{n+1}. The graph $G_c(A)$ in the case of (6.28) is shown in Fig. 6.12b.

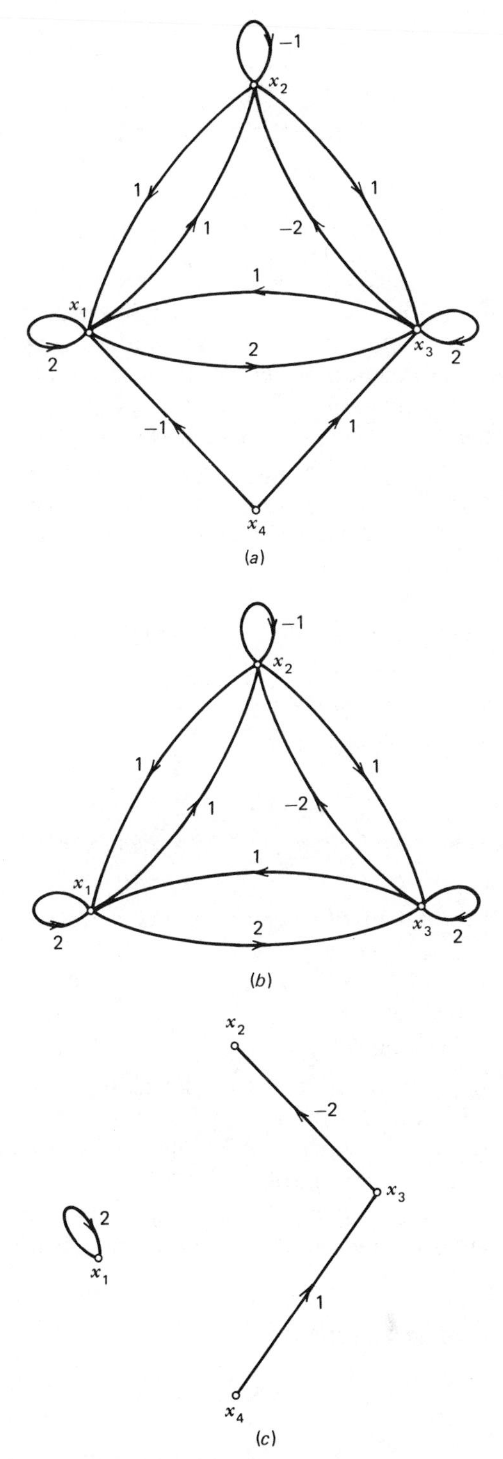

Figure 6.12. (*a*) The Coates graph $G_c(A')$. (*b*) The graph $G_c(A)$. (*c*) A 1-factorial connection $H_{4,2}$ of the graph $G_c(A')$.

Since A is the transpose of the adjacency matrix of $G_c(A)$ and since a matrix and its transpose have the same value of determinant, we get from Theorem 6.27

$$\det A = (-1)^n \sum_H (-1)^{L_H} w(H), \tag{6.29}$$

where H is a 1-factor in $G_c(A)$, $w(H)$ is the weight product of H, and L_H is the number of directed circuits in H. Thus we can evaluate the denominator of (6.27) in terms of the weight products of the 1-factors of $G_c(A)$. To derive a similar expression for the numerator of (6.27), we need to evaluate Δ_{ik}.

A *1-factorial connection* H_{ij} from x_i to x_j in $G_c(A)$ is a spanning subgraph of $G_c(A)$ which contains (a) a directed path P from x_i to x_j and (b) a set of vertex-disjoint directed circuits which include all the vertices of $G_c(A)$ except those contained in P. As an example, a 1-factorial connection from x_4 to x_2 of the graph $G_c(A')$ of Fig. 6.12*a* is shown in Fig. 6.12*c*.

THEOREM 6.28. Let $G_c(A)$ be the Coates graph associated with an $n \times n$ matrix A. Then:

1. $\Delta_{ii} = (-1)^{n-1} \Sigma_H (-1)^{L_H} w(H)$;
2. $\Delta_{ij} = (-1)^{n-1} \Sigma_{H_{ij}} (-1)^{L'_H} w(H_{ij})$, $\quad i \neq j$;

where

H is a 1-factor in the graph obtained by removing vertex x_i from $G_c(A)$,
H_{ij} is a 1-factorial connection in $G_c(A)$ from vertex x_i to vertex x_j, and
L_H and L'_H are the numbers of directed circuits in H and H_{ij}, respectively.

Proof

1. Follows from Theorem 6.27.
2. Note that Δ_{ij} is the determinant of the matrix obtained from A by replacing the jth column of A by a column of zeros except for the element of the ith row, which is 1. Let the resulting matrix be denoted by A_α. Then the Coates graph $G_c(A_\alpha)$ is obtained from $G_c(A)$ by removing all the edges incident out of x_j (including self-loops at x_j) and then adding an edge directed from x_j to x_i with weight 1. Now from Theorem 6.27 we get

$$\begin{aligned} \Delta_{ij} &= \det A_\alpha \\ &= (-1)^n \sum_{H_\alpha} (-1)^{L_\alpha} w(H_\alpha), \end{aligned} \tag{6.30}$$

where H_α is a 1-factor in $G_c(A_\alpha)$ and L_α is the number of directed circuits in H_α.

Each 1-factor H_α must necessarily contain the added edge with weight 1. Removing this edge from H_α we get a 1-factorial connection H_{ij} of $G_c(A)$. Furthermore, $w(H_\alpha)=w(H_{ij})$. We can also see that there is a one-to-one correspondence between H_α in $G_c(A_\alpha)$ and H_{ij} in $G_c(A)$ such that $w(H_\alpha)=w(H_{ij})$. Since the number of directed circuits in H_{ij} is one less than that in H_α, we have $L'_H=L_\alpha-1$. So we get from (6.30)

$$\Delta_{ij}=(-1)^{n-1}\sum_{H_{ij}}(-1)^{L'_H}w(H_{ij}). \quad \blacksquare$$

Consider now the term $\sum_{i=1}^{n}b_i\Delta_{ik}$, the numerator of (6.27). This term is equal to the determinant of the matrix obtained from A by replacing the jth column by B. We can easily relate this to $\det A'_{n+1,k}$ (where $A'_{n+1,k}$ is the matrix obtained by removing from A' the $(n+1)$th row and the kth column) as follows:

$$\sum_{i=1}^{n} b_i\Delta_{ik}=(-1)^{(n+1)-k-1}(-1)\det(A'_{n+1,k}). \tag{6.31}$$

From part 2 of Theorem 6.28 we get

$$(-1)^{n+1+k}\det(A'_{n+1,k})=(-1)^n \sum_{H_{n+1,k}} (-1)^{L'_H}w(H_{n+1,k}), \tag{6.32}$$

where L'_H is the number of directed circuits in the 1-factorial connection $H_{n+1,k}$ of $G_c(A')$. Combining (6.31) and (6.32),

$$\sum_{i=1}^{n} b_i\Delta_{ik}=(-1)^n \sum_{H_{n+1,k}} (-1)^{L'_H}w(H_{n+1,k}). \tag{6.33}$$

From (6.29) and (6.33) we get the following theorem.

THEOREM 6.29. If the coefficient matrix A is nonsingular, then the solution of (6.26) is given by

$$\frac{x_k}{x_{n+1}}=\frac{\sum\limits_{H_{n+1,k}} (-1)^{L'_H}w(H_{n+1,k})}{\sum\limits_{H}(-1)^{L_H}w(H)} \tag{6.34}$$

for $k=1,2,\ldots,n$, where

1. $H_{n+1,k}$ is a 1-factorial connection of $G_c(A')$ from vertex x_{n+1} to vertex x_k,
2. H is a 1-factor of $G_c(A)$, and
3. L'_H and L_H are the numbers of directed circuits in $H_{n+1,k}$ and H, respectively. $\blacksquare$

Equation (6.34) is called *Coates' gain formula*. We now illustrate an application of this formula by solving (6.28) for x_2/x_4.

The 1-factors of the graph $G_c(A)$ (Fig. 6.12*b*) associated with the matrix A of (6.28) are given below along with their weight products. The different directed circuits in a 1-factor are distinguished by enclosing the vertices in each directed circuit within parentheses.

1-Factor	Weight Product
$(x_1)(x_2)(x_3)$	-4
$(x_1)(x_2, x_3)$	-4
$(x_2)(x_1, x_3)$	-2
$(x_3)(x_1, x_2)$	2
(x_1, x_2, x_3)	1
(x_1, x_3, x_2)	-4

From the above we get the denominator in (6.34) as

$$\begin{aligned}\sum_H (-1)^{L_H} w(H) &= (-1)^3(-4) + (-1)^2(-4) + (-1)^2(-2) \\ &\quad + (-1)^2(2) + (-1)(1) + (-1)(-4) \\ &= 4 - 4 - 2 + 2 - 1 + 4 \\ &= 3.\end{aligned}$$

The 1-factorial connections from x_4 to x_2 of $G_c(A')$ (Fig. 6.12*a*) are given below. The vertices which lie in a directed path are also now included within parentheses.

1-Factorial Connection	Weight Product
$(x_4, x_1, x_2)(x_3)$	-2
(x_4, x_1, x_3, x_2)	4
$(x_4, x_3, x_2)(x_1)$	-4
(x_4, x_3, x_1, x_2)	1

From the above we get the numerator of (6.34) as

$$\begin{aligned}\sum_{H_{n+1,k}} (-1)^{L'_H} w(H_{n+1,k}) &= (-1)(-2) + 4 + (-1)(-4) + 1 \\ &= 2 + 4 + 4 + 1 \\ &= 11.\end{aligned}$$

Thus we get

$$\frac{x_2}{x_4} = \frac{11}{3}.$$

6.11.2 Mason's Method

To develop Mason's method, first we rewrite (6.26) as

$$x_j=(a_{jj}+1)x_j+\sum_{\substack{k=1\\k\neq j}}^{n} a_{jk}x_k-b_jx_{n+1}, \qquad j=1,2,\ldots,n$$

$$x_{n+1}=x_{n+1}.$$

We may write the above in matrix form as follows:

$$(A'+U_{n+1})X'=X',$$

where A' is the $(n+1)\times(n+1)$ matrix defined earlier, U_{n+1} is the unit matrix of order $n+1$, and X' is the column vector of variables $x_1, x_2, \ldots, x_{n+1}$.

The Coates graph $G_c(A'+U_{n+1})$ is called *Mason's signal flow graph* or simply the *Mason graph* associated with the matrix A' and it is denoted by $G_m(A')$. For example, the Mason graphs $G_m(A')$ and $G_m(A)$ associated with the system of equations (6.28) are shown in Fig. 6.13.

We may regard each vertex in $G_m(A')$ as representing a variable. If there is a directed edge from vertex x_i to x_j, then we may consider the variable x_i as making a contribution of $(a_{ji}x_i)$ to the variable x_j. So x_j is equal to the sum of the products of the weights of the edges incident into vertex x_j and the variables corresponding to the vertices from which these edges emanate. Thus the Mason graph is a convenient pictorial display of the flow of variables in a system.

Note that to obtain the Coates graph from a given Mason graph, we simply subtract one from the weight of each self-loop, and for each vertex of a Mason graph without a self-loop we add one with weight -1. This is equivalent to saying that the Coates graph G_c can be obtained from a Mason graph G_m simply by adding a self-loop of weight -1 at each vertex. The set of such self-loops of weight -1 added to construct G_c will be denoted by S. Note that the graph G_c so constructed will have at most two and at least one self-loop at each of its vertices.

Consider now the Mason graph $G_m(A)$ associated with a matrix A, and let G_c be the corresponding Coates graph. Let H be a 1-factor of G_c having j self-loops from the set S. Let L_Q+j be the total number of directed circuits in H. On removing from H the j added self-loops of the set S, we get a unique subgraph Q of G_m which is a collection of L_Q vertex-disjoint directed circuits. Further, for $j<n$,

$$w(H)=(-1)^j w(Q). \tag{6.35}$$

Note, that if $j=n$, then Q is an empty subgraph of G_m which by definition has

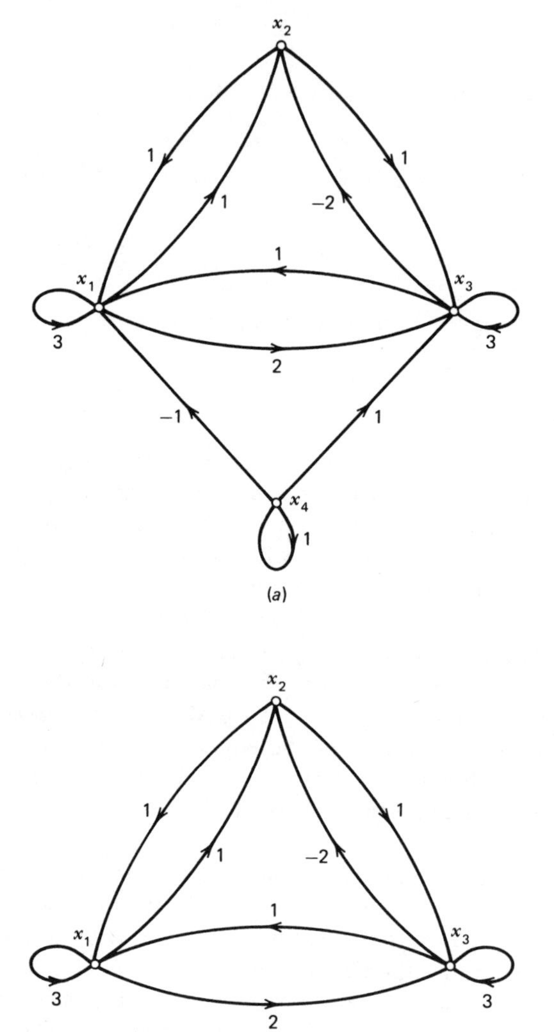

Figure 6.13. (*a*) The Mason graph $G_m(A')$. (*b*) The graph $G_m(A)$.

weight $w(Q)=1$. So, in such a case,

$$w(H)=(-1)^n. \tag{6.36}$$

We can also see that for each subgraph Q of G_m, which is a collection of L_Q vertex-disjoint directed circuits, there corresponds a unique 1-factor of G_c which is obtained by adding self-loops (from the set S) of weight -1 to the vertices not in Q.

From Theorem 6.27 we have

$$\det A = (-1)^n \sum_H (-1)^{L_Q + j} w(H)$$

$$= (-1)^n \left[1 + \sum_Q (-1)^{L_Q} w(Q) \right]. \tag{6.37}$$

The last step in the above follows from (6.35) and (6.36).

We may also rewrite (6.37) as

$$\det A = (-1)^n \left[1 - \sum_j Q_{j1} + \sum_j Q_{j2} - \sum_j Q_{j3} + \cdots \right] \tag{6.38}$$

where Q_{jk} is the weight product of the jth subgraph of G_m, which is a collection of k vertex-disjoint directed circuits.

Thus we have expressed the denominator of (6.27) in terms of the weight products of appropriate subgraphs of the Mason graph.

Let us refer to $(-1)^n \det A$ as the determinant of the graph $G_m(A)$ and denote it by Δ. Then using (6.33) and a reasoning exactly similar to that employed to derive (6.38), we can express the numerator of (6.27) as

$$\sum_{i=1}^{n} b_i \Delta_{ik} = (-1)^n \sum_j w(P^j_{n+1,k}) \Delta_j,$$

where $P^j_{n+1,k}$ is the jth directed path from x_{n+1} to x_k in $G_m(A')$ and Δ_j is determinant of the subgraph of $G_m(A')$ which is vertex-disjoint from the jth directed path $P^j_{n+1,k}$. Thus we get the following.

THEOREM 6.30. If the coefficient matrix A in (6.26) is nonsingular, then

$$\frac{x_k}{x_{n+1}} = \frac{\sum_j w(P^j_{n+1,k}) \Delta_j}{\Delta}, \qquad k = 1, 2, \ldots, n, \tag{6.39}$$

where $P^j_{n+1,k}$ is the jth directed path from x_{n+1} to x_k in $G_m(A')$, Δ_j is the determinant of the subgraph of $G_m(A')$ which is vertex-disjoint from the jth directed path $P^j_{n+1,k}$, and Δ is the determinant of the graph $G_m(A)$. ■

Equation (6.39) is known as *Mason's gain formula*. In systems theory literature $P^j_{n+1,k}$ is referred to as a *forward path* from vertex x_{n+1} to vertex x_k. Furthermore the directed circuits of $G_m(A')$ are called the *feedback loops*.

We now illustrate application of (6.39) by solving (6.28) for x_2/x_4.

The different collections of vertex-disjoint directed circuits of $G_m(A)$ (Fig. 6.13*b*) and their weight products are:

Directed Circuit	Weight Product
(x_1)	3
(x_3)	3
(x_1, x_2)	1
(x_2, x_3)	-2
(x_1, x_3)	2
(x_1, x_2, x_3)	1
(x_1, x_3, x_2)	-4
$(x_1)(x_3)$	9
$(x_1)(x_2, x_3)$	-6
$(x_3)(x_1, x_2)$	3

From the above we get the denominator in (6.39) as

$$\Delta = 1 - (3+3+1-2+2+1-4) + (9-6+3)$$
$$= 3.$$

The different directed paths of $G_m(A')$ (Fig. 6.13*a*) from x_4 to x_2 along with their weight products are:

j	$P_{4,2}^j$	Weight Product
1	(x_4, x_1, x_2)	-1
2	(x_4, x_1, x_3, x_2)	4
3	(x_4, x_3, x_2)	-2
4	(x_4, x_3, x_1, x_2)	1

The directed circuits which are vertex-disjoint from $P_{4,2}^1$ and $P_{4,2}^3$ are (x_3) and (x_1), respectively. So

$$\Delta_1 = 1 - 3 = -2,$$
$$\Delta_3 = 1 - 3 = -2.$$

There are no directed circuits which are vertex-disjoint from $P_{4,2}^2$ and $P_{4,2}^4$. So

$$\Delta_2 = 1,$$
$$\Delta_4 = 1.$$

Thus the numerator in (6.39) is

$$P_{4,2}^1\Delta_1 + P_{4,2}^2\Delta_2 + P_{4,2}^3\Delta_3 + P_{4,2}^4\Delta_4 = 2+4+4+1 = 11.$$

So

$$\frac{x_2}{x_4} = \frac{11}{3}.$$

6.12 FURTHER READING

The results of this chapter will form the basis for most of our discussions in Part II of the book. Seshu and Reed [6.13], Chen [6.2], Mayeda [6.14], and Deo [6.15] are other references for the topics covered in this chapter. Chen also gives historical details regarding the results presented here.

The textbook by Harary and Palmer [6.16] is devoted exclusively to the study of enumeration problems in graph theory, in particular, those related to unlabeled graphs.

The problem of counting spanning trees has received considerable attention in electrical network theory literature. Recurrence relations for counting spanning trees in special classes of graphs are available. For example, see Myers [6.17], [6.18], Bedrosian [6.19], Bose, Feick, and Sun [6.20], and Swamy and Thulasiraman [6.21]. Berge [6.22] contains formulas for the number of spanning trees having certain specified properties.

Chen [6.2] gives an extensive discussion of several aspects concerning the signal flow graph approach. See Robichaud, Boisvert, and Robert [6.23] for several applications of signal flow graphs. Balabanian and Bickart [6.24] may be consulted for the application of signal flow graphs in the study of electrical networks.

6.13 EXERCISES

6.1 Let $Q_f = [q_{ij}]$ be a fundamental cutset matrix and $B_f = [b_{ij}]$ be a fundamental circuit matrix of a connected directed graph. Prove the following:

(a) If the products $(q_{ij}q_{kj})$ and $(q_{ir}q_{kr})$ are both nonzero, then they are equal.

(b) Repeat (a) for B_f.

6.2 Let B be a matrix formed by any $\mu(G)$ independent rows of the circuit matrix of a connected directed graph G having m edges. Show that if F is any matrix of order $\rho(G)\times m$ and of rank $\rho(G)$, having entries 1, -1, and 0 and satisfying $BF^t=0$, then each row of F is a row of the cut matrix of G.

6.3 Let Q be a matrix formed by any $\rho(G)$ independent rows of the cut matrix of a connected directed graph G having m edges. Show that if F is any matrix of order $\mu(G)\times m$ and of rank $\mu(G)$, having entries 1, -1, and 0 and satisfying $QF^t=0$, then each row of F corresponds to a circuit or an edge-disjoint union of circuits of G.

6.4 Let $Q(x, y)$ be the submatrix of the cut matrix Q_c of a connected nonseparable directed graph G containing only those rows of Q_c which correspond to cuts separating vertices x and y. Show that $Q(x, y)$

contains a fundamental cutset matrix of G. (See Kajitani and Ueno [6.25] for a more general result.)

6.5 Let B be a submatrix of the circuit matrix B_c of a planar graph G containing only those rows of B_c which correspond to the meshes of G. Show that B is unimodular.

6.6 Let B be a matrix formed by any $\mu(G)$ independent rows of the circuit matrix of a connected directed graph G. Show that the determinants of all the major nonsingular sub matrices of B have the same magnitude.

6.7 Prove Theorem 4.11.

6.8 Prove that for a connected directed graph G

$$\tau(G)=\det(B_f B_f^t)=\det(Q_f Q_f^t),$$

where B_f is a fundamental circuit matrix and Q_f is a fundamental cutset matrix of G.

6.9 Let Q be a matrix formed by any $\rho(G)$ independent rows of the cut matrix of a connected directed graph G, and let B be a matrix formed by any $\mu(G)$ independent rows of the circuit matrix of G. Show that

$$\begin{bmatrix} Q \\ B \end{bmatrix}$$

is nonsingular.

6.10 Prove that the circuit and cutset subspaces W_C and W_S over $GF(2)$ of a connected undirected graph G are orthogonal complements of the vector space W_G if and only if the number of spanning trees of G is odd.

6.11 For any edge e of a graph G, let $G \cdot e$ denote the graph obtained from G by contracting the edge e. If G is connected, show that

$$\tau(G)=\tau(G-e)+\tau(G \cdot e).$$

6.12 Use the recursion formula in Exercise 6.11 to find the number of spanning trees in a wheel on n vertices. (See Section 8.4 for the definition of a wheel.)

6.13 Show that if e is an edge of K_n, then

$$\tau(K_n-e)=(n-2)n^{n-3}.$$

6.14 (a) Let H be an n-vertex graph in which there are k parallel edges between every pair of adjacent vertices and let G be the underlying simple graph of H. Show that

$$\tau(H)=k^{n-1}\tau(G).$$

(b) Let H be the graph obtained from an n-vertex connected graph G when each edge of G is replaced by a path of length k. Show that

$$\tau(H)=k^{m-n+1}\tau(G),$$

where m is the number of edges in G.

6.15 Establish Theorem 6.17 as a special case of Theorem 6.24.

6.16 A graph G with weights associated with its edges is called a *weighted graph*. Recall that the weight product of a subgraph of G is equal to the product of the weights of the edges of the subgraph and that if the subgraph has only isolated vertices, then its weight product is defined to be 1.

Let A be the incidence matrix of a connected directed graph G, with vertex v_n as reference. Assume that the ith row of A corresponds to vertex v_i. Let W be the diagonal matrix representing the weights of G and assume that the columns of A and W are arranged in the same edge order. Prove that

(a) $\det(AWA^t)$ = sum of the weight products of all the spanning trees of G.

(b) Δ_{ij} = sum of the weight products of all the spanning 2-trees $T_{ij,n}$ of G, where Δ_{ij} is the (i, j) cofactor of AWA^t.

6.17 The *out-degree matrix* $K=[k_{ij}]$ of a directed graph G with no self-loops is defined as follows:

$$k_{ij}=\begin{cases} -p, & \text{if } i\neq j \text{ and there are } p \text{ parallel edges directed from } v_i \text{ to } v_j. \\ d^+(v_p), & \text{if } p=q. \end{cases}$$

Also define an *in-going directed tree* as a directed graph which has a vertex v_r to which there is a directed path from every other vertex, and its underlying undirected graph is a tree. Show that the number of in-going spanning directed trees of a graph G can be computed in a way similar to Theorem 6.24.

6.18 The *out-degree matrix* $K=[k_{ij}]$ of a weighted directed graph G without self-loops is defined as follows:

(a) $k_{ij}=-$(sum of the weights of the edges directed from v_i to v_j), $i\neq j$.

(b) k_{ii} = sum of the weights of the edges incident out of v_i.

Prove that $\det(K_{ii})$ is the sum of the weight products of all the in-going spanning directed trees of G having v_i as root.

Note Here by a root we mean a vertex v_i such that there is a directed path to v_i from every other vertex.

6.19 Solve the following system of equations:

$$\begin{bmatrix} 3 & -2 & 1 \\ -1 & 2 & 0 \\ 3 & -2 & 2 \end{bmatrix} \begin{bmatrix} x_1 \\ x_2 \\ x_3 \end{bmatrix} = \begin{bmatrix} 3 \\ 1 \\ -2 \end{bmatrix} x_4. \tag{6.40}$$

6.20 Let G_m be the Mason graph associated with a system of linear equations. Prove the following:

(a) We can remove the self-loop of weight $a_{kk} \neq 1$ at vertex x_k simply by multiplying the weight of every edge incident into x_k by the factor $1/(1-a_{kk})$.

(b) We can remove a vertex x_p with no self-loop by doing the following: For all $i \neq p$ and $k \neq p$, add $(a_{pi}a_{kp})$ to the weight of the edge (x_i, x_k).

6.21 Using the method in Exercise 6.20, reduce the Mason graph associated with (6.40) to a graph containing only the vertices x_4 and x_2 and solve for x_2/x_4.

6.14 REFERENCES

6.1 F. E. Hohn, *Elementary Matrix Algebra*, Macmillan, New York, 1958.

6.2 W. K. Chen, *Applied Graph Theory*, North Holland, Amsterdam, 1971.

6.3 G. Kirchhoff, "Über die Auflösung der Gleichungen, auf welche man bei der Untersuchung der linearen Verteilung galvanischer Ströme geführt wird," *Ann. Phys. Chem.*, Vol. 72, 497–508 (1847).

6.4 A. Cayley, "A Theorem on Trees," *Quart. J. Math.*, Vol. 23, 376–378 (1889).

6.5 J. W. Moon, "Various Proofs of Cayley's Formula for Counting Trees," in *A Seminar on Graph Theory* (Ed. F. Harary and L. W. Beinke), Holt, Rinehart and Winston, New York, 1967, pp. 70–78.

6.6 H. Prüfer, "Neuer Beweis eines Satzes über Permutationen," *Arch. Math. Phys.*, Vol. 27, 742–744 (1918).

6.7 K. Sankara Rao, V. V. Bapeswara Rao, and V. G. K. Murti, "Two-Tree Admittance Products," *Electron. Lett.*, Vol. 6, 834–835 (1970).

6.8 W. T. Tutte, "The Dissection of Equilateral Triangles into Equilateral Triangles," *Proc. Cambridge Phil. Soc.*, Vol. 44, 203–217 (1948).

6.9 F. Harary, "The Determinant of the Adjacency Matrix of a Graph," *SIAM Rev.*, Vol. 4, 202–210 (1962).

6.10 C. L. Coates, "Flow-Graph Solutions of Linear Algebraic Equations," *IRE Trans. Circuit Theory*, Vol. CT-6, 170–187 (1959).

6.11 S. J. Mason, "Feedback Theory: Some Properties of Signal Flow Graphs," *Proc. IRE.*, Vol. 41, 1144–1156 (1953).

6.12 S. J. Mason, "Feedback Theory: Further Properties of Signal Flow Graphs," *Proc. IRE.*, Vol. 44, 920–926 (1956).

6.13 S. Seshu and M. B. Reed, *Linear Graphs and Electrical Networks*, Addison-Wesley, Reading, Mass., 1961.

6.14 W. Mayeda, *Graph Theory*, Wiley-Interscience, New York, 1972.

6.15 N. Deo, *Graph Theory with Applications to Engineering and Computer Science*, Prentice-Hall, Englewood Cliffs, N.J., 1974.

6.16 F. Harary and E. M. Palmer, *Graphical Enumeration*, Academic Press, New York, 1973.

6.17 B. R. Myers, "Number of Trees in a Cascade of 2-Port Networks," *IEEE Trans. Circuit Theory*, Vol. CT-14, 284–290 (1967).

6.18 B. R. Myers, "Number of Spanning Trees in a Wheel," *IEEE Trans. Circuit Theory*, Vol. CT-18, 280–282 (1971).

6.19 S. D. Bedrosian, "Number of Spanning Trees in Multigraph Wheels," *IEEE Trans. Circuit Theory*, Vol. CT-19, 77–78 (1972).

6.20 N. K. Bose, R. Feick, and F. K. Sun, "General Solution to the Spanning Tree Enumeration Problem in Multigraph Wheels," *IEEE Trans. Circuit Theory*, Vol. CT-20, 69–70 (1973).

6.21 M. N. S. Swamy and K. Thulasiraman, "A Theorem in the Theory of Determinants and the Number of Spanning Trees of a Graph," *Proc. IEEE Int. Symp. on Circuits and Systems*, 153–156 (1976).

6.22 C. Berge, *Graphs and Hypergraphs*, North Holland, Amsterdam, 1973.

6.23 L. P. A. Robichaud, M. Boisvert, and J. Robert, *Signal Flow Graphs and Applications*, Prentice-Hall, Englewood Cliffs, N.J., 1962.

6.24 N. Balabanian and T. A. Bickart, *Electrical Network Theory*, Wiley, New York, 1969.

6.25 Y. Kajitani and S. Ueno, "On the Rank of Certain Classes of Cutsets and Tie-Sets of a Graph," *IEEE Trans. Circuits and Sys.*, Vol. CAS-26, 666–668 (1979).

Chapter 7

Planarity and Duality

In this chapter we discuss two important concepts in graph theory, namely, planarity and duality. First we consider planar graphs and derive some properties of these graphs. Characterizations of planar graphs due to Kuratowski, Wagner, Harary and Tutte, and to MacLane are also discussed. We then discuss Whitney's definition of duality of graphs which is given in terms of circuits and cutsets and relate this concept to the seemingly unrelated concept of planarity.

Duality has been of considerable interest to electrical network theorists. This interest is due to the fact that the voltages and currents in an electrical network are dual variables. Duality of these variables arises as a result of Kirchhoff's laws. Kirchhoff's voltage law is in terms of circuits and Kirchhoff's current law is in terms of cutsets. Duality between circuits and cutsets was observed in Chapters 2 and 4. This duality will become obvious in Chapter 10, where we show that the set of circuits and the set of cutsets of a graph have the same algebraic structure.

7.1 PLANAR GRAPHS

A graph G is said to be *embeddable* on a surface S if it can be drawn on S so that its edges intersect only at their end vertices. A graph is said to be *planar* if it can be embedded on a plane. Such a drawing of a planar graph G is called a *planar embedding* of G.

Two planar embeddings of a graph are shown in Fig. 7.1. In one of these (Fig. 7.1*a*) all the edges are drawn as straight line segments, while in the other (Fig. 7.1*b*) one of the edges is drawn as a curved line. Note that the edge

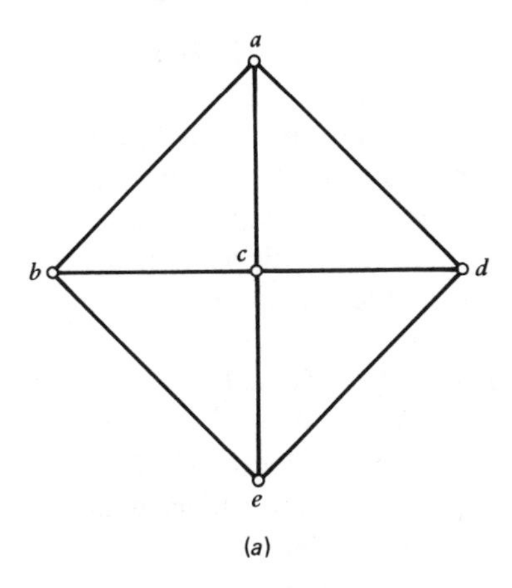

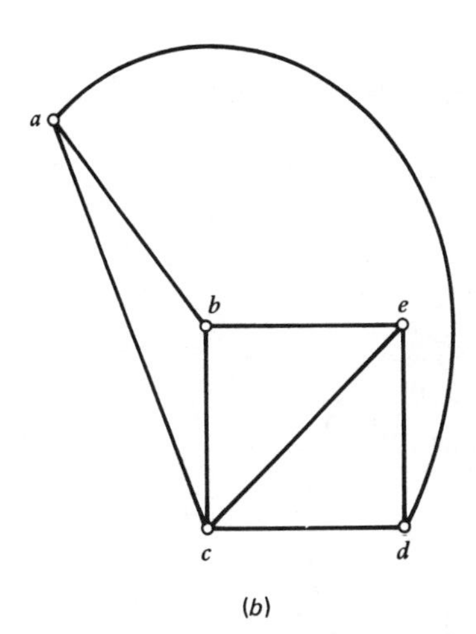

Figure 7.1. Two planar embeddings of a graph.

connecting vertices a and d in Fig. 7.1b cannot be drawn as a straight line if all the remaining edges are drawn as shown.

Obviously, if a graph has self-loops or parallel edges, then in none of its planar embeddings all the edges can be drawn as straight line segments. This naturally raises the question whether for every simple planar graph G there exists a planar embedding in which all the edges of G can be drawn as straight line segments. The answer to this question is in the affirmative, as stated in the following theorem.

THEOREM 7.1. For every simple planar graph there exists a planar embedding in which all the edges of the graph can be drawn as straight line segments. ∎

This result was proved independently by Wagner [7.1], Fary [7.2], and Stein [7.3].

If a graph is not embeddable on a plane, then it may be embeddable on some other surface. However, we now show that embeddability on a plane and embeddability on a sphere are equivalent; that is, if a graph is embeddable on a plane, then it is also embeddable on a sphere and vice versa. The proof of this result uses what is called the stereographic projection of a sphere onto a plane, which is described below.

Suppose that we place a sphere on a plane (Fig. 7.2) and call the point of contact the south pole and the diametrically opposite point on the sphere the north pole N. Let P be any point on the sphere. Then the point P' at which the straight line joining N and P, when extended, meets the plane, is called the *stereographic projection* of P onto the plane. It is clear that there is a one-to-one correspondence between the points on a sphere and their stereographic projections on the plane.

THEOREM 7.2. A graph G is embeddable on a plane if and only if G is embeddable on a sphere.

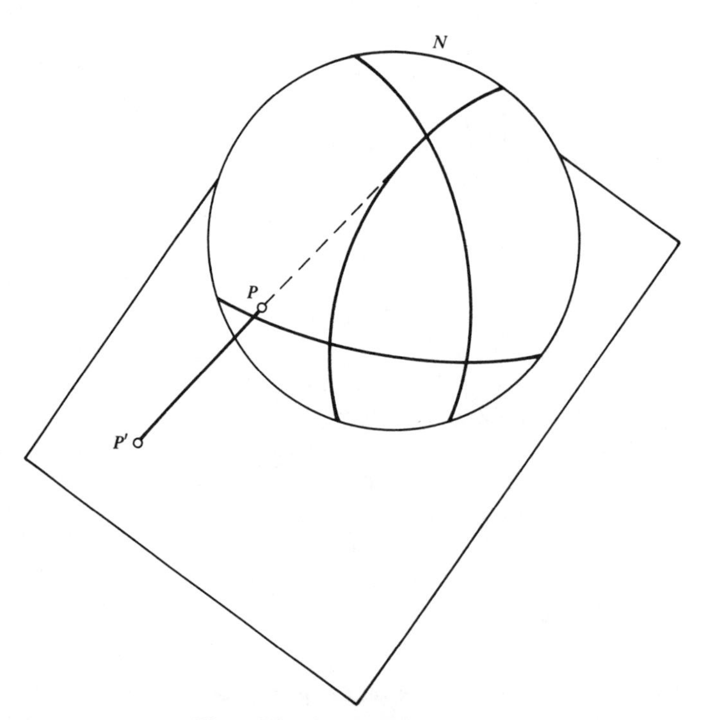

Figure 7.2. Stereographic projection.

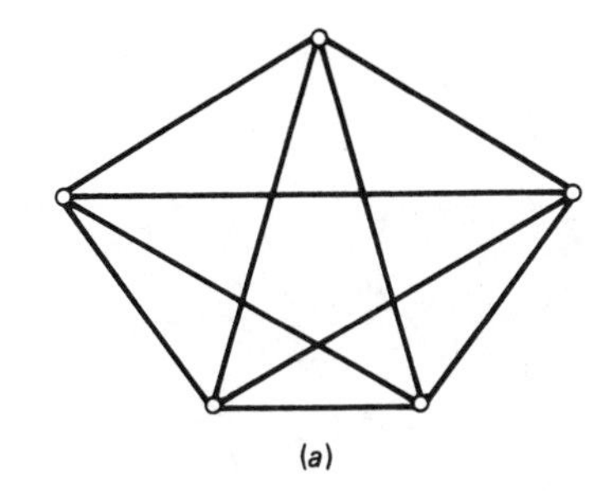

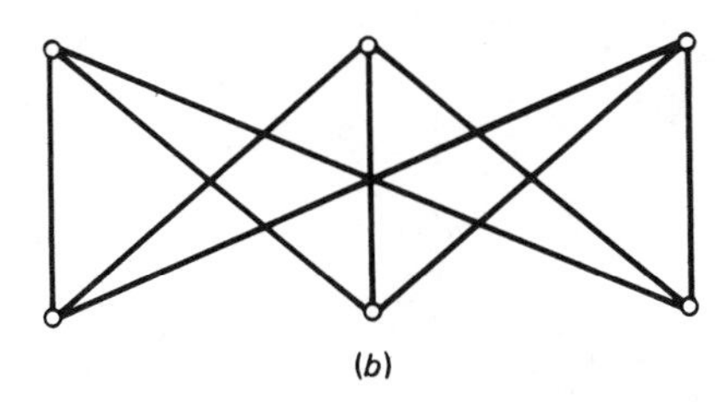

Figure 7.3. Basic nonplanar graphs. (*a*) K_5. (*b*) $K_{3,3}$.

Proof

Let G' be an embedding of G on a sphere. Place the sphere on the plane so that the north pole is neither a vertex of G' nor a point on an edge of G'.

Then the image of G' under the stereographic projection is an embedding of G on the plane because edges of G' intersect only at their end vertices and there is a one-to-one correspondence between points on the sphere and their images under stereographic projection.

The converse is proved similarly. ■

Two basic nonplanar graphs called *Kuratowski's graphs* are shown in Fig. 7.3. One of these is K_5, the complete graph on five vertices, and the other is $K_{3,3}$. We call these graphs basic nonplanar graphs because they play a fundamental role in an important characterization of planarity due to Kuratowski (Section 7.3). The nonplanarity of these two graphs is established in the next section.

Before we conclude this section, we would like to point out that Whitney [7.4] has proved that a separable graph is planar if and only if its blocks are planar. So while considering questions relating to the embedding on a plane, it is enough if we concern ourselves with only nonseparable graphs.

7.2 EULER'S FORMULA

An embedding of a planar graph on a plane divides the plane into *regions*. A region is finite if the area it encloses is finite; otherwise it is infinite.

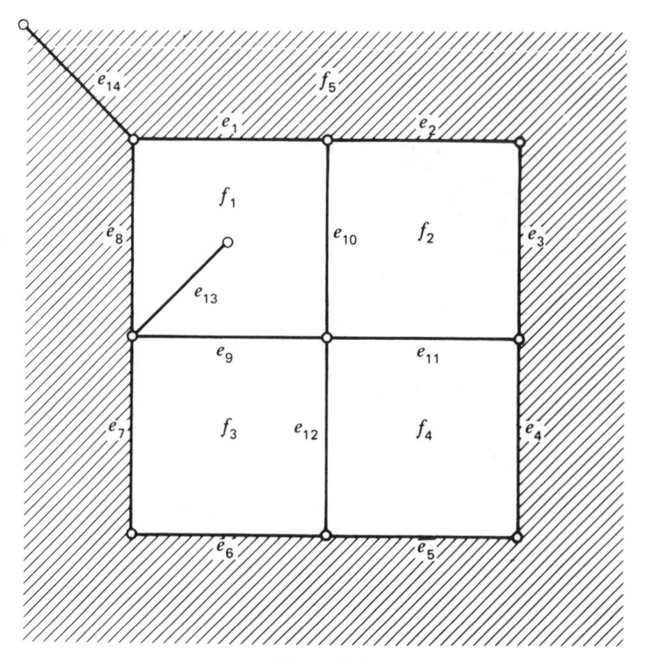

Figure 7.4.

For example, in the planar graph shown in Fig. 7.4, the hatched region f_5 is the infinite region; f_1, f_2, f_3, and f_4 are the finite regions.

Clearly, the edges on the boundary of a region contain exactly one circuit, and this circuit is said to enclose the region. For example, the edges e_1, e_8, e_9, e_{10}, and e_{13} form the region f_1 in the graph of Fig. 7.4, and they contain the circuit $\{e_1, e_8, e_9, e_{10}\}$.

Note that in any spherical embedding of a planar graph, every region is finite.

Suppose we embed a planar graph on a sphere and place the sphere on a plane so that the north pole is inside any chosen region, say, region f. Then under the stereographic projection the image of f will be the infinite region. Thus a planar graph can always be embedded on a plane so that any chosen region becomes the infinite region.

Let $f_1, f_2, \ldots, f_r$ be the regions of a planar graph with f_r as the infinite region. We denote by C_i, $1 \leqslant i \leqslant r$, the circuit on the boundary of region f_i.

The circuits $C_1, C_2, \ldots, C_{r-1}$, corresponding to the finite regions, are called *meshes*.

It is easy to verify that the ring sum of any $k \geqslant 2$ meshes, say, $C_1, C_2, \ldots, C_k$, is a circuit or union of edge-disjoint circuits enclosing the regions $f_1, f_2, \ldots, f_k$. Since each mesh encloses only one region, it follows that no mesh can be obtained as the ring sum of some of the remaining meshes. Thus the meshes $C_1, C_2, \ldots, C_{r-1}$ are linearly independent.

Suppose that any element C of the circuit subspace of G encloses the finite regions $f_1, f_2, \ldots, f_k$. Then it can be verified that

$$C = C_1 \oplus C_2 \oplus \cdots \oplus C_k.$$

For example, in the graph of Fig. 7.4, the set $C = \{e_1, e_2, e_3, e_4, e_5, e_6, e_7, e_8\}$ encloses the regions f_1, f_2, f_3, f_4. Therefore,

$$C = C_1 \oplus C_2 \oplus C_3 \oplus C_4.$$

Thus every element of the circuit subspace of G can be expressed as the ring sum of some or all of the meshes of G. Since the meshes are themselves independent, we get the following theorem.

THEOREM 7.3. The meshes of a planar graph G form a basis of the circuit subspace of G. ■

Following is an immediate consequence of this Theorem.

Corollary 7.3.1 (Euler's Formula) If a connected planar graph G has m edges, n vertices, and r regions, then

$$n - m + r = 2.$$

Proof

The proof follows if we note that by Theorem 7.3, the nullity μ of G is equal to $r-1$. ■

In general it is difficult to test whether a graph is planar or not. We now use Euler's formula to derive some properties of planar graphs. These properties can be of help in detecting nonplanarity in certain cases, as we shall see soon.

Corollary 7.3.2 If a connected simple planar graph G has m edges and $n \geq 3$ vertices, then

$$m \leq 3n - 6.$$

Proof

Let $F = \{f_1, f_2, \ldots, f_r\}$ denote the set of regions of G.

Let the *degree* $d(f_i)$ *of region* f_i denote the number of edges on the boundary of f_i, bridges* being counted twice. (For example, in the graph of Fig. 7.4, the degree of region f_1 is 6.) Noting the similarity between the definitions of the degree of a vertex and the degree of a region, we get from

*For the definition of a bridge, see Exercise 1.28.

Theorem 1.1,

$$\sum_{f_i \in F} d(f_i) = 2m.$$

Since G has neither parallel edges nor self-loops and $n \geqslant 3$, it follows that $d(f_i) \geqslant 3$, for all i. Hence

$$\sum_{f_i \in F} d(f_i) \geqslant 3r.$$

Thus $2m \geqslant 3r$, that is,

$$r \leqslant \frac{2}{3}m.$$

Using this inequality in Euler's formula, we get

$$n - m + \frac{2}{3}m \geqslant 2$$

or

$$m \leqslant 3n - 6. \quad \blacksquare$$

Corollary 7.3.3 K_5 is nonplanar.

Proof

For K_5, $n=5$ and $m=10$. If it were planar, then by Corollary 7.3.2,

$$m = 10 \leqslant 3n - 6 = 9;$$

a contradiction. Thus K_5 must be nonplanar. $\blacksquare$

Corollary 7.3.4 $K_{3,3}$ is nonplanar.

Proof

For $K_{3,3}$, $m=9$ and $n=6$. If it were planar, then by Euler's formula it has $r = 9 - 6 + 2 = 5$ regions.

In $K_{3,3}$ there is no circuit of length less than 4. Hence the degree of every region is at least 4. Thus,

$$2m = \sum_{i=1}^{r} d(f_i) \geqslant 4r$$

or

$$r \leqslant \frac{2m}{4},$$

that is, $r \leqslant 4$; a contradiction. Hence $K_{3,3}$ is nonplanar. $\blacksquare$

Corollary 7.3.5 In a simple planar graph G there is at least one vertex of degree less than or equal to 5.

Proof

Let G have m edges and n vertices. If every vertex of G has degree greater than 5, then by Theorem 1.1,

$$2m \geqslant 6n$$

or

$$m \geqslant 3n.$$

But by Corollary 7.3.2,

$$m \leqslant 3n-6.$$

These two inequalities contradict one another. Hence the result. ■

7.3 KURATOWSKI'S THEOREM AND OTHER CHARACTERIZATIONS OF PLANARITY

Characterizations of planarity given by Kuratowski's Wagner, Harary and Tutte, and MacLane are presented in this section.

To explain Kuratowski's characterization, we need the definition of the concept of homeomorphism between graphs.

The two edges incident on a vertex of degree 2 are called *series edges*.

Let $e_1=(u,v)$ and $e_2=(v,w)$ be the series edges incident on a vertex v. Removal of vertex v and replacing e_1 and e_2 by a simple edge (u,w) is called *series merger* (Fig. 7.5*a*).

Adding a new vertex v on an edge (u,w), thereby creating the edges (u,v) and (v,w), is called *series insertion* (Fig. 7.5*b*).

Two graphs are said to be *homeomorphic* if they are isomorphic or can be made isomorphic by repeated series insertions and/or mergers.

It is clear that if a graph G is planar, then any graph homeomorphic to G is also planar, that is, planarity of a graph is not affected by series insertions or mergers.

We proved in the previous section that K_5 and $K_{3,3}$ are nonplanar. Therefore, a planar graph does not contain a subgraph homeomorphic to K_5 or $K_{3,3}$. It is remarkable that Kuratowski [7.5] could prove that the converse of this result is also true. In the following theorem, we state this celebrated characterization of planarity. Proof of this may also be found in Harary [7.6].

THEOREM 7.4 (KURATOWSKI). A graph is planar if and only if it does not contain a subgraph homeomorphic to K_5 or $K_{3,3}$. ■

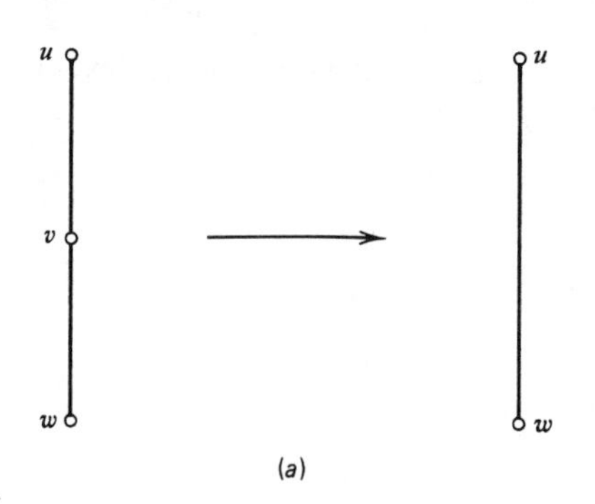

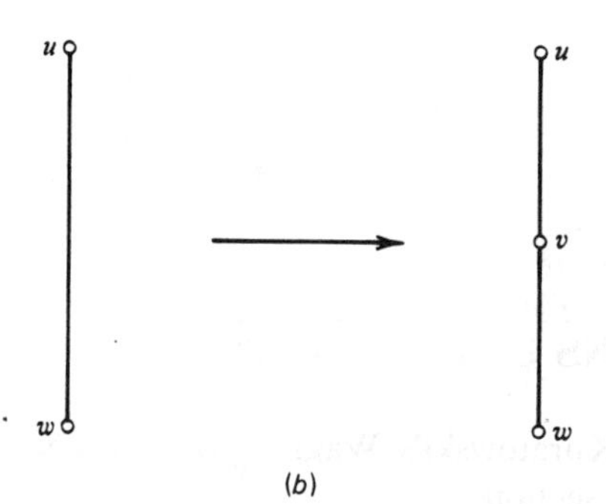

Figure 7.5. (*a*) Series merger. (*b*) Series insertion.

See also Tutte [7.7] for a constructive proof of this theorem.

As an illustration of Kuratowski's theorem, consider the graph G shown in Fig. 7.6*a*. It contains a subgraph (Fig. 7.6*b*) which is homeomorphic to $K_{3,3}$. Therefore G is nonplanar.

We now present another characterization of planarity independently proved by Wagner [7.8] and by Harary and Tutte [7.9].

THEOREM 7.5. A graph is planar if and only if it does not contain a subgraph contractible to K_5 or $K_{3,3}$. ∎

Consider now the graph (known as the *Petersen graph*) shown in Fig. 7.7. This graph does not contain any subgraph isomorphic to K_5 or $K_{3,3}$, but it is known to be nonplanar. So if we wish to use Kuratowski's criterion to establish the nonplanar character of the Petersen graph, then we need to locate a subgraph homeomorphic to K_5 or $K_{3,3}$. However, the nonplanarity of the graph follows easily from the above characterization, because the graph reduces to K_5 after contracting the edges e_1, e_2, e_3, e_4, and e_5.

MacLane's characterization of planar graphs is stated next.

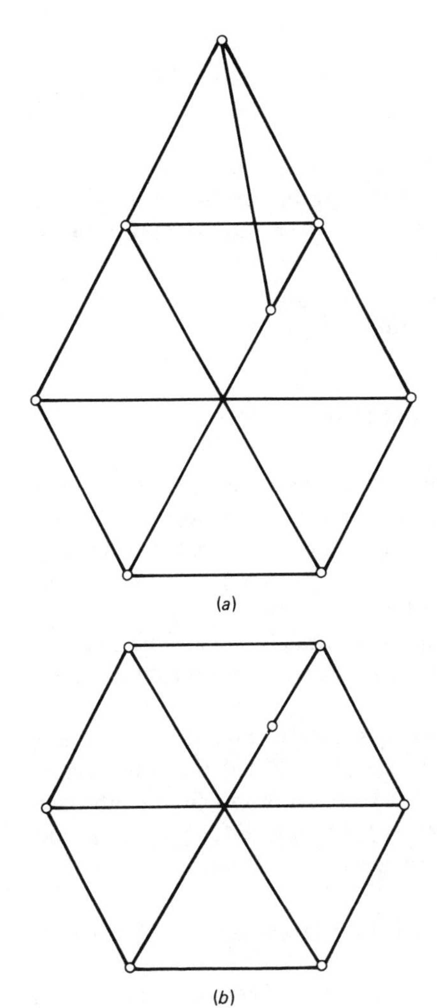

Figure 7.6. (a) Graph G. (b) Subgraph of G homeomorphic to $K_{3,3}$.

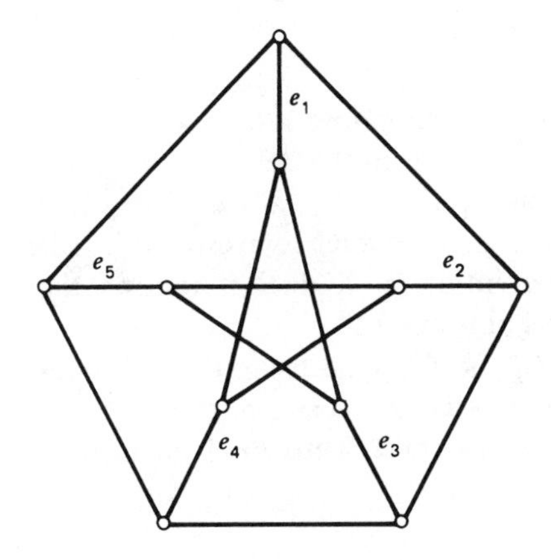

Figure 7.7. The Petersen graph.

THEOREM 7.6. A graph G is planar if and only if there exists in G a set of basis circuits such that no edge appears in more than two of these circuits. ■

We know that the meshes of a planar graph form a basis of the circuit subspace of the graph, and that no edge of the graph appears in more than two of the meshes. This proves the necessity of Theorem 7.6. The proof of the sufficiency may be found in MacLane [7.10].

Another important characterization of planar graphs in terms of the existence of dual graphs is discussed in Section 7.5.

7.4 DUAL GRAPHS

A graph G_2 is a *dual* of a graph G_1 if there is a one-to-one correspondence between the edges of G_2 and those of G_1 such that a set of edges in G_2 is a circuit vector of G_2 if and only if the corresponding set of edges in G_1 is a cutset vector of G_1. Duals were first defined by Whitney [7.11], though his original definition was given in a different form.

Clearly, to prove that G_2 is a dual of G_1 it is enough if we show that the vectors forming a basis of the circuit subspace of G_2 correspond to the vectors forming a basis of the cutset subspace of G_1.

For example, consider the graphs G_1 and G_2 shown in Fig. 7.8. The edge e_i of G_1 corresponds to the edge e_i^* of G_2. It may be verified that the circuits $\{e_1, e_2, e_3\}$, $\{e_3, e_4, e_5, e_6\}$, and $\{e_6, e_7, e_8\}$ form a basis of the circuit subspace of G_1, and the corresponding sets of edges $\{e_1^*, e_2^*, e_3^*\}$, $\{e_3^*, e_4^*, e_5^*, e_6^*\}$, and $\{e_6^*, e_7^*, e_8^*\}$ form a basis of the cutset subspace of G_2. Thus G_2 is a dual of G_1.

We now study some properties of the duals of a graph.

THEOREM 7.7. Let G_2 be a dual of a graph G_1. Then a circuit in G_2 corresponds to a cutset in G_1 and vice versa.

Proof

Let C^* be a circuit in G_2 and C be the corresponding set of edges in G_1. Suppose that C is not a cutset. Then it follows from the definition of a dual that C must be the union of disjoint cutsets $C_1, C_2, \ldots, C_k$, $k \geq 2$.

Let $C_1^*, C_2^*, \ldots, C_k^*$ be the sets of edges in G_2 which correspond to the cutsets $C_1, C_2, \ldots, C_k$. Again from the definition of a dual it follows that $C_1^*, C_2^*, \ldots, C_k^*$ are circuits or unions of disjoint circuits.

Since C^* is the union of $C_1^*, C_2^*, \ldots, C_k^*$, $k \geq 2$, it is clear that C^* must contain more than one circuit. However, this is not possible, since C^* is a circuit, and no proper subset of a circuit is a circuit. Thus $k = 1$, or in other words, C is a cutset of G_1.

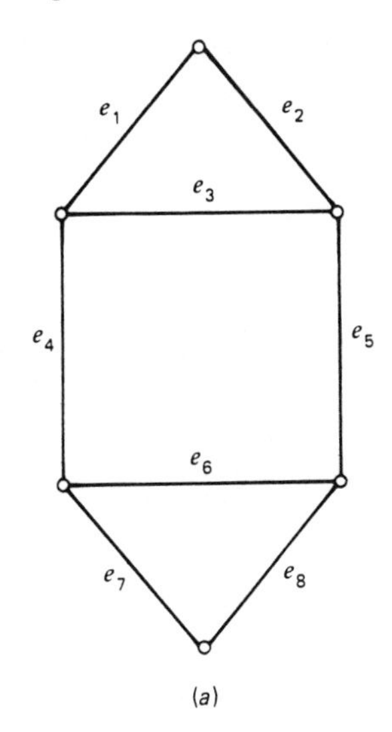

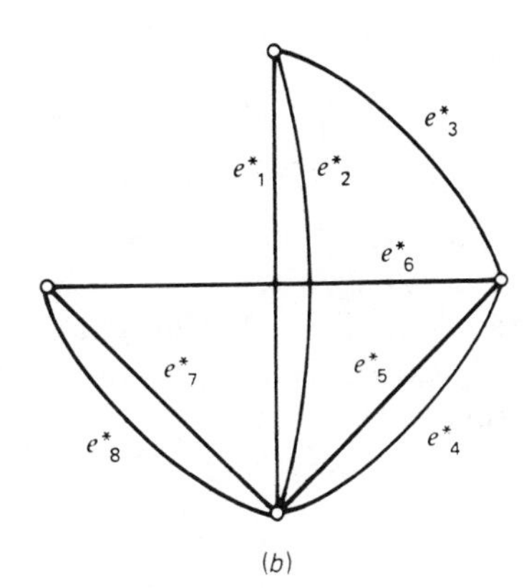

Figure 7.8. Dual graphs. (*a*) Graph G_1. (*b*) Graph G_2.

In a similar way we can show that each cutset in G_1 corresponds to a circuit in G_2. ■

THEOREM 7.8. If G_2 is a dual of G_1, then G_1 is a dual of G_2.

Proof

To prove the theorem we need to show that each circuit vector of G_1 corresponds to a cutset vector of G_2 and vice versa.

Let C be a circuit vector in G_1, with C^* denoting the corresponding set of edges in G_2.

By Theorem 4.7, C has an even number of common edges with every cutset vector of G_1. Since G_2 is a dual of G_1, C^* has an even number of common edges with every circuit vector of G_2. Therefore, by Theorem 4.8, C^* is a cutset vector of G_2.

In a similar way, we can show that each cutset vector of G_2 corresponds to a circuit vector of G_1. Hence the theorem. ■

In view of the above theorem, we refer to graphs G_1 and G_2 as simply duals if any one of them is dual of the other.

The following result is a consequence of Theorem 7.8 and the definition of a dual.

THEOREM 7.9. If G_1 and G_2 are dual graphs, then the rank of one is equal to the nullity of the other; that is,

$$\rho(G_1)=\mu(G_2)$$

and

$$\rho(G_2)=\mu(G_1). \quad ■$$

Suppose a graph G has a dual. Then the question arises whether every subgraph of G has a dual. To answer this question we need the following result.

THEOREM 7.10. Consider two dual graphs G_1 and G_2. Let $e=(v_1, v_2)$ be an edge in G_1, and $e^*=(v_1^*, v_2^*)$ be the corresponding edge in G_2. Let G_1' be the graph obtained by removing the edge e from G_1; let G_2' be the graph obtained by contracting e^* in G_2. Then G_1' and G_2' are duals, the one-to-one correspondence between their edges being the same as in G_1 and G_2.

Proof

Let C and C^* denote corresponding sets of edges in G_1 and G_2, respectively.

Suppose C is a circuit in G_1'. Since it does not contain e, it is also a circuit in G_1. Hence C^* is a cutset, say, $\langle V_a^*, V_b^* \rangle$, in G_2. Since C^* does not contain e^*, the vertices v_1^* and v_2^* are both in V_a^* or in V_b^*. Therefore C^* is also a cutset in G_2'. Thus every circuit in G_1' corresponds to a cutset in G_2'.

Suppose C^* is a cutset in G_2'. Since C^* does not contain e^*, it is also a cutset in G_2. Hence C is a circuit in G_1. Since it does not contain e, it is also a circuit in G_1'. Thus every cutset in G_2' corresponds to a circuit in G_1'. ■

In view of this theorem, we may say, using the language of electrical network theory, that "open-circuiting" an edge in a graph G corresponds to "short-circuiting" the corresponding edge in a dual of G.

A useful corollary of Theorem 7.10 now follows.

Corollary 7.10.1 If a graph G has a dual, then every edge-induced subgraph of G also has a dual.

Proof

The result follows from Theorem 7.10, if we note that every edge-induced subgraph H of G can be obtained by removing from G the edges not in H. ■

To illustrate the above corollary, consider the two dual graphs G_1 and G_2 of Fig. 7.8. The graph G_1' shown in Fig. 7.9*a* is obtained by removing from G_1 the edges e_3 and e_6. The graph G_2' of Fig. 7.9*b* is obtained by contracting the edges e_3^* and e_6^* of G_2. It may be verified that G_1' and G_2' are duals.

Observing that series edges in a graph G correspond to parallel edges in a dual of G, we get the following corollary of Theorem 7.10.

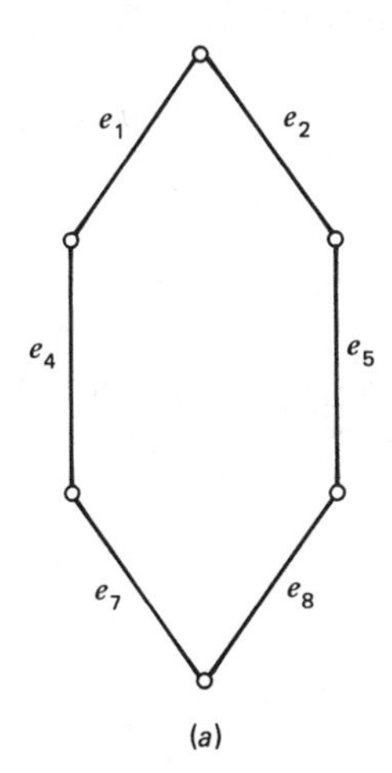

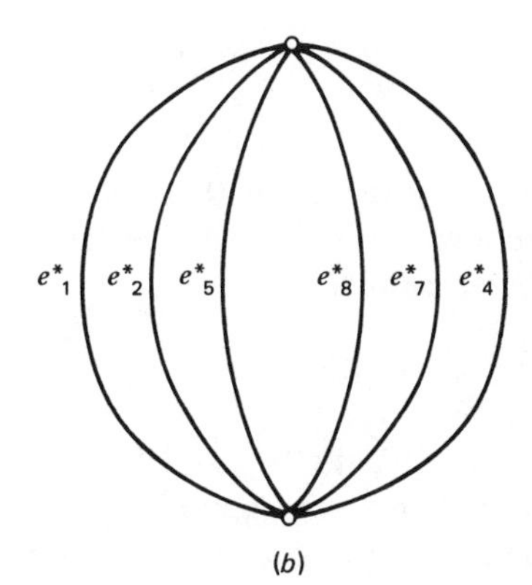

Figure 7.9. Dual graphs. (*a*) Graph G_1'. (*b*) Graph G_2'.

Corollary 7.10.2 If a graph G has a dual, then every graph homeomorphic to G also has a dual. ■

We now proceed to develop an equivalent characterization of a dual.

Let G be an n-vertex graph. We may assume without loss of generality that it is connected. Let K^* be a subgraph of G, and let G' be the graph obtained by contracting the edges of K^*. Note that G' is also connected.

If K^* has n^* vertices and p connected components, then G' will have $n-(n^*-p)$ vertices. Therefore, the rank of G' is given by

$$\begin{aligned}\rho(G') &= n-(n^*-p)-1\\ &= \rho(G)-\rho(K^*).\end{aligned} \tag{7.1}$$

THEOREM 7.11. Let G_1 and G_2 be two graphs with a one-to-one correspondence between their edges. Let H be any subgraph of G_1 and H^* the corresponding subgraph in G_2. Let K^* be the complement of H^* in G_2. Then G_1 and G_2 are dual graphs if and only if

$$\mu(H)=\rho(G_2)-\rho(K^*). \tag{7.2}$$

Proof

Necessity Let G_1 and G_2 be dual graphs. Let G_2' be the graph obtained from G_2 by contracting the edges of K^*. Then by Theorem 7.10, H and G_2' are dual graphs. Therefore, by Theorem 7.9,

$$\mu(H)=\rho(G_2').$$

But by (7.1),

$$\rho(G_2')=\rho(G_2)-\rho(K^*).$$

Hence

$$\mu(H)=\rho(G_2)-\rho(K^*).$$

Sufficiency Assume that (7.2) is satisfied for every subgraph H of G_1. We now show that each circuit in G_1 corresponds to a cutset in G_2 and vice versa.

Let H be a circuit in G_1. Then $\mu(H)=1$. Therefore by (7.2),

$$\rho(K^*)=\rho(G_2)-1.$$

Since H is a minimal subgraph of G with nullity equal to 1, and K^* is the complement of H^* in G_2, it is clear that K^* is a maximal subgraph of G_2 with rank equal to $\rho(G_2)-1$. It now follows from the definition of a cutset that H^* is a cutset in G_2.

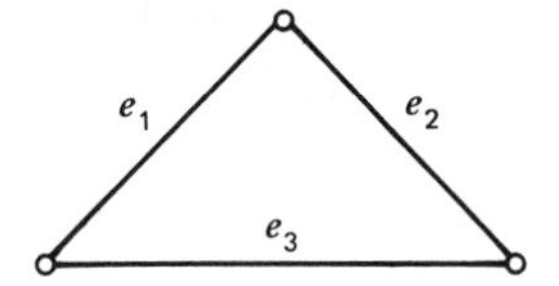

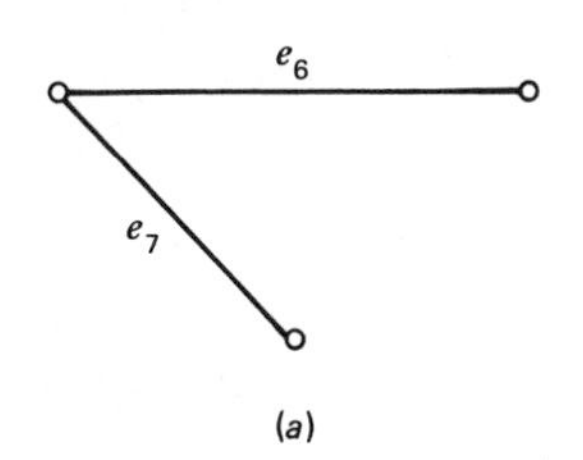

(a)

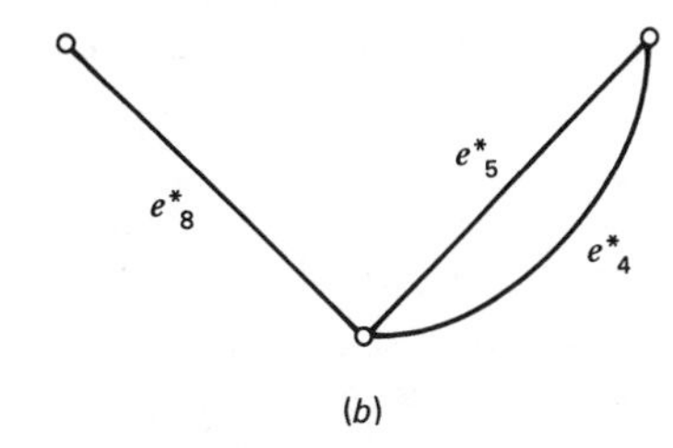

(b)

Figure 7.10. Illustration of Whitney's definition of duality. (*a*) Graph H, $\mu(H)=1$. (*b*) Graph K^*, $\rho(K^*)=2$.

In a similar way, we can show that a cutset in G_2 corresponds to a circuit in G_1.

Thus G_1 and G_2 are dual graphs. ■

Whitney's [7.11] original definition of duality was stated as in Theorem 7.11.

To illustrate this definition, consider the dual graphs G_1 and G_2 of Fig. 7.8. A subgraph H of G_1 and the complement K^* of the corresponding subgraph in G_2 are shown in Fig. 7.10. We may now verify that

$$\mu(H)=\rho(G_2)-\rho(K^*).$$

7.5 PLANARITY AND DUALITY

In this section we characterize the class of graphs which have duals. While doing so, we relate the two seemingly unrelated concepts, planarity and duality.

First we prove that every planar graph has a dual. The proof is based on a procedure for constructing a dual of a given planar graph.

Consider a planar graph and let G be a planar embedding of this graph. Let $f_1, f_2, \ldots, f_r$ be the regions of G. Construct a graph G^* defined as follows:

1. G^* has r vertices $v_1, v_2, \ldots, v_r$, vertex v_i, $1 \leqslant i \leqslant r$, corresponding to region f_i.
2. G^* has as many edges as G has.
3. If an edge e of G is common to the regions f_i and f_j (not necessarily distinct), then the corresponding edge e^* in G^* connects vertices v_i^* and v_j^*. (Note that each edge e of G is common to at most two regions, and it is possible that an edge may be in exactly one region.)

A simple way to construct G^* is to first place the vertices $v_1^*, v_2^*, \ldots, v_r^*$, one in each region of G. Then, for each edge e common to regions f_i and f_j, draw a line connecting v_i^* and v_j^* so that it crosses the edge e. This line represents the edge e^*.

The procedure for constructing G^* is illustrated in Fig. 7.11. The continuous lines represent the edges of the given planar graph G and the dashed lines represent those of G^*.

We now prove that G^* is a dual of G.

Let $C_1, C_2, \ldots, C_{r-1}$ denote the meshes of G, and $C_1^*, C_2^*, \ldots, C_{r-1}^*$ denote the corresponding sets of edges in G^*. It is clear from the procedure used to construct G^* that the edges in C_i^* are incident on the vertex v_i^* and form a cut whose removal will separate v_i^* from the remaining vertices of G^*.

By Theorem 7.3, $C_1, C_2, \ldots, C_{r-1}$ form a basis of the circuit subspace of G, and we know that the incidence vectors $C_1^*, C_2^*, \ldots, C_{r-1}^*$ form a basis of the cutset subspace of G^*. Since there is a one-to-one correspondence between

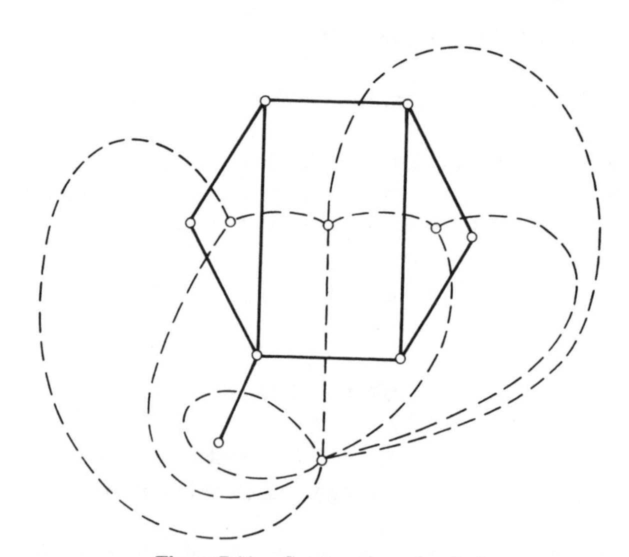

Figure 7.11. Construction of a dual.

C_i's and C_i^*'s, G and G^* are dual graphs. Thus we have the following theorem.

THEOREM 7.12. Every planar graph has a dual. ■

The question that immediately arises now is whether a nonplanar graph has a dual. The answer is "no," and it is based on the next two lemmas.

LEMMA 7.1. $K_{3,3}$ has no dual.

Proof

First observe that

1. $K_{3,3}$ has no cutsets of two edges.
2. $K_{3,3}$ has circuits of length four or six only.
3. $K_{3,3}$ has nine edges.

Suppose $K_{3,3}$ has a dual G. Then these observations would, respectively, imply the following for G:

1. G has no circuits of two edges; that is, G has no parallel edges.
2. G has no cutsets with less than four edges. Thus every vertex in G is of degree at least 4.
3. G has nine edges.

The first two of the above imply that G has at least five vertices, each of degree at least 4. Thus G must have at least $\frac{1}{2}\times 5\times 4=10$ edges. However, this contradicts observation 3. Hence $K_{3,3}$ has no dual. ■

LEMMA 7.2. K_5 has no dual.

Proof

First observe that

1. K_5 has no circuits of length one or two.
2. K_5 has cutsets with four or six edges only.
3. K_5 has ten edges.

Suppose K_5 has a dual G. Then by observation 2, G has circuits of lengths four and six only. In other words, all circuits of G are of even length. So G is bipartite.

Since a bipartite graph with six or fewer vertices cannot have more than nine edges, it is necessary that G has seven vertices. But by observation 1 the degree of every vertex of G is at least 3. Hence G must have at least $\frac{1}{2}\times 7\times 3>10$ edges. This, however, contradicts observation 3. Hence K_5 has no dual. ■

The main result of this section now follows.

THEOREM 7.13. A graph has a dual if and only if it is planar.

Proof

The sufficiency part of the theorem is the same as Theorem 7.12.

We can prove the necessity by showing that a nonplanar graph G has no dual. By Kuratowski's theorem, G has a subgraph H homeomorphic to $K_{3,3}$ or K_5. If G has a dual, then by Corollary 7.10.1, H has a dual. But then, by Corollary 7.10.2, $K_{3,3}$ or K_5 should have a dual. This, however, will contradict the fact that neither of these graphs has a dual. Hence G has no dual. ■

The above theorem gives a characterization of planar graphs in terms of the existence of dual graphs and was originally proved by Whitney. The proof given here is due to Parsons [7.12]. Whitney's original proof, which does not make use of Kuratowski's theorem, may be found in [7.13].

From the procedure given earlier in this section it is clear that different (though isomorphic) planar embeddings of a planar graph may lead to nonisomorphic duals (Exercise 7.6). The following theorem presents a property of the duals of a graph.

THEOREM 7.14. All duals of a graph G are 2-isomorphic; every graph 2-isomorphic to a dual of G is also a dual of G. ■

The proof of this theorem follows in a straightforward manner from the definition of a dual and Theorem 1.7.

7.6 FURTHER READING

Whitney [7.4], [7.11], [7.14] and books by Seshu and Reed [7.13], Ore [7.15], Harary [7.6], and Bondy and Murty [7.16] are highly recommended for further reading. For an algorithm for testing the planarity of a graph see Hopcroft and Tarjan [7.17].

Two properties of a nonplanar graph G which are of interest are:

1. The minimum number of planar subgraphs whose union is G; this is called the *thickness* of G.
2. The minimum number of crossings (or intersections) in order to draw a graph on a plane; this is called the *crossing number* of G.

For several results on the thickness and the crossing numbers of a nonplanar graph see Harary [7.6]. See also Bose and Prabhu [7.18].

7.7 EXERCISES

7.1 Show that if G is a connected planar graph with m edges, n vertices, and with girth* $k \geqslant 3$, then $m \leqslant k(n-2)/(k-2)$. Using this result deduce that the Petersen graph is nonplanar.

7.2 Prove that a simple planar graph with $n \geqslant 4$ vertices has at least four vertices with degree 5 or less.

7.3 Prove or disprove: Every connected simple nonplanar graph is contractible to K_5 or $K_{3,3}$.

7.4 Let G be a simple graph with at least eleven vertices. Show that G and its complement $\overline{G}$ cannot both be planar. (In fact, a similar result can be proved with eleven replaced by nine. See Tutte [7.19].) Give an example of a graph G on eight vertices with the property that both G and $\overline{G}$ are planar.

7.5 Using Kuratowski's theorem prove that the Petersen graph is nonplanar.

7.6 Find two nonisomorphic duals of the graph in Fig. 7.12.

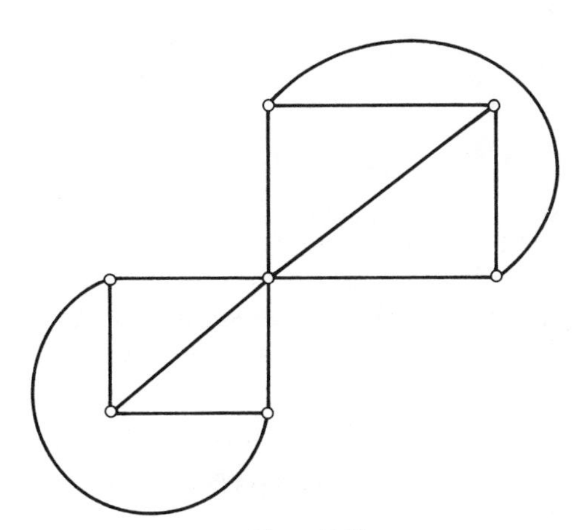

Figure 7.12.

7.7 Prove that a planar graph without self-loops is nonseparable if and only if its dual is nonseparable.

*For the definition of girth, see Exercise 1.18.

7.8 Show that the dual of a nonseparable planar graph is Eulerian if and only if the graph is bipartite.

7.9 A planar graph is *self-dual* if it is isomorphic to its dual. Show that if a graph G with n vertices and m edges is self-dual, then $m=2n-2$.

7.10 A *one-terminal-pair* graph is a graph with two vertices specially designated as the terminals of the graph. A *planar one-terminal-pair graph* is a one-terminal-pair graph that is planar and remains planar when an edge joining the two terminals is added to it.
A *series–parallel graph* is a one-terminal-pair graph defined recursively as follows:

(a) A single edge is a series-parallel graph. If G' and G'' are series-parallel, then:

(b) The series combination of G' and G'' is a series-parallel graph. By the *series combination* of G' and G'' we mean the joining of one of the terminals of G' with one of the terminals of G''. (Fig. 7.13*a*).

(c) The parallel combination of G' and G'' is a series-parallel graph. By the *parallel combination* of G' and G'' we mean the joining of the two terminals of G' with the two terminals of G''. (Fig. 7.13*b*).

Show that a dual of a graph G is a series-parallel graph if and only if G is series–parallel.

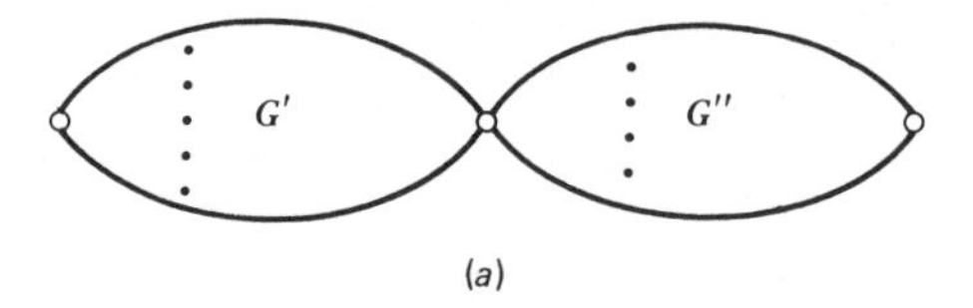

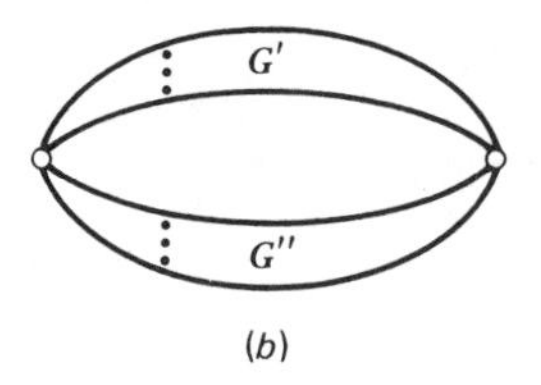

Figure 7.13. (*a*) Series combination of G' and G''. (*b*) Parallel combination of G' and G''.

7.8 REFERENCES

7.1 K. Wagner, "Bemerkungen zum Vierfarbenproblem," *Über. Deutsch. Math. Verein*, Vol. 46, 26–32 (1936).

7.2 I. Fary, "On Straight Line Representation of Planar Graphs," *Acta Sci. Math. Szeged*, Vol. 11, 229–233 (1948).

7.3 S. K. Stein, "Convex Maps," *Proc. Am. Math. Soc.*, Vol. 2, 464–466 (1951).

7.4 H. Whitney, "Non-Separable and Planar Graphs," *Trans. Am. Math. Soc.*, Vol. 34, 339–362 (1932).

7.5 C. Kuratowski, "Sur le problème des courbes gauches en topologie," *Fund. Math.*, Vol. 15, 271–283 (1930).

7.6 F. Harary, *Graph Theory*, Addison-Wesley, Reading, Mass., 1969.

7.7 W. T. Tutte, "How to Draw a Graph," *Proc. London Math. Soc.*, Vol. 13, 743–767 (1963).

7.8 K. Wagner, "Über eine Eigenschaft der ebenen Komplexe," *Math. Ann.*, Vol. 114, 570–590 (1937).

7.9 F. Harary and W. T. Tutte, "A Dual Form of Kuratowski's Theorem," *Canad. Math. Bull.*, Vol. 8, 17–20 (1965).

7.10 S. MacLane, "A Structural Characterization of Planar Combinatorial Graphs," *Duke Math. J.*, Vol. 3, 340–372 (1937).

7.11 H. Whitney, "Planar Graphs," *Fund. Math.*, Vol. 21, 73–84 (1933).

7.12 T. D. Parsons, "On Planar Graphs," *Am. Math. Monthly*, Vol. 78, 176–178 (1971).

7.13 S. Seshu and M. B. Reed, *Linear Graphs and Electrical Networks*, Addison-Wesley, Reading, Mass., 1961.

7.14 H. Whitney, "A Set of Topological Invariants for Graphs," *Am. J. Math.*, Vol. 55, 231–235 (1933).

7.15 O. Ore, *The Four Colour Problem*, Academic Press, New York, 1967.

7.16 J. A. Bondy and U. S. R. Murty, *Graph Theory with Applications*, Macmillan, London, 1976.

7.17 J. E. Hopcroft and R. E. Tarjan, "Efficient Planarity Testing," *J. ACM*, Vol. 21, 549–568 (1974).

7.18 N. K. Bose and K. A. Prabhu, "Thickness of Graphs with Degree Constrained Vertices," *IEEE Trans. Circuits and Sys.*, Vol. CAS-24, 184–190 (1975).

7.19 W. T. Tutte, "On the Nonbiplanar Character of the Complete 9-graph," *Canad. Math. Bull.*, Vol. 6, 319–330 (1963).

Chapter 8

Connectivity and Matching

In Chapter 1 we defined a graph to be connected if there exists a path between any two vertices of the graph. Suppose a graph G is connected. Then we may be interested in finding how "well-connected" G is. In other words, we may like to know the minimum number of vertices or edges whose removal will disconnect G. This leads us to the concepts of vertex and edge connectivities of a graph. In this chapter we first develop several results relating to vertex and edge connectivities of a graph. We also discuss a classical result in graph theory, namely, Menger's theorem, which relates connectivities to the number of vertex-disjoint and edge-disjoint paths.

A matching in a graph is a set of edges no two of which have common vertices. In the latter part of this chapter we develop the theory of matchings, starting our development with Hall's marriage theorem.

Connectivity and matching are two extensively studied topics in graph theory. Many deep results in graph theory belong to these two areas.

8.1 CONNECTIVITY OR VERTEX CONNECTIVITY

The *connectivity* $\kappa(G)$ of a graph G is the minimum number of vertices whose removal from G results in a disconnected graph or a trivial graph.* $\kappa(G)$ is also called *vertex connectivity* to distinguish it from edge connectivity which is introduced in the next section.

*Recall (Section 1.7) that a trivial graph is a graph with just a single vertex.

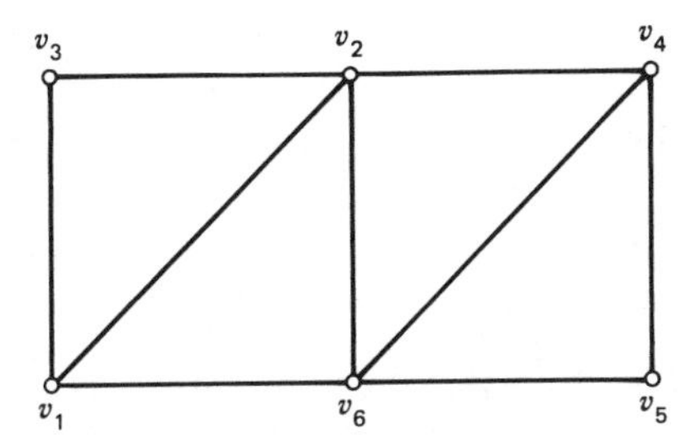

Figure 8.1.

For example, the connectivity of the graph of Fig. 8.1 is 2 since the removal of vertices v_1 and v_2 from this graph results in a disconnected graph, whereas the removal of no single vertex will disconnect the graph.

Obviously, the connectivity of a disconnected graph is 0.

Consider a graph on n vertices. It is clear that $\kappa(G)=n-1$ if G is complete. If G is not complete, it will have at least two nonadjacent vertices v_1 and v_2. The removal of the remaining $n-2$ vertices from G will then result in a graph in which v_1 and v_2 are not connected. Thus if G is not complete, we have

$$\kappa(G) \leqslant n-2.$$

A *disconnecting set* in a graph G is a set of vertices whose removal from G results in a disconnected graph or a trivial graph.

A graph G is said to be *k-connected* if $\kappa(G) \geqslant k$. So, a k-connected graph has no disconnecting set S with $|S| \leqslant k-1$.

If a graph is connected, its connectivity is greater than or equal to 1. So connected graphs are 1-connected.

If a connected graph has no cut-vertices, then its connectivity is greater than 1. So, connected graphs with no cut-vertices are 2-connected.

In the following theorem we present a simple upper bound on the connectivity of a graph.

THEOREM 8.1. For a simple connected graph G, $\kappa(G) \leqslant \delta(G)$, where $\delta(G)$ is the minimum degree in G.

Proof

Consider any vertex v in a simple connected graph $G=(V,E)$. Let $\Gamma(v)$ denote the set of vertices adjacent to v. It is clear that $\Gamma(v)$ is a disconnecting set, since the removal of the vertices in $\Gamma(v)$ results in a trivial graph or a disconnected graph in which v is not connected to any of the remaining vertices. Thus

$$\kappa(G) \leqslant |\Gamma(v)|, \qquad \text{for all } v \in V.$$

Since G is simple,

$$|\Gamma(v)| = d(v)$$

and hence

$$\kappa(G) \leqslant \min_{v \in V} \{d(v)\}$$
$$= \delta(G). \quad \blacksquare$$

If a graph G has m edges and n vertices $v_1, v_2, \ldots, v_n$, then we know from Theorem 1.1,

$$d(v_1) + d(v_2) + \cdots + d(v_n) = 2m.$$

So,

$$n \cdot \delta(G) \leqslant 2m$$

and

$$\delta(G) \leqslant \left\lfloor \frac{2m}{n} \right\rfloor, \tag{8.1}$$

where $\lfloor x \rfloor$ denotes the largest integer less than or equal to x.

Combining (8.1) with the result of Theorem 8.1, we get the following.

THEOREM 8.2. For a simple connected graph G having m edges and n vertices,

$$\kappa(G) \leqslant \left\lfloor \frac{2m}{n} \right\rfloor. \quad \blacksquare$$

Let $f(k, n)$ denote the least number of edges that a k-connected graph on n vertices must have. Of course, we assume that $k < n$. From Theorem 8.2 it is clear that

$$f(k, n) \geqslant \left\lceil \frac{kn}{2} \right\rceil, \tag{8.2}$$

where $\lceil x \rceil$ denotes the smallest integer greater than or equal to x.

Harary [8.1] has proved that equality holds in (8.2) by giving a procedure for constructing a k-connected graph $H_{k,n}$ which has exactly $\lceil kn/2 \rceil$ edges. This procedure is as follows.

Case 1 k even.

Let $k = 2r$. Then $H_{2r,n}$ has vertices $v_0, v_1, v_2, \ldots, v_{n-1}$ and two vertices v_i and v_j are adjacent if $i - r \leqslant j \leqslant i + r$, where addition is modulo n.

$H_{6,8}$ is shown in Fig. 8.2a.

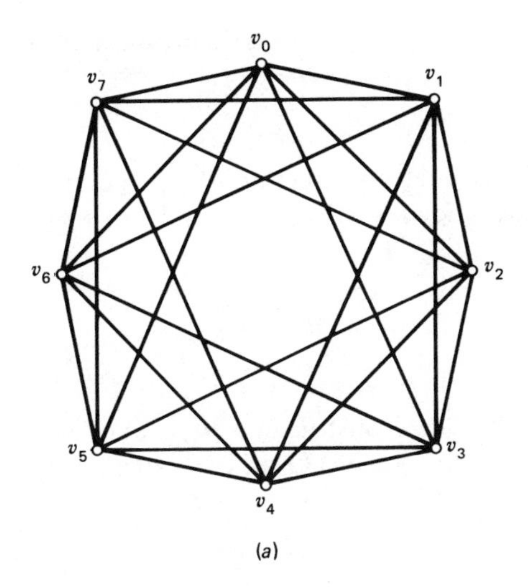

(a)

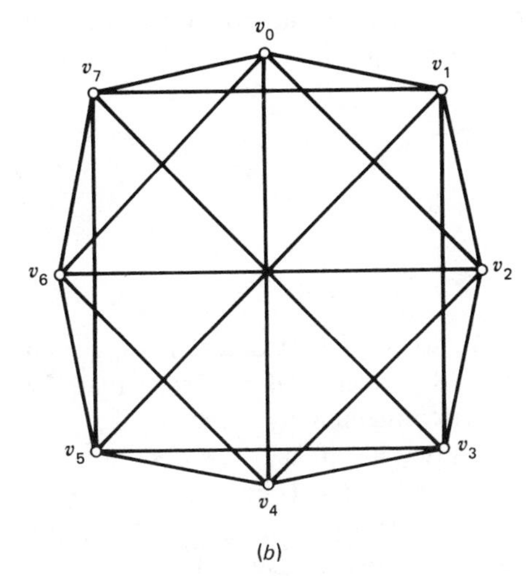

(b)

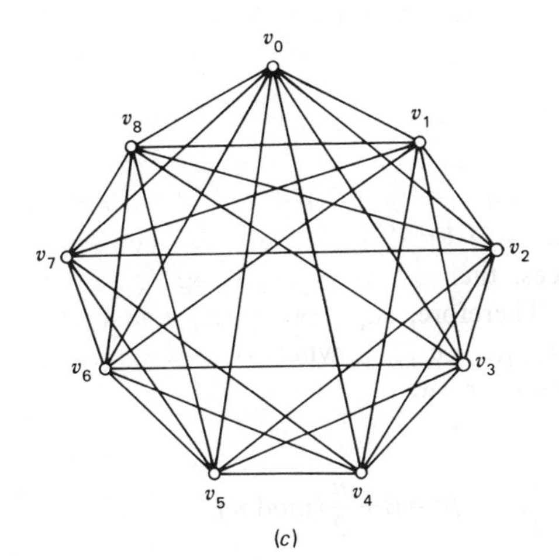

(c)

Figure 8.2. (a) $H_{6,8}$. (b) $H_{5,8}$. (c) $H_{7,9}$.

Case 2 k odd, n even.

Let $k=2r+1$. $H_{2r+1,n}$ is constructed by first constructing $H_{2r,n}$ and then adding edges joining vertex v_i to vertex v_j for $i=1,2,\ldots,n/2$ and $j=i+n/2 \pmod n$.

$H_{5,8}$ is shown in Fig. 8.2*b*.

Case 3 k odd, n odd.

Let $k=2r+1$. $H_{2r+1,n}$ is constructed by first constructing $H_{2r,n}$ and then adding edges joining v_i and v_j for $i=0,1,2,\ldots,(n-1)/2$ and $j=i+(n+1)/2 \pmod n$.

$H_{7,9}$ is shown in Fig. 8.2*c*.

It is easy to verify that the graph $H_{k,n}$ constructed as above has exactly $\lceil kn/2 \rceil$ edges. We now prove that $H_{k,n}$ is k-connected.

THEOREM 8.3. Graph $H_{k,n}$ is k-connected.

Proof

Case 1 $k=2r$.

From the symmetry of $H_{2r,n}$ it is sufficient to show that the vertices v_0 and v_α, $\alpha=1,2,\ldots,n-1$ cannot be disconnected by removal of fewer than $2r$ vertices. Suppose that v_0 and v_α can be disconnected by removal of $2r-1$ vertices $v_{i_1}, v_{i_2}, \ldots, v_{i_{2r-1}}$. One of the two intervals $[0,\alpha],[\alpha,n]$ contains at most $r-1$ of these indices. Suppose it is interval $[0,\alpha]$. Then two consecutive vertices of the sequence obtained by removing $v_{i_1}, v_{i_2}, \ldots, v_{i_{2r-1}}$ from the sequence $v_0, v_1, v_2, \ldots, v_\alpha$ are connected by an edge (because the difference between their indices is $\leqslant r$). Hence there is a path from v_0 to v_α, a contradiction.

Case 2 $k=2r+1$, n even.

Suppose the vertices v_0 and v_α are disconnected by the removal of $2r$ vertices $v_{i_1}, v_{i_2}, \ldots, v_{i_{2r}}$. If one of the two intervals $[0,\alpha],[\alpha,n]$ does not contain a consecutive r number of these indices, then a path from v_0 to v_α can be constructed as shown under case 1. Therefore, suppose that v_0 and v_α are disconnected by the removal of $v_i, v_{i+1}, \ldots, v_{i+r-1}$, where $1 \leqslant i \leqslant \alpha-r$, and $v_{\alpha+j}, v_{\alpha+j+1}, \ldots, v_{\alpha+j+r-1}$, where $1 \leqslant j \leqslant n-r-\alpha$.
Let

$$\beta=\left\lfloor \frac{\alpha+i+j+r+n-1}{2} \right\rfloor, \qquad \beta'=\beta+\frac{n}{2} \pmod n.$$

Then

$$\beta \in [\alpha+j+r, n+i-1], \qquad \beta' \in [i+r, \alpha+j-1].$$

There is a path from vertex v_0 to vertex v_β and from vertex $v_{\beta'}$ to vertex v_α. Hence there is a path from vertex v_0 to vertex v_α, since in $H_{2r+1,n}$ there is an edge connecting v_β and $v_{\beta'}$.

Case 3 $k=2r+1$, n odd.

Proof follows as in case 2. ■

In the next theorem we present a sufficient condition for a graph to be k-connected. This result is due to Bondy [8.2].

THEOREM 8.4. Let G be a simple graph of order n. Suppose we order the vertices of G so that

$$d(v_1) \leqslant d(v_2) \leqslant \cdots \leqslant d(v_n).$$

Then G is k-connected if

$$d(v_r) \geqslant r+k-1, \qquad \text{for } 1 \leqslant r \leqslant n-1-d(v_{n-k+1})$$

Proof

Suppose a simple graph G satisfies the conditions of the theorem. If it is not k-connected, then there exists a disconnecting set S such that $|S|=s<k$.

Consider now the graph $G-S$ which is not connected. Let H be a component of $G-S$ of minimum order h. Then it is clear that the degree in H of each vertex of H is at most $h-1$. Therefore, in G, the degree of each vertex of H is at most $h+s-1$. Thus

$$d(v_h) \leqslant h+s-1 < h+k-1. \tag{8.3}$$

Therefore, by the conditions of the theorem,

$$h > n-1-d(v_{n-k+1}). \tag{8.4}$$

Since $G-S$ has $n-s$ vertices and H is a component of $G-S$ of minimum order, we have

$$h \leqslant n-s-h$$

or

$$h+s \leqslant n-h.$$

Therefore,

$$d(v) \leqslant h+s-1 \leqslant n-h-1, \qquad v \in V(H), \tag{8.5}$$

where $V(H)$ is the vertex set of H. Since every vertex $u \in V(G)-V(H)-S$ is adjacent to at most $n-h-1$ vertices, we have

$$d(u) \leqslant n-h-1, \qquad u \in V(G)-V(H)-S. \tag{8.6}$$

From (8.5) and (8.6) we conclude that all the vertices of degrees exceeding $n-h-1$ are in S. Thus there are at most s vertices of degrees exceeding $n-h-1$. Therefore

$$d(v_{n-s}) \leqslant n-h-1. \tag{8.7}$$

Using (8.4) in (8.7), we get

$$d(v_{n-s}) < d(v_{n-k+1}).$$

Therefore

$$n-s < n-k+1$$

or

$$s \geqslant k,$$

a contradiction. ■

For example, the degrees of the graph in Fig. 8.3 satisfy the conditions of Theorem 8.4 for $k=3$. Hence it is 3-connected.

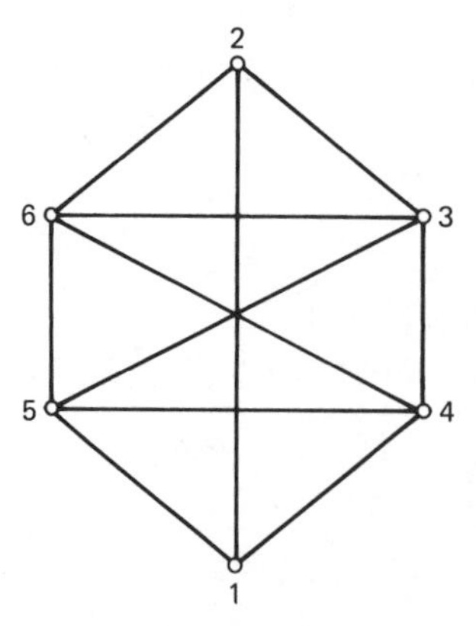

Figure 8.3. A 3-connected graph which satisfies the conditions of Theorem 8.4 for $k=3$.

8.2 EDGE CONNECTIVITY

Edge connectivity $\kappa'(G)$ of a graph G is the minimum number of edges whose removal from G results in a disconnected graph or a trivial graph. In other words, $\kappa'(G)$ is the number of edges in a cut having the minimum number of edges. For example, for the graph of Fig. 8.4 the edge connectivity is equal to 2, since the removal of the two edges e_1 and e_2 disconnects the graph, and removal of no single edge results in a disconnected graph.

A graph G is said to be *k-edge-connected* if $\kappa'(G) \geqslant k$. Thus at least k edges have to be removed to disconnect a k-edge-connected graph.

Since the edges incident on any vertex v of G form a cut of G, it follows that

$$\kappa'(G) \leqslant \delta(G).$$

In the following theorem we relate $\kappa(G)$, $\kappa'(G)$, and $\delta(G)$.

THEOREM 8.5. For a simple graph G,

$$\kappa(G) \leqslant \kappa'(G) \leqslant \delta(G).$$

Proof

We have already proved the second inequality. The first inequality can be proved as follows.

If G is not connected, then $\kappa(G) = \kappa'(G) = 0$. Thus in such a case the condition $\kappa(G) \leqslant \kappa'(G)$ is satisfied.

If G is connected and $\kappa'(G) = 1$, then G has a bridge e. In this case, if we remove one of the vertices on which e is incident, then a disconnected graph or the trivial graph results. Thus $\kappa(G) \leqslant \kappa'(G)$ in this case too.

Suppose $\kappa'(G) \geqslant 2$. Then G has $\kappa'(G)$ edges whose removal disconnects it. Removal of any $\kappa'(G)-1$ of these edges results in a graph with a bridge $e = (v_1, v_2)$. For each of these $\kappa'(G)-1$ edges select an end vertex different from v_1 and v_2. Removal of these vertices will remove from G the $\kappa'(G)-1$

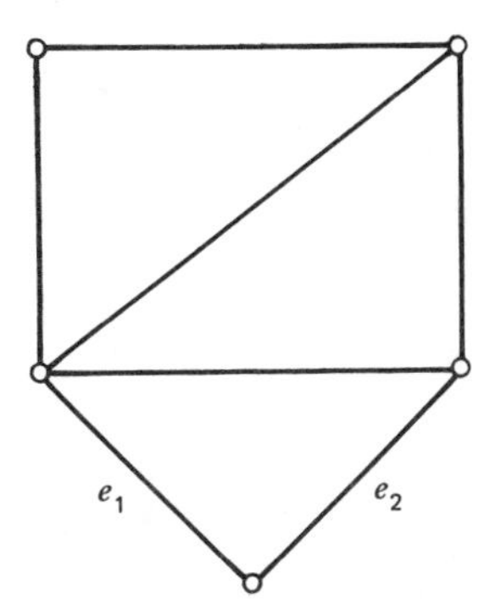

Figure 8.4. A 2-edge-connected graph.

edges and possibly some more. Suppose the resulting graph is disconnected. Then $\kappa(G) \leqslant \kappa'(G)-1$. Otherwise this graph will have e as a bridge, and so, removal of v_1 or v_2 will result in a disconnected graph or the trivial graph. In such a case, $\kappa(G) \leqslant \kappa'(G)$.

Thus in all cases,

$$\kappa(G) \leqslant \kappa'(G). \quad \blacksquare$$

Next we present a sufficient condition for $\kappa'(G)$ to be equal to $\delta(G)$. This result is due to Chartrand [8.3].

THEOREM 8.6. Let G be an n-vertex simple graph. If $\delta(G) \geqslant \lfloor n/2 \rfloor$ then $\kappa'(G) = \delta(G)$.

Proof

It can be shown that G is connected (see Exercise 1.14). Therefore $\kappa'(G) > 0$.

Since $\kappa'(G) \leqslant \delta(G)$, the proof will follow if we show that $\kappa'(G) \geqslant \delta(G)$.

Suppose $\kappa'(G) < \delta(G)$. Then there exists a cut $S = \langle V_1, \overline{V}_1 \rangle$ such that $\kappa'(G) = |S| < \delta(G)$. Let the edges of S be incident with q vertices in V_1 and p vertices in $\overline{V}_1$.

Suppose $|V_1| = q$. Then each vertex of V_1 is an end vertex of at least one edge in S. If we denote by G_1 the induced subgraph of G on the vertex set V_1, then G_1 has at least

$$m_1 = \tfrac{1}{2}(q\delta(G) - \kappa'(G))$$

edges. Since $\kappa'(G) < \delta(G)$, we get

$$\begin{aligned} m_1 &> \tfrac{1}{2}(q\delta(G) - \delta(G)) \\ &= \tfrac{1}{2}\delta(G)(q-1) \\ &> \tfrac{1}{2}q(q-1), \end{aligned}$$

since $\delta(G) > \kappa'(G) \geqslant q$. This is a contradiction since in a simple graph there cannot be more than $q(q-1)/2$ edges connecting q vertices. Therefore $|V_1| > q$. In a similar way we can prove that $|\overline{V}_1| > p$.

If $|V_1| > q$ and $|\overline{V}_1| > p$, then there are vertices in both V_1 and $\overline{V}_1$ which are adjacent to only vertices in V_1 and $\overline{V}_1$, respectively. Thus each of V_1 and $\overline{V}_1$ contains at least $\delta(G)+1$ vertices. Therefore G has at least $2\delta(G)+2$ vertices. But

$$\begin{aligned} 2\delta(G)+2 &\geqslant 2 \cdot \left\lfloor \frac{n}{2} \right\rfloor + 2 \\ &> n, \end{aligned}$$

leading to a contradiction. Therefore there is no cut S with $|S| < \delta(G)$. $\blacksquare$

8.3 GRAPHS WITH PRESCRIBED DEGREES

Recall that a sequence $(d_1, d_2, \ldots, d_n)$ of nonnegative integers is graphic if there exists an n-vertex graph with vertices $v_1, v_2, \ldots, v_n$ such that vertex v_i has degree d_i.

In this section we first describe an algorithm to construct a simple graph, if one exists, having a prescribed degree sequence. We then use this algorithm to establish Edmonds' theorem on the existence of k-edge-connected simple graphs having prescribed degree sequences.

Consider a graphic sequence $(d_1, d_2, \ldots, d_n)$ with $d_1 \geqslant d_2 \geqslant \cdots \geqslant d_n$. Let d_i be the degree of vertex v_i. "To lay off d_k" means to connect the corresponding vertex v_k to the vertices

$$v_1, v_2, \ldots, v_{d_k}, \qquad \text{if } d_k < k$$

or to the vertices

$$v_1, v_2, \ldots, v_{k-1}, v_{k+1}, \ldots, v_{d_k+1}, \qquad \text{if } d_k \geqslant k.$$

The sequence

$$\left(d_1 - 1, \ldots, d_{d_k} - 1, d_{d_k+1}, \ldots, d_{k-1}, 0, d_{k+1}, \ldots, d_n\right), \qquad \text{if } d_k < k$$

or

$$\left(d_1 - 1, \ldots, d_{k-1} - 1, 0, d_{k+1} - 1, \ldots, d_{d_k+1} - 1, d_{d_k+2}, \ldots, d_n\right), \qquad \text{if } d_k \geqslant k$$

is called the *residual sequence* after laying off d_k or simply the residual sequence.

Hakimi [8.4] and Havel [8.5] have given an algorithm for constructing a simple graph, if one exists, having a prescribed degree sequence. This algorithm is based on a result which is a special case (where $k=1$) of the following theorem due to Wang and Kleitman [8.6].

THEOREM 8.7. If a sequence $(d_1, d_2, \ldots, d_n)$ with $d_1 \geqslant d_2 \geqslant \cdots \geqslant d_n$ is the degree sequence of a simple graph, then so is the residual sequence after laying off d_k.

Proof

To prove the theorem we have to show that a graph having $(d_1, d_2, \ldots, d_n)$ as its degree sequence exists such that vertex v_k is adjacent to the first d_k vertices other than itself. If otherwise, select from among the graphs with the degree sequence $(d_1, d_2, \ldots, d_n)$ a simple graph G in which v_k is adjacent to the maximum number of vertices among the first d_k vertices other than itself. Let v_m be a vertex not adjacent to v_k in G such that

$$m \leqslant d_k, \qquad \text{if } d_k < k$$

or

$$m \leqslant d_k + 1, \qquad \text{if } k \leqslant d_k.$$

In other words, v_m is among the first d_k vertices other than v_k. So v_k is adjacent in G to some vertex v_q which is not among these first d_k vertices. Then $d_m > d_q$ (if equality, the order of q and m can be interchanged), and hence v_m is adjacent to some vertex v_t, $t \neq q$, $t \neq m$, such that v_t and v_q are not adjacent in G. If we now remove the edges (v_m, v_t) and (v_k, v_q) and replace them by (v_m, v_k) and (v_t, v_q), we obtain a graph G' with one more vertex adjacent to v_k among the first d_k vertices other than itself violating the definition of G. ■

From the above theorem we get the following algorithm which is a generalization of Hakimi's algorithm for realizing a sequence $D = (d_1, d_2, \ldots, d_n)$ with $d_1 \geqslant d_2 \geqslant \cdots \geqslant d_n$, by a simple graph.

Choose any $d_k \neq 0$. "Lay off" d_k by connecting v_k to the first d_k vertices other than itself. Compute the residual degree sequence. Reorder the vertices so that the residual degrees in the resulting sequence are in nonincreasing order. Repeat this process until one of the following occurs:

1. All the residual degrees are zero. In this case the resulting graph has D as its degree sequence.
2. One of the residual degrees is negative. This means that the sequence D is not graphic.

To illustrate the above algorithm, consider the sequence

$$D = \begin{matrix} v_1 & v_2 & v_3 & v_4 & v_5 \\ (4 & 3 & \textcircled{3} & 2 & 2) \end{matrix}.$$

After laying off d_3 (which is circled), we get the sequence

$$D' = \begin{matrix} v_1 & v_2 & v_3 & v_4 & v_5 \\ (3 & 2 & 0 & 1 & 2) \end{matrix}$$

which, after reordering of the residual degrees, becomes

$$D_1 = \begin{matrix} v_1 & v_2 & v_5 & v_4 & v_3 \\ (3 & 2 & \textcircled{2} & 1 & 0) \end{matrix}.$$

We next lay off the degree corresponding to v_5 and get

$$D_1' = \begin{matrix} v_1 & v_2 & v_5 & v_4 & v_3 \\ (2 & 1 & 0 & 1 & 0) \end{matrix}.$$

Reordering the residual degrees in D_1' we get

$$\begin{array}{ccccc} v_1 & v_2 & v_4 & v_5 & v_3 \end{array}$$
$$D_2 = (\textcircled{2} \quad 1 \quad 1 \quad 0 \quad 0).$$

Now laying off the degree corresponding to v_1, we get

$$\begin{array}{ccccc} v_1 & v_2 & v_4 & v_5 & v_3 \end{array}$$
$$D_2' = (0 \quad 0 \quad 0 \quad 0 \quad 0).$$

The algorithm terminates here. Since all the residual degrees are equal to zero, the sequence (4,3,3,2,2) is graphic. The required graph (Fig. 8.5) is obtained by the following sequence of steps which corresponds to the order in which the degrees were laid off:

1. Connect v_3 to v_1, v_2, and v_4.
2. Connect v_5 to v_1 and v_2.
3. Connect v_1 to v_2 and v_4.

Erdös and Gallai [8.7] have given a necessary and sufficient condition (not of an algorithmic type) for a sequence to be graphic. See also Harary [8.8].

Suppose that in the above algorithm, we lay off at each step the smallest nonzero residual degree. Then using induction we can easily show that the resulting graph is connected if

$$d_i \geqslant 1, \qquad \text{for all } i \tag{8.8}$$

and

$$\sum_{i=1}^{n} d_i \geqslant 2(n-1). \tag{8.9}$$

Note that (8.8) and (8.9) are necessary for the graph to be connected. In fact we prove in the following theorem a much stronger result. This result is due to Edmonds [8.9]. The proof given here is due to Wang and Kleitman [8.10].

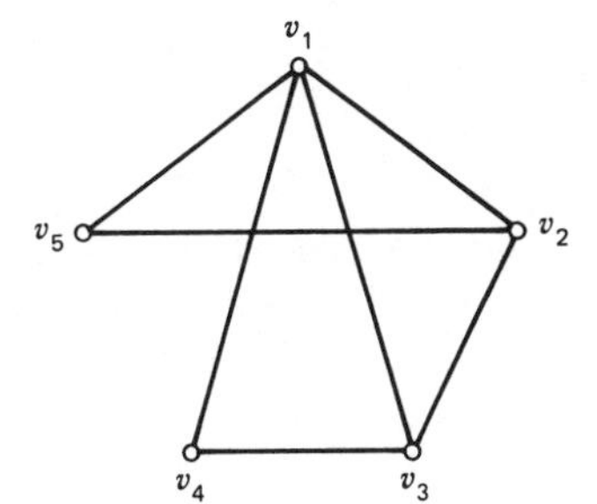

Figure 8.5. A graph with degree sequence (4,3,3,2,2).

THEOREM 8.8 (EDMONDS). A necessary and sufficient condition for a graphic sequence $(d_1, d_2, \ldots, d_n)$ to be the degree sequence of a simple k-edge-connected graph, for $k \geqslant 2$, is that each degree $d_i \geqslant k$.

Proof

Necessity is obvious.

We prove the sufficiency by showing that the algorithm we have just described results in a k-edge-connected graph when all the degrees in the given graphic sequence are greater than or equal to k. Note that at each step of the algorithm, we should lay off the smallest nonzero residual degree.

Proof is by induction. Assume that the algorithm is valid for all sequences in which each degree $d_i \geqslant p$ with $p \leqslant k-1$.

To prove the theorem we have to show that in the graph constructed by the algorithm every cutset $\langle A, \bar{A}\rangle$ has at least k edges. Proof is trivial if $|A|=1$ or $|\bar{A}|=1$. So assume that $|A| \geqslant 2$ and $|\bar{A}| \geqslant 2$.

We claim that in the connection procedure of the algorithm, one of the following three cases will eventually occur at some step, say, step r. (Step r means the step when the rth vertex is fully connected.)

Case 1 All the nonzero residual degrees are at least k.

Case 2 All the nonzero residual degrees are at least $k-1$, and there is at least one edge constructed by steps $1, \ldots, r$ which lies in $\langle A, \bar{A}\rangle$.

Case 3 All the nonzero residual degrees are at least $k-2$, and there are at least two edges constructed by steps $1, 2, \ldots, r$ which lie in $\langle A, \bar{A}\rangle$.

All these three cases imply that the cutset $\langle A, \bar{A}\rangle$ has at least k edges by induction.

We prove this claim as follows:

Let v_i be the vertex which is connected at step i. Without loss of generality we may assume that v_1 is in A. We now show that at some step in the connection procedure case 3 must occur if cases 1 and 2 never occur.

At step 1, v_1 is fully connected. Then

a. The smallest nonzero residual degrees are $k-1$, because case 1 has not occurred; and

b. The degrees of none of the vertices in $\bar{A}$ are decreased by 1 when v_1 is connected, because case 2 has not occurred.

Hence v_2, the vertex to be connected at step 2, must be in A. (All vertices in $\bar{A}$ must have degree at least k.) Now if v_2 is connected and case 1 does not occur and no edge connects A with $\bar{A}$, then we shall still have (a) and (b) as before. Hence the next vertex to be connected will still be in A.

Since the residual degrees are decreased at each step of the connection procedure, sooner or later there must exist an r with v_r in A, such that when v_r is connected, an edge will connect A with $\bar{A}$. If case 2 does not occur, then one of the nonzero residual degrees among the vertices of A not yet fully connected becomes $k-2$. This means, by our connection procedure, that v_r must connect to every vertex in $\bar{A}$, because vertices in $\bar{A}$ all have residual degree equal to k, and since we connect v_r to the vertices of the largest residual degree. Since $|\bar{A}| \geqslant 2$, at this step case 3 occurs. ■

Wang and Kleitman [8.6] have also established necessary and sufficient conditions for a graphic sequence to be the degree sequence of a simple k-vertex-connected graph.

8.4 MENGER'S THEOREM

We present in this section a classical result in graph theory, namely, Menger's theorem [8.11]. This theorem helps to relate the connectivity of a graph to the number of vertex-disjoint paths between any two distinct vertices in the graph.

THEOREM 8.9 (MENGER). The minimum number of vertices whose removal from a graph disconnects two nonadjacent vertices s and t is equal to the maximum number of vertex-disjoint s–t paths in the graph. ■

Proof of this theorem is given in Section 15.7.

THEOREM 8.10. A necessary and sufficient condition that a simple graph $G=(V, E)$, with $|V| \geqslant k+1$, be k-connected is that there are k vertex-disjoint s–t paths between any two vertices s and t of G.

Proof

Obviously, the theorem is true for $k=1$. So we need to prove the theorem for $k \geqslant 2$.

Necessity If s and t are not adjacent, then the necessity of the theorem follows from Theorem 8.9.

Suppose that s and t are adjacent and that there are at most $k-1$ vertex-disjoint s–t paths in G. Let $e=(s, t)$. Consider now the graph $G'=G-e$. Since there are at most $k-1$ vertex-disjoint s–t paths in G, there cannot be more than $k-2$ vertex-disjoint s–t paths in G'. Thus there exists a set $A \subseteq V-\{s, t\}$ of vertices, with $|A| \leqslant k-2$, whose removal disconnects s and t

in G'. Then

$$|V-A|=|V|-|A|\geqslant k+1-(k-2)=3,$$

and therefore, there is a vertex u in $V-A$ different from s and t.

Now we show that there exists a s–u path in G' which does not contain any vertex of A. Clearly, this is true if s and u are adjacent. If s and u are not adjacent, then there are k vertex-disjoint s–u paths in G, and hence there are $k-1$ vertex-disjoint s–u paths in G'. Since $|A|\leqslant k-2$, at least one of these $k-1$ paths will not contain any vertex of A.

In a similar way, we can show that in G' there exists a u–t path which does not contain any vertex of A.

Thus there exists in G' a s–t path which does not contain any vertex of A. This, however, contradicts that A is a s–t disconnecting set in G'.

Hence the necessity.

Sufficiency G is connected because there are k vertex-disjoint paths between any two distinct vertices of G. Further, not more than one of these paths can be of length 1, since there are no parallel edges in G. The union of the remaining $k-1$ paths must contain at least $k-1$ distinct vertices other than s and t. Hence

$$|V|\geqslant(k-1)+2>k.$$

Suppose in G there is a disconnecting set A with $|A|<k$. Then consider the subgraph G' of G on the vertex set $V-A$. This graph contains at least two distinct components. If we select two vertices s and t from any two different components of G', then there are at most $|A|<k$ vertex-disjoint s–t paths in G. This contradicts that any two vertices are connected by k vertex-disjoint paths in G.

Hence the sufficiency. ∎

The above result is due to Whitney [8.12]. Since it is only a variation of Theorem 8.9, we shall refer to this also as Menger's theorem.

Consider next two special classes of k-connected graphs, namely, the 2-connected graphs and the 3-connected graphs. There are several equivalent characterizations of 2-connected graphs. Some of them we have already listed as exercises in Chapter 1.

Tutte has given a characterization of 3-connected graphs. This characterization is given in terms of a special class of graphs called *wheels*.

Consider a circuit C of length n. If we add a new vertex and connect it to all the vertices of C, then we get the $(n+1)$-wheel W_{n+1}. For example, W_7 is shown in Fig. 8.6.

Tutte's [8.13] characterization of 3-connected graphs is stated in the next theorem.

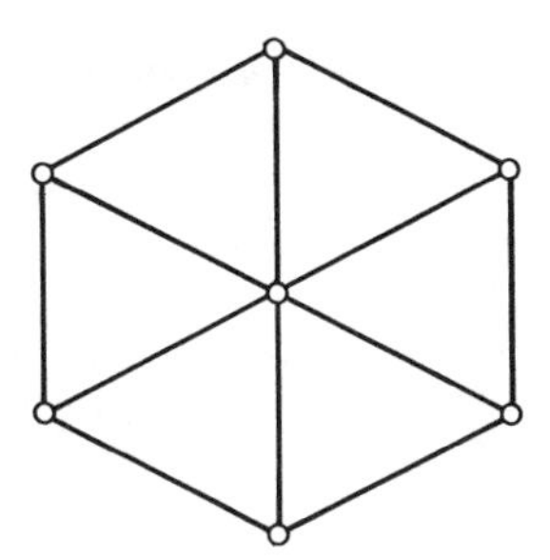

Figure 8.6. Wheel W_7.

THEOREM 8.11. A simple graph G is 3-connected if and only if G is a wheel or can be obtained from a wheel by a sequence of operations of the following types:

1. The addition of a new edge.
2. The replacement of a vertex v of degree ≥ 4 by two adjacent vertices v' and v'' of degrees ≥ 3 such that each vertex formerly adjacent to v is now adjacent to exactly one of v' and v''. ■

We conclude this section with a characterization of k-edge-connected graphs. This characterization is analogous to Theorem 8.9 and is also known as Menger's theorem, though it was independently discovered by Ford and Fulkerson [8.14], and by Elias, Feinstein, and Shannon [8.15].

THEOREM 8.12. The minimum number of edges whose removal from a connected graph G disconnects two distinct vertices s and t is equal to the maximum number of edge-disjoint s–t paths in G. ■

Proof of the above theorem is also given in Section 15.7.

8.5 MATCHINGS

The discussions of this and the remaining sections of this chapter concern the following problem, known as the *marriage problem*, and related ones.

We have a finite set of boys, each of whom has several girlfriends. Under what conditions can we marry off the boys in such a way that each boy marries one of his girlfriends? (Of course, no girl should marry more than one boy!)

This problem can be posed in graph-theoretical terminology as follows:

Construct a bipartite graph G in which the vertices $x_1, x_2, \ldots, x_n$ represent the boys and the vertices $y_1, y_2, \ldots, y_m$ represent the girls. An edge (x_i, y_j) is present in G if and only if y_j is a girlfriend of x_i. The marriage problem is then equivalent to finding in G a set of edges such that no two of them have a common vertex and each x_i is an end vertex of one of these edges.

For example, suppose there are four boys b_1, b_2, b_3, and b_4 and four girls g_1, g_2, g_3, and g_4, with their relationships as shown below:

$$b_1 \to \{g_1\},$$

$$b_2 \to \{g_2\},$$

$$b_3 \to \{g_1, g_2\},$$

$$b_4 \to \{g_3, g_4\}.$$

The bipartite graph representing this situation is shown in Fig. 8.7. It is easy to verify that it is not possible to marry off all the four boys such that each one marries one of his girlfriends. However, we can marry off as many as three boys without violating the requirement of the marriage problem. For example, two sets of such appropriate pairing are

1. (b_1, g_1), (b_2, g_2), (b_4, g_3),
2. (b_1, g_1), (b_3, g_2), (b_4, g_4).

The above discussions motivate the definition of a matching in a graph.

Two edges in a graph are said to be *independent* if they do not have any common vertex. Edges $e_1, e_2, \ldots$ are said to be *independent* if no two of them have a common vertex.

A *matching* in a graph is a set of independent edges. For example, in the graph of Fig. 8.8 $\{e_1, e_4\}$ is a matching.

Obviously, a maximal matching is a maximal set of independent edges. Thus in the graph of Fig. 8.8 $\{e_1, e_4\}$ is a maximal matching, whereas $\{e_5, e_6\}$ is not.

A matching with the largest number of edges is called a *maximum matching*. The set $\{e_1, e_5, e_6\}$ is a maximum matching in the graph of Fig. 8.8. The

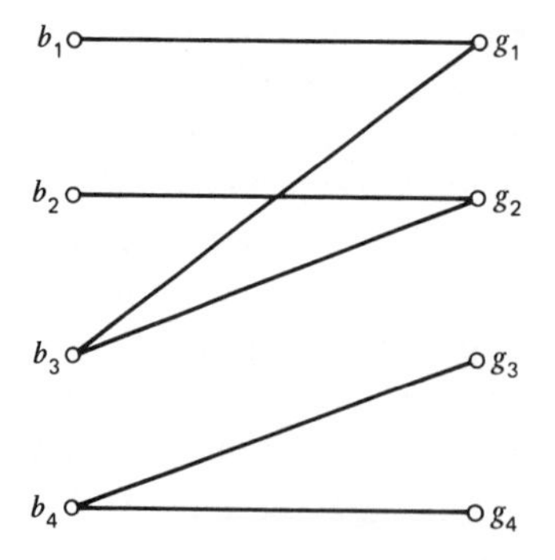

Figure 8.7.

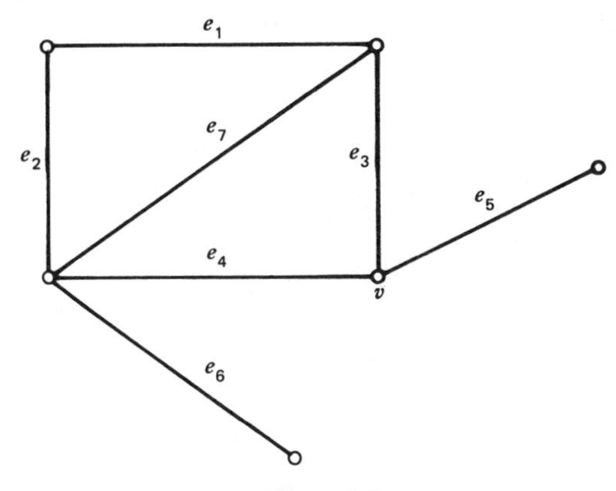

Figure 8.8.

number of edges in a maximum matching of G will be called the *matching number* of G and will be denoted by $\alpha_1(G)$.

A vertex is said to be *saturated* in a matching M if it is an end vertex of an edge in M. For example, the vertex v is saturated in the matching $\{e_1, e_5, e_6\}$ in the graph of Fig. 8.8.

In the following a bipartite graph $G=(V, E)$ with bipartition (X, Y) will be denoted by the triplet (X, Y, E).

We say that the set X can be *matched into* Y in the bipartite graph (X, Y, E) if there is a matching M such that every vertex of X is saturated in M. The matching M will then be called a *complete matching* of X into Y.

With this new terminology the marriage problem becomes equivalent to finding necessary and sufficient conditions for the existence of a complete matching of X into Y, in a bipartite graph (X, Y, E). In the next section we present several results concerning matchings in bipartite graphs, in addition to providing an answer to the marriage problem.

8.6 MATCHINGS IN BIPARTITE GRAPHS

Consider a bipartite graph $G=(X, Y, E)$. Let S be any subset of X and let $\Gamma(S)$ denote the set of vertices adjacent to the vertices in S. Suppose $|S|>|\Gamma(S)|$. Then it is clear that there is no complete matching of S into $\Gamma(S)$, and any matching M in G will saturate at most $|\Gamma(S)|$ of the vertices of S. So

$$|M| \leqslant |X| - (|S| - |\Gamma(S)|). \tag{8.10}$$

Since (8.10) is valid for any subset S of X, we can conclude that for any matching M,

$$|M| \leqslant |X| - \max_{S \subseteq X} (|S| - |\Gamma(S)|). \tag{8.11}$$

We now define the *deficiency* $\sigma(S)$ of any subset S of X and the deficiency $\sigma(G)$ of G as follows:

$$\sigma(S)=|S|-|\Gamma(S)| \tag{8.12}$$

and

$$\sigma(G)=\max_{S\subseteq X}(|S|-|\Gamma(S)|). \tag{8.13}$$

Now using (8.13), we can rewrite (8.11) as

$$|M|\leqslant|X|-\sigma(G). \tag{8.14}$$

The main result of this section is that in a bipartite graph $G=(X,Y,E)$ there exists a matching with $|X|-\sigma(G)$ edges. In other words, we show that the number of edges in a maximum matching of G is equal to $|X|-\sigma(G)$. The result will be proved in two parts. We consider the case $\sigma(G)=0$ first, and the case $\sigma(G)>0$ later. (Note that $\sigma(G)\geqslant 0$, since for the empty set $\varnothing$, $\sigma(\varnothing)=0$.)

Note that the case $\sigma(G)=0$ occurs if for every subset S of X

$$|S|\leqslant|\Gamma(S)|,$$

and the case $\sigma(G)>0$ occurs when for some subset S of X

$$|S|>|\Gamma(S)|.$$

The following result is due to Hall [8.16]. Our proof is based on Halmos and Vaughan [8.17].

THEOREM 8.13 (HALL). In a bipartite graph $G=(X,Y,E)$ there exists a complete matching of X into Y if and only if, for any $S\subseteq X$,

$$|S|\leqslant|\Gamma(S)|.$$

Proof

Necessity From (8.14) it is clear that complete matching of X into Y requires that $\sigma(G)=0$ or, in other words,

$$|S|\leqslant|\Gamma(S)|, \qquad S\subseteq X.$$

Sufficiency Proof is by induction on $|X|$, the number of vertices in X.

If $|X|=1$, then obviously there is a complete matching of X into Y, since the only vertex in X should be adjacent to at least one vertex in Y.

Assume as the induction hypothesis that the sufficiency is true for any bipartite graph with $|X|\leqslant m-1$. Consider then a bipartite graph $G=(X,Y,E)$

with $|X|=m$. Let for every subset S of X, $|S| \leqslant |\Gamma(S)|$. We now show that there is a complete matching of X into Y by examining the following cases.

Case 1 For every nonempty proper subset S of X, $|S| < |\Gamma(S)|$.

Select any edge (x_0, y_0), $x_0 \in X$ and $y_0 \in Y$. Let G' be the graph obtained from G by removing x_0 and y_0 and the edges incident on them. Since the vertices of every subset S of $X-\{x_0\}$ are adjacent to more than $|S|$ vertices in Y, they are adjacent to $|S|$ or more vertices in $Y-\{y_0\}$. Therefore by the induction hypothesis, there exists a complete matching of $X-\{x_0\}$ into $Y-\{y_0\}$. This matching together with the edge (x_0, y_0) is a complete matching of X into Y.

Case 2 There is a nonempty proper subset S_0 of X such that $|S_0| = |\Gamma(S_0)|$.

Let G' be the subgraph of G containing the vertices in the sets S_0 and $\Gamma(S_0)$ and the edges connecting these vertices.

Let G'' be the subgraph of G containing the vertices in the sets $X-S_0$ and $Y-\Gamma(S_0)$ and the edges connecting these vertices.

We show that in G' there is a complete matching of S_0 into $\Gamma(S_0)$, and in G'' there is a complete matching of $X-S_0$ into $Y-\Gamma(S_0)$. These two matchings together will form a complete matching of X into Y.

Consider first the graph G'. Let, for any subset S of S_0, $\Gamma'(S)$ denote the set of vertices in $\Gamma(S_0)$ adjacent to those in S. It is then clear that

$$\Gamma'(S) = \Gamma(S), \qquad S \subseteq S_0.$$

Since, for every subset S of S_0,

$$|S| \leqslant |\Gamma(S)| = |\Gamma'(S)|,$$

it follows from the induction hypothesis that there is a complete matching of S_0 into $\Gamma(S_0)$.

Consider next the graph G''. Let, for any subset of $X-S_0$, $\Gamma''(S)$ denote the set of vertices in $Y-\Gamma(S_0)$ adjacent to those in S. Now for $S \subseteq X-S_0$,

$$\begin{aligned} |S \cup S_0| &= |S| + |S_0| \\ &\leqslant |\Gamma(S \cup S_0)| \\ &= |\Gamma''(S)| + |\Gamma(S_0)|. \end{aligned}$$

Since $|S_0| = |\Gamma(S_0)|$, we get from the above

$$|S| \leqslant |\Gamma''(S)|, \qquad S \subseteq X - S_0.$$

Thus again by the induction hypothesis there is a complete matching of $X-S_0$ into $Y-\Gamma(S_0)$.

This completes the proof of sufficiency. ∎

As we observed before, Theorem 8.13 provides an answer to the marriage problem, and it is stated next.

THEOREM 8.14 (HALL). A necessary and sufficient condition for a solution of the marriage problem is that every set of k boys collectively have at least k girlfriends, $1 \leqslant k \leqslant m$, where m is the number of boys. ∎

Theorem 8.13 is simply a translation into graph-theoretical terminology of Theorem 8.14.

Next we prove that when $\sigma(G)>0$, the number of edges in a maximum matching is equal to $|X|-\sigma(G)$. To do so, we need the following two lemmas.

LEMMA 8.1. Let S_1 and S_2 be any two subsets of X. Then

$$\sigma(S_1 \cup S_2)+\sigma(S_1 \cap S_2) \geqslant \sigma(S_1)+\sigma(S_2).$$

Proof

It is a simple exercise to verify that

$$|S_1 \cup S_2|+|S_1 \cap S_2|=|S_1|+|S_2|. \tag{8.15}$$

Since

$$|\Gamma(S_1 \cup S_2)|=|\Gamma(S_1) \cup \Gamma(S_2)|$$

and

$$|\Gamma(S_1 \cap S_2)| \leqslant |\Gamma(S_1) \cap \Gamma(S_2)|,$$

we get

$$\begin{aligned} |\Gamma(S_1 \cup S_2)|+|\Gamma(S_1 \cap S_2)| &\leqslant |\Gamma(S_1) \cup \Gamma(S_2)|+|\Gamma(S_1) \cap \Gamma(S_2)| \\ &=|\Gamma(S_1)|+|\Gamma(S_2)|. \end{aligned} \tag{8.16}$$

Now subtracting (8.16) from (8.15) we get

$$\begin{aligned} \sigma(S_1 \cup S_2)+\sigma(S_1 \cap S_2) &\geqslant |S_1|+|S_2|-|\Gamma(S_1)|-|\Gamma(S_2)| \\ &=\sigma(S_1)+\sigma(S_2). \end{aligned}$$ ∎

LEMMA 8.2. Let S_1 and S_2 be two subsets of X such that $\sigma(S_1)=\sigma(S_2)=\sigma(G)$. Then

$$\sigma(S_1 \cup S_2)=\sigma(S_1 \cap S_2)=\sigma(G).$$

Proof

From Lemma 8.1,

$$\sigma(S_1 \cup S_2)+\sigma(S_1 \cap S_2) \geqslant \sigma(S_1)+\sigma(S_2)=2\sigma(G).$$

Since neither $\sigma(S_1 \cup S_2)$ nor $\sigma(S_1 \cap S_2)$ is greater than $\sigma(G)$, we get

$$\sigma(S_1 \cup S_2)=\sigma(S_1 \cap S_2)=\sigma(G). \quad \blacksquare$$

THEOREM 8.15. In a bipartite graph $G=(X, Y, E)$ with $\sigma(G)>0$, the number of edges in a maximum matching is equal to $|X|-\sigma(G)$.

Proof

Let $S_1, S_2, \ldots, S_k$ be all the subsets of X with their deficiencies equal to $\sigma(G)$. Let

$$S_0=S_1 \cap S_2 \cap \cdots \cap S_k.$$

By Lemma 8.2,

$$\sigma(S_0)=\sigma(G)>0.$$

Thus S_0 is nonempty. We may now note that every subset of X having its deficiency equal to $\sigma(G)$ contains S_0.

Consider any vertex x_0 in S_0. Let G' be the graph which is obtained by removing from G the vertex x_0 and all the edges incident on it. It is clear that $\sigma(G')<\sigma(G)$, since no subset of the set $X-\{x_0\}$ contains S_0. We now show that $\sigma(G')=\sigma(G)-1$.

Consider the set $S_0'=S_0-\{x_0\}$. We have

$$\sigma(S_0')=|S_0'|-|\Gamma(S_0')|$$

$$=|S_0|-1-|\Gamma(S_0')|.$$

Since $\sigma(S_0')<\sigma(G)$, we obtain

$$|S_0|-1-|\Gamma(S_0')|<|S_0|-|\Gamma(S_0)|,$$

that is,

$$|\Gamma(S_0')| \geqslant |\Gamma(S_0)|. \tag{8.17}$$

On the other hand, since S_0' is a subset of S_0,

$$|\Gamma(S_0')| \leqslant |G(S_0)|. \tag{8.18}$$

Combining (8.17) and (8.18), we get

$$|\Gamma(S_0')| = |\Gamma(S_0)|.$$

Therefore

$$\sigma(S_0') = |S_0| - 1 - |\Gamma(S_0)| = \sigma(G) - 1.$$

Since $\sigma(G') < \sigma(G)$, we can conclude that $\sigma(G') = \sigma(G) - 1$.

If we repeat the above argument until an appropriate set of $\sigma(G)$ vertices is removed from X along with the edges incident on these vertices, we get a subgraph of G with deficiency equal to zero. By Theorem 8.13, there is a complete matching in this subgraph. This will be a matching for G containing $X - \sigma(G)$ edges. It is clear from (8.14) that such a matching will be a maximum matching of G. ■

We can combine Theorems 8.13 and 8.15 into one, and because of its importance, we present it below. See König [8.18] and Ore [8.19].

THEOREM 8.16 (KÖNIG). The number of edges in a maximum matching of a bipartite graph $G = (X, Y, E)$ is equal to $|X| - \sigma(G)$, where $\sigma(G)$ is the deficiency of G. ■

Corollary 8.16.1 In a nonempty bipartite graph $G = (X, Y, E)$ there is a complete matching of X into Y if

$$\min_{x \in X} \{d(x)\} \geqslant \max_{y \in Y} \{d(y)\}.$$

Proof

Let

$$\min_{x \in X} \{d(x)\} = d_1$$

and

$$\max_{y \in Y} \{d(y)\} = d_2.$$

Consider any subset A of X. Let E_1 be the set of edges incident on the vertices in A, and E_2 the set of edges incident on the vertices in $\Gamma(A)$. Then we have

$$|E_1| \geqslant |A|\, d_1$$

and

$$|E_2| \leqslant |\Gamma(A)|\, d_2.$$

Since $E_1 \subseteq E_2$, we have

$$|\Gamma(A)|\, d_2 \geqslant |E_2| \geqslant |E_1| \geqslant |A|\, d_1.$$

So

$$|\Gamma(A)| \geqslant |A|, \qquad A \subseteq X.$$

Thus by Hall's theorem there is a complete matching of X into Y. ■

Next we consider two results which are related to Hall's theorem.

Our first result is from transversal theory.

Let M be a nonempty finite set and $S = \{S_1, S_2, \ldots, S_r\}$ be a family of (not necessarily distinct) nonempty subsets of M. Then a *transversal* (or *system of distinct representatives*) of S is a set of r distinct elements of M, one from each set S_i.

For example, if $M = \{1,2,3,4,5,6\}$ and $S_1 = \{1,3,4\}$, $S_2 = \{1,3,4\}$, $S_3 = \{1,2,5\}$, and $S_4 = \{5,6\}$, then $\{1,3,2,6\}$ is a transversal of the family $S = \{S_1, S_2, S_3, S_4\}$. On the other hand, if $S_1 = S_2 = \{1,3\}$, $S_3 = \{3,4\}$, and $S_4 = \{1,4\}$, then there is no transversal for the family $\{S_1, S_2, S_3, S_4\}$.

The following question now arises:

What are the necessary and sufficient conditions for a family of subsets of a set to have a transversal?

Suppose we construct a bipartite graph $G = (X, Y, E)$ such that

1. The vertex x_i of X corresponds to the set S_i in S;
2. The vertex y_i of Y corresponds to element i of M; and
3. Edge $(x_i, y_j) \in E$ if and only if $j \in S_i$.

We can now see that the above question becomes equivalent to finding a complete matching of X into Y in the bipartite graph G constructed as above.

Thus we get the following theorem, which is simply a restatement of Hall's theorem in the language of transversal theory.

THEOREM 8.17. Let M be a nonempty set and $S=\{S_1, S_2, \ldots, S_r\}$ a family of subsets of M. Then S has a transversal if and only if the union of any k, $1 \leqslant k \leqslant r$, of the subsets S_i contain at least k elements of M. ∎

A purely combinatorial proof of the above theorem without using concepts from graph theory was given by Rado. The proof is very elegant and may be found in Rado [8.20].

The next result relates to matrices whose elements are 0's and 1's. Such matrices are called *(0, 1)–matrices*. In the following, a line of a matrix refers to a row or a column of the matrix.

THEOREM 8.18 (KÖNIG AND EGERVÁRY). The minimum number of lines which contain all the 1's in a (0, 1)–matrix is equal to the maximum number of 1's no two of which are in the same line of M. ∎

Given a (0, 1)–matrix M of order $m \times m$. Suppose we construct a bipartite graph $G=(X, Y, E)$ such that

1. The vertices $x_1, x_2, \ldots, x_m$ of X correspond to the m rows of M;
2. The vertices $y_1, y_2, \ldots, y_n$ of Y correspond to the n columns of Y; and
3. The edge $(x_i, y_j) \in E$ if the (i, j) entry of M is equal to 1.

If we now consider a vertex as covering all the edges incident on the vertex, then the König-Egerváry theorem can be restated as follows:

In a bipartite graph the minimum number of vertices which cover all the edges is equal to the number of edges in any maximum matching of the graph.

In Chapter 9 (Theorem 9.2) we prove the above form of the König-Egerváry theorem.

8.7 MATCHINGS IN GENERAL GRAPHS

In this section we establish some results relating to matchings in a general graph.

Consider a graph $G=(V, E)$ and a matching M in G. An *alternating chain* in G is a trail whose edges are alternately in M and $(E-M)$. For example, the sequence of edges $e_1, e_2, e_3, e_4, e_7, e_6$ is an alternating chain relative to the matching $M=\{e_2, e_4, e_6\}$ in the graph of Fig. 8.9. The edges in the alternating chain which belong to M are called *dark edges* and those which belong to $E-M$ are called *light edges*. Thus e_1, e_3, e_7 are light edges, whereas e_2, e_4, e_6 are dark edges in the alternating chain considered above.

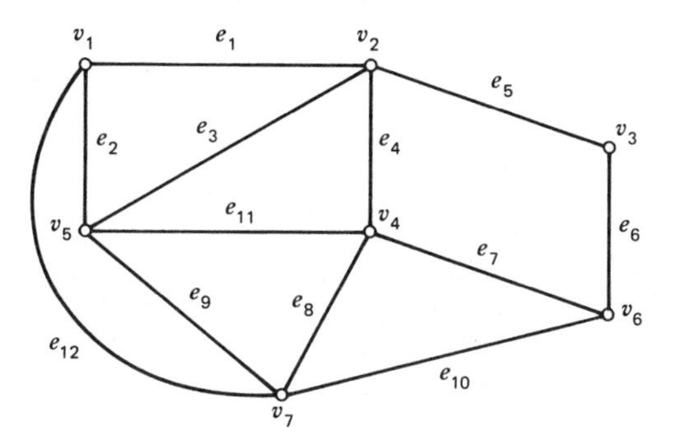

Figure 8.9.

THEOREM 8.19. Let M_1 and M_2 be two matchings in a simple graph $G=(V, E)$. Let $G'=(V', E')$ be the induced subgraph of G on the edge set

$$M_1 \oplus M_2 = (M_1 - M_2) \cup (M_2 - M_1).$$

Then each component of G' is of one of the following types:

1. A circuit of even length whose edges are alternately in M_1 and M_2.
2. A path whose edges are alternately in M_1 and M_2 and whose end vertices are unsaturated in one of the two matchings.

Proof

Consider any vertex $v \in V'$.

Case 1 $v \in V(M_1 - M_2)$ and $v \notin V(M_2 - M_1)$, where $V(M_i - M_j)$ denotes the set of vertices of the edges in $M_i - M_j$.

In this case v is the end vertex of an edge in $M_1 - M_2$. Since M_1 is a matching, no other edge of $M_1 - M_2$ is incident on v. Further, no edge of $M_2 - M_1$ is incident on v because $v \notin V(M_2 - M_1)$. Thus in this case the degree of v in G' is equal to 1.

Case 2 $v \in V(M_1 - M_2)$ and $v \in V(M_2 - M_1)$.

In this case a unique edge of $M_1 - M_2$ is incident on v and a unique edge of $M_2 - M_1$ is incident on v. Thus the degree of v is equal to 2.

Since the two cases considered are exhaustive, it follows that the maximum degree in G' is 2. Therefore the connected components will be of one of the two types described in the theorem. ■

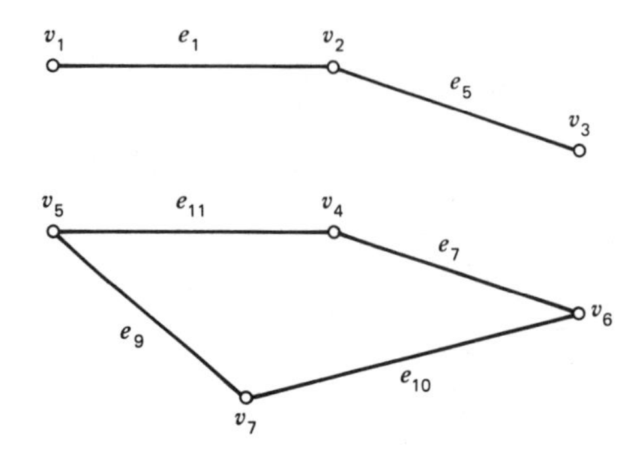

Figure 8.10.

For example, consider the two matchings $M_1=\{e_5,e_7,e_9\}$ and $M_2=\{e_1,e_{10},e_{11}\}$ of the graph G of Fig. 8.9. Then

$$M_1 \oplus M_2=\{e_1,e_5,e_7,e_9,e_{10},e_{11}\},$$

and the graph G' will be as in Fig. 8.10. It may be seen that the components of G' are of the two types described in Theorem 8.19.

In the following theorem we establish Berge's [8.21] characterization of a maximum matching in terms of an alternating chain.

THEOREM 8.20 (BERGE). A matching M is maximum if and only if there exists no alternating chain between any two unsaturated vertices.

Proof

Necessity Suppose there is an alternating chain P between two unsaturated vertices. Then replacing the dark edges in the chain by the light edges will give a matching M_1 with

$$|M_1|=|M|+1.$$

Note that $M_1=(M-P)\cup(P-M)$.

For example, in the graph of Fig. 8.9, consider the matching $M=\{e_2,e_4\}$. There is an alternating chain e_5, e_4, e_{11}, e_2, and e_{12} between the unsaturated vertices v_3 and v_7. If we replace in M the dark edges e_4 and e_2 by the light edges e_5, e_{11}, and e_{12}, we get the matching $\{e_5,e_{11},e_{12}\}$ which has one more edge than M.

Sufficiency Suppose M satisfies the conditions of the theorem. Let M' be a maximum matching. Then it follows from the necessity part of the theorem that M' also satisfies the conditions of the theorem, namely, there is no alternating chain between any vertices which are not saturated in M'. We now show that $|M|=|M'|$, thereby proving the sufficiency.

Since $M=(M\cap M')\cup(M-M')$ and $M'=(M\cap M')\cup(M'-M)$, it is clear that $|M|=|M'|$ if and only if $|M-M'|=|M'-M|$.

Let G' be the graph on the edge set $M\oplus M'=(M-M')\cup(M'-M)$.

Consider first a circuit in G'. By Theorem 8.19 such a circuit is of even length, and the edges in this circuit are alternately in $M-M'$ and $M'-M$. Therefore each circuit in G' has the same number of edges from both $M-M'$ and $M'-M$.

Consider next a component of G' which is a path. Again, by Theorem 8.19, the edges in this path are alternately in $M-M'$ and $M'-M$. Further the end vertices of this path are unsaturated in M or M'. Suppose the path is of odd length, then the end vertices of the path will be both unsaturated in the same matching. This would mean that with respect to one of these two matchings there is an alternating chain between two unsaturated vertices. But this is a contradiction because both M and M' satisfy the condition of the theorem. So each component of G' which is a path has an even number of edges, and hence it has the same number of edges from $M-M'$ and $M'-M$.

Thus each component of G' has an equal number of edges from $M-M'$ and $M'-M$. Since the edges of G' constitute the set $(M-M')\cup(M'-M)$, we get

$$|M-M'|=|M'-M|,$$

and so, $|M|=|M'|$. ■

Given a matching M in a graph G. Let P be an alternating chain between any two vertices which are not saturated in M. Then as we have seen before, $M\oplus P$ is a matching with one more edge than M. For this reason the path P is called an *augmenting path* relative to M.

Next we prove two interesting results on bipartite graphs using the theory of alternating chains.

Consider a bipartite graph $G=(X, Y, E)$ with maximum degree Δ. Let X_1 denote the set of all the vertices in X of degree Δ. If G' is the bipartite graph $(X_1, \Gamma(X_1), E')$, where E' is the set of edges connecting X_1 and $\Gamma(X_1)$, then it can be seen from Corollary 8.16.1 that there exists in G' a complete matching of X_1 into $\Gamma(X_1)$. Such a matching clearly saturates all the vertices in X_1. Thus there exists a matching in G which saturates all the vertices in X of degree Δ. Similarly, there exists a matching in G which saturates all the vertices in Y of degree Δ. The question that now arises is whether in a bipartite graph there exists a matching which saturates all the maximum degree vertices in both X and Y. To answer this question, we need the following result due to Mendelsohn and Dulmage [8.22].

THEOREM 8.21 (MENDELSOHN AND DULMAGE). Let $G=(X, Y, E)$ be a bipartite graph, and let M_i be a matching which matches $X_i\subseteq X$ with

$Y_i \subseteq Y$ $(i=1,2)$. Then there exists a matching $M' \subseteq M_1 \cup M_2$ which saturates X_1 and Y_2.

Proof

Consider the bipartite graph $G'=(X_1 \cup X_2, Y_1 \cup Y_2, M_1 \cup M_2)$. Each vertex of this graph has degree 1 or 2; hence each component of this graph is either a path or a circuit whose edges are alternately in M_1 and M_2. (See proof of Theorem 8.19.)

Each vertex $y \in Y_2 - Y_1$ has degree 1 in G'. So it is in a connected component which is a path P_y from y to a vertex $x \in X_2 - X_1$ or to a vertex $z \in Y_1 - Y_2$. In the former case the last edge of P_y is in M_2 and so $M_1 \oplus P_y$ matches $X_1 \cup \{x\}$ with $Y_1 \cup \{y\}$. In the latter case the last edge of P_y is in M_1 and so $M_1 \oplus P_y$ matches X_1 with $(Y_1 - z) \cup \{y\}$. In either case $M_1 \oplus P_y$ saturates $Y_1 \cap Y_2$. Thus $M_1 \oplus P_y$ saturates $y \in Y_2 - Y_1$ and all the vertices in X_1 and $Y_1 \cap Y_2$.

If we let

$$P = \bigcup_{y \in Y_2 - Y_1} P_y,$$

then we can see that $M_1 \oplus P$ is a matching which saturates X_1 and Y_2. This is a required matching $M' \subseteq M_1 \cup M_2$. ■

THEOREM 8.22. In a bipartite graph there exists a matching which saturates all the maximum degree vertices.

Proof

Consider a bipartite graph $G=(X, Y, E)$. Let $X' \subseteq X$ and $Y' \subseteq Y$ contain all the vertices of maximum degree in G. As we have seen before, there exists a matching M_1 which saturates all the vertices in X' and a matching M_2 which saturates all the vertices in Y'. So by Theorem 8.21, there exists a matching $M' \subseteq M_1 \cup M_2$ which saturates all the vertices in X' and Y'. This is a required matching saturating all the maximum degree vertices in G. ■

Corollary 8.22.1 The set of edges of a bipartite graph with maximum degree Δ can be partitioned into Δ matchings.

Proof

Consider a bipartite graph $G=(X, Y, E)$ with maximum degree Δ. By Theorem 8.22 there exists a matching M_1 which saturates all the vertices of degree Δ. Then the bipartite graph $G'=(X, Y, E-M_1)$ has maximum degree $\Delta - 1$. This graph contains a matching M_2 which saturates every vertex of degree $\Delta - 1$. By repeating this process we can construct a sequence of disjoint matchings $M_1, M_2, \ldots, M_\Delta$ which form a partition of E. ■

An application of Theorems 8.21 and 8.22 and Corollary 8.22.1 is considered in Section 15.6.

Now we return to our discussion of matchings in general graphs.

A matching saturating all the vertices of a graph G is called a *perfect matching* of G.

We conclude this section with a theorem due to Tutte [8.23] on the conditions for the existence of a perfect matching in a graph. Our proof here is due to Anderson [8.24].

A component of a graph is *odd* if it has an odd number of vertices; otherwise it is *even*. If S is a subset of vertices of a graph G, then we shall denote by $p_0(S)$ the number of odd components of $G-S$.

THEOREM 8.23 (TUTTE). A graph $G=(V, E)$ has a perfect matching if and only if

$$p_0(S) \leqslant |S|, \qquad \text{for all } S \subset V. \tag{8.19}$$

Proof

Necessity Suppose that G has a perfect matching M. For some $S \subset V$, let $G_1, G_2, \ldots, G_k$ be the odd components of $G-S$. Since each G_i is odd, some vertex v_i of G_i must be matched under M with some vertex v_j of S. Thus S has at least k vertices, and hence

$$p_0(S) \leqslant |S|.$$

Sufficiency First note that if a graph G satisfies (8.19), then choosing $S = \varnothing$ we get $p_0(\varnothing) \leqslant 0$. So in such a case there will be no odd components in G. In other words, G will have an even number of vertices.

Proof of sufficiency is by induction on l, where $2l = |V|$. We use Hall's theorem (Theorem 8.13) and the simple fact that

$$p_0(S) = |S| (\text{mod}\, 2). \tag{8.20}$$

The case $l=1$ is trivial. Assume that the result is true for all graphs with less than $2l$ vertices. Consider then a graph G with $2l$ vertices and which satisfies (8.19). Now we have two cases.

Case 1 Suppose that $p_0(S) < |S|$, for all S, $2 \leqslant |S| < 2l$.

Consider any edge $e=(a, b)$ in G. Let $A=\{a, b\}$ and $G_A=G-A$. For any subset T of vertices of G_A, let $p_0'(T)$ denote the number of odd components in G_A-T. Then $p_0'(T)\leqslant|T|$, for if $p_0'(T)>|T|$, then we would get $p_0(T\cup A)=p_0'(T)>|T|=|T\cup A|-2$, contradicting (8.19). Therefore, by induction, G_A, and hence G, has a perfect matching.

Case 2 Let there be a set S such that $p_0(S)=|S|\geqslant 2$. Assume that S is a maximal such set.

First observe that there are no even components in $G-S$. For if there were any, we would remove a vertex v from one and add it to S. This would necessarily give us at least one more odd component. So $p_0(S\cup v)\geqslant p_0(S)+1=|S|+1$. Condition (8.19) requires that $p_0(S\cup v)\leqslant|S|+1$. So $p_0(S\cup v)=|S|+1$. But this would contradict the maximality of S. Hence there are no even components in $G-S$.

Let $|S|=s$, and let $G_1, G_2, \ldots, G_s$ be the s odd components of $G-S$. We now show that we can take a vertex from each one of these odd components and match it with a vertex in S. If this is not possible, then by Hall's theorem, there are k odd components which are connected in G to only $h<k$ vertices of S. But if T denotes such a set of h vertices, we would then have

$$p_0(T)\geqslant k>h=|T|,$$

contradicting (8.19). Thus we can take a vertex v_i from each G_i, $1\leqslant i\leqslant s$, and match it with a vertex in S.

Now each $G_i'=G_i-v_i$ has an even number of vertices. The proof will be completed if we show that each G_i' has a perfect matching.

If G_i' contains a set R of vertices such that $p_0''(R)>|R|$, where $p_0''(R)$ is the number of odd components in $G_i'-R$, then by (8.20), $p_0''(R)\geqslant|R|+2$, so that

$$\begin{aligned} p_0(R\cup S\cup\{v_i\}) &= p_0''(R)+p_0(S)-1 \\ &\geqslant |R|+|S|+1 \\ &= |R\cup S\cup\{v_i\}|. \end{aligned} \tag{8.21}$$

But condition (8.19) requires that

$$p_0(R\cup S\cup\{v_i\})\leqslant|R\cup S\cup\{v_i\}|.$$

So

$$p_0(R\cup S\cup\{v_i\})=|R\cup S\cup\{v_i\}|$$

contradicting the maximality of S. Thus by induction, G_i' has a perfect matching. ■

Lovász [8.25] gives an alternative proof of the above theorem using Theorem 8.20.

8.8 FURTHER READING

Berge [8.26] and Harary [8.8] are two general references recommended for further reading on connectivity and matching. Berge also presents a detailed discussion on the realizability of degree sequences for the cases of both directed and undirected graphs. Harary gives an historical account of Menger's theorem and its several variations.

A communication network can be modeled by a graph. The concept of vulnerability arises in the study of a network modeled by a graph. By vulnerability we mean the susceptibility of the network to attack. A network may be considered "destroyed" if after removal of some vertices or edges the resulting graph is not connected. Thus vulnerability of a network is related to the vertex and edge connectivities of the network. For example, a network N_1 may be considered more vulnerable than a network N_2 if the vertex connectivity of N_1 is less than that of N_2.

Boesch [8.27] has several papers on the design of graphs having specified connectivity and reliability properties. Some of the other papers which deal with this topic include Hakimi [8.28], Boesch and Thomas [8.29], and Amin and Hakimi [8.30]. Frank and Frisch [8.31] is recommended for further reading on this topic.

The problem of testing the connectivity of a graph is closely related to that of finding a maximum flow in a transport network. For further discussions and related references see Section 15.7 where we also prove Menger's theorems.

The papers by Mirsky and Perfect [8.32], Brualdi [8.33], and the book by Mirsky [8.34] are highly recommended for further reading on matchings and transversal theory. Applications involving the theory of matchings (in particular, the optimal assignment problem and a time-tabling problem) and related algorithms are discussed in Sections 15.4 through 15.6.

8.9 EXERCISES

8.1 Let $d_1 \leqslant d_2 \leqslant \cdots \leqslant d_n$ denote the degrees of a simple graph G. Show that G is k-connected, $k<n$, if

(a) $d(v_r) \geqslant r+k-1$, $r \leqslant \dfrac{n-k}{2}$; and

(b) $d(v_{n-k+1}) \geqslant \dfrac{n+k-2}{2}$.

8.2 Show that a simple n-vertex graph G is k-connected if

$$\delta(G) \geqslant \frac{n+k-2}{2}.$$

8.3 Let $G=(V, E)$ be a simple k-connected graph. Let $B=\{b_1, b_2, \ldots, b_k\}$ be a set of vertices with $|B|=k$. If $a \in V-B$, show that there exist k vertex-disjoint a–b_i paths from a to B.

8.4 Let G be a simple k-connected graph with $k \geqslant 2$. Show that there exists a circuit which passes through an arbitrary set of two edges e_1 and e_2 and $k-2$ vertices.

8.5 Show that $\kappa(H_{k,n})=\kappa'(H_{k,n})=k$.

8.6 Find a 5-connected graph with 7 vertices and 18 edges.

8.7 Show that the edges of a simple graph G can be oriented to form a strongly connected graph if and only if G is 2-edge-connected.

8.8 The *associated directed graph* $D(G)$ of an undirected graph G is the graph obtained by replacing each edge e of G by two oppositely oriented edges having the same end vertices as e. Show that

(a) There is a one-to-one correspondence between paths in G and directed paths in $D(G)$; and

(b) $D(G)$ is k-edge-connected if and only if G is k-edge-connected.

Note A directed graph G is *k-edge-connected* if and only if at least k edges have to be removed to destroy all directed s–t paths for any two vertices s and t of G.

8.9 Find a simple graph having the degree sequence $(5,4,4,4,3,3,3,3,3)$ and with the largest possible edge connectivity.

8.10 Show that a sequence $(d_1, d_2, \ldots, d_n)$ of nonnegative integers is the degree sequence of a tree if and only if

$$d_i \geqslant 1, \qquad \text{for all } i,$$

and

$$\sum_{i=1}^{n} d_i = 2(n-1).$$

8.11 Show that a sequence $(d_1, d_2, \ldots, d_n)$ of nonnegative integers is the degree sequence of a graph if and only if $\sum_{i=1}^{n} d_i$ is even (Hakimi [8.4]).

8.12 Show that a sequence $(d_1, d_2, \ldots, d_n)$ of nonnegative integers is the degree sequence of a simple graph if and only if

(a) $\sum_{i=1}^{n} d_i$ is even; and

(b) $\sum_{i=1}^{k} d_i \leqslant k(k-1) + \sum_{i=k+1}^{n} \min\{k, d_i\}$, for $1 \leqslant k \leqslant n-1$

(Erdös and Gallai [8.7]).

8.13 Show that a sequence $(d_1, d_2, \ldots, d_n)$ (realizable by a simple graph) is the degree sequence of a simple k-connected graph if and only if

(a) $d_i \geqslant k$, for $1 \leqslant i \leqslant n$,

(b) $m - \sum_{i=1}^{k-1} d_i + \frac{(k-1)(k-2)}{2} \geqslant n-k$,

where m is the number of edges of G. (Wang and Kleitman [8.6]).

8.14 Prove or disprove: For every matching M there exists a maximum matching M' such that $M \subseteq M'$.

8.15 A nonnegative real square matrix P is *doubly stochastic* if the sum of the entries in each row of P is 1 and the sum of the entries in each column of P is 1. A *permutation matrix* is a (0, 1)-matrix which has exactly one 1 in each row and each column. Show that a doubly stochastic matrix P can be expressed as

$$P = c_1 P_1 + c_2 P_2 + \cdots + c_k P_k,$$

where each P_i is a permutation matrix, each c_i is a nonnegative real number, and $\sum_{i=1}^{k} c_i = 1$.

8.16 Let G be a bipartite graph with bipartition (X, Y). If G has a complete matching of X into Y, show that there exists an $x_0 \in X$, such that for each $y \in \Gamma(x_0)$ at least one maximum matching uses the edge (x_0, y).

8.17 A *k-factor* of a graph G is a k-regular spanning subgraph of G. Clearly a 1-factor is a perfect matching. G is *k-factorable* if G is the union of some edge-disjoint k-factors. Show that $K_{n,n}$ and K_{2n} are 1-factorable.

8.18 (a) Show that K_{2n+1} can be expressed as the union of n connected 2-factors $(n \geqslant 1)$.

Note A connected 2-factor is a Hamilton circuit.

(b) Show that K_{2n} is the union of a 1-factor and $n-1$ connected 2-factors.

8.19 Show that a connected graph G is 2-factorable if and only if it is regular with even degree.

8.20 Let M and N be two edge-disjoint matchings of a graph G with $|M|>|N|$. Show that there are disjoint matchings M' and N' with $|M'|=|M|-1$ and $|N'|=|N|+1$, and $M' \cup N' = M \cup N$.

8.21 If $G=(V, E)$ is a bipartite graph and $k \geqslant \Delta$, then show that there exist k disjoint matchings $M_1, M_2, \ldots, M_k$ of G such that

$$E = M_1 \cup M_2 \cup \cdots \cup M_k$$

and for $1 \leqslant i \leqslant k$,

$$\left\lfloor \frac{m}{k} \right\rfloor \leqslant |M_i| \leqslant \left\lceil \frac{m}{k} \right\rceil,$$

where m is the number of edges of G.

8.22 Show that a tree T has a perfect matching if and only if $p_0(v)=1$ for all vertices v in T, where $p_0(v)$ is the number of odd components of $T-v$.

8.23 Derive Hall's theorem from Tutte's theorem (see Exercise 5.3.1 in Bondy and Murty [8.35]).

8.24 Derive Hall's theorem from Menger's theorem (see Wilson [8.36], Theorem 28d).

8.10 REFERENCES

8.1 F. Harary, "The Maximum Connectivity of a Graph," *Proc. Nat. Acad. Sci., U.S.*, Vol. 48, 1142–1146 (1962).

8.2 J. A. Bondy, "Properties of Graphs with Constraints on Degrees," *Studia Sci. Math. Hung.*, Vol. 4, 473–475 (1969).

8.3 G. Chartrand, "A Graph Theoretic Approach to a Communication Problem," *SIAM J. Appl. Math.*, Vol. 14, 778–781 (1966).

8.4 S. L. Hakimi, "On the Realizability of a Set of Integers as Degrees of the Vertices of a Graph," *SIAM J. Appl. Math.*, Vol. 10, 496–506 (1962).

8.5 V. Havel, "A Remark on the Existence of Finite Graphs" (in Hungarian), *Časopois Pěst. Mat.*, Vol. 80, 477–480 (1955).

8.6 D. L. Wang and D. J. Kleitman, "On the Existence of n-Connected Graphs with Prescribed Degrees ($n \geqslant 2$)," *Networks*, Vol. 3, 225–239 (1973).

8.7 P. Erdös and T. Gallai, "Graphs with Prescribed Degrees of Vertices" (in Hungarian), *Mat. Lapok*, Vol. 11, 264–274 (1960).

8.8 F. Harary, *Graph Theory*, Addison-Wesley, Reading, Mass., 1969.

8.9 J. Edmonds, "Existence of k-Edge-Connected Ordinary Graphs with Prescribed Degrees," *J. Res. Nat. Bur. Stand. B.*, Vol. 68, 73–74 (1964).

8.10 D. L. Wang and D. J. Kleitman, "A Note on *n*-Edge Connectivity," *SIAM J. Appl. Math.*, Vol. 26, 313–314 (1974).

8.11 K. Menger, "Zur allgemeinen Kurventheorie," *Fund. Math.*, Vol. 10, 96–115 (1927).

8.12 H. Whitney, "Congruent Graphs and the Connectivity of Graphs," *Am. J. Math.*, Vol. 54, 150–168 (1932).

8.13 W. T. Tutte, "A Theory of 3-Connected Graphs," *Indag. Math.*, Vol. 23, 441–455 (1961).

8.14 L. R. Ford and D. R. Fulkerson, "Maximal Flow Through a Network," *Canad. J. Math.*, Vol. 8, 399–404 (1956).

8.15 P. Elias, A. Feinstein, and C. E. Shannon, "A Note on the Maximum Flow Through a Network," *IRE Trans. Inform. Theory*, IT-2, 117–119 (1956).

8.16 P. Hall, "On Representatives of Subsets," *J. London Math. Soc.*, Vol. 10, 26–30 (1935).

8.17 P. R. Halmos and H. E. Vaughan, "The Marriage Problem," *Am. J. Math.*, Vol. 72, 214–215 (1950).

8.18 D. König, "Graphs and Matrices" (in Hungarian), *Mat. Fiz. Lapok*, Vol. 38, 116–119 (1931).

8.19 O. Ore, "Graphs and Matching Theorems," *Duke Math. J.*, Vol. 22, 625–639 (1955).

8.20 R. Rado, "Note on the Transfinite Case of Hall's Theorem on Representatives," *J. London Math. Soc.*, Vol. 42, 321–324 (1967).

8.21 C. Berge, "Two Theorems in Graph Theory," *Proc. Nat. Acad. Sci. U.S.*, Vol. 43, 842–844 (1957).

8.22 N. S. Mendelsohn and A. L. Dulmage, "Some Generalizations of the Problem of Distinct Representatives," *Canad. J. Math.*, Vol. 10, 230–241 (1958).

8.23 W. T. Tutte, "The Factorization of Linear Graphs," *J. London Math. Soc.*, Vol. 22, 107–111 (1947).

8.24 I. Anderson, "Perfect Matchings of a Graph," *J. Combinatorial Theory B*, Vol. 10, 183–186 (1971).

8.25 L. Lovász, "Three Short Proofs in Graph Theory," *J. Combinatorial Theory B*, Vol. 19, 111–113 (1975).

8.26 C. Berge, *Graphs and Hypergraphs*, North Holland, Amsterdam, 1973.

8.27 F. T. Boesch (Ed.), *Large-Scale Networks: Theory and Design*, IEEE Press, New York, 1976.

8.28 S. L. Hakimi, "An Algorithm for Construction of Least Vulnerable Communication Networks or the Graph with the Maximum Connectivity," *IEEE Trans. Circuit Theory*, Vol. CT-16, 229–230 (1969).

8.29 F. T. Boesch and R. E. Thomas, "On Graphs of Invulnerable Communication Nets," *IEEE Trans. Circuit Theory*, Vol. CT-17, 183–192 (1970).

8.30 A. T. Amin and S. L. Hakimi, "Graphs with Given Connectivity and Independence Number or Networks with Given Measures of Vulnerability and Survivability," *IEEE Trans. Circuit Theory*, Vol. CT-20, 2–10 (1973).

8.31 H. Frank and I. T. Frisch, *Communication, Transmission, and Transportation Networks*, Addison-Wesley, Reading, Mass. 1971.

8.32 L. Mirsky and H. Perfect, "Systems of Representatives," *J. Math. Anal. Applic.*, Vol. 15, 520–568 (1966).

8.33 R. A. Brualdi, "Transversal Theory and Graphs," in *Studies in Graph Theory*, Part II, MAA Press, 1975, pp. 23–88.

8.34 L. Mirsky, *Transversal Theory*, Academic Press, New York, 1971.

8.35 J. A. Bondy and U. S. R. Murty, *Graph Theory with Applications*, Macmillan, London, 1976.

8.36 R. J. Wilson, *Introduction to Graph Theory*, Oliver and Boyd, Edinburgh, 1972.

Chapter 9

Covering and Coloring

In the previous chapters we defined several useful parameters associated with a graph, namely, rank, nullity, connectivity, matching number, and so on. As we mentioned earlier, rank and nullity arise in the study of electrical networks. Connectivity arises in the study of communication nets. In this chapter we study several other useful parameters of a graph—vertex independence number, vertex and edge covering numbers, chromatic index, and chromatic number. We begin with a discussion of the vertex independence number and the vertex and edge covering numbers. We relate these numbers to the matching number defined in the previous chapter and develop equivalent formulations of Hall's theorem. Then we study the chromatic index and chromatic number which relate to the vertex and edge colorability properties of a graph.

The parameters to be discussed in this chapter arise in the study of several practical problems such as time-table scheduling and communication net studies.

9.1 INDEPENDENT SETS AND VERTEX COVERS

Consider a graph $G=(V, E)$. A subset S of V is an *independent set* of G if no two vertices of S are adjacent in G. An independent set is also called a *stable set*.

An independent set S of G is *maximum* if G has no independent set S' with $|S'|>|S|$. The number of vertices in a maximum independent set of G is called the *independence number* (*stability number*) of G and is denoted by $\alpha_0(G)$.

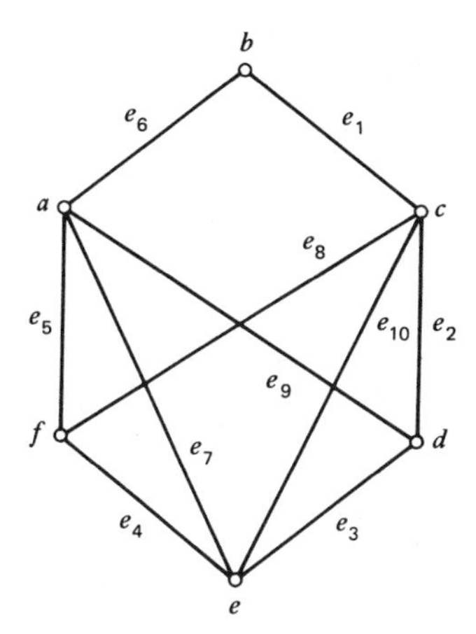

Figure 9.1.

For example, in the graph of Fig. 9.1, the sets $\{b,d\}$, $\{b,f\}$, $\{a,c\}$, and $\{b,d,f\}$ are independent sets. Of these, $\{b,d\}$ and $\{b,f\}$ are not maximal independent sets; $\{a,c\}$ is maximal but it is not maximum; $\{b,d,f\}$ is maximum.

A subset K of V is a *vertex cover* of G if every edge of G has at least one end vertex in K. If we consider a vertex as covering all the edges incident on it, then a vertex cover of G is a subset of V which covers all the edges of G.

A vertex cover K of G is *minimum* if G has no vertex cover K' with $|K'|<|K|$. The number of vertices in a minimum vertex cover of G is called the *vertex covering number* of G and is denoted by $\beta_0(G)$.

For example, in the graph of Fig. 9.1, the sets $\{a,c,e,f\}$, $\{a,c,d,e\}$, $\{b,d,e,f\}$, and $\{a,c,e\}$ are vertex covers. Of these, $\{a,c,e,f\}$ and $\{a,c,d,e\}$ are not minimal; $\{b,d,e,f\}$ is minimal but not minimum; $\{a,c,e\}$ is minimum.

$\alpha_0(G)$ and $\beta_0(G)$ will be denoted simply as α_0 and β_0, respectively, whenever the graph G under consideration is clear from the context. Recall that $\alpha_1(G)$ is the number of edges in a maximum matching of G.

Independent sets and vertex covers are closely related concepts, as we show in the following theorem.

THEOREM 9.1. Consider a graph $G=(V,E)$. A subset S of V is an independent set of G if and only if the complement $\bar{S}$ of S in V (i.e., $\bar{S}=V-S$) is a vertex cover of G.

Proof

By definition S is an independent set of G if and only if no edge of G has both its end vertices in S. In other words, S is an independent set if and only if every edge of G has at least one end vertex in $\bar{S}$, the complement of S in V. The theorem now follows from the definition of a vertex cover. ∎

Corollary 9.1.1 For a simple n-vertex graph, $\alpha_0+\beta_0=n$.

Proof

Consider a maximum independent set S^* and a minimum vertex cover K^* of a graph $G=(V, E)$. Then

$$|S^*|=\alpha_0$$

and

$$|K^*|=\beta_0.$$

By Theorem 9.1, $\bar{S}^*=V-S^*$ is a vertex cover and $\bar{K}^*=V-K^*$ is an independent set. Therefore,

$$|\bar{S}^*|=|V-S^*|=n-\alpha_0 \geqslant \beta_0$$

and

$$|\bar{K}^*|=|V-K^*|=n-\beta_0 \leqslant \alpha_0.$$

Combining the above inequalities, we get

$$\alpha_0+\beta_0=n. \quad \blacksquare$$

Consider any maximum matching M^* and any minimum vertex cover K^* in a graph G. Since at least $|M^*|$ vertices are required to cover the edges of M^*, the vertex cover K^* must contain at least $|M^*|$ vertices. Therefore,

$$|M^*| \leqslant |K^*|. \tag{9.1}$$

In general, equality does not hold in (9.1). However, we show in the following theorem that $|M^*|=|K^*|$ if G is a bipartite graph.

THEOREM 9.2. For a bipartite graph the number of edges in a maximum matching equals the number of vertices in a minimum vertex cover, that is, $\alpha_1=\beta_0$.

Proof

Let M^* be a maximum matching and K^* a minimum vertex cover in a bipartite graph $G=(X, Y, E)$.

Consider any subset A of X. Each edge e of G is incident on either a vertex in A or a vertex in $X-A$. Further, any edge incident on a vertex in A is also incident on a vertex in $\Gamma(A)$, the set of vertices adjacent to those in A. Thus the set $(X-A)\cup\Gamma(A)$ is a vertex cover of G, and hence

$$|(X-A)|+|\Gamma(A)| \geqslant |K^*|.$$

But from Hall's theorem (Theorem 8.13),

$$|M^*| = \min_{A \subseteq X} \{|X-A|+|\Gamma(A)|\} \geqslant |K^*|. \tag{9.2}$$

Now combining (9.1) and (9.2), we get

$$|M^*| = |K^*|. \quad \blacksquare$$

The above theorem is due to König [9.1].

We established in Theorem 8.20 a characterization of a maximum matching using the concept of an alternating chain. Since an independent set is the vertex analog of a matching, a similar characterization may be expected to exist in the case of a maximum independent set too. This is indeed true, and in the following we develop such a characterization. We follow the treatment given in Berge [9.2].

Let us first define an alternating sequence, the vertex analog of an alternating chain.

Let Y be an independent set of a graph $G=(V, E)$. An *alternating sequence* relative to Y is a sequence

$$\sigma = \{x_1, y_1, x_2, y_2, \ldots\}$$

of distinct vertices alternately belonging to $X=\overline{Y}=V-Y$ and Y such that the following conditions are satisfied:

1. $x_i \in X$, $y_i \in Y$.
2. y_i is adjacent to at least one vertex in the set $\{x_1, x_2, \ldots, x_i\}$.
3. x_{i+1} is not adjacent to any vertex in $\{x_1, x_2, \ldots, x_i\}$.
4. x_{i+1} is adjacent to at least one vertex in $\{y_1, y_2, \ldots, y_i\}$.

An alternating sequence is maximal if no more vertices can be added to it without violating the above conditions.

For example, in the graph of Fig. 9.1, the sequence $\{f, a, b, c\}$ is a maximal alternating sequence relative to the independent set $\{a, c\}$.

Consider a tree T. Recall (Exercise 2.2) that T is bipartite, that is, its vertex V set can be partitioned into two subsets X and Y such that

1. $V = X \cup Y$;
2. $X \cap Y = \varnothing$; and
3. X and Y are both independent sets in the tree.

We shall refer to (X, Y) as a *bipartition* of the vertex set of T, and the vertices of X and Y as X- and Y-vertices, respectively.

As a first step toward the development of a characterization of a maximum independent set in terms of an alternating sequence we now show that if $|X|>|Y|$ in a tree, then there is an alternating sequence relative to Y using all the vertices of Y. To do so we need the following lemma.

LEMMA 9.1. Let (X, Y) be a bipartition of the vertex set of a tree T. If $|X|=|Y|+p$, $p \geqslant 0$, then X contains at least $p+1$ pendant vertices of T.

Proof

Proof is by induction on the number n of vertices of T. Clearly, the lemma is true for $n=2$, 3, 4. Let the result be valid for all trees having fewer than n vertices, $n \geqslant 5$.

Consider an n-vertex tree T with $|X|=|Y|+p$. Let v be a pendant vertex of T and let

$$T' = T - \{v\}.$$

Note that T' is a tree with $n-1$ vertices.

Suppose v is an X-vertex. If $p=0$, then v is the required X-vertex of T. Otherwise, by the induction hypothesis, at least p pendant vertices of T' are X-vertices. These p vertices are also pendant in T and hence they, along with v, are the required X-vertices for T.

Proof is immediate if v is a Y-vertex. ■

LEMMA 9.2. Let (X, Y) be a bipartition of the vertex set of a tree T.

1. If $|X|=|Y|$ or $|X|=|Y|+1$, then there is an alternating sequence $\{x_1, y_1, x_2, y_2, \ldots\}$ which uses each vertex of T exactly once.
2. If $|X|>|Y|+1$, then there is a maximal alternating sequence of odd length $\{x_1, y_1, \ldots\}$ which uses each vertex of Y exactly once.

Proof

1. The result is true for a tree with 2 vertices. Assume the result to be true for any tree with $2k$ vertices. Consider then a tree T with $2k+1$ vertices and $|X|=|Y|+1$. Then, by the previous lemma, there exists in T a pendant vertex $x_{k+1} \in X$. Let $T'=T-\{x_{k+1}\}$. By the induction hypothesis, there is in T' an alternating sequence $\{x_1, y_1, \ldots, y_k\}$ which uses all the vertices of T'. Therefore the sequence $\{x_1, y_1, \ldots, y_k, x_{k+1}\}$ is a required sequence for T. (Note that this sequence is of odd length.)

   Next assume the result to be valid for any tree with $2k+1$ vertices. Consider a tree T with $2k+2$ vertices and $|X|=|Y|$. Then by the previous lemma, there exists in T a pendant vertex $y_{k+1} \in Y$. Let $T'=T-\{y_{k+1}\}$. Now, by the induction hypothesis, there is an alternating sequence $\{x_1, y_1, \ldots, x_{k+1}\}$ which uses all the vertices of T'. Therefore the sequence $\{x_1, y_1, \ldots, x_{k+1}, y_{k+1}\}$ is a required sequence for T.

2. If $|X|>|Y|+1$, remove from T as many pendant X-vertices as necessary until we get a tree T' in which the number of X-vertices is one more than the number of Y-vertices. By the previous lemma, this is always possible.

   Now, by the result of part 1 of the lemma, there exists a maximal alternating sequence of odd length which uses all the vertices of T' and hence all the Y-vertices of T. ■

For example, in the tree T of Fig. 9.2, $\{x_1, y_1, x_2, y_2, x_3, y_3, x_4, y_4, x_5, y_5, x_6, y_6, x_7\}$ is an alternating sequence which uses all the Y-vertices of T.

We now state and prove a characterization of a maximum independent set in terms of an alternating sequence.

THEOREM 9.3. An independent set Y in a graph G is maximum if and only if in G there exists no maximal alternating sequence of odd length relative to Y.

Proof

Necessity Consider a graph $G=(V, E)$. Let Y be a maximum independent set in G and $X=V-Y$.

Suppose there exists a maximal alternating sequence σ of odd length relative to Y, and let

$$\sigma=\sigma_1 \cup \sigma_2,$$

where

$$\sigma_2=\sigma \cap Y$$

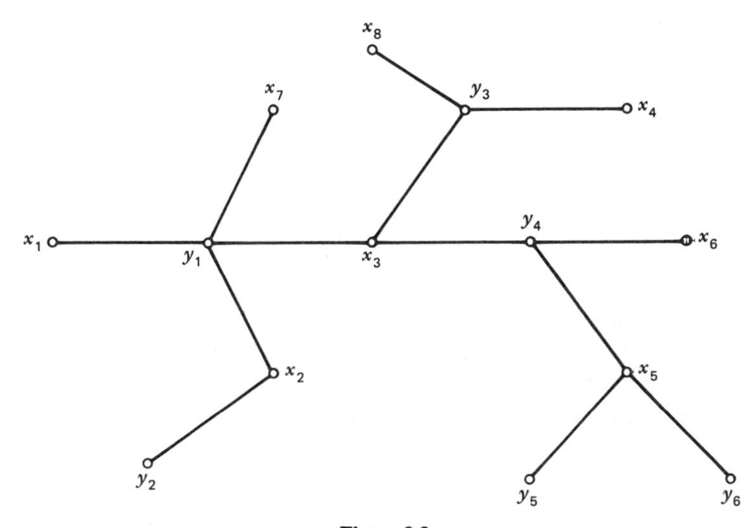

Figure 9.2.

and

$$\sigma_1 = \sigma \cap X.$$

Note that since σ is of odd length, $|\sigma_1| > |\sigma_2|$, and the last vertex of σ is an X-vertex. Furthermore, since σ is maximal, no vertex $y_j \in Y - \sigma_2$ can be added to σ without violating condition (2) in the definition of an alternating sequence.

Thus no vertex $y_j \in Y - \sigma_2$ is adjacent to any vertex $x_i \in \sigma_1$. Since σ_1 and $(Y - \sigma_2)$ are themselves independent sets, it follows that $(Y - \sigma_2) \cup \sigma_1$ is an independent set. So,

$$|(Y - \sigma_2) \cup \sigma_1| > |(Y - \sigma_2)| + |\sigma_2| = |Y|,$$

contradicting that Y is a maximum independent set.

Sufficiency Let X be a maximum independent set and Y an independent set with $|Y| < |X|$. We now show that there exists an alternating sequence of odd length relative to Y. Let

$$Y_0 = Y - (X \cap Y),$$

$$X_0 = X - (X \cap Y),$$

and let G_0 be the induced subgraph of G on the vertex set $(X_0 \cup Y_0)$. If G_0 is not connected, then let $G_1, G_2, \ldots, G_k$ be the connected components of G_0 with $(X_i \cup Y_i)$ as the vertex set of G_i. Note that

$$\bigcup_{i=1}^{k} X_i = X_0$$

and

$$\bigcup_{i=1}^{k} Y_i = Y_0.$$

Since $|X| > |Y|$, $|X_0| > |Y_0|$. Therefore for some i, $|X_i| > |Y_i|$. Let, without loss of generality, $i = 1$.

Now let T be a spanning tree of G_1. Clearly, (X_1, Y_1) is a bipartition of the vertex set of T. Since $|X_1| > |Y_1|$, by the previous lemma there exists in G_1 a maximal alternating sequence σ (relative to Y_1) of odd length which uses all the vertices of Y_1. Clearly σ is an alternating sequence of odd length relative to Y for G also. We now show that σ is a maximal such sequence in G.

Note that the X-vertices and Y-vertices of σ belong to X_0 and Y_0, respectively.

Consider any Y-vertex y_j which is not in σ. Two cases now arise.

Case 1 $y_j \in Y \cap X$.

In this case y_j is not adjacent to X-vertices. In particular y_j is not adjacent to any X-vertex of σ.

Case 2 $y_j \in Y_0$.

In this case y_j is not in G_1 because σ contains all the Y-vertices of G_1. Hence y_j is not adjacent to any of the vertices of G_1. In particular y_j is not adjacent to any X-vertex of σ.

Thus in both the cases, y_j cannot be used to extend σ. Therefore, for G, σ is a maximal alternating sequence of odd length relative to Y. ■

For example, consider the independent set $Y = \{a, c\}$ in the graph of Fig. 9.1. This is not maximum and it may be verified that $\{d, c, b, a, f\}$ is a maximal alternating sequence of odd length relative to Y.

9.2 EDGE COVERS

Consider a graph $G = (V, E)$. A subset P of E is an *edge cover* of G if every vertex of G is an end vertex of at least one of the edges of P. If we consider an edge as covering its end vertices, then an edge cover is a subset of edges covering all the vertices of G.

An edge cover P of G is *minimum* if G has no edge cover P' with $|P'| < |P|$. The number of edges in a minimum edge cover of G is called the *edge covering number* of G and is denoted by $\beta_1(G)$.

For example, in the graph of Fig. 9.1, the set $\{e_1, e_3, e_5\}$ is a minimum edge cover.

Recall that the vertex covering number and the edge covering number are denoted by β_0 and β_1, respectively; similarly the independence number and matching number are denoted by α_0 and α_1, respectively.

In the following we use the same symbol to denote an edge cover as well as the induced subgraph on the edges of the cover.

Suppose that an edge cover P is minimal. Then it is easy to verify that P has neither circuits nor paths of length greater than 2. This means that each component of P is a tree in which all the edges are incident on a common vertex.

In the next theorem, which is the edge analog of Corollary 9.1.1, we relate α_1 and β_1. This result is due to Gallai [9.3].

THEOREM 9.4. For a simple n-vertex graph G without isolated vertices,

$$\alpha_1 + \beta_1 = n.$$

Proof

Let M^* be a maximum matching and P^* a minimum edge cover of G. Let $N(M^*)$ denote the set of vertices which are not saturated in M^*. Then $|N(M^*)|=n-2\alpha_1$. Now, for each vertex v in $N(M^*)$, select an edge which is incident on v and adjacent to an edge in M^*. Let P_a be the set of $(n-2\alpha_1)$ edges so selected. Then it is clear that the set $P=M^*\cup P_a$ is an edge cover and

$$|P|=\alpha_1+n-2\alpha_1=n-\alpha_1.$$

Thus

$$\beta_1=|P^*|\leqslant|P|=n-\alpha_1,$$

so that

$$\alpha_1+\beta_1\leqslant n. \tag{9.3}$$

Now let P^* have r connected components so that

$$\beta_1=|P^*|=n-r.$$

Select one edge from each one of the r components of P. Let M be the set of r edges so chosen. It is clear that M is a matching, and so

$$\alpha_1=|M^*|\geqslant|M|=r=n-\beta_1.$$

Thus

$$\alpha_1+\beta_1\geqslant n. \tag{9.4}$$

Combining (9.3) and (9.4), we get

$$\alpha_1+\beta_1=n. \quad \blacksquare$$

Consider a maximum independent set S^* and a minimum edge cover P^* in a graph G with no isolated vertices. Since $|S^*|$ edges are required to cover the vertices of S^*, any edge cover must contain at least $|S^*|$ edges. Thus

$$|S^*|\leqslant|P^*|. \tag{9.5}$$

In general, equality does not hold in (9.5). However, when a graph G is bipartite, $|S^*|=|P^*|$, and this we prove in the next theorem which is analogous to Theorem 9.2.

THEOREM 9.5. In a bipartite graph G, without isolated vertices, the number of vertices in a maximum independent set is equal to the number of edges in a minimum edge cover, that is,

$$\alpha_0=\beta_1.$$

Proof

Let G have n vertices. Then by Corollary 9.1.1, $\alpha_0 = n - \beta_0$, and by Theorem 9.4, $\beta_1 = n - \alpha_1$. But by Theorem 9.2, $\alpha_1 = \beta_0$. So we get $\alpha_0 = \beta_1$. ■

Theorems 9.2 and 9.5 are both equivalent formulations of Hall's theorem (Exercises 9.5 and 9.6).

9.3 EDGE COLORING AND CHROMATIC INDEX

A *k-edge coloring* of a graph is the assignment of k distinct colors to the edges of the graph. An edge coloring is *proper* if in the coloring no two adjacent edges receive the same color. A 3-edge coloring and a proper 4-edge coloring of a graph are shown in Fig. 9.3. A graph is *k-edge colorable* if it has a proper k-edge coloring.

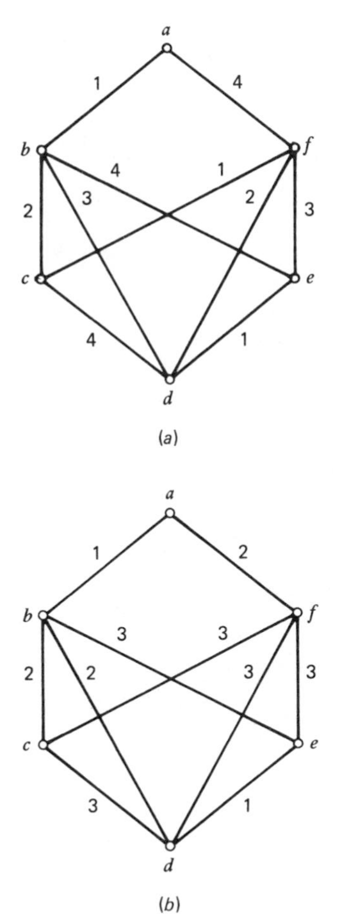

Figure 9.3. Edge coloring (numbers indicate colors). (*a*) A proper 4-edge coloring. (*b*) A 3-edge coloring.

We shall assume, without any loss of generality, that the graphs to be considered in this section have no self-loops.

The *chromatic index* or the *edge chromatic number* $\chi'(G)$ of a graph G is the minimum k for which G has a proper k-edge coloring. Graph G is *k-edge chromatic* if $\chi'(G)=k$. It may be verified that the chromatic index of the graph of Fig. 9.3*a* is 4.

It may be seen that a k-edge coloring of a graph $G=(V,E)$ induces the partition $(E_1, E_2, \ldots, E_k)$ of E, where E_i denotes the subset of edges which are assigned the color i in the coloring. Similarly every partition of E into k subsets corresponds to a k-edge coloring of G. So we shall often denote an edge coloring by the partition of G it induces.

If a coloring $\mathcal{E}=(E_1, E_2, \ldots, E_k)$ is proper, then each E_i is a matching. Therefore $\chi'(G)$ may be regarded as the smallest number of matchings into which the edge set of G can be partitioned. This interpretation of $\chi'(G)$ will be helpful in the proof of certain useful results.

Since in any proper coloring the edges incident on a vertex should receive different colors, it follows that

$$\chi'(G) \geqslant \Delta, \tag{9.6}$$

where Δ is the maximum degree in G.

In general $\chi'(G) \neq \Delta$. However, in the case of a bipartite graph $\chi'=\Delta$.

THEOREM 9.6. For a bipartite graph $\chi'=\Delta$.

Proof

By Corollary 8.22.1, the edge set of a bipartite graph G can be partitioned into Δ matchings. Thus

$$\chi'(G) \leqslant \Delta.$$

Combining this with (9.6) proves the theorem. ■

Though Δ colors are not, in general, sufficient to properly color the edges of a graph, Vizing [9.4] has shown that for any simple graph $\Delta+1$ colors are sufficient. To prove Vizing's theorem we need to establish a few basic results.

We shall say that, in a coloring, color i is represented at vertex v if in the coloring at least one of the edges incident on v is assigned the color i. The number of distinct colors represented at vertex v will be denoted by $c(v)$. For example, in the coloring shown in Fig. 9.3*b*

$$c(f)=2.$$

We now consider the question of existence of 2-edge colorings of a graph in which $c(v)=2$ for every vertex v in the graph.

LEMMA 9.3. Let G be a connected graph which is not a circuit of odd length. Then G has a 2-edge coloring in which both colors are represented at each vertex of degree at least 2.

Proof

Case 1 G is Eulerian.

If G is a circuit, then by the hypothesis of the theorem it must be of even length. In such a case it is easy to verify that G has a 2-edge coloring having the required property.

If G is not a circuit, then it must have a vertex v_0 of degree at least 4. Let

$$v_0, e_1, v_1, e_2, v_2, e_3, \ldots, e_m, v_0$$

be an Euler trail and let

$$E_1 = \{e_i | i \text{ odd}\} \qquad \text{and} \qquad E_2 = \{e_i | i \text{ even}\}. \tag{9.7}$$

Then (E_1, E_2) is a 2-edge coloring in which at every vertex both colors are represented because in the Euler trail considered every vertex including v_0 appears as an internal vertex.

Case 2 G is not Eulerian.

In this case G must have an even number of odd vertices. Now construct an Eulerian graph G' by adding a new vertex v_0 and connecting it to every vertex of odd degree in G. Let $v_0, e_1, v_1, \ldots, e_m, v_0$ be an Euler trail in G'. If E_1 and E_2 are defined as in (9.7), then (E_1, E_2) is a 2-edge coloring of G' in which both colors are represented at each vertex of G. It may now be verified that $(E_1 \cap E, E_2 \cap E)$ is a 2-edge coloring of G in which both colors are represented at each vertex of degree at least 2. ∎

A k-edge coloring $\mathfrak{S}$ of a graph $G=(V, E)$ is *optimum* if there is no other k-edge coloring $\mathfrak{S}'$ of G such that

$$\sum_{v \in V} c(v) < \sum_{v \in V} c'(v),$$

where $c(v)$ and $c'(v)$ are the number of distinct colors represented at vertex v in the colorings $\mathfrak{S}$ and $\mathfrak{S}'$, respectively.

Clearly, in any coloring $\mathfrak{S}$, $c(v) \leq d(v)$ for every vertex v. Further, for every v, $c(v) = d(v)$ if and only if $\mathfrak{S}$ is a proper coloring.

A useful result on the nature of an optimum coloring is presented next.

LEMMA 9.4. Let $\mathfrak{S} = (E_1, E_2, \ldots, E_k)$ be an optimum k-edge coloring of a graph $G=(V, E)$. Suppose that there is a vertex u in G and there are two colors i and j such that i is not represented at u and j is represented twice at u. Let G' be the induced subgraph of G on the edge set $E_i \cup E_j$. Then the component H of G' containing the vertex u is a circuit of odd length.

Proof

If H is not a circuit of odd length, then, by Lemma 9.3, there is a 2-edge coloring of H in which both colors i and j are represented at each vertex of degree at least 2 in H.

If we use this 2-edge coloring to recolor the edges of H and leave the colors of the other edges of G unaltered as they are in $\mathcal{E}$, then we get a new k-edge coloring $\mathcal{E}'$ in which both colors i and j are represented at vertex u. Hence,

$$c'(u) = c(u) + 1,$$

where $c'(u)$ is the number of distinct colors represented at vertex u in the coloring $\mathcal{E}'$. Further, for every vertex $v \neq u$,

$$c'(v) \geqslant c(v).$$

Hence

$$\sum_{v \in V} c'(v) > \sum_{v \in V} c(v),$$

contradicting that $\mathcal{E}$ is an optimum edge coloring. Thus it follows that H is a circuit of odd length. ■

A k-edge coloring $\mathcal{E}'$ is an *improvement* on the k-edge coloring $\mathcal{E}$ if

$$\sum_{v \in V} c'(v) > \sum_{v \in V} c(v).$$

We now prove Vizing's theorem [9.4]. Our proof is due to Fournier [9.5]. We follow the treatment given in Bondy and Murty [9.6].

THEOREM 9.7 (VIZING). If $G = (V, E)$ is a simple graph, then either $\chi'(G) = \Delta$ or $\chi'(G) = \Delta + 1$.

Proof

Since $\chi'(G) \geqslant \Delta$, it is enough if we prove that $\chi'(G) \leqslant \Delta + 1$.

Let $\mathcal{E} = \{E_1, E_2, \ldots, E_{\Delta+1}\}$ be an optimal $(\Delta+1)$-edge coloring of G. Assume that $\chi'(G) > \Delta + 1$, that is, $\mathcal{E}$ is not a proper $(\Delta+1)$-edge coloring. Then there exists a vertex u in G such that $c(u) < d(u)$. This implies that there exist colors i_0 and i_1 such that i_0 is not represented at u and i_1 is represented twice at u. Let edge (u, v_1) have color i_1 (Fig. 9.4a).

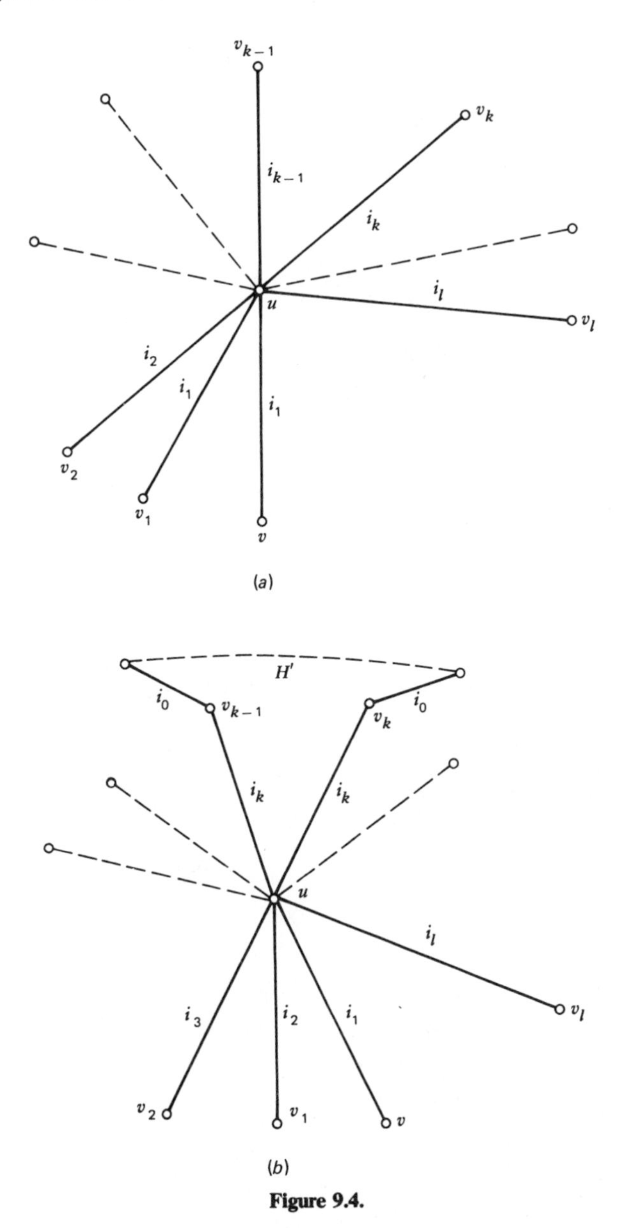

Figure 9.4.

Since the maximum degree in G is Δ, it follows that in the coloring $\mathfrak{S}$, not all the $\Delta+1$ colors are represented at v_1.

Let color i_2 be not represented at vertex v_1. Then i_2 must be represented at vertex u, for otherwise we can recolor edge (u, v_1) with color i_2 and obtain a new coloring of G which is an improvement on $\mathfrak{S}$. Let edge (u, v_2) have color i_2. Again not all the colors are represented at v_2.

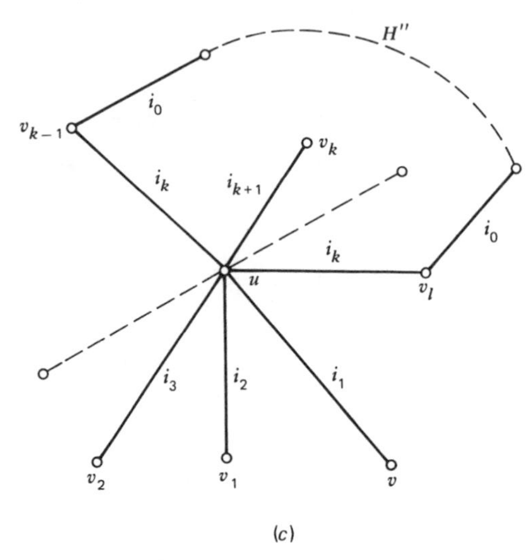

(c)

Figure 9.4. (*continued*)

Suppose that color i_3 is not represented at v_2. Then it must be represented at u, for otherwise we can recolor edge (u, v_1) with i_2 and edge (u, v_2) with i_3, thereby obtaining an improvement on $\mathcal{E}$. Let edge (u, v_3) have color i_3.

Repeating the above arguments, we can construct a sequence $v_1, v_2, \dots$ of vertices and a sequence $i_1, i_2, \dots$ of colors such that

1. (u, v_j) has color i_j, and
2. i_{j+1} is not represented at v_j.

We can now easily see that there exists a smallest integer l such that, for some $k < l$,

$$i_{l+1} = i_k \qquad \text{(Fig. 9.4}a\text{)}.$$

To establish a contradiction, we next recolor, in two different ways, some of the edges incident on u and obtain two new optimal $(\Delta+1)$-edge colorings $\mathcal{E}'$ and $\mathcal{E}''$.

$\mathcal{E}' = \{E_1', E_2', \dots, E_{\Delta+1}'\}$ is obtained by recoloring (u, v_j) with color i_{j+1}, $1 \leq j \leq k-1$. Each one of the remaining edges of G receives in $\mathcal{E}'$ the same color as it does in $\mathcal{E}$ (Fig. 9.4b).

$\mathcal{E}'' = \{E_1'', E_2'', \dots, E_{\Delta+1}''\}$ is obtained by recoloring (u, v_j) with color i_{j+1}, $1 \leq j \leq l-1$, and (u, v_l) with color i_k. Each one of the remaining edges of G receives in $\mathcal{E}''$ the same color as it does in $\mathcal{E}$ (Fig. 9.4c).

Let G' be the subgraph of G on the edge set $(E_{i_0}' \cup E_{i_k}')$ with H' denoting its component containing u. Similarly, G'' is the subgraph of G on the edge set $(E_{i_0}'' \cup E_{i_k}'')$ with H'' denoting its component containing u.

In both $\mathcal{E}'$ and $\mathcal{E}''$, color i_0 is not represented at u and color i_k is represented twice at this vertex. Therefore it follows from Lemma 9.4 that H' and H'' are both circuits.

Now (u, v_k) is the only edge of H' which is not in H''. Therefore u and v_k are connected in G'' too. Thus they lie in the same component of G'', namely, the component H''. The degree of v_k in H'' is then 1, for otherwise its degree in H' will be greater than 2. This contradicts that H'' is also a circuit. Thus $\mathfrak{S}$ is a proper $(\Delta+1)$-edge coloring. ■

For a more general result than the above theorem see Exercise 9.8.

9.4 VERTEX COLORING AND CHROMATIC NUMBER

A *k-vertex coloring* of a graph is the assignment of k-distinct colors to the vertices of the graph. A vertex coloring is *proper* if in the coloring no two adjacent vertices receive the same color. A proper 3-vertex coloring is shown in Fig. 9.5. A graph is *k-vertex colorable* if it has a proper k-vertex coloring.

We may assume, without any loss of generality, that the graphs to be considered in this section are simple.

The *chromatic number* $\chi(G)$ of a graph G is the minimum number k for which G is k-vertex colorable. G is *k-chromatic* if $\chi(G)=k$. For example, the chromatic number of the graph in Fig. 9.5 is 3.

We shall hereafter refer to a "proper k-vertex coloring" as simply a "k-coloring." Similarly, "k-vertex colorable" will be abbreviated to "k-colorable."

Note that a k-coloring of a graph $G=(V, E)$ induces a partition $(V_1, V_2, \ldots, V_k)$ of V, where each V_i is the subset of vertices assigned the color i and therefore is an independent set. Similarly each partition of V into k independent sets corresponds to a k-coloring of G.

We proved in the previous section (Theorem 9.7) that at most $\Delta+1$ colors are required to properly color the edges of a simple graph. We now prove an analogous result for vertex coloring.

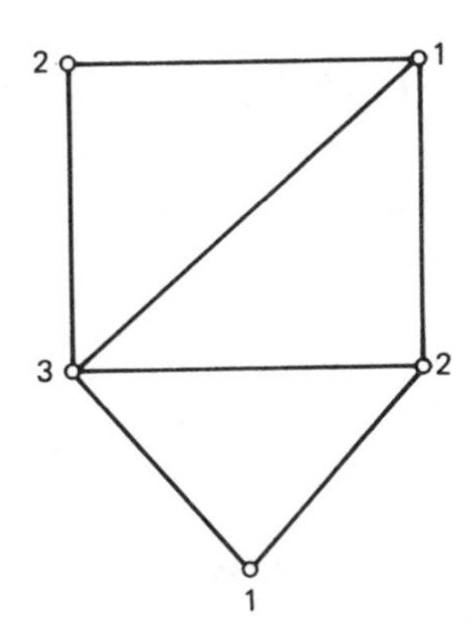

Figure 9.5. A proper 3-vertex coloring.

THEOREM 9.8. If a graph G is simple, then it is $(\Delta+1)$-colorable.

Proof

Given $\Delta+1$ distinct colors, we can obtain a $(\Delta+1)$-coloring of G as follows.

Pick any vertex v_0 and assign to it any one of the $\Delta+1$ colors. Then select an uncolored vertex, say v_1. Assign to v_1 a color which has not been assigned to the vertices adjacent to it. This is always possible, because $d(v_1) \leqslant \Delta$, and hence at most Δ colors will have been assigned to the vertices adjacent to v_1. Repeat this process until all the vertices are colored. The resulting coloring is clearly a proper $(\Delta+1)$-coloring. ■

Clearly, $\chi = \Delta+1$ for complete graphs and for circuits of odd length. A very interesting result is that for all other graphs $\chi \leqslant \Delta$. This result due to Brooks [9.7] is proved next. Our proof here is due to Melnikov and Vizing [9.8]. For an alternate proof see Lovász [9.9].

THEOREM 9.9 (BROOKS). Let G be a connected simple graph. If G is neither a circuit of odd length nor a complete graph, then $\chi(G) \leqslant \Delta$.

Proof

Clearly the theorem is valid for $\Delta = 0, 1, 2$.

To prove the theorem for $\Delta \geqslant 3$, assume the contrary, that is, there exist graphs which are not complete and for which $\Delta \geqslant 3$ and $\chi = \Delta+1$. Then select such a graph $G=(V, E)$ with the minimum number of vertices.

Let $v_0 \in V$, and G' be the graph that results after removing v_0 from G. It follows from the choice of G that G' is Δ-colorable. This implies that $d(v_0) = \Delta$, for otherwise one of the Δ colors used to color G' can be used to color v_0, contradicting that $\chi(G) = \Delta+1$. Another important implication is the following.

Property 1 In any Δ-coloring of G' the vertices adjacent to v_0 are colored differently.

Let $u_1, u_2, \ldots, u_\Delta$ be the vertices adjacent to v_0. Let $u_1, u_2, \ldots, u_\Delta$ receive the colors $1, 2, \ldots, \Delta$, respectively, in a coloring of G'. Let $G(i, j)$ denote the induced subgraph of G' on the vertices which are assigned colors i and j. Then

Property 2 Vertices u_i and u_j are in the same connected component of $G(i, j)$.

For otherwise, by interchanging the colors i and j in the component containing u_i, we can get a new Δ-coloring of G' in which u_i and u_j are assigned the same color, contradicting Property 1.

Let C_{ij} be the component of $G(i, j)$ containing u_i and u_j. Then

Property 3 C_{ij} is a path from u_i to u_j.

Suppose that the degree of u_i in C_{ij} is greater than 1. Then u_i is adjacent to at least two vertices of color j. Since in G', $d(u_i) \leqslant \Delta - 1$, we can recolor u_i with a color $k \neq i, j$ so that in the resulting new coloring u_i and u_k receive the same color, contradicting Property 1.

Similarly, the degree of u_j in C_{ij} is 1.

The degree in C_{ij} of all other vertices is 2. For otherwise, let u be the first vertex of degree (in C_{ij}) greater than 2 as we move on a path from u_i to u_j. If u is colored with color i, then it is adjacent to at least three vertices of color j. Since $d(u) \leqslant \Delta$, we can then recolor u with a color $k \neq i, j$ so that in the new coloring u_i and u_j will be in different components, contradicting Property 2.

Thus C_{ij} is a path from u_i to u_j.

Property 4 C_{ij} and C_{ik} have no common vertex except u_i.

Let $u \neq u_i$ be common to C_{ij} and C_{ik}. Then u is colored with color i, and it is adjacent to at least two vertices of color j and two vertices of color k. Since $d(u) \leqslant \Delta$, there exists a color $l \neq i, j, k$ with which we can recolor u. But this would disconnect u_i and u_j, contradicting Property 2.

Now we proceed to establish a contradiction of Property 4.

Since G is not a $(\Delta + 1)$-vertex complete graph, there exist two vertices, say u_1 and u_2, which are not adjacent. Then the path C_{12} contains a vertex $u \neq u_2$ adjacent to u_1. Suppose that we interchange colors 1 and 3 in the path C_{13} (which exists because $\Delta \geqslant 3$), so that in the new coloring C', u_1 receives color 3 and u_3 receives color 1. But then the new components C'_{12} and C'_{23} contain the common vertex $u \neq u_2$, contradicting Property 4.

This completes the proof of the theorem. ■

9.5 CHROMATIC POLYNOMIALS

In this section we discuss the question of counting the number of distinct proper λ-colorings of a graph.

If a graph is λ-colorable, then it may be possible to color it with λ colors in more than one way. Two colorings of a graph are considered distinct if at least one vertex of the graph is assigned different colors in the two colorings.

The *chromatic polynomial* $P(G, \lambda)$ expresses for each integer λ the number of distinct λ-colorings possible for a graph G.

For example, consider the graph shown in Fig. 9.6. Given λ colors, we can select any one of them to color vertex a. Vertex b can then be colored with any one of the remaining $\lambda - 1$ colors. For each coloring of vertex b, there are $\lambda - 1$ different ways of coloring vertex c. Thus the graph of Fig. 9.6 can be colored in $\lambda(\lambda - 1)^2$ different ways. In other words, the chromatic polynomial

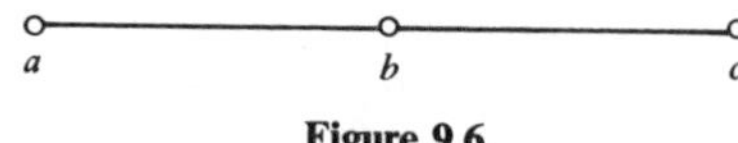

Figure 9.6.

of the graph is $\lambda(\lambda-1)^2$. In fact, we can repeat these arguments to show that the chromatic polynomial of a path on n vertices is equal to $\lambda(\lambda-1)^{n-1}$.

As another example, consider next K_n, the complete graph on n vertices $v_1, v_2, \ldots, v_n$. Given λ colors, the vertex v_1 can be colored with any one of the λ colors, vertex v_2 with any one of the remaining $\lambda-1$ colors, vertex v_3 with any one of the remaining $\lambda-2$ colors, and so on. Thus

$$P(K_n, \lambda)=\lambda(\lambda-1)\cdots(\lambda-n+1).$$

Next we present a formula for determining the chromatic polynomial of a graph G.

THEOREM 9.10. Let u and v be two nonadjacent vertices in a simple graph G. Let $e=(u, v)$. If $G\cdot e$ denotes the simple graph obtained from G by short-circuiting the vertices u and v and replacing the resulting sets of parallel edges with single edges, and if $G+e$ denotes the graph obtained by adding the edge e to G, then

$$P(G, \lambda)=P(G+e, \lambda)+P(G\cdot e, \lambda).$$

Proof

Each λ-coloring of G in which the vertices u and v are assigned different colors corresponds to a λ-coloring of $G+e$ and vice versa. Similarly, each λ-coloring of G in which u and v are assigned the same color corresponds to a λ-coloring of $G\cdot e$ and vice versa. Hence

$$P(G, \lambda)=P(G+e, \lambda)+P(G\cdot e, \lambda). \quad \blacksquare$$

The above result can be stated in a different form as follows.

Corollary 9.10.1 If $e=(u, v)$ is an edge of a simple graph G, then

$$P(G, \lambda)=P(G-e, \lambda)-P(G\cdot e, \lambda),$$

where $G-e$ is obtained by removing e from G and $G\cdot e$ is as defined in Theorem 9.10. ■

If we repeatedly apply the formula given in Theorem 9.10 on a graph G, then the process will terminate in complete graphs, say, $H_1, H_2, \ldots, H_k$, so that

$$P(G, \lambda)=P(H_1, \lambda)+P(H_2, \lambda)+\cdots+P(H_k, \lambda).$$

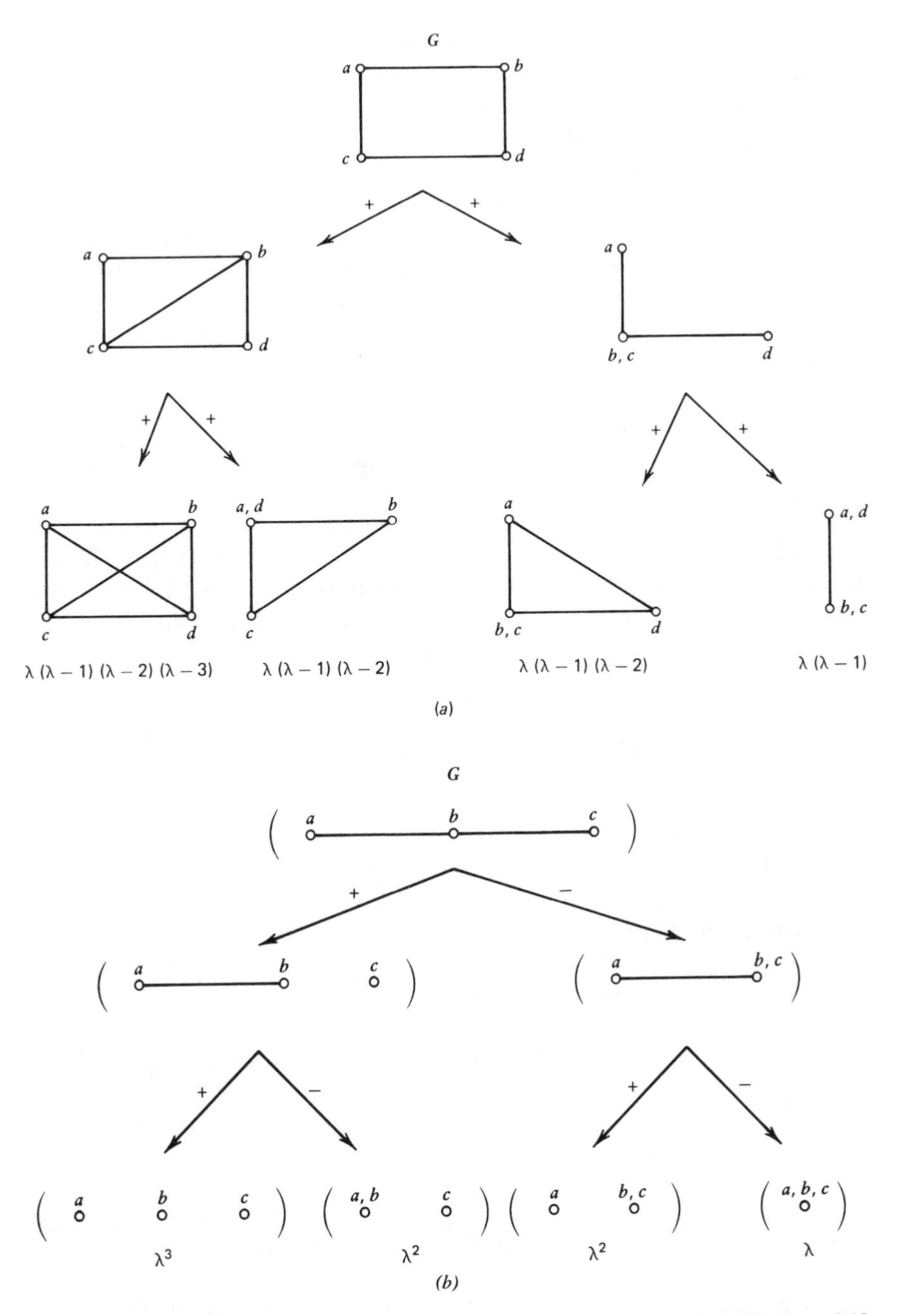

Figure 9.7. Calculation of the chromatic polynomial of a graph G. (a) Using Theorem 9.10. $P(G, \lambda) = \lambda(\lambda-1)(\lambda-2)(\lambda-3) + 2\lambda(\lambda-1)(\lambda-2) + \lambda(\lambda-1)$. ($b$) Using Corollary 9.10.1. $P(G, \lambda) = \lambda^3 - 2\lambda^2 + \lambda$.

On the other hand, if we use the formula given in Corollary 9.10.1, the process will terminate in empty graphs (i.e., graphs that have no edges) so that the chromatic polynomial is a linear combination of the chromatic polynomials of empty graphs.

Both these procedures are illustrated in Fig. 9.7.

THEOREM 9.11. The chromatic polynomial $P(G, \lambda)$ of an n-vertex graph G is of degree n, with leading term λ^n and constant term zero. Furthermore, the coefficients are all integers and alternate in sign.

Proof

The proof is by induction on the number m of edges. Clearly the theorem is valid for $m=0$, since the chromatic polynomial of an n-vertex empty graph is λ^n.

Assume that the theorem holds for all graphs with fewer than m edges. Consider now an n-vertex graph G with m edges. Let e be an edge of G. Then $G-e$ is an n-vertex graph with $m-1$ edges and $G\cdot e$ is an $(n-1)$-vertex graph with $m-1$ or less edges.

By the induction hypothesis it now follows that there are nonnegative integer coefficients $a_1, a_2, \ldots, a_{n-1}$ and $b_1, b_2, \ldots, b_{n-2}$, such that

$$P(G-e, \lambda) = \lambda^n - a_{n-1}\lambda^{n-1} + a_{n-2}\lambda^{n-2} - \cdots + (-1)^{n-1}a_1\lambda$$

and

$$P(G\cdot e, \lambda) = \lambda^{n-1} - b_{n-2}\lambda^{n-2} + b_{n-3}\lambda^{n-3} - \cdots + (-1)^{n-2}b_1\lambda.$$

By Corollary 9.10.1,

$$\begin{aligned} P(G, \lambda) &= P(G-e, \lambda) - P(G\cdot e, \lambda) \\ &= \lambda^n - (a_{n-1}+1)\lambda^{n-1} + \sum_{i=2}^{n-1} (-1)^i (a_{n-i} + b_{n-i})\lambda^{n-i}. \end{aligned}$$

Thus G too satisfies the theorem. ■

9.6 THE FOUR-COLOR PROBLEM

In map making, to distinguish the different regions, one is interested in coloring the regions so that no two adjacent regions receive the same color. This problem of properly coloring the regions of a planar graph is the same as that of properly coloring the vertices of the dual of the graph. It is easy to construct an example to show that three colors are, in general, not sufficient to properly color a planar graph. In the following theorem, known as the *five-color theorem*, we show that five colors are sufficient.

THEOREM 9.12. Every planar graph is 5-colorable.

Proof

We shall assume that the theorem is true for all planar graphs of order less than n and then show that it is also true for a planar graph $G=(V, E)$ of order n.

By Corollary 7.3.5, there exists in G a vertex v_0 of degree $\leqslant 5$. Let G' be the induced subgraph of G on $V-\{v_0\}$.

Color G' with the five colors $\alpha_1, \alpha_2, \ldots, \alpha_5$. (This is possible by the induction hypothesis.) Clearly, one of these colors can be assigned to v_0 in a proper 5-coloring of G, unless $d(v_0)=5$, and all the 5 vertices $v_1, v_2, \ldots, v_5$ adjacent to v_0 are assigned different colors.

Assume that v_i, $1 \leqslant i \leqslant 5$, is assigned the color α_i. Further, let $v_1, v_2, \ldots, v_5$ be arranged in clockwise order as shown in Fig. 9.8.

Let $G(\alpha_i, \alpha_j)$ be the subgraph of G on the vertices which have been assigned the color α_i or α_j.

In $G(\alpha_1, \alpha_3)$, the component that contains v_1 should contain v_3; for otherwise the colors α_1 and α_3 could be interchanged in this component and v_0 could be colored with α_1.

Similarly, in $G(\alpha_1, \alpha_4)$ the component that contains v_1 should also contain v_4. (See Fig. 9.8.)

Then in $G(\alpha_2, \alpha_5)$, v_2 and v_5 will not be connected. If now colors α_2 and α_5 are interchanged in the component of $G(\alpha_2, \alpha_5)$ that contains v_2, then v_0 can be colored with α_2. Thus G is 5-colorable. ■

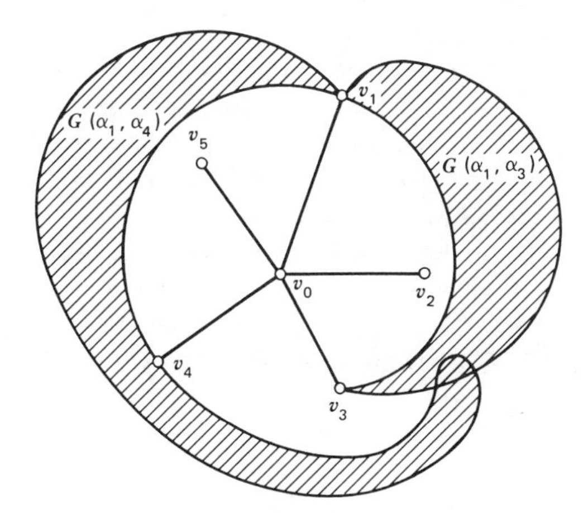

Figure 9.8.

The above result is due to Heawood [9.10].

The question now arises as to whether the five-color theorem is the best possible. It was conjectured that every planar graph is 4-colorable. This was known as the *four-color conjecture*. This conjecture remained unsolved for more than 100 years and was the subject of extensive research. Recently, Appel and Haken [9.11] proved this conjecture true. This is stated in the following theorem.

THEOREM 9.13 (FOUR-COLOR THEOREM). Every planar graph is 4-colorable. ■

There is a vast literature on the four-color problem. See Ore [9.12], Saaty [9.13], and Saaty and Kainen [9.14] for an extensive survey of the problem with historical details. See also Whitney and Tutte [9.15].

9.7 FURTHER READING

Berge [9.2], Bondy and Murty [9.6], and Harary [9.16] are general references for the topics covered in this chapter.

A classical result in graph theory due to Turán [9.17] determines the minimum number of edges that an n-vertex graph with independence number $\leqslant k$ must have (Exercise 9.2). See Erdös [9.18]. This result is the starting point for a branch of graph theory known as *extremal graph theory*. The book by Bollobás [9.19] is devoted exclusively to the study of extremal graph problems. Extremal results form the basis of the design of communication nets having specified vulnerability and reliability properties. See Frank and Frisch [9.20], Boesch [9.21], and Karivaratharajan and Thulasiraman [9.22], [9.23].

See Berge [9.2] (Chapter 16), Liu [9.24] (Chapter 9), and Lovász [9.25] for an application of the independence number in information theory.

Read [9.26] gives an excellent survey of results on chromatic polynomials. See also Birkhoff [9.27] and Tutte [9.28], [9.29]. Liu [9.24] (pp. 245–246) discusses an application of coloring in the decomposition of a graph into planar subgraphs. This problem arises in the design of printed circuits. See Wood [9.30] for an application of coloring in time-tabling problems, and Christofides [9.31] for coloring algorithms.

A set of vertices D in a graph G is a *dominating set* in G if every vertex not in D is adjacent to a vertex in D. For a review of results concerning dominating sets, see Cockayne and Hedetniemi [9.32]. An extremal result concerning dominating sets is given in Cockayne and Hedetniemi [9.33].

9.8 EXERCISES

9.1 Prove or disprove: Every vertex cover of a graph contains a minimum vertex cover.

9.2 Given any two integers n and k with $n \geq k > 0$. Let $q = \lfloor n/k \rfloor$ and let r be an integer such that

$$n = kq + r, \qquad 0 \leq r < k.$$

Let $G_{n,k}$ be the simple graph that consists of k disjoint complete graphs, of which r have $q+1$ vertices and $k-r$ have q vertices. Then show that every graph G with n vertices and $\alpha_0(G) \leq k$ that has the minimum possible number of edges is isomorphic to $G_{n,k}$ (Turán [9.17]). An easy corollary of this result is:

If G is a simple graph with n vertices and m edges and with $\alpha_0(G) = k$, then

$$m \geq q\left(n - \frac{k}{2} - \frac{kq}{2}\right).$$

9.3 If G is a simple graph with n vertices and m edges, then show that

$$\alpha_0(G) \geq \frac{n^2}{2m+n}.$$

9.4 Show that the maximum number of edges that an n-vertex simple graph G with no triangles can have is $\lfloor n^2/4 \rfloor$. Construct a graph with this property.

9.5 Derive Hall's theorem, Theorem 8.13, from Theorem 9.2.

9.6 Derive Hall's theorem, Theorem 8.13, from Theorem 9.5.

9.7 Show that

$$\chi'(K_n) = \begin{cases} n-1, & \text{if } n \text{ is even.} \\ n, & \text{if } n \text{ is odd.} \end{cases}$$

9.8 Show that if G is a graph without self-loops, then

$$\chi'(G) \leq \Delta + k,$$

where k is the maximum number of parallel edges between any two vertices of G (Vizing [9.4]).

9.9 Let G be a graph without self-loops. If $\Delta(G) = 3$, then show that $\chi'(G) = 3$ or 4.

9.10 Show that, for a regular nonempty simple graph with odd number of vertices,

$$\chi'(G) = \Delta + 1.$$

9.11 Show that for any arbitrary orientation of the edges of a simple graph G, there exists a directed path of length $\chi(G) - 1$.

9.12 Show that if any two circuits of odd length of a graph G have a vertex in common, then

$$\chi(G) \leqslant 5.$$

9.13 Using Corollary 7.3.5, show that every planar graph is 6-colorable.

9.14 Show that if a simple graph G has degree sequence $(d_1, d_2, \ldots, d_n)$ with $d_1 \geqslant d_2 \geqslant \cdots \geqslant d_n$, then

$$\chi(G) \leqslant \max_i \min\{d_i + 1, i\}$$

(Welsh and Powell [9.34]).

9.15 Show that for an n-vertex simple graph G,

(a) $\chi(G)\alpha_0(G) \geqslant n$;
(b) $\chi(G) + \alpha_0(G) \leqslant n+1$;
(c) $\chi(G) + \chi(\bar{G}) \leqslant n+1$.

9.16 Let G be a simple n-vertex regular graph of degree k. Show that

$$\chi(G) \geqslant \frac{n}{n-k}.$$

9.17 Let G be a planar graph in which each region is bounded by exactly three edges. Show that G is 3-colorable if and only if G is Eulerian.

9.18 Show that a simple graph is nonplanar if it has seven vertices and the degrees of its vertices are all equal to 4 (Liu [9.24]).

9.19 (a) Show that a graph is 2-colorable if and only if all its circuits are of even length.
(b) Show that the regions of a planar graph G can be properly colored with two colors if and only if G is Eulerian.

9.20 Find the chromatic polynomial of the graph shown in Fig. 9.9.

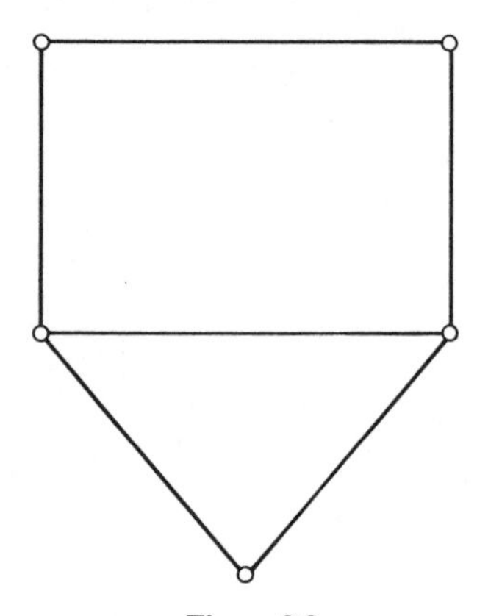

Figure 9.9.

9.21 (a) Show that an n-vertex graph G is a tree if and only if

$$P(G,\lambda)=\lambda(\lambda-1)^{n-1}.$$

(b) Show that if an n-vertex graph G is connected, then

$$P(G,\lambda)\leqslant\lambda(\lambda-1)^{n-1}.$$

9.22 Show that if G is a circuit of length n, then

$$P(G,\lambda)=(\lambda-1)^{n}+(-1)^{n}(\lambda-1).$$

9.23 Show that if G is a wheel on $n+1$ vertices, then

$$P(G,\lambda)=\lambda(\lambda-2)^{n}+(-1)^{n}\lambda(\lambda-2).$$

9.24 Show that if a simple graph G has n vertices, m edges, and k components, then

(a) The coefficient of λ^{n-1} in $P(G,\lambda)$ is $-m$.

(b) The smallest exponent of λ in $P(G,\lambda)$ with a nonzero coefficient is k.

9.25 A *clique* of a graph $G=(V,E)$ is a set $S\subseteq V$ such that the induced subgraph of G on S is a complete graph. Let $P=\{P_1, P_2,\ldots, P_k\}$ be a partition of V into cliques. Let

$$\theta(G)=\min_{P}\{|P|\}.$$

Thus $\theta(G)$ is the smallest possible number of cliques into which V can be partitioned.

Show that, for a simple graph G, $\alpha_0(G)\leqslant\theta(G)$. Further, if S is an independent set and P is a partition of V into cliques such that $|S|=|P|$, then show that S is a maximum independent set and P is a minimum partition of V into cliques.

9.9 REFERENCES

9.1 D. König, "Graphs and Matrices," (in Hungarian), *Mat. Fiz. Lapok*, Vol. 38, 116–119 (1931).

9.2 C. Berge, *Graphs and Hypergraphs*, North Holland, Amsterdam, 1973.

9.3 T. Gallai, "Über extreme Punkt und Kantenmengen," *Ann. Univ. Sci. Budapest, Eötvös Sect. Math.*, Vol. 2, 133–138 (1959).

9.4 V. G. Vizing, "On an Estimate of the Chromatic Class of a p-Graph," (in Russian), *Diskret. Analiz.*, Vol. 3, 25–30 (1964).

9.5 J. C. Fournier, "Colorations des arêtes d'un graphe," *Cahiers du CERO*, Vol. 15, 311–314 (1973).

9.6 J. A. Bondy and U. S. R. Murty, *Graph Theory with Applications*, Macmillan, London, 1976.

9.7 R. L. Brooks, "On Colouring the Nodes of a Network," *Proc. Cambridge Phil. Soc.*, Vol. 37, 194–197 (1941).

9.8 L. S. Melnikov and V. G. Vizing, "New Proof of Brooks' Theorem," *J. Combinatorial Theory*, Vol. 7, 289–290 (1969).

9.9 L. Lovász, "Three Short Proofs in Graph Theory," *J. Combinatorial Theory B*, Vol. 19, 111–113 (1975).

9.10 P. J. Heawood, "Map Colour Theorems," *Quart. J. Math.*, Vol. 24, 332–338 (1890).

9.11 K. I. Appel and W. Haken, "Every Planar Map is Four-Colorable," *Bull. Am. Math. Soc.*, Vol. 82, 711–712 (1976).

9.12 O. Ore, *The Four-Color Problem*, Academic Press, New York, 1967.

9.13 T. L. Saaty, "Thirteen Colorful Variations on Guthrie's Four-Color Conjecture," *Am. Math. Monthly*, Vol. 79, 2–43 (1972).

9.14 T. L. Saaty and P. C. Kainen, *The Four-Color Problem: Assaults and Conquests*, McGraw-Hill, New York, 1977.

9.15 H. Whitney and W. T. Tutte, "Kempe Chains and the Four Colour Problem," in *Studies in Graph Theory*, Part II, MAA Press, 1975, pp. 378–413.

9.16 F. Harary, *Graph Theory*, Addison-Wesley, Reading, Mass., 1969.

9.17 P. Turán, "An Extremal Problem in Graph Theory," (in Hungarian), *Mat. Fiz. Lapok*, Vol. 48, 436–452 (1941).

9.18 P. Erdös, "On the Graph Theorem of Turán," (in Hungarian), *Mat. Lapok*, Vol. 21, 249–251, (1970).

9.19 B. Bollobás, *Extremal Graph Theory*, Academic Press, New York, 1978.

9.20 H. Frank and I. T. Frisch, *Communication, Transmission, and Transportation Networks*, Addison-Wesley, Reading, Mass., 1971.

9.21 F. T. Boesch (Ed.), *Large-Scale Networks: Theory and Design*, IEEE Press, New York, 1976.

9.22 P. Karivaratharajan and K. Thulasiraman, "K-Sets of a Graph and Vulnerability of Communication Nets," *Matrix and Tensor Quart.*, Vol. 25, 63–66, (1974), 77–86 (1975).

9.23 P. Karivaratharajan and K. Thulasiraman, "An Extremal Problem in Graph Theory and Its Applications," *Proc. IEEE Intl. Symp. on Circuits and Systems*, Tokyo, Japan, 1979.

9.24 C. L. Liu, *Introduction to Combinatorial Mathematics*, McGraw-Hill, New York, 1968.

9.25 L. Lovász, "On the Shannon Capacity of a Graph," *IEEE Trans. Inform. Theory*, Vol. IT-25, 1–7 (1979).

9.26 R. C. Read, "An Introduction to Chromatic Polynomials," *J. Combinatorial Theory*, Vol. 4, 52–71 (1968).

9.27 G. D. Birkhoff, "A Determinant Formula for the Number of Ways of Coloring a Map," *Ann. Math.*, Vol. 14, 42–46 (1912).

9.28 W. T. Tutte, "On Chromatic Polynomials and the Golden Ratio," *J. Combinatorial Theory*, Vol. 9, 289–296 (1970).

9.29 W. T. Tutte, "Chromials," in *Studies in Graph Theory*, Part II, MAA Press, 1975, pp. 361–377.

9.30 D. C. Wood, "A Technique for Colouring a Graph Applicable to Large-Scale Timetabling Problems," *Computer J.*, Vol. 12, 317 (1969).

9.31 N. Christofides, *Graph Theory: An Algorithmic Approach*, Academic Press, New York, 1975.

9.32 E. J. Cockayne and S. T. Hedetniemi, "Towards a Theory of Domination in Graphs," *Networks*, Vol. 7, 247–267 (1977).

9.33 E. J. Cockayne and S. T. Hedetniemi, "Optimal Domination in Graphs," *IEEE Trans. Circuits and Sys.*, Vol. CAS-22, 41–44 (1975).

9.34 D. J. A. Welsh and M. B. Powell, "An Upper Bound for the Chromatic Number of a Graph and Its Application to Timetabling Problems," *Computer J.*, Vol. 10, 85–87 (1967).

Chapter 10

Matroids

Consider a finite set S of vectors over an arbitrary field. It is well known that each subset of vectors of S is either linearly dependent or linearly independent. Further, the collection of the independent sets of vectors possesses several interesting properties. For example,

1. Any subset of an independent set is independent.
2. If I_p and I_{p+1} are any two independent sets with $|I_{p+1}|=|I_p|+1$, then I_p together with some element of I_{p+1} forms an independent set of $|I_{p+1}|$ elements.

It is interesting to note that there are several algebraic systems which possess the above properties. For instance, the collection of subsets of edges of a graph which do not contain any circuit possesses these properties. It was while studying the properties of such systems that Whitney [10.1] introduced the concept of a matroid.

In this chapter we give an introduction to the theory of matroids. We discuss several fundamental properties of a matroid. Of special interest to us is the result that the "duality" which we observed between the circuits and the cutsets of a graph is not accidental. As we shall see, this duality is a consequence of the fact that the collection of subgraphs which have no circuits and the collection of subgraphs which have no cutsets both have the matroid structure. We also study the "painting" theorem and the arc coloring lemma which find an application in network analysis. We conclude the chapter with a discussion of the "greedy" algorithm which is a generalization of a well-known algorithm due to Kruskal for finding a minimum cost spanning tree of a weighted connected graph.

10.1 BASIC DEFINITIONS

There are several equivalent axiom systems for characterizing a matroid. We begin our discussion with what are known as the independence axioms. In Section 10.3 we derive some of the equivalent axiom systems.

A *matroid* M is a finite set S and a collection $\mathcal{I}$ of subsets of S such that the following axioms, called *independence axioms*, are satisfied:

I-1. $\emptyset \in \mathcal{I}$.

I-2. If $X \in \mathcal{I}$ and $Y \subseteq X$, then $Y \in \mathcal{I}$.

I-3. If X and Y are members of $\mathcal{I}$ with $|X| = |Y| + 1$, there exists $x \in X - Y$ such that $Y \cup x \in \mathcal{I}$.

The elements of S are called the *elements of the matroid* M. Members of $\mathcal{I}$ are called the *independent sets* of M. A maximal independent set of M is called a *base* of M. The collection of the bases of M is denoted by $\mathcal{B}(M)$ or simply $\mathcal{B}$.

A subset of S not belonging to $\mathcal{I}$ is called *dependent*. A minimal dependent subset of S is called a *circuit* of M. An element x of S is a *loop* of M if $\{x\}$ is dependent. The collection of circuits of M is denoted by $\mathcal{C}(M)$ or simply $\mathcal{C}$.

The *rank function* ρ of M associates to each subset A of S a nonnegative integer $\rho(A)$ defined by

$$\rho(A) = \max\{|X| : X \subseteq A, X \in \mathcal{I}\}.$$

$\rho(A)$ is called the *rank* of A. The *rank of the matroid* M denoted by $\rho(M)$ is the rank of the set S.

We now consider some examples.

Let S be a finite subset of a vector space. As we observed earlier, the family of all the subsets of linearly independent vectors in S satisfy the independence axioms I-1 through I-3. Hence these subsets of S form the collection of independent sets of a matroid on S. In this matroid, the rank of a subset X of S is equal to the dimension of the vector space generated by X.

Let G be an undirected graph with edge set E. We can define two matroids on E.

First consider the collection $\mathcal{I}$ of all the subsets of E which do not contain any circuits. Clearly $\mathcal{I}$ satisfies I-1 and I-2. We can easily show (Exercise 4.8) that $\mathcal{I}$ satisfies I-3. Thus $\mathcal{I}$ is the collection of independent sets of a matroid M on E. Each base of M is a spanning forest of G. In this matroid the rank of any subset X of E is equal to the rank of the subgraph of G induced by X. Furthermore, each circuit of M is a circuit of G. For this reason M is called the *circuit matroid* of G.

Consider next the family $\mathcal{I}^*$ of all the subsets of E which do not contain any cutset of G. We can show (Exercise 4.9) that $\mathcal{I}^*$ satisfies the axioms I-1 through I-3, and hence it is the family of independent sets of a matroid M^* on E. Each base of M^* is a cospanning forest of G. In this matroid the rank of any subset X of E is equal to the nullity of the subgraph of G induced by X. Furthermore, each circuit of M^* is a cutset of G. The matroid M^* is called the *cutset matroid* or the *bond matroid* of G.

The two matroids M and M^* defined above possess the interesting property that the bases of one are the complements in E of the bases of the other. This result is true for any matroid on any finite S (not necessarily the edge set of a graph). In other words, for any matroid M on a set S there exists a matroid M^* on S with the bases of M^* being the complements of the bases of M. We discuss this result in Section 10.4.

Another example of a matroid is the *matching matroid* defined on the vertex set of a graph.

THEOREM 10.1. Let G be an undirected graph with vertex set V. Let $\mathcal{I}$ be the collection of all the subsets $I \subseteq V$ such that the elements of I are saturated in some matching of G. Then $\mathcal{I}$ is the collection of independent sets of a matroid on V.

Proof

Clearly $\mathcal{I}$ satisfies axioms I-1 and I-2.

To show that $\mathcal{I}$ satisfies axiom I-3, consider any two members I_p and I_{p+1} of $\mathcal{I}$ containing p and $p+1$ vertices, respectively, and let X_p and X_{p+1} be any two matchings saturating the elements of I_p and I_{p+1}, respectively. Now two cases arise.

Case 1 Suppose that some element $x \in I_{p+1} - I_p$ is saturated in X_p. Then X_p saturates $I_p \cup x$ and axiom I-3 is satisfied.

Case 2 Suppose that no element of $I_{p+1} - I_p$ is saturated in X_p. Then consider the subgraph G' on the edge set $X_p \oplus X_{p+1} = (X_p - X_{p+1}) \cup (X_{p+1} - X_p)$. By Theorem 8.19 each component of G' is either

1. A circuit whose edges are alternately in X_p and X_{p+1}; or
2. A path whose edges are alternately in X_p and X_{p+1} and whose end vertices are not saturated in one of the two matchings.

Since $|I_{p+1} - I_p| > |I_p - I_{p+1}|$, there is a path P in G' from a vertex $v \in I_{p+1} - I_p$ to a vertex not in $I_p - I_{p+1}$. Then $X_p \oplus P$ will be a matching saturating v and all the elements of I_p. Thus $I_p \cup v$ is a member of $\mathcal{I}$, and axiom I-3 is satisfied. ■

As an example, consider the graph G shown in Fig. 10.1*a*. The sets $I_4 = \{v_1, v_2, v_3, v_4\}$ and $I_6 = \{v_1, v_2, v_4, v_5, v_6\}$ are saturated in the matchings $X_4 = \{e_1, e_2\}$ and $X_6 = \{e_2, e_3, e_4\}$, respectively. The subgraph on the edge set $X_4 \oplus X_6 = \{e_1, e_3, e_4\}$ is shown in Fig. 10.1*b*. There is a path P from the vertex $v_5 \in I_6 - I_4$ to the vertex $v_6 \notin I_4 - I_6$. The matching $X_4 \oplus P = \{e_2, e_3, e_4\}$ saturates the set $I_4 \cup v_5$.

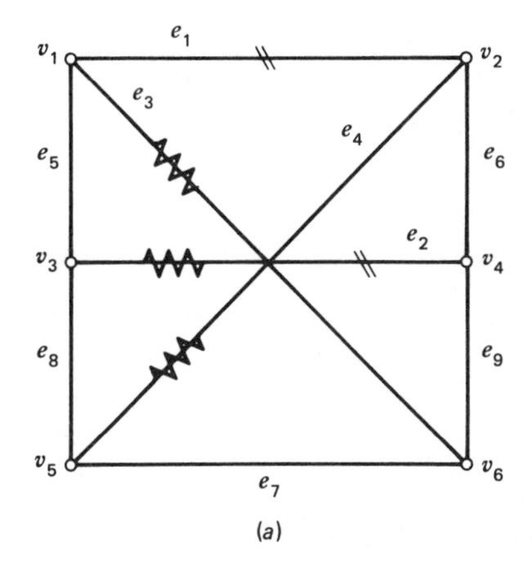

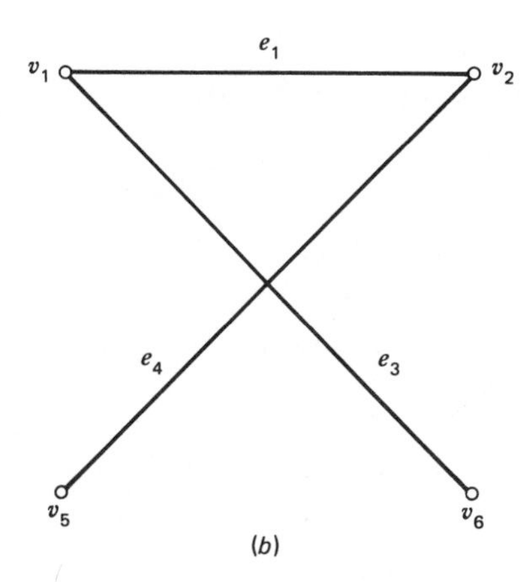

Figure 10.1. (*a*) A graph G with the matching X_4 and X_6 indicated. (*b*) Subgraph of G on the edge set $X_4 \oplus X_6$.

Two matroids M_1 and M_2 on the sets S_1 and S_2, respectively, are *isomorphic* if there is a one-to-one correspondence between the elements of S_1 and S_2 which preserves independence.

A matroid M on S is called *graphic* if it is isomorphic to the circuit matroid of a graph G. M is called *cographic* if it is isomorphic to the cutset matroid of a graph G.

10.2 FUNDAMENTAL PROPERTIES

We establish in this section several fundamental properties of a matroid. All these properties are familiar in graph theory.

We first consider the independent sets and the bases of a matroid M.

THEOREM 10.2 (AUGMENTATION THEOREM). Let X and Y be any two independent subsets in a matroid M. If $|X|<|Y|$, then there exists $Z \subseteq Y-X$ such that $|X \cup Z|=|Y|$ and $X \cup Z$ is independent in M.

Proof

Let Z_0 be a set such that

1. $Z_0 \subseteq Y-X$;
2. $X \cup Z_0$ is independent in M; and
3. If $X \cup Z$ is independent in M for any $Z \subseteq Y-X$, then $|X \cup Z_0| \geqslant |X \cup Z|$.

Suppose that $|X \cup Z_0|<|Y|$. Then there exists a set $Y_0 \subseteq Y$ with $|Y_0|=|X \cup Z_0|+1$. Since Y_0 is independent in M, by axiom I-3 there exists an element $y \in Y_0-(X \cup Z_0)$ such that $X \cup Z_0 \cup y$ is independent in M. The set $Z_0 \cup y$ contradicts the choice of Z_0. Hence $|X \cup Z_0| \geqslant |Y|$, and the proof now follows. ■

Corollary 10.2.1 All the bases of a matroid M on a set S have the same cardinality equal to the rank of M.

Proof

If not, let B_1, B_2 be bases with $|B_1|<|B_2|$. Then by the augmentation theorem there exists $Z \subseteq B_2-B_1$ such that $B_1 \cup Z$ is independent. This, however, contradicts the maximality of B_1 in $\mathcal{I}$. ■

The following generalization of the above result can also be proved using the augmentation theorem.

Corollary 10.2.2 Let M be a matroid on a set S and $A \subseteq S$. Then all the maximal independent subsets of A have the same cardinality. ■

Another corollary of Theorem 10.2 is stated next without proof.

Corollary 10.2.3 If B_1 and B_2 are the bases of a matroid M and $x \in B_1-B_2$, then there exists $y \in B_2-B_1$ such that $(B_1 \cup y)-x$ is a base of M. ■

We next obtain some properties of the rank function of a matroid.

THEOREM 10.3. The rank function ρ of a matroid M on a set S satisfies the following properties:

1. $0 \leqslant \rho(A) \leqslant |A|$.
2. If $A \subseteq B$, then $\rho(A) \leqslant \rho(B)$.
3. $\rho(A) \leqslant \rho(A \cup x) \leqslant \rho(A)+1$.

4. The submodular inequality
$$\rho(A)+\rho(B)\geqslant\rho(A\cup B)+\rho(A\cap B).$$
5. If $\rho(A\cup y)=\rho(A\cup x)=\rho(A)$, then $\rho(A\cup x\cup y)=\rho(A)$.

Proof

The first three properties follow easily from the definition of the rank function.

To establish property 4, let X be a maximal independent subset of $A\cap B$. By the augmentation theorem there exists a maximal independent subset Y of $A\cup B$ containing X. Let $Y=X\cup V\cup W$, where $V\subseteq A-B$ and $W\subseteq B-A$. Since $X\cup V$ is an independent subset of A, and $X\cup W$ is an independent subset of B, we get

$$\rho(A)\geqslant|X\cup V|$$

and

$$\rho(B)\geqslant|X\cup W|.$$

Hence

$$\begin{aligned}\rho(A)+\rho(B)&\geqslant|X\cup V|+|X\cup W|\\&=2|X|+|V|+|W|\\&=|X|+(|X|+|V|+|W|)\\&=\rho(A\cap B)+\rho(A\cup B).\end{aligned}$$

Property 5 is an easy consequence of property 4. ■

We next consider the circuits of a matroid. Some properties which follow directly from the definition of a circuit are stated below.

1. Every proper subset of a circuit is independent. So, if C_1 and C_2 are distinct circuits, then $C_1\not\subseteq C_2$.
2. If C is a circuit, then $\rho(C)=|C|-1$.
3. A matroid M on a set S has no circuits if and only if all the subsets of S are independent. Thus S is the only base of such a matroid M.

THEOREM 10.4. If C_1 and C_2 are distinct circuits of a matroid M and $x\in C_1\cap C_2$, then there exists a circuit C_3 such that $C_3\subseteq(C_1\cup C_2)-x$.

Proof

Let $C'=(C_1\cup C_2)-x$. We prove the theorem by showing that C' is dependent, that is, $\rho(C')<|C'|$.

Since C_1 and C_2 are distinct circuits, $C_1 \cap C_2$ is a proper subset of both C_1 and C_2. Hence $C_1 \cap C_2$ is independent. Thus

$$\rho(C_1 \cap C_2) = |C_1 \cap C_2|.$$

Furthermore,

$$\rho(C_1) = |C_1| - 1$$

and

$$\rho(C_2) = |C_2| - 1.$$

Now, using the above and the submodularity of the rank function (Theorem 10.3), we get

$$\begin{aligned}\rho(C_1 \cup C_2) &\leqslant \rho(C_1) + \rho(C_2) - \rho(C_1 \cap C_2)\\ &= |C_1| - 1 + |C_2| - 1 - |C_1 \cap C_2|\\ &= |C_1 \cup C_2| - 2\\ &< |C'|. \end{aligned} \tag{10.1}$$

Since $C' \subseteq C_1 \cup C_2$, we also have

$$\rho(C') \leqslant \rho(C_1 \cup C_2). \tag{10.2}$$

Combining (10.1) and (10.2), we get

$$\rho(C') < |C'|. \quad \blacksquare$$

Corollary 10.4.1 If A is an independent set in a matroid M on a set S, then for any $x \in S$, $A \cup x$ contains at most one circuit.

Proof

Let, for some $x \in S$, there exist two distinct circuits C_1 and C_2 such that $x \in C_1 \cap C_2$, $C_1 \subseteq A \cup x$, and $C_2 \subseteq A \cup x$. Then by the previous theorem there exists a circuit $C' \subseteq (C_1 \cup C_2) - x$. Hence $(C_1 \cup C_2) - x$ is dependent. But $(C_1 \cup C_2) - x \subseteq A$, contradicting the independence of A. $\blacksquare$

Corollary 10.4.2 If B is a base of a matroid M on a set S and $x \in S - B$, then there exists a unique circuit $C = C(x, B)$ such that $x \in C \subseteq B \cup x$.

Proof

Since B is a maximal independent set $B \cup x$ contains a circuit. By the previous corollary, such a circuit is unique. $\blacksquare$

The circuit $C(x, B)$ defined in the above corollary is called the *fundamental circuit* of x with respect to base B.

We next prove a much stronger result than Theorem 10.4.

THEOREM 10.5. If C_1 and C_2 are distinct circuits of a matroid and $x \in C_1 \cap C_2$, then for any element $y \in C_1 - C_2$ there exists a circuit C such that

$$y \in C \subseteq (C_1 \cup C_2) - x.$$

Proof

Suppose that C_1, C_2, x, y are such that the theorem is false and that $|C_1 \cup C_2|$ is minimum among all pairs of distinct circuits violating the theorem.

By Theorem 10.4 there exists a circuit $C_3 \subseteq (C_1 \cup C_2) - x$. From our choice of x and y it is clear that $y \notin C_3$. Furthermore, $C_3 \cap (C_2 - C_1) \neq \emptyset$, for otherwise $C_3 \subseteq C_1$, contradicting the fact that C_1 is a minimal dependent set. Let then $z \in C_3 \cap (C_2 - C_1)$.

Now, for the circuits C_2 and C_3 we have the following:

1. $z \in C_2 \cap C_3$.
2. $x \in C_2 - C_3$.
3. $C_2 \cup C_3$ is a proper subset of $C_1 \cup C_2$ because $y \notin C_2 \cup C_3$.

From our choice of C_1 and C_2 it follows that there exists a circuit C_4 such that

$$x \in C_4 \subseteq (C_2 \cup C_3) - z.$$

Consider next the circuits C_1 and C_4 for which we have:

1. $x \in C_1 \cap C_4$.
2. $y \in C_1 - C_4$, because $y \notin C_2 \cup C_3$.
3. $C_1 \cup C_4$ is a proper subset of $C_1 \cup C_2$, because $z \notin C_1 \cup C_4$.

Again from our choice of C_1 and C_2 it follows that there exists a circuit C_5 such that $y \in C_5 \subseteq (C_1 \cup C_4) - x$.

Since $C_1 \cup C_4$ is a proper subset of $C_1 \cup C_2$, we have found a circuit C_5 such that $y \in C_5 \subseteq (C_1 \cup C_2) - x$, contradicting our choice of C_1 and C_2. ■

10.3 EQUIVALENT AXIOM SYSTEMS

In this section we establish some alternative axiom systems for defining a matroid. Such equivalent characterizations would help us in gaining more insight into the structure of matroids.

We begin with an equivalent set of independence axioms.

THEOREM 10.6 (INDEPENDENCE AXIOMS). A collection $\mathcal{I}$ of subsets of a set S is the set of independent sets of a matroid on S if and only if $\mathcal{I}$ satisfies axioms I-1, I-2, and

I-3′. If A is any subset of S, then all the maximal subsets Y of A, with $Y \in \mathcal{I}$, have the same cardinality.

Proof

The proof is left as an exercise. See Corollary 10.2.2. ■

THEOREM 10.7 (BASE AXIOMS). Let $\mathcal{B}$ be the set of bases of a matroid. Then

B-1. $\mathcal{B} \neq \varnothing$, and no set in $\mathcal{B}$ contains another properly.

B-2. If $B_1, B_2 \in \mathcal{B}$ and $x \in B_1$, then there exists $y \in B_2$ such that $(B_1 - x) \cup y \in \mathcal{B}$.

Conversely, if $\mathcal{B}$ is a collection of subsets of a set S satisfying B-1 and B-2, then

$$\mathcal{I} = \{I \mid I \subseteq B, \text{ for some } B \in \mathcal{B}\}$$

is the collection of independent sets of a matroid on S.

Proof

We have already proved that the bases of a matroid satisfy B-1 and B-2. (Recall the definition of a base and see Corollary 10.2.3.)

Given $\mathcal{B}$ satisfying B-1 and B-2, we now show that $\mathcal{I}$ as defined in the theorem is the collection of independent sets of a matroid on S.

First note that B-1 and B-2 imply that for $x \in B_1 - B_2$ there exists $y \in B_2 - B_1$ such that $(B_1 - x) \cup y$ is a member of $\mathcal{B}$.

Clearly, $\mathcal{I}$ satisfies axioms I-1 and I-2. To prove axiom I-3, first we show that all the members of $\mathcal{B}$ have the same cardinality.

Suppose there exist $B_1, B_2 \in \mathcal{B}$ with $|B_1| > |B_2|$. Then by repeated application of B-2 we can remove $B_2 - B_1$ from B_2 and replace it by a set $A \subset B_1 - B_2$ with $|A| = |B_2 - B_1|$ and such that $B_3 = (B_2 \cap B_1) \cup A$ is a member of $\mathcal{B}$. But $B_3 \subset B_1$, contradicting B-1. Thus all the members of $\mathcal{B}$ have the same cardinality.

Consider next any two distinct members X and Y of $\mathcal{I}$ with $|Y| = |X| + 1$. Let $X \subseteq B_1$, $Y \subseteq B_2$, and $B_1, B_2 \in \mathcal{B}$. Let

$$X = \{x_1, x_2, \ldots, x_k\},$$

$$B_1 = \{x_1, x_2, \ldots, x_k, b_1, b_2, \ldots, b_q\},$$

$$Y = \{y_1, y_2, \ldots, y_k, y_{k+1}\},$$

$$B_2 = \{y_1, y_2, \ldots, y_k, y_{k+1}, c_1, c_2, \ldots, c_{q-1}\}.$$

Now consider B_1-b_q. By B-2 there exists $z\in B_2$ such that $B'=(B_1-b_q)\cup z$ is a member of $\mathcal{B}$. If $z\in Y$, then $(X\cup z)\subseteq B'$, and hence it is a member of $\mathcal{I}$. So I-3 is satisfied in this case.

If $z\notin Y$, then consider $B''=B'-b_{q-1}$. Again by B-2 there exists $z'\in B_2$ such that $B''=(B'-b_{q-1})\cup z'$ is a member of $\mathcal{B}$. If $z'\in Y$, then $X\cup z'\subseteq B''$. Hence $(X\cup z')\in\mathcal{I}$, and axiom I-3 is satisfied.

If $z'\notin Y$, remove b_{q-2} from B'', and so on. Since $|\{b_1,b_2,\ldots,b_q\}|>|\{c_1,c_2,\ldots,c_{q-1}\}|$, after at most q steps we arrive at a situation where we replace some b_i by an element of Y, thereby satisfying axiom I-3. ■

THEOREM 10.8 (RANK AXIOMS). The rank function ρ of a matroid M on a set S satisfies:

R-1. $0\leqslant\rho(A)\leqslant|A|$, for all $A\subseteq S$.

R-2. If $A\subseteq B\subseteq S$, then $\rho(A)\leqslant\rho(B)$.

R-3. For any $A, B\subseteq S$,
$\rho(A)+\rho(B)\geqslant\rho(A\cup B)+\rho(A\cap B)$.

Conversely, if an integer-valued function ρ defined on a finite set S satisfies R-1, R-2, and R-3, then the set

$$\mathcal{I}=\{I\,|\,\rho(I)=|I|,\ I\subseteq S\}$$

is the collection of independent sets of a matroid on S.

Proof

We have already proved in Theorem 10.3 that the rank function of a matroid satisfies R-1, R-2, and R-3.

To prove the converse, suppose that ρ satisfies R-1, R-2, and R-3. It follows from R-1 that $\rho(\varnothing)=0$. So $\varnothing\in\mathcal{I}$, and axiom I-1 is satisfied.

Let $B\in\mathcal{I}$ and $A\subseteq B$. Then by R-3, we get

$$\begin{aligned}\rho(A)+\rho(B-A)&\geqslant\rho(A\cup(B-A))+\rho(A\cap(B-A))\\&=\rho(B)+\rho(\varnothing)\\&=|B|.\end{aligned}\qquad(10.3)$$

If $\rho(A)<|A|$, then by R-1,

$$\begin{aligned}\rho(A)+\rho(B-A)&<|A|+|B-A|\\&=|B|,\end{aligned}$$

contradicting (10.3). Hence $\rho(A)=|A|$, and thus A is also a member of $\mathcal{I}$, thereby satisfying axiom I-2.

To prove I-3, first note that R-2 and R-3 imply the following:

R-3′. If $\rho(A\cup x)=\rho(A\cup y)=\rho(A)$, then $\rho(A\cup x\cup y)=\rho(A)$.

Now let X and Y be two distinct members of $\mathcal{I}$ with $|X|=k$ and $|Y|=k+1$. Suppose that, for every $y\in Y-X$, $\rho(X\cup y)=k$. Then by repeated application of R-3′ we get

$$\rho(X\cup(Y-X))=\rho(X\cup Y)=k,$$

contradicting that $\rho(Y)=k+1$. So there exists an element $y\in Y-X$ with the property $\rho(X\cup y)=k+1$, and thus axiom I-3 is satisfied. ■

THEOREM 10.9 (CIRCUIT AXIOMS). Let $\mathcal{C}$ be the set of circuits of a matroid on a set S. Then

C-1. $\varnothing\notin\mathcal{C}$ and no set in $\mathcal{C}$ contains another properly.

C-2. If $C_1, C_2\in\mathcal{C}$, $C_1\neq C_2$, and $x\in C_1\cap C_2$, then there exists $C_3\in\mathcal{C}$ such that $C_3\subseteq(C_1\cup C_2)-x$.

Conversely, if $\mathcal{C}$ is a collection of subsets of a set S satisfying C-1 and C-2, then the set $\mathcal{I}=\{I|C\not\subseteq I$, for all $C\in\mathcal{C}\}$ is the collection of independent sets of a matroid on S.

Proof

We have proved earlier that the circuits of a matroid satisfy C-1 and C-2 (see Theorem 10.4.)

Given $\mathcal{C}$ satisfying C-1 and C-2, we now show that the set $\mathcal{I}$ defined as in the theorem is the collection of independent sets of a matroid on S. We do so by showing that $\mathcal{I}$ satisfies axioms I-1, I-2, and I-3′ (see Theorem 10.6.)

First note that $\mathcal{I}$ is the collection of all the subsets of S which do not contain any member belonging to $\mathcal{C}$. Therefore it satisfies axioms I-1 and I-2.

To prove that $\mathcal{I}$ satisfies axiom I-3′, consider any subset A of S. Let S_1 and S_2 be distinct maximal subsets of A which belong to $\mathcal{I}$. Assuming that $|S_2|>|S_1|$, we establish a contradiction.

Since S_1 and S_2 are distinct, $S_1-S_2\neq\varnothing$ and $S_2-S_1\neq\varnothing$. Let $x\in S_2-S_1$. Then clearly $S_1\cup x\notin\mathcal{I}$. Hence there exists $C\in\mathcal{C}$ with $x\in C\subseteq S_1\cup x$. Further, $C\cap(S_1-S_2)\neq\varnothing$, as otherwise $C\subset S_2$, contradicting that $S_2\in\mathcal{I}$.

Let then $x'\in C\cap(S_1-S_2)$ and $S_3=(S_1-x')\cup x$. Note that $|S_3|=|S_1|$. We first prove:

1. $S_3\in\mathcal{I}$.

Clearly $S_1-x'\in\mathcal{I}$. If $S_3\notin\mathcal{I}$, then there exists $C'\in\mathcal{C}$ with $x\in C'\subseteq S_3\subset S_1\cup x$. Further $C\neq C'$ as $x'\notin C'$. Thus C and C' are distinct members of $\mathcal{C}$ such that

(a) $x\in C\cap C'$.

(b) $C\subset S_1\cup x$ and $C'\subset S_1\cup x$.

Then it follows from C-2 that there exists $C''\in\mathcal{C}$ with $C''\subseteq(C\cup C')-x$

$\subseteq S_1 \in \mathcal{I}$. This contradicts that no member of $\mathcal{I}$ contains a member of $\mathcal{C}$. Therefore, $S_3 \in \mathcal{I}$.

We next prove:

2. *$S_3 \in \mathcal{I}$ is a maximal subset of A.*

   If not, let $S' \in \mathcal{I}$ be a maximal subset of A with $S_3 \subset S'$. Now $x' \notin S'$, as otherwise $S_1 \subset S'$, contradicting the maximality of S_1 in A. Then $S' \cup x' \notin \mathcal{I}$. Hence there exists $C''' \in \mathcal{C}$ with $x' \in C''' \subseteq S' \cup x'$. Further, $C''' \cap (S' - S_1) \neq \emptyset$, as otherwise $C''' \subseteq S_1$. Let $x'' \in C''' \cap (S' - S_1)$. Then we can show (as in the proof of 1) that $(S' - x'') \cup x' \in \mathcal{I}$. But $S_1 \subset (S' - x'') \cup x' \in \mathcal{I}$, contradicting the maximality of S_1 in A. Thus S_3 is a maximal subset of A.

Note that S_3 is constructed from S_1 by replacing an element $x' \in S_1 - S_2$ by an element $x \in S_2 - S_1$. Since $|S_2| > |S_1|$, we can repeat this construction a finite number of times to obtain a maximal subset S_n of A with $S_n \in \mathcal{I}$ and $S_n \subset S_2$, a contradicting that $\rho(Y) = k+1$. So there exists an element $y \in Y - X$ with the property $\rho(X \cup y) = k+1$, and thus axiom I-3 is satisfied. ∎

10.4 MATROID DUALITY AND GRAPHOIDS

In this section we first introduce the concept of matroid duality and discuss some results relating a matroid and its dual. We then establish the "painting" theorem which would lead to the definition of a graphoid. We conclude with the development of Minty's self-dual axiom system for matroids which is based on graphoids.

We begin with a theorem due to Whitney [10.1], which defines the dual of a matroid.

THEOREM 10.10. Let $\mathcal{B}$ be the set of bases of a matroid M on a set S. Then $\mathcal{B}^* = \{S - B \mid B \in \mathcal{B}\}$ is the set of bases of a matroid M^* on S.

Proof

Clearly, $\mathcal{B}^*$ satisfies axiom B-1. To show that axiom B-2 is also satisfied, consider any two members B_1^* and B_2^* of $\mathcal{B}^*$ with $B_1^* = S - B_1$ and $B_2^* = S - B_2$. Let $x \in B_1^* - B_2^*$. Then $x \in B_2 - B_1$. By Exercise 10.4 there exists $y \in B_1 - B_2$ such that $B_3 = (B_1 - y) \cup x$ is a base of M. Now $y \in B_2^* - B_1^*$ and

$$\begin{aligned}(B_1^* - x) \cup y &= S - ((B_1 - y) \cup x)\\ &= S - B_3 = B_3^*.\end{aligned}$$

Thus axiom B-2 is satisfied and hence $\mathcal{B}^*$ is the set of bases of a matroid M^* on S. ∎

The matroid M^* defined in the above theorem is called the *dual matroid* or simply the *dual* of M. It is easy to see that M is the dual of M^*. So we shall refer to M and M^* as *dual matroids*.

The circuit matroid and the cutset matroid of a graph defined in Section 10.1 may now be seen to be dual matroids.

A base of M^* is called a *cobase* of M. Similarly a circuit of M^* is a *cocircuit* of M, a loop of M^* is a *coloop* of M, the rank function of M^* is the *corank function* of M, and so on. The rank function of M^* is denoted by ρ^*. If B is a base of M, then the cobase $S-B$ is denoted by B^*. The collection of cobases of M is denoted by $\mathscr{B}^*(M)$ and the collection of cocircuits is denoted by $\mathscr{C}^*(M)$.

Now we establish some results relating a matroid and its dual. Many of these results are well known in graph theory (see Chapter 2).

It is clear from the definition of a dual that

$$\rho^*(M)=|S|-\rho(M). \tag{10.4}$$

The following theorem generalizes this relationship between ρ and ρ^*.

THEOREM 10.11. If M and M^* are dual matroids on a set S, then for all $A \subseteq S$

$$\rho^*(A)=|A|+\rho(S-A)-\rho(M). \tag{10.5}$$

Proof

Consider $A \subseteq S$ and let B^* be a base of M^* such that $|B^* \cap A|$ is maximum. Then B is a base of M with $|B \cap (S-A)|$ maximum. It follows from the definition of a rank function that

$$\rho^*(A)=|B^* \cap A| \tag{10.6}$$

and

$$\rho(S-A)=|B \cap (S-A)|. \tag{10.7}$$

Now

$$|B^* \cap A|=|A|-|B \cap A|$$

and

$$\begin{aligned}|B \cap (S-A)| &= |B|-|B \cap A| \\ &= \rho(M)-|B \cap A|.\end{aligned}$$

So we get from (10.6) and (10.7)

$$\rho^*(A)=|A|+\rho(S-A)-\rho(M). \quad \blacksquare$$

Note that for each statement involving the circuits, bases, and so on, of a matroid there is a dual statement involving the cocircuits, cobases, and so on, of the matroid. This is because the cocircuits, cobases, and so on, are themselves the circuits, bases, and so on, of a matroid.

In all the lemmas and theorems of this section we include the dual statements wherever applicable. The proofs of these dual statements follow in the obvious manner.

LEMMA 10.1. Let M be a matroid on a set S. If $A \subseteq S$ is independent in M, then $S-A$ contains a cobase of M. Dually, if $A^* \subseteq S$ is independent in M^*, then $S-A^*$ contains a base of M.

Proof

There exists a base B of M with $A \subseteq B$. So $S-A$ contains the corresponding cobase B^*. ∎

THEOREM 10.12. Let M be a matroid on a set S. If A and A^* are subsets of S with $A \cap A^* = \emptyset$, A independent in M and A^* independent in M^*, then there exists a base B of M with $A \subseteq B$ and $A^* \subseteq B^*$.

Proof

By the previous lemma $S-A^*$ contains a base of M. Since $A \subseteq S-A^*$ is independent in M, there exists, by Theorem 10.2, a base B of M with $A \subseteq B \subseteq S-A^*$, and hence $A^* \subseteq B^*$. ∎

LEMMA 10.2. Let M be a matroid on a set S. Then each base B of M has nonnull intersection with every cocircuit of M.

Dually, each cobase of M has nonnull intersection with every circuit of M.

Proof

If, for any cocircuit C^* of $M, C^* \cap B = \emptyset$, then $B^* = S-B$ would contain C^*, contradicting the independence of B^* in M^*. ∎

Let B^* be any cobase of a matroid on a set S. By the dual of Corollary 10.4.2, for any $x \in B$, $B^* \cup x$ contains exactly one cocircuit. This cocircuit is called the *fundamental cocircuit* of x with respect to B. Note that this cocircuit contains exactly one element from B, namely, the element x.

LEMMA 10.3. Let M be a matroid. For any independent set A of M there exists a cocircuit having exactly one element from A. Further, if $|A| < \rho(M)$, then there exists a cocircuit having null intersection with A.

Dually, for any independent set A^* of M^* there exists a circuit having exactly one element from A^*. Further, if $|A^*|<\rho^*(M^*)$, then there exists a circuit having null intersection with A^*.

Proof

Let B be a base of M with $A \subseteq B$, and let C^* be the fundamental cocircuit of an element $x \in B$ with respect to B. Then $C^* \cap A = \{x\}$ if $x \in A$, and $C^* \cap A = \varnothing$ if $x \in B - A$. Thus follows the proof. ■

THEOREM 10.13. Let M be a matroid on a set S. A subset X of S is a base of M if and only if X is a minimal subset having nonnull intersection with every cocircuit of M.

Dually, a subset X^* of S is a cobase of M if and only if it is a minimal subset having nonnull intersection with every circuit of M.

Proof

Necessity follows easily from Lemmas 10.2 and 10.3.

To prove the sufficiency, let X be a minimal subset of S having nonnull intersection with every cocircuit of M. Then $S - X$ is a maximal subset having no cocircuit of M. So, by definition, $S - X$ is a cobase of M, and hence X is a base of M. ■

Next we obtain some new characterizations of circuits and cocircuits.

THEOREM 10.14. Let M be a matroid on a set S. A subset X of S is a circuit of M if and only if it is a minimal subset having nonnull intersection with every cobase of M.

Dually, a subset X^* of S is a cocircuit of M if and only if it is a minimal subset having nonnull intersection with every base of M.

Proof

See proof of Theorem 2.13. ■

Let C^* be a cocircuit of a matroid M on S. By the previous theorem, C^* is a minimal subset having nonnull intersection with every base of M. So $S - C^*$ is a maximal subset containing no base of M. Thus we have the following.

LEMMA 10.4. If C^* is a cocircuit of a matroid M on a set S, then for any $x \in C^*$, $(S - C^*) \cap x$ contains a base B of M.

Dually, if C is a circuit of a matroid M on a set S, then for any $x \in C$, $(S - C) \cup x$ contains a cobase of M. ■

THEOREM 10.15. Let M be a matroid on a set S. A subset X of S is a circuit of M if and only if it is a minimal subset with $|X \cap C^*| \neq 1$ for every cocircuit C^* of M.

Dually, a subset X^* of S is a cocircuit of M if and only if it is a minimal subset with $|X^* \cap C| \neq 1$ for every circuit C of M.

Proof

Necessity Proof is by contradiction. For any proper subset C' of a circuit C, there exists, by Lemma 10.3, a cocircuit C^* such that $|C^* \cap C'| = 1$.

Suppose that $|C \cap C^*| = 1$ for some circuit C and some cocircuit C^*, and let $C \cap C^* = \{x\}$. Consider now $S' = S - C^*$ and $C' = C - x$. Clearly, $C' \subseteq S'$. By Lemma 10.4, $S' \cup x$ contains a base. Let then $B \subseteq S' \cup x$ be a base with $C' \subseteq B$. Note that $x \in B$. So the circuit $C = C' \cup x$ is contained in B, a contradiction.

Sufficiency If X is a subset of S with $|X \cap C^*| \neq 1$ for every cocircuit C^*, then X should contain a circuit, for otherwise X would be independent in M, and so by Lemma 10.3 there would exist a cocircuit having exactly one element from X.

Let then C be a circuit contained in X. It is clear from the necessity part of the theorem that $|C \cap C^*| \neq 1$ for every cocircuit C^*. Therefore $X = C$, for otherwise the minimality of X would be contradicted. ∎

We now introduce the notion of a painting of a finite set S and establish the "painting" theorem.

A *painting* of a set S is a partitioning of S into three subsets R, G, and B such that $|G| = 1$. For easy visualization, we regard the elements in R as being "painted red," the element in G as being "painted green," and the elements in B as being "painted blue."

THEOREM 10.16 (PAINTING THEOREM). Let M be a matroid on a set S. For any painting of S, there is either

1. A circuit C of M consisting of the green element and no blue elements; or
2. A cocircuit C^* of M consisting of the green element and no red elements.

Proof

We shall assume that the green element is in some circuit, for otherwise it will be a coloop (why?), and the theorem will follow trivially.

Suppose, for some painting of S, statement 1 is not true. Then consider the subset $S' = R \cup G$. Clearly, any circuit contained in S' consists of only red elements. Let R' be a minimal subset of red elements such that the set $S'' = S' - R'$ contains no circuit, and let B be a base with $S'' \subseteq B$. Then we can easily see that the fundamental circuit of any element $y \in R'$ with respect to B consists of only red elements.

Now, let C^* be the fundamental cocircuit of the green element with respect to B. Suppose C^* contains a red element x. Then $x \in R'$ and $|C^* \cap C_x| = 1$, where C_x is the fundamental circuit of x with respect to B. This contradicts Theorem 10.15, and hence C^* is a cocircuit containing the green element and no red elements. Thus statement 2 is true. ■

Theorems 10.15 and 10.16 lead us to the definition of a graphoid introduced by Minty [10.2].

A *graphoid* is a structure $(S, \mathcal{C}, \mathcal{D})$ consisting of a finite set S and two collections $\mathcal{C}$ and $\mathcal{D}$ of nonempty subsets of S satisfying the following conditions:

G-1. If $C \in \mathcal{C}$ and $D \in \mathcal{D}$, then $|C \cap D| \neq 1$.

G-2. For any painting of S, there is either

- **a.** A member of $\mathcal{C}$ consisting of the green element and no blue elements; or
- **b.** A member of $\mathcal{D}$ consisting of the green element and no red elements.

G-3. No member of $\mathcal{C}$ contains another member of $\mathcal{C}$ properly; no member of $\mathcal{D}$ contains another member of $\mathcal{D}$ properly.

THEOREM 10.17. Let M be a matroid on a set S. Then $(S, \mathcal{C}(M), \mathcal{C}^*(M))$ is a graphoid.

Proof

Follows from Theorems 10.15 and 10.16. ■

We now proceed to establish the converse of the above theorem.

THEOREM 10.18. Let $(S, \mathcal{C}, \mathcal{D})$ be a graphoid. Then $B \subseteq S$ is a maximal subset containing no member of $\mathcal{C}$ if and only if B contains no member of $\mathcal{C}$ and $S - B$ contains no member of $\mathcal{D}$.

Proof

Necessity Let B be a maximal subset of S containing no member of $\mathcal{C}$. Suppose $S-B$ contains a member D of $\mathcal{D}$. Let $x \in D$. Then $B \cup x$ contains a member C of $\mathcal{C}$ and $x \in C$. So $C \cap D = \{x\}$, contradicting G-1. Thus $S-B$ contains no member of $\mathcal{D}$.

Sufficiency Let $B \subseteq S$ contain no member of $\mathcal{C}$ and $S-B$ contain no member of $\mathcal{D}$. We now show that, for any $x \in S-B$, $B \cup x$ contains a member of $\mathcal{C}$.

Consider a painting of S in which all the elements of B are red, x is green, and all the remaining elements of $S-B$ are blue. Clearly, there is no member of $\mathfrak{D}$ containing only green and blue elements. Thus by G-2, $B \cup x$ contains a member of $\mathcal{C}$. ■

THEOREM 10.19. Let $(S, \mathcal{C}, \mathfrak{D})$ be a graphoid. If B is a maximal subset of S containing no member of $\mathcal{C}$, then $S-B$ is a maximal subset of S containing no member of $\mathfrak{D}$.

Proof

Follows from Theorem 10.18 and its dual. ■

THEOREM 10.20. Let $(S, \mathcal{C}, \mathfrak{D})$ be a graphoid. Then $\mathcal{C}$ is the collection of circuits of a matroid M on S and $\mathfrak{D}$ is the collection of cocircuits of M.

Proof

We first show that $\mathcal{C}$ is the set of circuits of a matroid M_1 on S.

Clearly, axiom C-1 is satisfied.

To show that C-2 is also satisfied, let C_1 and C_2 be any two distinct members of $\mathcal{C}$ with $x \in C_1 \cap C_2$ and $y \in C_1 - C_2$, and consider a painting of S in which y is green, x is blue, the rest of $C_1 \cup C_2$ are red, and the rest of S are blue (see Fig. 10.2).

Now there exists no member D of $\mathfrak{D}$ consisting of the green element and no red elements. For if such a member D of $\mathfrak{D}$ exists, then we can easily show that either $|C_1 \cap D| = 1$ or $|C_2 \cap D| = 1$, thereby contradicting G-1. So by G-2 there exists a member C_3 of $\mathcal{C}$ consisting of the green element and no blue elements. By G-3, $\{y\}$ is not a member of $\mathcal{C}$. So it follows that $y \in C_3 \in (C_1 \cup C_2) - x$. Thus axiom C-2 is satisfied, and hence $\mathcal{C}$ is the collection of circuits of a matroid M_1 on S.

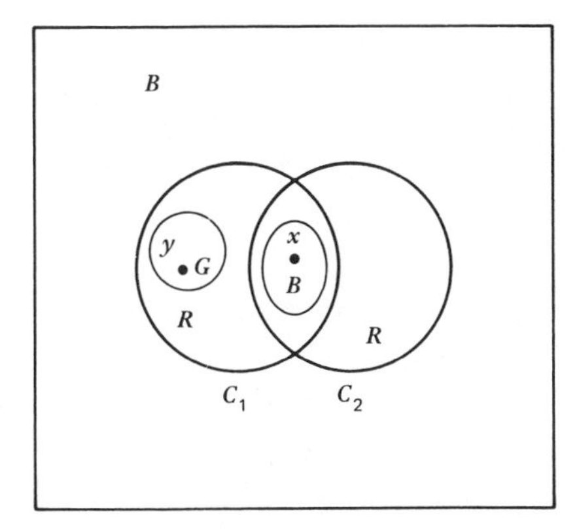

Figure 10.2.

In a similar way, we can show that $\mathfrak{D}$ is the collection of circuits of a matroid M_2 on S.

By Theorem 10.19 the bases of M_1 are the complements of the bases of M_2. So M_1 and M_2 are dual matroids. ■

The equivalence of the graphoid axiom system and the circuit axiom system follows from Theorems 10.17 and 10.20. This result is due to Minty [10.2] who has given an elegant exposition of the theory of matroids based on graphoids.

10.5 RESTRICTION, CONTRACTION, AND MINORS OF A MATROID

Consider a graph G with edge set E. Relative to any subset T of E, we, can define two graphs denoted by $G|T$ and $G \cdot T$. The graph $G|T$, called The *restriction* of G to T, is obtained from G by deleting (open-circuiting) the edges belonging to $E-T$ and then removing the resulting isolated vertices. In fact $G|T$ is the induced subgraph of G on T. The graph $G \cdot T$, called the *contraction* of G to T, is obtained by contracting the edges belonging to $E-T$. Fig. 10.3 shows a graph G and the graphs $G|T$ and $G \cdot T$ for $T=\{e_1, e_6, e_7, e_8\}$.

Now it is easy to verify the following:

1. A subset X of edges is an acyclic subgraph of $G|T$ if and only if it is an acyclic subgraph of G.
2. A subset X of edges is an acyclic subgraph of $G \cdot T$ if and only if there exists a subset Y of $E-T$ such that Y is a spanning forest of $G|(E-T)$ and $X \cup Y$ is an acyclic subgraph of G.

The above ideas motivate the introduction of two submatroids induced on a matroid by a subset of the elements of the matroid.

If $\mathcal{I}(M)$ is the set of independent sets of a matroid M on S and $T \subseteq S$, let

$$\mathcal{I}(M|T)=\{X \mid X \subseteq T, X \in \mathcal{I}(M)\}.$$

We can easily show that $\mathcal{I}(M|T)$ is the set of independent sets of a matroid on T. This matroid is denoted by $M|T$ and is called the *restriction* of M to T. The following observations are obvious:

1. $X \subseteq T$ is independent in $M|T$ if and only if X is independent in M.
2. $X \subseteq T$ is a circuit of $M|T$ if and only if it is a circuit of M.
3. If λ is the rank function of $M|T$, then for any $X \subseteq T$,

$$\lambda(X)=\rho(X). \tag{10.8}$$

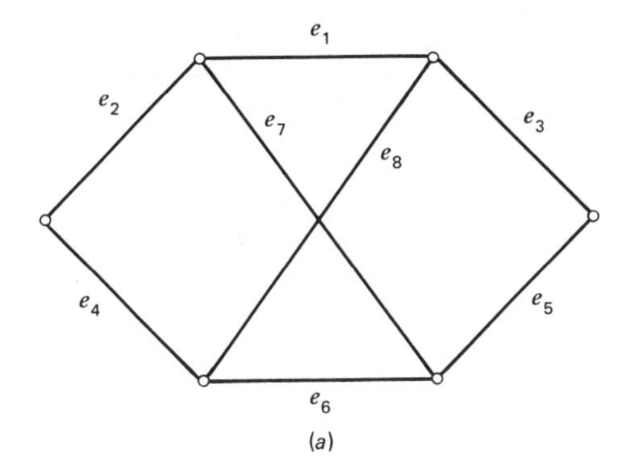

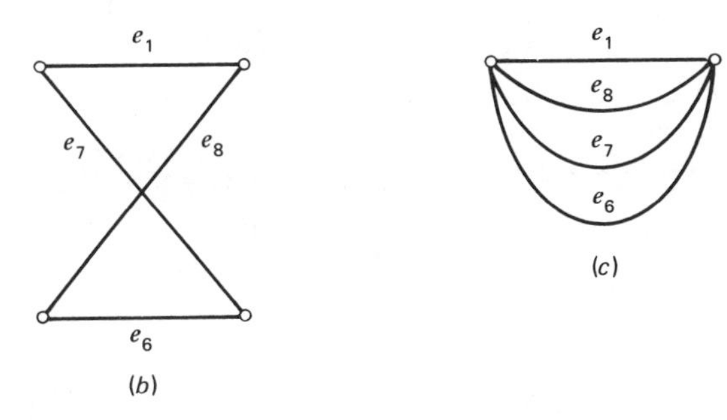

Figure 10.3. (*a*) Graph G. (*b*) $G|T$, restriction of G to T; $T=\{e_1, e_6, e_7, e_8\}$. (*c*) $G\cdot T$, contraction of G to T.

4. If $M(G)$ is the circuit matroid of a graph G with edge set E, then for any $T\subseteq E$, $M(G)|T=M(G|T)$.

Next we define the submatroid $M\cdot T$. Let $\mathcal{I}(M\cdot T)$ be the collection of all those subsets X of T such that there exists a maximal independent subset Y of $S-T$ with $X\cup Y\in\mathcal{I}(M)$.

THEOREM 10.21. $\mathcal{I}(M\cdot T)$ is the collection of independent sets of a matroid on T.

Proof

Let X_1 and X_2 be two members of $\mathcal{I}(M\cdot T)$ such that $|X_2|=|X_1|+1$. Then there exist maximal independent subsets Y_1 and Y_2 of $S-T$ with $X'=X_1\cup Y_1$ and $X''=X_2\cup Y_2$ independent in M. Now $|X''|=|X'|+1$. So there exists, by axiom I-3, an element $x\in X''-X'$ such that $X'\cup x$ is independent in M. Note that $X''-X'=(X_2-X_1)\cup(Y_2-Y_1)$. Further, $x\notin Y_2-Y_1$ because Y_1 is a maximal independent subset of $S-T$. So $x\in X_2-X_1$ and $(X_1\cup x)\subseteq T$ is a member of $\mathcal{I}(M\cdot T)$. Thus $\mathcal{I}(M\cdot T)$ satisfies axiom I-3. It is easy to see that axioms I-1 and I-2 are also satisfied. ■

The matroid defined in the above theorem is called the *contraction* of M to T and is denoted by $M\cdot T$. Note that if $M(G)$ is the circuit matroid of a graph G with edge set E then, for any $T\subseteq E$, $M(G\cdot T)=M(G)\cdot T$.

Let ρ^T be the rank function of $M\cdot T$. Then from the definition of $M\cdot T$ we get, for any $A\subseteq T$,

$$\rho^T(A)=\rho(A\cup(S-T))-\rho(S-T). \tag{10.9}$$

We observed in Chapter 7 that open-circuiting or removing of an edge and the contracting of an edge of a graph are dual operations. In particular, if G and G' are dual graphs and T and T' are the corresponding subsets of edges of G and G', respectively, then $G|T$ is the dual of $G'\cdot T'$ (see Thoerem 7.10). The following theorem is the matroid analog of this relationship.

THEOREM 10.22. If M is a matroid on S and $T\subseteq S$, then

1. $(M|T)^*=M^*\cdot T$.
2. $(M\cdot T)^*=M^*|T$.

Proof

1. Let λ be the rank function of $M|T$ and λ^* be the rank function of $(M|T)^*$. Then from (10.5) we have, for any $X\subseteq T$, $\lambda^*(X)=|X|-\lambda(T)+\rho(T-X)$.

   Let ρ^T be the rank function of $M\cdot T$ and $(\rho^*)^T$ be the rank function of $M^*\cdot T$. Then from (10.9) and (10.5) we get

$$\begin{aligned}(\rho^*)^T(X)&=\rho^*(X\cup(S-T))-\rho^*(S-T)\\&=\rho^*(S-(T-X))-\rho^*(S-T)\\&=|S|-|T|+|X|-\rho(S)+\rho(T-X)-(|S|-|T|-\rho(S)+\rho(T))\\&=|X|-\rho(T)+\rho(T-X)\\&=|X|-\lambda(T)+\lambda(T-X),\qquad\text{by (10.8)}\\&=\lambda^*(X).\end{aligned}$$

   Since $(M|T)^*$ and $M^*\cdot T$ have the same rank function, we get

$$(M|T)^*=M^*\cdot T.$$

2. Putting M for M^* in 1 and taking the duals, we get 2. ■

If M is a matroid on S and $T \subseteq S$, a matroid N on T is called a *minor* of M if N is obtained by a succession of restrictions and/or contractions of M. This topic has been extensively developed by Tutte [10.3].

Using the terminology developed thus far, we can state in matroid terminology many of the results relating to planar graphs and dual graphs. For example, Theorem 7.5 can be stated as follows:

A graph is planar if and only if the circuit matroid $M(G)$ of the graph does not contain either of the matroids $M(K_5)$ or $M(K_{3,3})$ as a minor.

10.6 REPRESENTABILITY OF A MATROID

Let M be a matroid on a set S. Let $B=\{b_1, b_2, \ldots, b_r\}$ be a base of M and let $B^* = S - B = \{e_1, e_2, \ldots, e_q\}$ be the corresponding cobase of M. Recall that $B^* \cap C(e_j) = e_j$, $j=1,\ldots,q$, where $C(e_j)$ is the fundamental circuit of e_j. Similarly $B \cap C^*(b_j) = b_j, j=1,\ldots,r$, where $C^*(b_j)$ is the fundamental cocircuit of b_j. As in the case of undirected graphs (Chapter 6), we may now define the *fundamental cocircuit matrix* D_f and the *fundamental circuit matrix* D_f^* of M with respect to the base B. Clearly, these two matrices will be of the form

$$[U \quad D],$$

where U is a unit matrix whose columns correspond to the elements of B. A matrix of this form is called a *standard representation* of M with respect to the base B.

Now consider a graph G. Suppose we arbitrarily assign orientations to the edges of G and let G' be the resulting directed graph. Then by Theorem 6.9 the incidence matrix of G' is a representation of the circuit matroid $M(G)$ over any field F. We can also see (Theorem 6.10) that any fundamental cutset matrix of G' is a standard representation of $M(G)$ over any field F. Further by Theorem 6.11 any fundamental circuit matrix of G' is a standard representation of the cutset matroid $M^*(G)$ of G over any field F. Thus we have the following.

THEOREM 10.23. The circuit matroid and the cutset matroid of a graph G are both representable over any field F. ∎

We now prove the following main theorem of this section.

THEOREM 10.24. If a matroid M on S is representable over a field F, then the dual matroid M^* on S is also representable over F.

Proof

Suppose M is of rank r and has n elements. Let the $r \times n$ matrix A be a representation of M over F.

Let X be the set of all those column vectors x such that $Ax=0$. X is called the *null space* of A. It is well known in linear algebra that the dimension of X is equal to $n-r$. Now select from X a set of $n-r$ linearly independent column vectors, and with these vectors as columns form an $n\times(n-r)$ matrix B. Note that $AB=0$.

We now show that B^t is a representation over F of the dual matroid M^*. We do so by proving that any r columns of A are linearly dependent if and only if the complementary set of $n-r$ columns of B^t are linearly dependent. For this purpose we shall choose the first r columns of A. Clearly, this involves no loss of generality.

The first r columns of A are linearly dependent if and only if there exists a nonzero column vector $x=[x_1, x_2, \ldots, x_r, 0, 0, \ldots, 0]^t$ belonging to X. Such a vector $x\in X$ exists if and only if there exists a column vector $y\neq 0$ with $n-r$ entries and such that

$$x=By.$$

Now writing B as

$$\begin{bmatrix} B_1 \\ B_2 \end{bmatrix}$$

where B_1 is $r\times(n-r)$ and B_2 is $(n-r)\times(n-r)$, we can see from the above equation that $B_2 y=0$. Since $y\neq 0$, it follows that B_2 is singular. Thus the rows of B_2, and hence the last $n-r$ columns of B^t, are dependent, and the theorem is proved. ■

An easy corollary of the above theorem is the following result.

Corollary 10.24.1 Let M be a matroid of rank r on a set $S=\{s_1, s_2, \ldots, s_n\}$. If M has the standard representation

$$\begin{array}{ccc} s_1\, s_2 \cdots s_r & & s_{r+1} \cdots s_n \\ \left[\quad U_r \right. & | & \left. A \quad\right] \end{array}$$

then M^* has the standard representation

$$\begin{array}{ccc} s_1\, s_2 \cdots s_r & & s_{r+1} \cdots s_n \\ \left[\quad -A^t \right. & | & \left. U_{n-r} \quad\right] \end{array}$$

where U_k is the $k\times k$ identity matrix. ■

The relationship (6.13) between the fundamental circuit matrix and the fundamental cutset matrix of a connected graph follows easily from the above corollary.

For further discussions on the matroid representability problem see Welsh [10.4].

10.7 BINARY MATROIDS

A matroid is *binary* if it is representable over $GF(2)$, the field of integers modulo 2.

Clearly, the circuit matroid $M(G)$ and the cutset matroid $M^*(G)$ of a graph G are binary. We have proved in Theorem 4.6 that each circuit of $M(G)$ can be expressed as a ring sum of some of the fundamental circuits of G. A similar result is true in the case of the cutsets of G. This property of circuits and cutsets (cocircuits) holds true in the case of any binary matroid. However, it is not, in general, true in the case of arbitrary matroids. As an example, let $S=\{1,2,3,4\}$ and let M have as its circuits all subsets of three elements of S. Then the ring sum of the circuits $\{1,2,3\}$ and $\{1,2,4\}$ is $\{3,4\}$, which is an independent set of M.

In this section we establish several properties of a binary matroid, which would lead us to the development of alternative characterizations of such a matroid.

Let M be a matroid on a set S. Let $B=\{b_1, b_2, \ldots, b_r\}$ be a base of M and let $B^*=S-B=\{e_1, e_2, \ldots, e_q\}$ be the corresponding cobase of M. Recall that $B^* \cap C(e_j)=e_j$, $j=1,\ldots,q$, where $C(e_j)$ is the fundamental circuit of e_j. Similarly $B \cap C^*(b_j)=b_j, j=1,\ldots,r$, where $C^*(b_j)$ is the fundamental cocircuit of b_j. As in the case of undirected graphs (Chapter 6), we may now define the *fundamental cocircuit matrix* D_f and the *fundamental circuit matrix* D_f^* of M with respect to the base B. Clearly, these two matrices will be of the form

$$\begin{array}{rcccc} & & b_1\, b_2 \cdots b_r & & e_1\, e_2 \cdots e_q \\ D_f = & [& U_r & | & F \quad] \\ D_f^* = & [& G & | & U_q \quad]. \end{array}$$

Note that $D_f=[d_{ij}]$ and $D_f^*=[d_{ij}^*]$ are both (0, 1) matrices.

Suppose the matrix M is binary. Then it has a standard representation over $GF(2)$ of the form

$$\begin{array}{cccc} & b_1\, b_2 \cdots b_r & & e_1\, e_2 \cdots e_q \\ [& U_r & | & A \quad]. \end{array} \tag{10.10}$$

Let the fundamental circuit $C(e_j)$ be $\{e_j, b_1, b_2, \ldots, b_k\}$. Then the modulo 2 sum of the columns of the matrix in (10.10) corresponding to the elements of $C(e_j)$ is zero. Since the modulo 2 sum of vectors corresponds to the ring sum of the corresponding sets, we can see that

$$a_{ij}=\begin{cases}1, & \text{if } b_i \in C(e_j).\\ 0, & \text{if } b_i \notin C(e_j).\end{cases}$$

In other words,

$$A=G^t. \tag{10.11}$$

Thus the matrix $[U_r \quad G^t]$ is a standard representation over $GF(2)$ of M with respect to the base B.

Starting from a standard representation for the dual matroid M^*, we can show in a similar manner that $[F^t \quad U_q]$ is a standard representation over $GF(2)$ of M^*.

Since a matroid has a unique standard representation with respect to a given base, it follows from Corollary 10.24.1 that

$$F=G^t. \tag{10.12}$$

Thus we have proved the following theorem.

THEOREM 10.25. Let M be a binary matroid on a set S.

1. The fundamental cocircuit matrix of M with respect to any base is a standard representation of M.
2. The fundamental circuit matrix of M with respect to any base is a standard representation of M^*. ■

From (10.12) we also get the following important result.

THEOREM 10.26. Let M be a binary matroid. Let D_f and D_f^* be the fundamental cocircuit matrix and the fundamental circuit matrix of M with respect to a common base. Then

$$D_f(D_f^*)^t=0. \quad ■ \tag{10.13}$$

Given a circuit C of a matroid M, we can associate with C a (0,1) row vector, each entry of which corresponds to an element of M, and an entry in this vector is 1 if the corresponding element is in C. For example, the rows of D_f^* are the vectors which correspond to the fundamental circuits. The matrix of all the circuit vectors of M is called the *circuit matrix* of M and is denoted by $D^*(M)$. In a similar way the *cocircuit matrix* $D(M)$ of M is defined.

Suppose x is the vector corresponding to a circuit C of a binary matroid M. Since D_f is a standard representation of M, the ring sum of the columns of D_f corresponding to the elements of C is zero. In other words,

$$D_f x^t=0. \tag{10.14}$$

Similarly if x is a cocircuit vector, we have

$$D_f^* x^t = 0. \tag{10.15}$$

Note that in the above we have assumed that the columns of D_f^*, D_f, and x are arranged in the same element order.

THEOREM 10.27. Let M be a binary matroid on S. For any base B and circuit C of M, if $C-B=\{x_1, x_2, \ldots, x_k\}$ and if $C(x_i)$ is the fundamental circuit of x_i with respect to B, then

$$C = C(x_1) \oplus C(x_2) \oplus \cdots \oplus C(x_k).$$

Proof

Proof is based on (10.12) and (10.14). See proof of Theorem 6.7. ∎

To prove the converse of the above theorem, we need the following.

THEOREM 10.28. Let M be a matroid. For any base B and circuit C of M, let $C=C(x_1)\oplus C(x_2)\oplus\cdots\oplus C(x_k)$, where $\{x_1, x_2, \ldots, x_k\} = C-B$ and $C(x_i)$ is the fundamental circuit of x_i in B. Then the ring sum $C_1 \oplus C_2$ of any two distinct circuits of M contains a circuit of M.

Proof

Suppose $C_1\oplus C_2$ does not contain a circuit and let $C_1 \cap C_2 = \{x_1, x_2, \ldots, x_p\}$. Then $C_1 \oplus C_2 = (C_1 \cup C_2) - \{x_1, \ldots, x_p\}$ is independent in M. So there exists a base $B \supseteq C_1 \oplus C_2$. But then $C_1 - B = C_2 - B = \{x_1, \ldots, x_p\}$, so that by the hypothesis of the theorem,

$$C_1 = C(x_1) \oplus C(x_2) \oplus \cdots \oplus C(x_p) = C_2,$$

contradicting the fact that $C_1 \neq C_2$. ∎

THEOREM 10.29. Let M be a matroid. For any base B and circuit C of M, let $C=C(x_1)\oplus C(x_2)\oplus\cdots\oplus C(x_k)$, where $\{x_1, x_2, \ldots, x_k\} = C-B$ and $C(x_i)$ is the fundamental circuit of x_i in B. Then M is binary.

Proof

Let $B=\{b_1, b_2, \ldots, b_r\}$ be any base of M and let $S-B=\{e_1, e_2, \ldots, e_q\}$. The fundamental circuit matrix $D_f^*(M)$ of M with respect to B is of the form

$$\begin{array}{ccc} b_1\, b_2 \cdots b_r & & e_1\, e_2 \cdots e_q \\ \left[\quad A \right. & \big| & \left. U_q \quad \right] \end{array}$$

We prove the theorem by showing that the matrix

$$\begin{array}{ccc} b_1\, b_2 \cdots b_r & & e_1\, e_2 \cdots e_q \\ [\quad U_r & | & A^t \quad] \end{array} \tag{10.16}$$

is a standard representation of M over $GF(2)$.

Let C be a circuit of M. If $C-B=\{e_{i_1}, e_{i_2}, \ldots, e_{i_k}\}$, then by the hypothesis of the theorem

$$C=C(e_{i_1})\oplus C(e_{i_2})\oplus \cdots \oplus C(e_{i_k}).$$

It now follows from the definition of A that the modulo 2 sum of the columns of the matrix in (10.16) corresponding to the members of C is zero. Thus these columns are linearly dependent over $GF(2)$.

Suppose $W=\{b_1, \ldots, b_t, e_1, \ldots, e_p\}$ is a circuit of the matroid M' induced on S by the linear dependence of the corresponding column vectors of the matrix in (10.16). Then

$$W=C(e_1)\oplus C(e_2)\oplus \cdots \oplus C(e_p).$$

By Theorem 10.28, W contains a circuit C' of M. But C' cannot be a proper subset of W, for then it would contradict that W is a circuit of M'. Thus $W=C'$.

Hence the matrix in (10.16) is a representation of M over $GF(2)$. ■

Further characterizations of a binary matroid are given in the following theorem.

THEOREM 10.30. For a matroid M the following statements are equivalent:

1. For any circuit C and any cocircuit C^* of M, $|C \cap C^*|$ is even.
2. The ring sum of any collection of distinct circuits of M is the union of disjoint circuits of M.
3. For any base B and circuit C of M, if $C-B=\{e_1, e_2, \ldots, e_q]$ and $C(e_i)$ is the fundamental circuit of e_i in the base B, then $C=C(e_1)\oplus C(e_2) \oplus \cdots \oplus C(e_q)$.
4. M is binary.

Proof

1⇒2 Let $C_1, C_2, \ldots, C_k$ be distinct circuits of M, and let $A=C_1\oplus C_2 \oplus \cdots \oplus C_k$. We may assume without loss of generality that A contains no loops.

Since for any cocircuit C^*, $|C^* \cap C_i|$ is even for $1 \leqslant i \leqslant k$, it follows easily that $|A \cap C^*|$ is also even. Suppose that A is not dependent. Then a contradiction results because by Lemma 10.3 there exists a cocircuit having exactly one element from A. Thus A is dependent and contains a circuit C.

If $A = C$, the theorem is proved. If not, let $A_1 = A \oplus C$. Note that for any cocircuit C^*, $|C^* \cap A_1|$ is even. So we may now repeat the above argument on A_1. Since A_1 is finite and $A_1 = A - C$, this process will eventually terminate, yielding a finite collection of disjoint circuits whose union is A.

2⇒3 See proof of Theorem 4.6.

3⇒4 Same as Theorem 10.29.

4⇒1 By Theorem 10.27 each circuit C can be expressed in terms of fundamental circuits as

$$C = C(x_1) \oplus C(x_2) \oplus \cdots \oplus C(x_k).$$

By (10.15), $|C^* \cap C(x_i)|$ is even for all $1 \leqslant i \leqslant k$. So it now readily follows that $|C^* \cap C|$ is even. ■

Obviously, alternative characterizations of a binary matroid in terms of cocircuits can be obtained by replacing circuits by cocircuits in Theorem 10.30.

10.8 ORIENTABLE MATROIDS

A matroid M is *orientable* if it is possible to assign negative signs to some of the nonzero entries in the circuit matrix $D^* = D^*(M)$ and in the cocircuit matrix $D = D(M)$ in such a way that the inner product of any row of the signed circuit matrix with any row of the signed cocircuit matrix is zero over the ring of integers.

Clearly, the circuit matroid and the cutset matroid of a graph are both orientable.

A *painting* of an *orientable matroid* M is a partitioning of the elements of M into three sets R, G, and B and the distinguishing of one element of the set G. We can visualize this as coloring of the elements of M with three colors, each element being painted red, green, or blue, and exactly one green element being colored dark green. Note that the dark green element is also to be treated as a green element.

The main result of this section is the "arc coloring lemma" due to Minty [10.2].

THEOREM 10.31 (ARC COLORING LEMMA). Let M be an orientable matroid. For any painting of the elements of M, exactly one of the following is true:

1. There exists a circuit containing the dark green element but no blue elements, in which all the green elements are similarly oriented (that is, all have the same sign in the signed circuit matrix).
2. There exists a cocircuit containing the dark green element but no red elements, in which all the green elements are similarly oriented.

Proof

Proof is by induction on the number of green elements.

If there is only one green element, the result follows by axiom G-2.

Suppose the result is true when the number of green elements is m. Consider then a painting in which there are $m+1$ green elements. Choose a green element x other than the dark green element (see Fig. 10.4).

Color the element x red. In the resulting painting there are m green elements. If now there is a cocircuit of type 2, the theorem is proved.

Suppose we color x blue. If in the resulting painting there is a circuit of type 1, the theorem is proved.

Suppose neither of the above two occurs. Then by the induction hypothesis we have the following:

a. There is a cocircuit of type 2 when x is colored blue.
b. There is a circuit of type 1 when x is colored red.

Now let the corresponding rows of the signed circuit and cocircuit matrices be as shown below:

	dg	R	B	G	x
cocircuit	$+1$	$0\ 0\cdots0\ \ 0$	$1\ -1\cdots0\ 1$	$1\ 1\cdots1\ 0$	?
circuit	$+1$	$-1\ 1\cdots0\ -1$	$0\ \ 0\cdots0\ 0$	$0\ 1\cdots1\ 0$	?

Here we have assumed without loss of generality that $+1$ appears in the dark green position of both vectors.

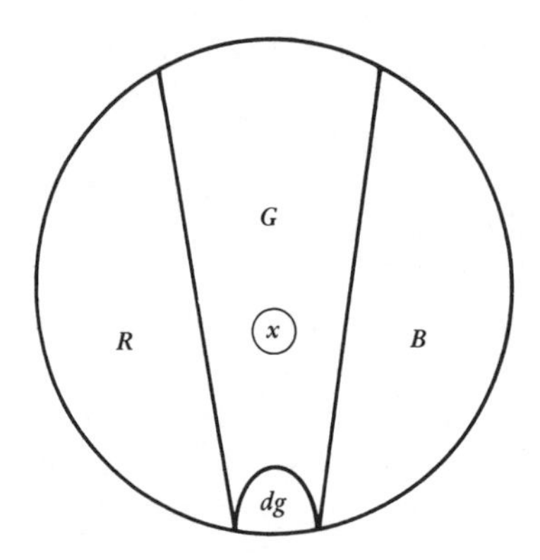

Figure 10.4.

By definition, the inner product of these two vectors is zero. The contribution to this inner product from the dark green element is 1; from all the red and blue elements, zero; from the green elements, a nonnegative integer p; and from x, an unknown integer q, which must be 0, 1, or -1. Thus we have $1+p+q=0$. This equation is satisfied only for $p=0$ and $q=-1$. Therefore in one of the two vectors the question mark is $+1$ and in the other it is -1. Choosing the vector in which the question mark is 1, we get the desired circuit or cocircuit.

Thus either statement 1 or statement 2 occurs. Both cannot occur simultaneously, for then the inner product of the corresponding circuit and cocircuit vectors will not be zero. ■

The arc coloring lemma in the special case of graphs [10.5], [10.6], is obvious, and it is used in Chapter 11 in proving the "no-gain" property of resistance networks.

10.9 MATROIDS AND THE GREEDY ALGORITHM

Consider a set S whose elements s_i have been assigned nonnegative weights $w(s_i)$. The weight of a subset of S is defined as equal to the sum of the weights of all the elements in the subset. Let $\mathcal{G}$ be a collection of subsets of S. Many problems in combinatorial optimization reduce to the following problem:

Determine a maximum weight member of $\mathcal{G}$.

For example, finding a maximum weight spanning tree of a weighted connected graph G reduces to the above problem, where $\mathcal{G}$ is the collection of all spanning trees of G.

The following algorithm, called the *greedy algorithm*, naturally suggests itself to solve such a combinatorial optimization problem.

Algorithm 10.1 The Greedy Algorithm

S1. Choose an element s_1 such that $\{s_1\} \in \mathcal{G}$ and such that $w(s_1) \geqslant w(s)$ for all s with $\{s\} \in \mathcal{G}$. If no such s_1 exists, stop.

S2. Choose an element s_2 such that $\{s_1, s_2\} \in \mathcal{G}$ and such that $w(s_2) \geqslant w(s)$ for all $s \neq s_1$ with $\{s_1, s\} \in \mathcal{G}$. If no such element s_2 exists, stop.

Sk. Choose an element s_k distinct from $s_1, s_2, \ldots, s_{k-1}$ such that $\{s_1, s_2, \ldots, s_{k-1}, s_k\} \in \mathcal{G}$ and such that $w(s_k)$ is a maximum over all such s. If no such s_k exists, stop. ■

Clearly, when the greedy algorithm terminates, it will have picked a maximal member of $\mathcal{G}$. But this member may not be of maximum weight in $\mathcal{G}$.

For example, if

$$S=\{a,b,c,d\},$$

$$w(a)=4, \qquad w(b)=3, \qquad w(c)=2, \qquad w(d)=2,$$

and

$$\mathcal{I}=\{\{a\},\{a,c\},\{b,c,d\},\{b,d\}\},$$

then the algorithm will pick $\{a,c\}$. But $\{b,c,d\}$ is a maximum weight member of $\mathcal{I}$. However, suppose we modify the weights so that

$$w(a)=6, \qquad w(b)=3, \qquad w(c)=2, \qquad w(d)=2.$$

The algorithm will again pick $\{a,c\}$, which is now a maximum weight member of $\mathcal{I}$.

Next we investigate the relationship between the greedy algorithm and the structure of $\mathcal{I}$.

Consider a matroid M on a set S. Let $\mathcal{I}$ be the collection of independent sets of M. Let

$$I_1=\{a_1,a_2,\ldots,a_m\}$$

and

$$I_2=\{b_1,b_2,\ldots,b_n\}$$

be two independent sets whose elements are arranged in the order of nonincreasing weights. Thus $w(a_1)\geqslant w(a_2)\geqslant\cdots\geqslant w(a_m)$ and $w(b_1)\geqslant w(b_2)\geqslant\cdots\geqslant w(b_n)$. Then I_1 is *lexicographically greater* than I_2 if there is some k such that $w(a_i)=w(b_i)$ for $1\leqslant i\leqslant k-1$ and $w(a_k)>w(b_k)$ or else $w(a_i)=w(b_i)$ for $1\leqslant i\leqslant n$ and $m>n$. A set which is not lexicographically less than any other set is said to be *lexicographically maximum*. From this definition it should be clear that a lexicographically maximum independent set must be a base.

A set $B\in\mathcal{I}$ is *Gale optimal* in $\mathcal{I}$ if for every $I\in\mathcal{I}$ there exists a one-to-one correspondence between I and B such that, for all $a\in I$, $w(a)\leqslant w(b)$, where b is the element of B which corresponds to a. Clearly, only bases can be Gale optimal. Also if a base is Gale optimal, then it must have maximum weight.

In the following we assume that the elements of each set are arranged in the order of nonincreasing weights.

THEOREM 10.32. Let $\mathcal{I}$ be the collection of independent sets of a matroid on S and B a base of M. For any nonnegative weighting of the elements of S, the following are equivalent:

1. B is lexicographically maximum in $\mathcal{I}$.
2. B is Gale optimal in $\mathcal{I}$.
3. B is a maximum weight member of $\mathcal{I}$.

Proof

1⇒2 Let $B=\{b_1, b_2,\ldots, b_r\}$ be a lexicographically maximum base of M. Suppose B is not Gale optimal. Then there exists an independent set $I=\{a_1, a_2,\ldots, a_k\}$ such that $w(a_i)=w(b_i)$ for $1\leqslant i\leqslant k-1$, and $w(a_k)>w(b_k)$. Consider then the independent sets $B_{k-1}=\{b_1, b_2,\ldots, b_{k-1}\}$ and $I=\{a_1, a_2,\ldots, a_k\}$. By axiom I-3 there exists $a_j\in I$ such that

$$I'=\{b_1, b_2,\ldots, b_{k-1}, a_j\}$$

is an independent set. But then I' is lexicographically greater than B because $w(a_j)\geqslant w(a_k)>w(b_k)$, which contradicts that B is lexicographically maximum in $\mathcal{I}$.

2⇒3 Obvious.

3⇒1 Let $B=\{b_1, b_2,\ldots, b_r\}$ be a lexicographically maximum base and $B'=\{b'_1, b'_2,\ldots, b'_r\}$ be a base of maximum weight. Since 1⇒2, it follows that

$$w(b_i)\geqslant w(b'_i) \qquad \text{for all } 1\leqslant i\leqslant r.$$

As B' is a maximum weight base, it follows that $w(b_i)=w(b'_i)$. Hence B' is also lexicographically maximum. ■

It is easy to show (as in the proof of 1⇒2 in the above theorem) that the greedy algorithm will pick a lexicographically maximum base which, by the above theorem, has maximum weight. Thus we have the following.

THEOREM 10.33. Let $\mathcal{I}$ be the collection of independent sets of a matroid on S whose elements have been assigned nonnegative weights. The greedy algorithm when applied on $\mathcal{I}$ will pick a maximum weight member of $\mathcal{I}$. ■

In view of the above theorem it is clear that a maximum weight base will be selected if we choose the elements of the matroid in the order of nonincreasing weights, rejecting an element only if its selection would destroy the independence of the set of chosen elements. The greedy algorithm to pick a minimum weight base is obvious.

As an example, consider the weighted graph G in Fig. 10.5. The weights of the edges are as shown in the figure. The greedy algorithm to select a maximum weight spanning tree of G will first arrange the edges in the order of nonincreasing weight. Thus the edges will be ordered as

$$a, b, e, f, d, c, g, h.$$

The algorithm will pick the first three edges a, b, and e because they do not contain any circuit. The edge f will be rejected because the set $\{a, b, e\}\cup\{f\}$

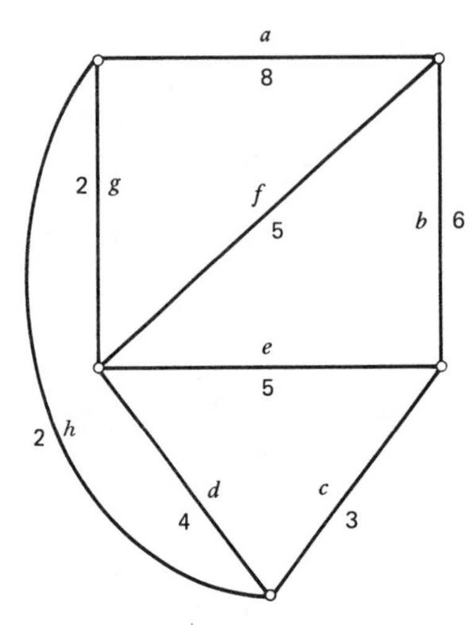

Figure 10.5. A weighted graph.

contains a circuit. The edge d will then be picked. c will be rejected because it forms a circuit with e and d which have already been picked. For the same reason g and h will also be rejected. Thus the greedy algorithm picks $\{a, b, e, d\}$, which is a maximum weight spanning tree of G.

We now prove the converse of Theorem 10.33.

THEOREM 10.34. Let $\mathcal{I}$ be the collection of subsets of a set S with the property that $A \in \mathcal{I}$ and $B \subseteq A$ implies that $B \in \mathcal{I}$. Then for all nonnegative weightings of the elements of S, the greedy algorithm when applied on $\mathcal{I}$ picks a maximum weight member of $\mathcal{I}$ only if $\mathcal{I}$ is the collection of independent sets of a matroid on S.

Proof

It is clear from the independence axioms that we have only to show that if $A = \{a_1, a_2, \ldots, a_k\} \in \mathcal{I}$ and $B = \{b_1, b_2, \ldots, b_{k+1}\} \in \mathcal{I}$, then there exists a $b_i \notin A$ such that $A \cup b_i \in \mathcal{I}$.

For this purpose let us define a weighting of the elements of S as follows:

$$w(a_i) = 1, \qquad 1 \leqslant i \leqslant k,$$

$$w(b_i) = x, \qquad b_i \in B - A,$$

$$w(e) = 0, \qquad e \in S - (A \cup B),$$

where $0 < x < 1$. Then the greedy algorithm first selects the elements $a_1, a_2, \ldots, a_k$. If no b_i exists with $\{b_i, a_1, \ldots, a_k\} \in \mathcal{I}$, the algorithm will thereafter select from members of $S - (A \cup B)$. So when it terminates, it will have picked a set whose weight is equal to the weight of A.

If $|B \cap A| = t$, then

$$w(A) = k$$

and

$$w(B) = t + (k+1-t)x.$$

Clearly, we can select $0 < x < 1$ such that $w(A) < w(B)$. But then the greedy algorithm will not have selected a maximum weight member of $\mathcal{I}$. This is a contradiction. ∎

The greedy algorithm for the problem of finding a maximum weight spanning tree was first discovered by Kruskal [10.7]. The extension of this algorithm to matroids was independently discovered by Rado [10.8], Edmonds [10.9], Gale [10.10], and Welsh [10.11].

10.10 FURTHER READING

Welsh [10.4] and Randow [10.12] are two excellent texts which contain a wealth of information on matroid theory. Berge [10.13] has a chapter on matroids. Wilson [10.14] gives an elegant introduction to matroids. He brings out clearly the power of the generality of a matroid by including simple proofs of two theorems on edge-disjoint spanning trees in a graph. In addition to Whitney's original paper [10.1] we highly recommend for further reading the papers by Rado [10.15], Lehman [10.16], Tutte [10.17] through [10.20], Mirsky [10.21], Harary and Welsh [10.22], Wilson [10.23], Edmonds [10.24], and Edmonds and Fulkerson [10.25] as well as the books by Mirsky [10.26] and Lawler [10.27].

Tutte [10.3], [10.17], [10.18] develops the theory of chain groups and matroids. He defines a matroid as regular if it is isomorphic to the matroid of a regular chain group. It can be shown that a matroid is regular if and only if it can be representable over every field. Another characterization is that a matroid is *regular* if and only if it is orientable [10.2]. Tutte also develops necessary and sufficient conditions for a matroid to be graphic. A binary matroid is graphic if and only if it does not contain as minor the Fano matroid or its dual or $M^*(K_5)$ or $M^*(K_{3,3})$. For more discussions on this question see [10.19] and [10.4]. (See Exercise 10.22 for a definition of the Fano matroid.)

For a theory of connectivity in matroids see Tutte [10.20]. For a theory of orientability which is applicable to general matroids, see Bland and Las Vergnas [10.28] and Folkman and Lawrence [10.29].

Lehman [10.16] gives a solution of the Shannon switching game using matroid concepts. See Edmonds [10.30], Bruno and Weinberg [10.31], and Welsh [10.4].

A beauty of matroid theory is its unifying nature which has yielded simple proofs of several results in transversal theory and graph theory. See Edmonds [10.24], [10.30], and Wilson [10.14].

Lawler [10.27] is an excellent text for matroid algorithms. See also Knuth [10.32], Edmonds [10.33], and Welsh [10.4].

Matroid theory is now being increasingly used in the study of electrical network problems. See Duffin [10.34], Duffin and Morley [10.35], Narayanan [10.36], Bruno and Weinberg [10.37], [10.38], [10.39], Iri and Tomizawa [10.40], Recski [10.41], [10.42], and Petersen [10.43]. Bruno and Weinberg [10.37] also give a good introduction to matroid theory.

10.11 EXERCISES

10.1 Let M be a matroid on S with $A \subseteq S$. Define $\mathcal{I}'$ to be the collection of subsets X of S such that X is independent in M and $X \cap A = \varnothing$. Prove that $\mathcal{I}'$ is the collection of independent sets of a matroid on S.

10.2 Let S be a set having n elements. Show that the collection $\mathcal{I}$ of all subsets of S having k or fewer elements is the set of independent sets of a matroid. This matroid is called the *uniform matroid* of rank k and is denoted by $U_{k,n}$.

10.3 Let M be a matroid on S, and let B_1, B_2 be distinct bases of M. Prove that there exists a one-to-one correspondence between B_1 and B_2 such that for all $e \in B_1$, $(B_2 - e') \cup e$ is a base of M, where $e' \in B_2$ corresponds to e.

10.4 If B_1, B_2 are bases of a matroid M and $X_1 \subseteq B_1$, prove that there exists $X_2 \subseteq B_2$ such that $(B_1 - X_1) \cup X_2$ and $(B_2 - X_2) \cup X_1$ are both bases of M (Greene [10.44]). (See also [10.12], Chap. V.)

10.5 Let $\mathcal{D}$ be a collection of nonnull subsets of S such that for any two distinct members X, Y of $\mathcal{D}$ with $x \in X \cap Y, y \in X - Y$ there exists $Z \in \mathcal{D}$ such that $y \in Z \subseteq (X \cup Y) - x$. Then prove that the collection $\mathcal{D}'$ of minimal members of $\mathcal{D}$ is the set of circuits of a matroid (Tutte [10.3]).

10.6 If C is a circuit of a matroid M and $a \in C$, prove that there exists a base B such that $C = C(a, B)$.

10.7 Prove that if B is a base of a matroid M and $x \in B$, there is exactly one cocircuit C^* of M such that $C^* \cap (B - x) = \varnothing$.

10.8 If C is a circuit of a matroid M and x, y are distinct elements of C, prove that there exists a cocircuit C^* containing x and y but no other element of C (Minty [10.2]).

10.9 Let M be a matroid on S and let x, y, z be distinct elements of S. If there is a circuit C_1 containing x and y and a circuit C_2 containing y and z, then prove that there exists a circuit C_3 containing x and z.

10.10 A set $A \subseteq S$ is *closed* in a matroid M on S if for all $x \in S-A$, $\rho(A \cup x)=\rho(A)+1$. Show that the intersection of two closed sets is a closed set.

10.11 Let M be a matroid on a set S. The *closure* $\sigma(A)$ of a subset A of S is the set of all elements x of S with the property that $\rho(A \cup x)=\rho(A)$. Prove the following:

(a) If x is contained in $\sigma(A \cup y)$ but not in $\sigma(A)$, then y is contained in $\sigma(A \cup x)$.

(b) An element x belongs to $\sigma(A)$ if and only if $x \in A$ or there exists a circuit C of M for which $C-A=\{x\}$.

10.12 A *hyperplane* of a matroid M on S is a maximal proper closed subset of S. Show that H is a hyperplane of the matroid M if and only if $S-H$ is a cocircuit of M (see Welsh [10.4] for a matroid axiom system in terms of hyperplanes).

10.13 Show that if M is a matroid on S and $A \subseteq S$, then the contraction $M \cdot T$ of M to T is the matroid whose cocircuits are precisely those cocircuits of M which are contained in A.

10.14 If M is a matroid on S and $T \subseteq X \subseteq S$, prove that

(a) $M|T=(M|X)|T$.

(b) $M \cdot T=(M \cdot X) \cdot T$.

(c) $(M|X) \cdot T=(M \cdot (S-(X-T)))|T$.

(d) $(M \cdot X)|T=(M|S-(X-T)) \cdot T$.

10.15 A matroid M on S is *connected* or *nonseparable* if for every pair of distinct elements x and y of S there is a circuit of M containing x and y. Otherwise it is *disconnected* or *separable*. Show that a matroid M is connected if and only if its dual M^* is connected.

Note If G is a graph, then $M(G)$ is connected if and only if G is 2-connected.

10.16 Show that a matroid M on S is not connected if and only if there exists a proper subset A of S such that

$$\rho(A)+\rho(S-A)=\rho(S)$$

(Whitney [10.1]).

10.17 Prove or disprove: Graph G is contractible to H if and only if $M(G)$ contains $M(H)$ as a contraction minor.

10.18 Prove that the uniform matroid $U_{2,4}$ is representable over every field except $GF(2)$.

10.19 Prove that a matroid is binary if and only if for any circuit C and cocircuit C^*, $|C \cap C^*| \neq 3$.

10.20 Let $\mathcal{P}$ be a family of finite nonempty subsets of a set S. A transversal of a subfamily of $\mathcal{P}$ is called a *partial transversal* of $\mathcal{P}$. Show that if $\mathcal{P}$ is a collection of finite nonempty subsets of a set S, then the collection of partial transversals of $\mathcal{P}$ is the set of independent sets of a matroid on S. (See Section 8.6 for the definition of a transversal.) Find the rank function of this matroid. A matroid M on S is called a *transversal matroid* if there exists some family $\mathcal{P}$ of subsets of S such that $\mathcal{I}(M)$ is the family of partial transversals of $\mathcal{P}$. Find the rank function of a transversal matroid (see Theorem 8.15).

10.21 Show that every k-uniform matroid is a transversal matroid.

10.22 The *Fano matroid* F is the matroid defined on the set $S = \{1,2,3,4,5,6,7\}$ whose bases are all those subsets of S containing three elements except $\{1,2,3\}$, $\{1,4,5\}$, $\{1,6,7\}$, $\{2,4,7\}$, $\{2,5,6\}$, $\{3,4,6\}$, and $\{3,5,7\}$. Show that F is

(a) binary,
(b) nontransversal,
(c) nongraphic,
(d) noncographic.

10.23 Show that the circuit matroid of K_4 is not a transversal matroid.

10.24 A matroid M on S is *Eulerian* if S can be expressed as the union of disjoint circuits. A matroid is *bipartite* if every circuit of M contains an even number of elements. Prove that a matroid is bipartite if and only if M^* is Eulerian.

10.25 Let D be a directed graph without self-loops and let X and Y be two disjoint sets of vertices of D. A subset A of X is called independent if there exist $|A|$ vertex-disjoint chains from A to Y. Show that these independent sets form the independent sets of a matroid on X. (Such a matroid is called a *gammoid*.) (Mason [10.45.])

10.26 A matroid M on S is *base orderable* if for any two bases B_1, B_2 of M there exists a one-to-one correspondence between B_1 and B_2 such that for each $x \in B_1$, both $(B_1 - x) \cup x'$ and $(B_2 - x') \cup x$ are bases of M where x' is the element of B_2 which corresponds to x. Show the following:

(a) $M(K_4)$ is not base orderable.
(b) If M is base orderable on S and $T \subseteq S$, then $M|T$ is base orderable.

(c) If M is base orderable, then any minor of M is base orderable.

10.27 Let M_1 and M_2 be two matroids on a set S.

(a) Show that the set of all unions $I \cup J$ of an independent set I of M_1 and an independent set J of M_2 form the independent sets of a new matroid. (This matroid is called the *union* of M_1 and M_2 and is denoted by $M_1 \cup M_2$.)

(b) If ρ_1 and ρ_2 denote the rank functions of matroids M_1 and M_2 on a set S, show that

$$\rho(A) = \min_{X \subseteq A} \{\rho_1(X) + \rho_2(X) + |A - X|\}$$

where $A \subseteq S$ and ρ is the rank function of $M_1 \cup M_2$ (Wilson [10.14], pp. 154–158).

10.28 Let M be a matroid on S. Prove the following:

(a) M contains k disjoint bases if and only if, for any $A \subseteq S$,

$$\rho(A) + |S - A| \geqslant k\rho(S).$$

(b) S can be expressed as the union of not more than k independent sets if and only if for any $A \subseteq S$,

$$k\rho(A) \geqslant |A|.$$

Hint Consider the union of k copies of M and use the result of Exercise 10.27 (Wilson [10.14], pp. 154–158).

10.29 Show that when the greedy algorithm has chosen k elements, these k elements are of maximum weight with respect to all independent sets of k or fewer elements.

10.30 A number of jobs are to be processed by a single machine. All jobs require the same processing time. Each job has assigned to it a deadline.

(a) Show that the collection of all subsets of jobs which can be completed on time forms the independent sets of a matroid.

(b) Suppose each job has a penalty which must be paid if it is not completed by its deadline. In what order should these jobs be processed so that the total penalty is minimum?

10.31 Let M be a matroid whose elements have been assigned nonnegative weights. Prove the following:

(a) No element of a maximum weight base is the smallest element of any circuit of M.

(b) Each element of a maximum weight base is the largest element of at least one cocircuit of M.

Using (b), design a procedure for constructing a maximum weight base of a matroid. (Prim [10.46] describes such a procedure for constructing a minimum weight spanning tree of a connected graph.)

10.32 Let M be a matroid on S with nonnegative weights assigned to the elements of S. Let $\mathcal{B}$ be the collection of bases of M and $\mathcal{C}^*$ the collection of cocircuits of M. Prove

$$\min_{B\in\mathcal{B}} \max_{e\in B} w(e) = \max_{C^*\in\mathcal{C}^*} \min_{e\in C^*} w(e).$$

10.12 REFERENCES

10.1 H. Whitney, "On the Abstract Properties of Linear Dependence," *Am. J. Math.*, Vol. 57, 509–533 (1935).

10.2 G. J. Minty, "On the Axiomatic Foundations of the Theories of Directed Linear Graphs, Electrical Networks and Network Programming," *J. Math. and Mech.*, Vol. 15, 485–520 (1966).

10.3 W. T. Tutte, *Introduction to the Theory of Matroids*, American Elsevier, New York, 1971.

10.4 D. J. A. Welsh, *Matroid Theory*, Academic Press, New York, 1976.

10.5 G. J. Minty, "Monotone Networks," *Proc. Roy. Soc., A*, Vol. 257, 194–212 (1960).

10.6 G. J. Minty, "Solving Steady-State Non-Linear Networks of 'Monotone' Elements," *IRE Trans. Circuit Theory*, Vol. CT-8, 99–104 (1961).

10.7 J. B. Kruskal, "On the Shortest Spanning Subgraph of a Graph and the Travelling Salesman Problem," *Proc. Am. Math. Soc.*, Vol. 7, 48–49 (1956).

10.8 R. Rado, "Note on Independence Functions," *Proc. London Math. Soc.*, Vol. 7, 300–320 (1957).

10.9 J. Edmonds, "Matroids and the Greedy Algorithm," *Math. Programming*, Vol. 1, 127–136 (1971).

10.10 D. Gale, "Optimal Assignments in an Ordered Set: An Application of Matroid Theory," *J. Combinatorial Theory*, Vol. 4, 176–180 (1968).

10.11 D. J. A. Welsh, "Kruskal's Theorem for Matroids," *Proc. Cambridge Phil. Soc.*, Vol. 64, 3–4 (1968).

10.12 Rabe von Randow, *Introduction to the Theory of Matroids*, Springer Lecture Notes in Mathematical Economics, Vol. 109, 1975.

10.13 C. Berge, *Graphs and Hypergraphs*, North Holland, Amsterdam, 1973.

10.14 R. J. Wilson, *Introduction to Graph Theory*, Oliver and Boyd, Edinburgh, 1972.

10.15 R. Rado, "A Theorem on Independence Relations," *Quart, J. Math. (Oxford)*, Vol. 13, 83–89 (1942).

10.16 A. Lehman, "A Solution of the Shannon Switching Game," *SIAM J. Appl. Math.*, Vol. 12, 687–725 (1964).

10.17 W. T. Tutte, "A Homotopy Theorem for Matroids-I and II," *Trans. Am. Math. Soc.*, Vol. 88, 144–174 (1958).

10.18 W. T. Tutte, "Lectures on Matroids," *J. Res. Nat. Bur. Stand.*, Vol. 69B, 1–48 (1965).

10.19 W. T. Tutte, "Matroids and Graphs," *Trans. Am. Math. Soc.*, Vol. 90, 527–552 (1959).

10.20 W. T. Tutte, "Connectivity in Matroids," *Canad. J. Math.*, Vol. 18, 1301–1324 (1966).

10.21 L. Mirsky, "Application of the Notion of Independence to Combinatorial Analysis," *J. Combinatorial Theory*, Vol. 2, 327–357 (1967).

10.22 F. Harary and D. J. A. Welsh, "Matroids versus Graphs," in *The Many Facets of Graph Theory*, Springer Lecture Notes, Vol. 110, 1969, pp. 155–170.

10.23 R. J. Wilson, "An Introduction to Matroid Theory," *Am. Math. Monthly*, Vol. 80, 500–525 (1973).

10.24 J. Edmonds, "Minimum Partition of a Matroid into Independent Subsets," *J. Res. Nat. Bur. Stand.*, Vol. 69B, 67–72 (1965).

10.25 J. Edmonds and D. R. Fulkerson, "Transversals and Matroids Partition," *J. Res. Nat. Bur. Stand.*, Vol. 69B, 147–153 (1965).

10.26 L. Mirsky, *Transversal Theory*, Academic Press, London, 1971.

10.27 E. L. Lawler, *Combinatorial Optimization: Networks and Matroids*, Holt, Rinehart and Winston, New York, 1976.

10.28 R. G. Bland and M. Las Vergnas," Orientability of Matroids," *J. Combinatorial Theory B*, Vol. 24, 94–123 (1978).

10.29 J. Folkman and J. Lawrence, "Oriented Matroids," *J. Combinatorial Theory B*, Vol. 25, 199–236 (1978).

10.30 J. Edmonds, "Lehman's Switching Game and a Theorem of Tutte and Nash-Williams," *J. Res. Nat. Bur. Stand.*, Vol. 69B, 73–77 (1965).

10.31 J. Bruno and L. Weinberg, "A Constructive Graph-Theoretic Solution of the Shannon Switching Game," *IEEE Trans. Circuit Theory*, Vol. CT-17, 74–81 (1970).

10.32 D. E. Knuth, "Matroid Partitioning," Stanford University Rep. STAN-CS-73-342, 1–12 (1973).

10.33 J. Edmonds, "Matroid Partition," in *Lectures in Appl. Math.*, Vol. 11: *Mathematics of Decision Sciences*, 1967, pp. 335–346.

10.34 R. J. Duffin, "Topology of Series-Parallel Networks," *J. Math. Anal. Appl.*, Vol. 10, 303–318 (1965).

10.35 R. J. Duffin and T. D. Morley, "Wang Algebra and Matroids," *IEEE Trans. Circuits and Syst.*, Vol. CAS-25, 755–762 (1978).

10.36 H. Narayanan, "Theory of Matroids and Network Analysis," Ph.D. Thesis, Indian Institute of Technology, Bombay, India, 1974.

10.37 J. Bruno and L. Weinberg, "Generalized Networks: Networks Embedded on a Matroid, Part 1," *Networks*, Vol. 6, 53–94 (1976).

10.38 J. Bruno and L. Weinberg, "Generalized Networks: Networks Embedded on a Matroid, Part 2," *Networks*, Vol. 6, 231–272 (1976).

10.39 L. Weinberg, "Matroids, Generalized Networks and Electric Network Synthesis," *J. Combinatorial Theory B*, Vol. 23, 106–126 (1977).

10.40 M. Iri and N. Tomizawa, "A Unifying Approach to Fundamental Problems in Network Theory by Means of Matroids," *Electron. Commun. in Japan*, Vol. 58-A, 28–35 (1975).

10.41 A. Recski, "On Partitional Matroids with Applications," in *Coll. Math. Soc. J. Bolyai*, Vol. 10: *Infinite and Finite Sets*, North-Holland-American Elsevier, Amsterdam, 1974, pp. 1169–1179.

10.42 A. Recski, "Matroids and Independent State Variables," *Proc. 2nd European Conf. Circuit Theory and Design*, Genova, 1976.

10.43 B. Petersen, "Investigating Solvability and Complexity of Linear Active Networks by Means of Matroids," *IEEE Trans. Circuits and Syst.*, Vol. CAS-26, 330–342 (1979).

10.44 C. Greene, "A Multiple Exchange Property for Bases," *Proc. Am. Math. Soc.*, Vol. 39, 45–50 (1973).

10.45 J. H. Mason, "On a Class of Matroids Arising from Paths in Graphs," *Proc. London Math. Soc.*, Vol. 25, 55–74 (1972).

10.46 R. C. Prim, "Shortest Connection Networks and Some Generalizations," *Bell Sys. Tech. J.*, Vol. 36, 1389–1402 (1957).

II

Electrical Network Theory

Chapter 11

Graphs and Networks

An electrical network is an interconnection of electrical network elements such as resistances, capacitances, inductances, and voltage and current sources. Each network element is associated with two variables, the voltage variable $v(t)$ and the current variable $i(t)$. We need to specify reference directions for these variables because they are functions of time and may take on positive and negative values in the course of time. This is done by assigning an arrow, called *orientation*, to each network element (Fig. 11.1). This arrow means that $i(t)$ is positive whenever the current is in the direction of the arrow. Further we assume that the positive polarity of the voltage $v(t)$ is at the tail end of the arrow. Thus $v(t)$ is positive whenever the voltage drop in a network element is in the direction of the arrow.

Network elements are characterized by the physical relationships between the associated voltage and current variables. Note that for some of the network elements the voltage variables may be required to have specified values, and for some others the current variables may be specified. Such elements are called, respectively, the *voltage* and *current sources*.

Two fundamental laws of network theory are *Kirchhoff's laws*, which can be stated as follows:

Kirchhoff's Current Law (KCL) The algebraic sum of the currents flowing out of a node is equal to zero.

Kirchhoff's Voltage Law (KVL) The algebraic sum of the voltages around any circuit is equal to zero.

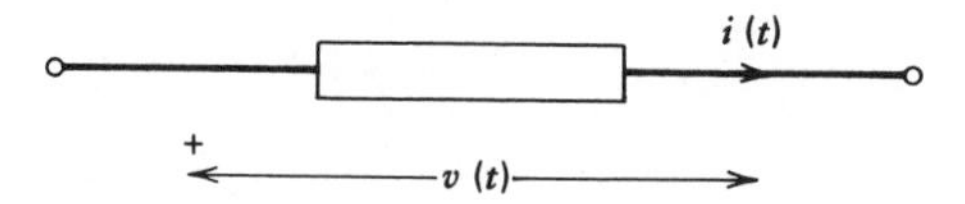

Figure 11.1. A network element (representation).

For instance, for the network shown in Fig. 11.2a the KCL and KVL equations are as given below:

KCL equations

node a $\quad i_1 - i_5 + i_6 = 0,$

node c $\quad -i_2 + i_4 - i_6 = 0,$

node b $\quad -i_1 + i_2 + i_3 = 0.$

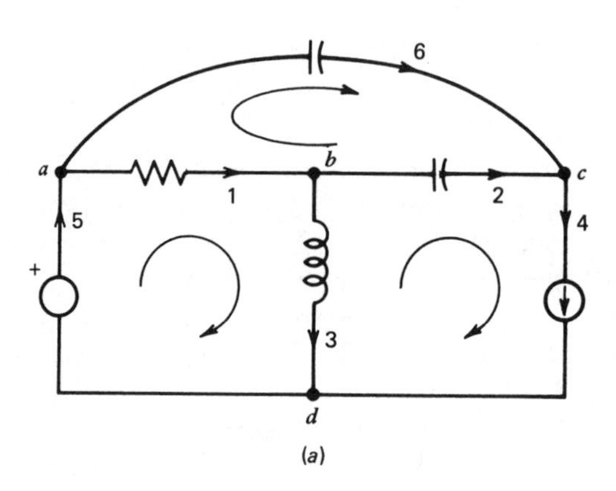

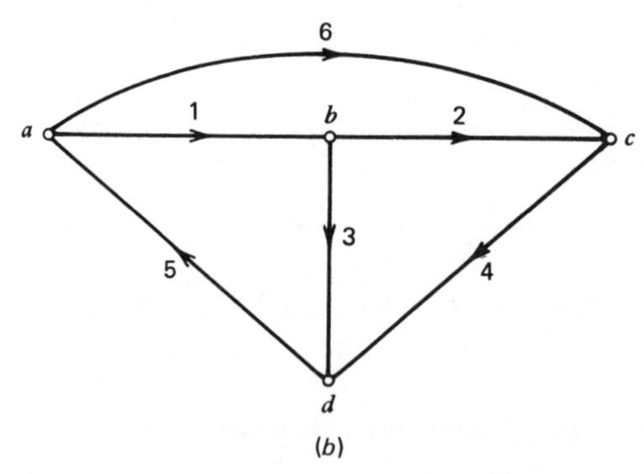

Figure 11.2. Directed graph representation of a network. (a) Network N. (b) Directed graph of N.

KVL equations

$$\text{circuit } \{1,3,5\} \qquad v_1+v_3+v_5=0,$$

$$\text{circuit } \{2,4,3\} \qquad v_2+v_4-v_3=0,$$

$$\text{circuit } \{1,6,2\} \qquad -v_1+v_6-v_2=0.$$

Given an electrical network N, the problem of network analysis is to determine the element voltages and currents that satisfy Kirchhoff's laws and the voltage-current relationships characterizing the different network elements constituting the network.

Notice that the equations which arise from an application of Kirchhoff's laws are algebraic in nature, and they depend only on the way the network elements are interconnected and not on the nature of the network elements. There are several properties of an electrical network which depend on the structure of the network. In studying such properties it will be convenient to treat each network element as a directed edge associated with the two variables $v(t)$ and $i(t)$. Thus we may consider an electrical network as a directed graph in which each edge is associated with the two variables $v(t)$ and $i(t)$, which are required to satisfy Kirchhoff's laws and certain specified physical relationships.

For example, the directed graph corresponding to the network of Fig. 11.2*a* is shown in Fig. 11.2*b*.

It is now easy to see that KCL and KVL equations for a network N can be written, respectively, as

$$Q_c I_e = 0$$

and

$$B_c V_e = 0,$$

where Q_c and B_c are the cut and circuit matrices of the directed graph associated with N, and I_e and V_e are, respectively, the column vectors of element currents and voltages of N.

In all our discussions in this and the subsequent chapters of this part of the book we denote both an electrical network and the associated directed graph by the same symbol. Most often a graph is also referred to as a network, and vice versa. Thus all the definitions relating to a graph, such as connectedness, nullity, and rank, are also applicable in the case of a network. We may also refer to a vertex as a *node*.

In this chapter we discuss some aspects of network analysis which depend heavily on the theory of graphs. Our main aim here is to highlight the usefulness of graph theory in the systematic formulation of network equations and in the discovery of certain fundamental properties of networks.

11.1 LOOP AND CUTSET TRANSFORMATIONS

In this section we investigate the relationship among the element currents and the relationship among the element voltages of an electrical network N. These relationships arise as a result of Kirchhoff's laws and the orthogonality relation between the circuit and cutset matrices of N. Without loss of generality, we may assume that N is connected.

Let T be a spanning tree of N, and let B_f and Q_f denote the fundamental circuit and cutset matrices of N with respect to T. Then Kirchhoff's current and voltage equations become

$$Q_f I_e = 0 \tag{11.1}$$

and

$$B_f V_e = 0. \tag{11.2}$$

Suppose we partition I_e and V_e as

$$I_e = \begin{bmatrix} I_c \\ I_t \end{bmatrix}$$

and

$$V_e = \begin{bmatrix} V_c \\ V_t \end{bmatrix},$$

where the vectors which correspond to the chords and the branches of T are distinguished by the subscripts c and t, respectively. Then (11.1) and (11.2) can be written as

$$\begin{bmatrix} Q_{fc} & U \end{bmatrix} \begin{bmatrix} I_c \\ I_t \end{bmatrix} = 0 \tag{11.3}$$

and

$$\begin{bmatrix} U & B_{ft} \end{bmatrix} \begin{bmatrix} V_c \\ V_t \end{bmatrix} = 0. \tag{11.4}$$

Recall that (see (6.13))

$$Q_{fc} = -B_{ft}^t. \tag{11.5}$$

Consider first (11.3). We get from this equation

$$\begin{aligned} I_t &= -Q_{fc} I_c \\ &= B_{ft}^t I_c. \end{aligned} \tag{11.6}$$

So we can express I_e as

$$I_e=\begin{bmatrix} I_c \\ I_t \end{bmatrix}$$

$$=\begin{bmatrix} U \\ B_{ft}^t \end{bmatrix} I_c$$

$$=B_f^t I_c. \tag{11.7}$$

Starting from (11.4) we can show in a similar manner that

$$V_e=Q_f^t V_t. \tag{11.8}$$

Thus we get the following theorem.

THEOREM 11.1.

1. All the element currents in an electrical network N can be expressed as linear combinations of chord currents, that is, the currents associated with the chords of a spanning tree of N.
2. All the element voltages of an electrical network N can be expressed as linear combinations of branch voltages, that is, the voltages associated with the branches of a spanning tree of N. ■

To illustrate (11.7) and (11.8), consider the network N shown in Fig. 11.2. The B_f and Q_f matrices with respect to the spanning tree T consisting of the elements 1, 4, and 5 are

$$\begin{array}{cc} & \begin{array}{cccccc} 2 & 3 & 6 & 1 & 4 & 5 \end{array} \\ B_f= & \left[\begin{array}{ccc|ccc} 1 & 0 & 0 & 1 & 1 & 1 \\ 0 & 1 & 0 & 1 & 0 & 1 \\ 0 & 0 & 1 & 0 & 1 & 1 \end{array}\right], \end{array}$$

$$\begin{array}{cc} & \begin{array}{cccccc} 2 & 3 & 6 & 1 & 4 & 5 \end{array} \\ Q_f= & \left[\begin{array}{ccc|ccc} -1 & -1 & 0 & 1 & 0 & 0 \\ -1 & 0 & -1 & 0 & 1 & 0 \\ -1 & -1 & -1 & 0 & 0 & 1 \end{array}\right]. \end{array}$$

Then we can express I_e and V_e as follows:

$$\begin{bmatrix} i_2 \\ i_3 \\ i_6 \\ i_1 \\ i_4 \\ i_5 \end{bmatrix} = \begin{bmatrix} 1 & 0 & 0 \\ 0 & 1 & 0 \\ 0 & 0 & 1 \\ 1 & 1 & 0 \\ 1 & 0 & 1 \\ 1 & 1 & 1 \end{bmatrix} \begin{bmatrix} i_2 \\ i_3 \\ i_6 \end{bmatrix},$$

$$\begin{bmatrix} v_2 \\ v_3 \\ v_6 \\ v_1 \\ v_4 \\ v_5 \end{bmatrix} = \begin{bmatrix} -1 & -1 & -1 \\ -1 & 0 & -1 \\ 0 & -1 & -1 \\ 1 & 0 & 0 \\ 0 & 1 & 0 \\ 0 & 0 & 1 \end{bmatrix} \begin{bmatrix} v_1 \\ v_4 \\ v_5 \end{bmatrix}.$$

THEOREM 11.2. Let N be an electrical network of rank ρ and nullity μ. Let B be a matrix formed by any μ independent rows of the circuit matrix of N, and let Q be a matrix formed by any ρ independent rows of the cut matrix of N.

1. *Loop Transformation* A column vector I_e satisfies KCL equations of N if and only if there exists a column vector I' of μ entries such that

$$I_e = B^t I'. \tag{11.9}$$

2. *Cutset Transformation* A column vector V_e satisfies KVL equations of N if and only if there exists a column vector V' of ρ entries such that

$$V_e = Q^t V'. \tag{11.10}$$

Proof

1. Let B_f be a fundamental circuit matrix of N. Then there exists a nonsingular matrix D such that

$$B_f = DB. \tag{11.11}$$

If I_e satisfies KCL equations, then we have from (11.7)

$$I_e = B_f^t I_c.$$

Substituting (11.11) into the above equation and letting $I' = D^t I_c$, we get

$$I_e = B^t I'.$$

Conversely, if there exists I' such that

$$I_e = B^t I',$$

then

$$QI_e = (QB^t)I'$$
$$= 0, \qquad \text{by Theorem 6.6,}$$

and KCL is satisfied.

2. Proof follows in a dual manner. ■

Equations (11.9) and (11.10) are known as the *loop transformation* and the *cutset transformation*, respectively. The elements of I' and V' are called, respectively, the *loop* and *cutset variables*. In general, the loop and cutset variables are linear combinations of the chord currents and branch voltages, respectively. However, if we use B_f (Q_f) in the loop (cutset) transformation, then the chord currents (branch voltages) become loop (cutset) variables.

The transformation

$$V_e = A^t V', \tag{11.12}$$

where A is an incidence matrix, is called *node transformation*. Clearly this is a special case of cutset transformation. If v_r is the reference node for A, then the entries of V' can be identified as the voltages of all the nodes (except v_r) with respect to v_r. These voltages are usually referred to as *node-to-datum* or simply as *node voltages* of the network.

We conclude this section with an interesting result due to Tellegen [11.1], which follows easily from the loop and cutset transformations.

THEOREM 11.3 (TELLEGEN). Consider two electrical networks N and $\hat{N}$ such that the graphs associated with them are identical. Let V_e and Ψ_e denote the element voltage vectors of N and $\hat{N}$, respectively, and let I_e and Λ_e be the corresponding element current vectors. Then

1. $V_e^t \Lambda_e = 0$.
2. $I_e^t \Psi_e = 0$.

Proof

1. Let B_f and Q_f denote the fundamental circuit and cutset matrices of N with respect to a spanning tree T. Since the graph of N is the same as

that of $\hat{N}$, it is clear that $\hat{N}$ also has the same B_f and Q_f matrices with respect to T.

We have from the loop and cutset transformations:

$$V_e = Q_f^t V_t$$

and

$$\Lambda_e = B_f^t \Lambda_c.$$

So

$$V_e^t \Lambda_e = V_t^t (Q_f B_f^t) \Lambda_c$$
$$= 0, \qquad \text{by Theorem 6.6.}$$

2. Proof follows in a dual manner. ■

Tellegen's theorem is a very profound result in network theory with several applications. See Penfield, Spence, and Duinker [11.2], [11.3]. We discuss in Chapter 13 an application of this theorem in the computation of network sensitivities using the concept of the adjoint of a network. See Bordewijk [11.4] and Director and Rohrer [11.5]; see also Kishi and Kida [11.6].

11.2 LOOP AND CUTSET SYSTEMS OF EQUATIONS

As we observed earlier, the problem of network analysis is to determine the voltages and currents associated with the elements of an electrical network. These voltages and currents can be determined from Kirchhoff's equations and the element voltage-current (in short, v–i) relations. However, these equations involve a large number of variables. As we have seen in Theorem 11.1, not all these variables are independent. Further, it is clear from Theorem 11.2 that in place of KCL equations we can use the loop transformation which involves only chord currents as variables. Similarly, KVL equations can be replaced by the cutset transformation which involves only branch voltage variables. We can take advantage of these transformations to establish different systems of network equations which involve only a subset of voltages and/or currents as variables. Two such systems of equations, known as the loop and cutset systems, are developed in this section. In deriving the loop system we use the loop transformation in place of KCL, and in this case the loop variables will serve as independent variables. In deriving the cutset system we use the cutset transformation in place of KVL, and the cutset variables will serve as the independent variables in this case.

Consider a connected electrical network N. We assume that N consists of only resistances(R), capacitances(C), inductances(L) including mutual inductances, and independent voltage and current sources. We also assume that all initial inductor currents and initial capacitor voltages have been replaced by appropriate sources. Further, the voltage and current variables are all Laplace transforms of the complex frequency variables.

In N there can be no circuit consisting of only independent voltage sources. For if such a circuit of sources were present, then, by KVL, there would be a linear relationship among the corresponding voltages, violating the independence of the voltage sources. For the same reason, in N there can be no cutset consisting of only independent current sources. So, by Theorem 10.12, there exists in N a spanning tree containing all the voltage sources but no current sources. Such a tree is the starting point for the development of both the loop and cutset systems of equations.

We first derive the loop system.

Let T be a spanning tree of the given network such that T contains all the voltage sources but no current sources. Let us partition the element voltage vector V_e and the element current vector I_e as follows:

$$V_e = \begin{bmatrix} V_1 \\ V_2 \\ V_3 \end{bmatrix}$$

and

$$I_e = \begin{bmatrix} I_1 \\ I_2 \\ I_3 \end{bmatrix},$$

where the subscripts 1, 2, and 3 refer to the vectors corresponding to the current sources, RLC elements, and voltage sources, respectively. Let B_f be the fundamental circuit matrix of N with respect to T. The KVL equations for N can be written as follows:

$$B_f V_e = \begin{bmatrix} U & B_{12} & B_{13} \\ 0 & B_{22} & B_{23} \end{bmatrix} \begin{bmatrix} V_1 \\ V_2 \\ V_3 \end{bmatrix} = [0],$$

that is,

$$V_1 = -B_{12}V_2 - B_{13}V_3 \tag{11.13}$$

and

$$B_{22}V_2 = -B_{23}V_3. \tag{11.14}$$

In place of KCL, we can use the loop transformation:

$$\begin{bmatrix} I_1 \\ I_2 \\ I_3 \end{bmatrix} = \begin{bmatrix} U & 0 \\ B_{12}^t & B_{22}^t \\ B_{13}^t & B_{23}^t \end{bmatrix} \begin{bmatrix} I_1 \\ I_l \end{bmatrix},$$

where I_l denotes the vector of currents associated with the nonsource chords of T. From this equation we get

$$I_2 = B_{12}^t I_1 + B_{22}^t I_l, \tag{11.15}$$

$$I_3 = B_{13}^t I_1 + B_{23}^t I_l. \tag{11.16}$$

Note that among the chord currents only those in I_l are to be determined.

If Z_2 is the impedance matrix of the RLC elements, then the v–i relations for these elements can be written as

$$V_2 = Z_2 I_2. \tag{11.17}$$

Using (11.17) in (11.14), we get

$$B_{22} Z_2 I_2 = -B_{23} V_3.$$

Using (11.15) in the above and rearranging the terms, we get

$$\left(B_{22} Z_2 B_{22}^t\right) I_l = -B_{23} V_3 - B_{22} Z_2 B_{12}^t I_1. \tag{11.18}$$

The above is the *loop system* of equations which involves only $\mu - n_c$ variables, where n_c is the number of current sources in N. Note that $\mu - n_c$ is equal to the nullity of the network obtained after removing from N all the current sources.

The matrix

$$Z_l = B_{22} Z_2 B_{22}^t$$

in (11.18) is called the *loop-impedance matrix* of N. If the network N has no mutual inductances, then Z_2 will be a diagonal matrix with no zero entries along the diagonal. Hence in this case Z_l will be nonsingular because B_{22} has maximum rank equal to $\mu - n_c$. If N has mutual inductances, then Z_l will be nonsingular only if Z_2 is positive definite.

Once we have determined I_l using (11.18), we can determine I_2 using (11.15), and we can determine V_2 using (11.17). We can then determine V_1 and I_3 using (11.13) and (11.16). (Note that I_1 and V_3 have specified values.) This would then complete the analysis of network N using the loop system of equations.

Note that the loop system is derived by first substituting the element v–i relations into KVL equations and then using the loop transformation. In an exactly dual manner we can derive as follows the cutset system of equations.

We first write the KCL equations in partitioned form:

$$\begin{bmatrix} Q_{11} & Q_{12} & 0 \\ Q_{21} & Q_{22} & U \end{bmatrix} \begin{bmatrix} I_1 \\ I_2 \\ I_3 \end{bmatrix} = [0],$$

where the coefficient matrix is the same as the fundamental cutset matrix of N with respect to T. From these equations we get

$$Q_{12}I_2 = -Q_{11}I_1, \tag{11.19}$$

$$I_3 = -Q_{21}I_1 - Q_{22}I_2. \tag{11.20}$$

In place of KVL equations we can use the cutset transformation:

$$\begin{bmatrix} V_1 \\ V_2 \\ V_3 \end{bmatrix} = \begin{bmatrix} Q_{11}^t & Q_{21}^t \\ Q_{12}^t & Q_{22}^t \\ 0 & U \end{bmatrix} \begin{bmatrix} V_b \\ V_3 \end{bmatrix},$$

where V_b is the vector of voltages associated with the nonsource branches of T. From these equations we get

$$V_1 = Q_{11}^t V_b + Q_{21}^t V_3, \tag{11.21}$$

$$V_2 = Q_{12}^t V_b + Q_{22}^t V_3. \tag{11.22}$$

If Y_2 is the admittance matrix of the RLC elements of N, then the element v–i relations can be written as

$$I_2 = Y_2 V_2. \tag{11.23}$$

It is now a straightforward exercise to derive the cutset system of equations which relate V_b to I_1 and V_3.

Using (11.23) in (11.19), we get

$$Q_{12}Y_2V_2 = -Q_{11}I_1.$$

Using (11.22) in the above, we get

$$(Q_{12}Y_2Q_{12}^t)V_b = -Q_{11}I_1 - Q_{12}Y_2Q_{22}^t V_3. \tag{11.24}$$

The above is the *cutset system of equations* which involve $\rho - n_v$ variables, where n_v is the number of voltage sources in N. Note that $\rho - n_v$ is equal to the rank of the network obtained from N by contracting all the voltage sources.

The matrix

$$Y_b = Q_{12} Y_2 Q_{12}^t$$

in (11.24) is called the *cutset admittance matrix* of N. This matrix will be nonsingular if N has no mutual inductances. If N has mutual inductances, then Y_b will be nonsingular only if Y_2 is positive definite.

Once we have determined V_b using (11.24), we can determine V_2 using (11.22) and then determine I_2 using (11.23). Finally we can determine I_3 and V_1 from (11.20) and (11.21), respectively.

Now we illustrate the formulation of loop and cutset systems of equations.

Consider the network shown in Fig. 11.3*a* where $u(t)$ is the unit step function. The graph of this network is shown in Fig. 11.3*b*. We choose the spanning tree T consisting of the edges 4, 5, and 6. Note that T contains the voltage source and no current source. The fundamental circuit and the fundamental cutset matrices with respect to T are given below in the required

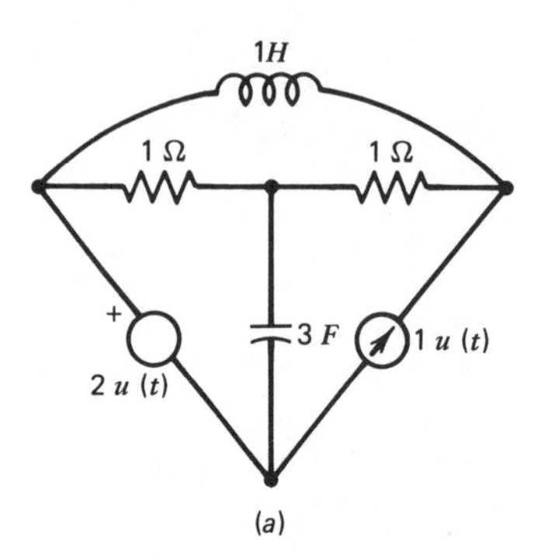

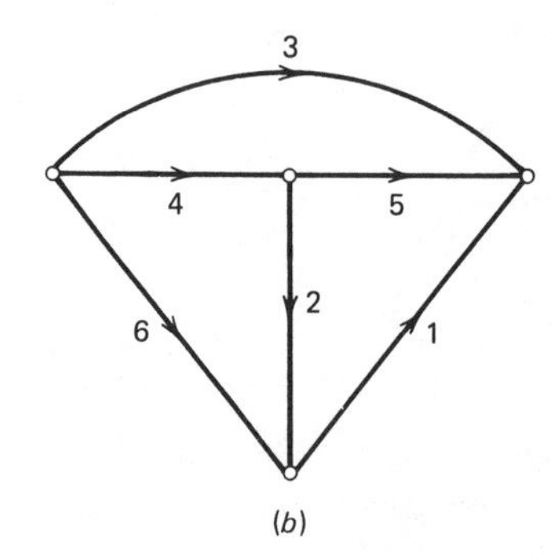

Figure 11.3. A network and its graph.

partitioned form:

$$
\begin{array}{c} \\ B_f= \\ \\ \end{array}
\left[\begin{array}{c|cccc|c}
1 & 2 & 3 & 4 & 5 & 6 \\
1 & 0 & 0 & -1 & -1 & 1 \\ \hline
0 & 1 & 0 & 1 & 0 & -1 \\
0 & 0 & 1 & -1 & -1 & 0
\end{array}\right],
$$

$$
Q_f=\left[\begin{array}{c|cccc|c}
1 & -1 & 1 & 1 & 0 & 0 \\
1 & 0 & 1 & 0 & 1 & 0 \\ \hline
-1 & 1 & 0 & 0 & 0 & 1
\end{array}\right].
$$

From these matrices we get

$$B_{12}=\begin{bmatrix} 0 & 0 & -1 & -1 \end{bmatrix},$$

$$B_{13}=\begin{bmatrix} 1 \end{bmatrix},$$

$$B_{22}=\begin{bmatrix} 1 & 0 & 1 & 0 \\ 0 & 1 & -1 & -1 \end{bmatrix},$$

$$B_{23}=\begin{bmatrix} -1 \\ 0 \end{bmatrix},$$

$$Q_{11}=\begin{bmatrix} 1 \\ 1 \end{bmatrix},$$

$$Q_{12}=\begin{bmatrix} -1 & 1 & 1 & 0 \\ 0 & 1 & 0 & 1 \end{bmatrix},$$

$$Q_{21}=\begin{bmatrix} -1 \end{bmatrix},$$

$$Q_{22}=\begin{bmatrix} 1 & 0 & 0 & 0 \end{bmatrix}.$$

We also have

$$Z_2=\begin{bmatrix} 1/3s & 0 & 0 & 0 \\ 0 & s & 0 & 0 \\ 0 & 0 & 1 & 0 \\ 0 & 0 & 0 & 1 \end{bmatrix},$$

$$Y_2=\begin{bmatrix} 3s & 0 & 0 & 0 \\ 0 & 1/s & 0 & 0 \\ 0 & 0 & 1 & 0 \\ 0 & 0 & 0 & 1 \end{bmatrix}$$

and

$$v_6(s)=\frac{2}{s},$$

$$i_1(s)=\frac{1}{s}.$$

Using the above in (11.18) and (11.24), we get the loop and the cutset systems of equations as given below:

loop system

$$\begin{bmatrix} 1+\frac{1}{3s} & -1 \\ -1 & s+2 \end{bmatrix}\begin{bmatrix} i_2(s) \\ i_3(s) \end{bmatrix}=\begin{bmatrix} \frac{3}{s} \\ -\frac{2}{s} \end{bmatrix}.$$

cutset system

$$\begin{bmatrix} 3s+1+\frac{1}{s} & \frac{1}{s} \\ \frac{1}{s} & 1+\frac{1}{s} \end{bmatrix}\begin{bmatrix} v_4(s) \\ v_5(s) \end{bmatrix}=\begin{bmatrix} 6-\frac{1}{s} \\ -\frac{1}{s} \end{bmatrix}.$$

Suppose a network N has no independent voltage sources. Then a convenient description of N with the node voltages as independent variables can be obtained as follows.

Let A be the incidence matrix of N with vertex v_r as reference. Let us also partition A as $A=[A_{11} \quad A_{12}]$, where the columns of A_{11} and A_{12} correspond, respectively, to the RLC elements and the current sources. If I_1 and I_2 denote the column vectors of RLC element currents and current source currents, then the KCL equations for N become

$$A_{11}I_1=-A_{12}I_2.$$

We also have

$$I_1=Y_1V_1,$$

where V_1 is the column vector of voltages of RLC elements and Y_1 is the corresponding admittance matrix. Furthermore, by the node transformation (11.12) we have

$$V_1=A_{11}^tV_n,$$

where V_n is the column vector of node voltages. So we get from the KCL equations

$$(A_{11}Y_1A_{11}^t)V_n=-A_{12}I_2.$$

The above equations are called *node equations*. The matrix $A_{11}Y_1A_{11}^t$ is called the *node-admittance matrix* of N.

11.3 MIXED-VARIABLE METHOD

In this section we discuss the mixed-variable method of network analysis. In this method, which is essentially a combination of both the loop and the cutset methods, some of the independent variables are voltages and the others are currents. We restrict our discussion to RLC networks (without mutual inductances) containing independent voltage and current sources. It could be extended in a straight forward manner to networks containing mutual inductances.

Consider a connected network N. Let us partition the elements of N into two sets E_1 and E_2, such that E_1 contains all the voltage sources and E_2 contains all the current sources. Let N_1' be the network obtained by removing E_2 from N and let N_2^* be obtained by contracting all the elements of E_1. Let T_1 be a spanning forest of N_1', and let T_2 be a spanning tree of N_2^*. Then $T = T_1 \cup T_2$ is a spanning tree of N. We select T_1 and T_2 such that T contains all voltage but no current sources. Let us define

T_v = Subgraph of T_1 containing all the elements of T_1 except the voltage sources.
$\hat{T}_1$ = Complement of T_1 in N_1'.
$\hat{T}_2$ = Complement of T_2 in N_2^*.
T_i = Subgraph of $\hat{T}_2$ containing all the elements of $\hat{T}_2$ except the current sources.

Let us partition the element voltage vector V_e and the element current vector I_e as follows:

$$V_e = \begin{bmatrix} V_E \\ V_1 \\ V_2 \\ V_3 \\ V_4 \\ V_J \end{bmatrix},$$

$$I_e = \begin{bmatrix} I_E \\ I_1 \\ I_2 \\ I_3 \\ I_4 \\ I_J \end{bmatrix},$$

where the subscripts E, J, 1, 2, 3, and 4 refer, respectively, to the voltage sources, current sources, elements of T_v, T_2, $\hat{T}_1$, and T_i.

We now seek a description of N in terms of the variables V_1 and I_4. Using the B_f and Q_f matrices with respect to T, we can write KVL and KCL equations as follows:

$$\begin{array}{c} \quad (T_v)(T_2)(\hat{T}_1)(T_i) \\ \begin{array}{c|cccccc} & E & 1 & 2 & 3 & 4 & J \\ \hline 3 & A & B & 0 & U & 0 & 0 \\ 4 & C & D & F & 0 & U & 0 \\ J & G & H & K & 0 & 0 & U \end{array} \end{array} \begin{bmatrix} V_E \\ V_1 \\ V_2 \\ V_3 \\ V_4 \\ V_J \end{bmatrix} = [0]. \tag{11.25}$$

(**Note** Explain the presence of the zero submatrix in the column corresponding to T_2.)

$$\begin{array}{c|cccccc} & E & 1 & 2 & 3 & 4 & J \\ \hline E & U & 0 & 0 & -A^t & -C^t & -G^t \\ 1 & 0 & U & 0 & -B^t & -D^t & -H^t \\ 2 & 0 & 0 & U & 0 & -F^t & -K^t \end{array} \begin{bmatrix} I_E \\ I_1 \\ I_2 \\ I_3 \\ I_4 \\ I_J \end{bmatrix} = [0]. \tag{11.26}$$

Consider the second sets of equations in (11.25) and (11.26):

$$FV_2 + V_4 = -CV_E - DV_1, \tag{11.27}$$

$$I_1 - B^t I_3 = D^t I_4 + H^t I_J. \tag{11.28}$$

Using the v–i relations

$$V_2 = Z_2 I_2,$$

$$V_4 = Z_4 I_4,$$

$$I_1 = Y_1 V_1,$$

$$I_3 = Y_3 V_3,$$

we can write (11.27) and (11.28) as

$$FZ_2 I_2 + Z_4 I_4 = -CV_E - DV_1, \tag{11.29}$$

$$Y_1 V_1 - B^t Y_3 V_3 = D^t I_4 + H^t I_J. \tag{11.30}$$

From the loop and cutset transformations we get

$$I_2 = F^t I_4 + K^t I_J, \tag{11.31}$$

$$V_3 = -AV_E - BV_1. \tag{11.32}$$

Now substituting from the above for I_2 and V_3 in (11.29) and (11.30) and rearranging the terms, we get

$$\begin{bmatrix} Z_4 + FZ_2F^t & D \\ -D^t & Y_1 + B^tY_3B \end{bmatrix} \begin{bmatrix} I_4 \\ V_1 \end{bmatrix} = \begin{bmatrix} -C & -FZ_2K^t \\ -B^tY_3A & H^t \end{bmatrix} \begin{bmatrix} V_E \\ I_J \end{bmatrix}.$$

The above is called the *hybrid* or *mixed-variable system* of equations.

It is an easy exercise to verify that once I_4 and V_1 are determined using the above equations, all other variables can be easily computed using (11.25), (11.26), and the loop and cutset transformations.

Note that the mixed-variable method reduces to the loop method if we choose $E_1 = \varnothing$ and $E_2 = E$, and it reduces to the cutset method if we choose $E_1 = E$ and $E_2 = \varnothing$, where E is the element set of N.

Further the mixed variable system of equations involves $\rho(N_1') + \mu(N_2^*) - n_v - n_c$ variables. Clearly, the number $\rho(N_1') + \mu(N_2^*)$ depends on the choice of E_1 and E_2. Thus arises the problem of determining a partition (E_1, E_2) of the element set E of a network N such that $\rho(N_1') + \mu(N_2^*)$ is as small as possible. A method for getting such a partition is discussed in the next section.

11.4 PRINCIPAL PARTITION OF A GRAPH

We describe in this section the principal partition of a graph introduced by Kishi and Kajitani [11.7]. As we shall see, the principal partition of a graph G defines a partition (E_1, E_2) of the edge set E of G which, when used for the mixed-variable method of analysis, leads to the minimum number of independent variables. Our discussion here is based on Kishi and Kajitani [11.7] and Ohtsuki, Ishizaki, and Watanabe [11.8].

Consider a connected graph G. All subgraphs of G to be considered in this section are edge-induced subgraphs. So a subgraph and its edge set will both be denoted by the same symbol.

The *distance* $d(T_1, T_2)$ between any two spanning trees T_1 and T_2 of G is defined as

$$d(T_1, T_2) = |T_1 - T_2| = |T_2 - T_1|.$$

Thus $d(T_1, T_2)$ is equal to the number of edges which are present in $T_1(T_2)$

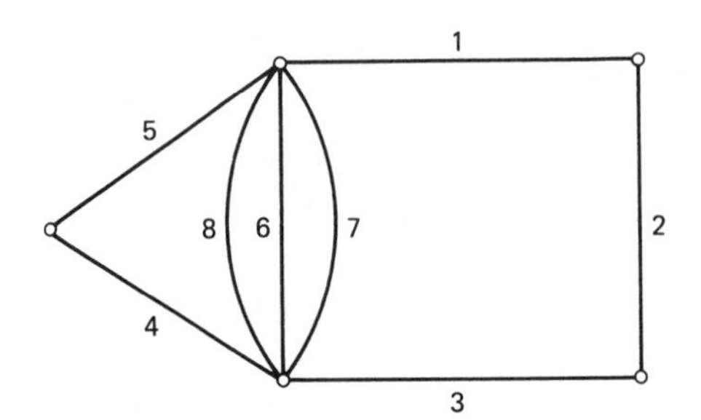

Figure 11.4.

and not in $T_2(T_1)$. We can easily show that

$$d(T_1, T_2) = \rho(G) - \text{number of common branches of } T_1 \text{ and } T_2$$
$$= \mu(G) - \text{number of common chords of } T_1 \text{ and } T_2.$$

Spanning trees T_1 and T_2 are said to be *maximally distant* if $d(T_1, T_2) \geqslant d(T_i, T_j)$ for every pair of spanning trees T_i and T_j of G.

For example, $T_1 = \{2,3,4,7\}$ and $T_2 = \{1,3,5,6\}$ form a pair of maximally distant spanning trees of the graph shown in Fig. 11.4.

THEOREM 11.4. Let T_1 and T_2 form a pair of maximally distant spanning trees of a connected graph G.

1. The fundamental circuit of G with respect to T_1 or T_2 defined by a common chord of T_1 and T_2 contains no common branches of these spanning trees.
2. The fundamental cutset of G with respect to T_1 or T_2 defined by a common branch of T_1 and T_2 contains no common chords of these spanning trees.

Proof

1. Suppose the fundamental circuit of G with respect to T_1 defined by a common chord c contains a common branch b. Then the distance $d(T_1', T_2)$ between T_2 and the spanning tree $T_1' = (T_1 - b) \cup c$ satisfies

$$d(T_1', T_2) = d(T_1, T_2) + 1,$$

   contradicting that T_1 and T_2 are maximally distant.
2. Proof follows in a dual manner. ■

For any two spanning trees T_1 and T_2, let c be a common chord and b be a common branch. Then the sequence

$$P: \quad c\, T_1 e_1 T_2 e_2 T_1 \cdots e_i T^* b,$$

where T^* is either T_1 or T_2, is called a *derivation sequence* of length i from common chord c to common branch b if P has the following properties:

1. T_1 and T_2 appear alternately in P.
2. e_1 is in the fundamental circuit with respect to T_1 defined by c.
3. b is in the fundamental circuit with respect to T^* defined by e_i.
4. If e_j and e_{j+1} appear in P as $e_j T_a e_{j+1}$, where $T_a = T_1$ or T_2, then e_j is a chord of T_a, e_{j+1} is a branch of T_a, and e_{j+1} is in the fundamental circuit with respect to T_a defined by e_j.

By replacing in the above definition circuit by cutset and branch by chord, we can define in an exactly dual manner a *derivation sequence* from a common branch to a common chord. In fact, if P is a derivation sequence from a common chord c to a common branch b, then the sequence P', which is the same as P written in the reverse order, will be a derivation sequence from b to c.

As an example consider the spanning trees $T_1 = \{1,2,3,7,10\}$ and $T_2 = \{2,6,8,11,12\}$ of the graph shown in Fig. 11.5. The edge 5 is a common chord of T_1 and T_2, and the edge 2 is a common branch of T_1 and T_2. Then

$$P: \quad 5T_1 7T_2 8T_1 2$$

is a derivation sequence from 5 to 2, and

$$P': \quad 2T_1 8T_2 7T_1 5$$

is a derivation sequence from 2 to 5.

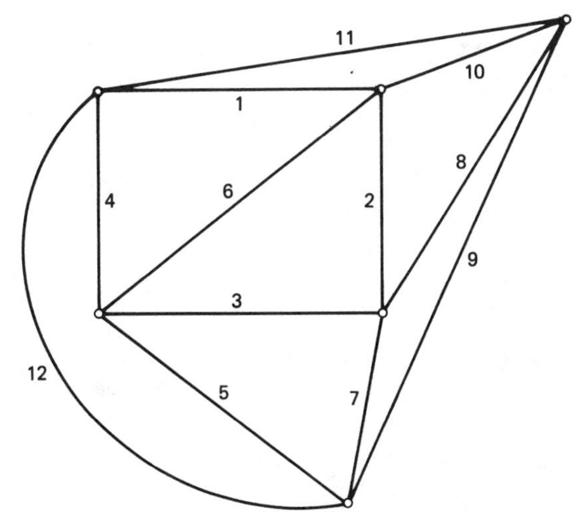

Figure 11.5.

Let T_1 and T_2 form a pair of maximally distant spanning trees. Suppose there exists a derivation sequence

$$P:\quad cT_1e_1T_2b$$

of length 1 from a common chord c to a common branch b. Then T_1 and $T_2' = (T_2 - b) \cup e_1$ will form a pair of maximally distant spanning trees for which c is a common chord and e_1 is a common branch. Further the common branch e_1 will be in the fundamental circuit with respect to T_1 defined by the common chord c. But this contradicts Theorem 11.4. So there exists no derivation sequence of length 1 from c to b. In a dual manner we can show that there is no derivation sequence of length 1 from b to c.

THEOREM 11.5. Let c be a common chord and let b be a common branch of a pair of spanning trees T_1 and T_2 of a connected graph. If T_1 and T_2 are maximally distant, then there exists no derivation sequence from c to b and no derivation sequence from b to c.

Proof

Proof is by induction on the length of a derivation sequence with respect to any pair of maximally distant spanning trees.

As we have just seen, there is no derivation sequence of length 1 from a common chord to a common branch of any pair of maximally distant spanning trees. Assume that there is no such sequence of length less than $k \geqslant 2$.

Suppose there exists a derivation sequence

$$P:\quad cT_1e_1T_2e_2T_1 \cdots e_kT^*b$$

of length k from a common chord c to a common branch b of a pair of maximally distant spanning trees T_1 and T_2. Then it should be a shortest such sequence because by assumption there is no such sequence of length less than k.

If $T^* = T_1$, then T_2 and $T_1' = (T_1 - b) \cup e_k$ form a pair of maximally distant spanning trees for which c is a common chord and e_k is a common branch. Since P is a shortest sequence of length k, it can be shown that

$$P':\quad cT_1'e_1T_2e_2T_1' \cdots e_{k-1}T_2e_k$$

is a derivation sequence from c to e_k. But P' is of length $k-1$, a contradiction because we have assumed there is no such sequence of length $k-1$.

Similarly if $T^* = T_2$, then T_1 and $T_2' = (T_2 - b) \cup e_k$ form a pair of maximally distant spanning trees for which c is a common chord and e_k is a common branch. Again it can be shown that

$$P'':\quad cT_1e_1T_2'e_2T_1 \cdots e_{k-1}T_1e_k$$

is a derivation sequence from c to e_k. Since P'' is of length $k-1$, a contradiction results.

Thus there is no derivation sequence of any length from a common chord to a common branch of T_1 and T_2. Similarly there is no derivation sequence of any length from a common branch to a common chord. ■

Given a pair of maximally distant trees T_1 and T_2, suppose c is a common chord of T_1 and T_2. The K-subgraph G_c of G with respect to c is constructed as follows:

1. Let L_1 be the set of all the edges in the fundamental circuit with respect to T_1 defined by c. By Theorem 11.4, L_1 has no common branches.
2. Let L_2 be the union of all the fundamental circuits with respect to T_2 defined by every edge in L_1. By Theorem 11.5, L_2 has no common branches.
3. Repeating the above, we can obtain a sequence of sets of edges $L_1, L_2, \ldots$ until we arrive at a set $L_{k+1} = L_k$. Then the induced subgraph on the edge set L_k is called the *K-subgraph* G_c with respect to c.

Replacing, in the above construction, circuit by cutset and chord by branch, we can define in an exactly dual manner the *K-subgraph* G_b with respect to a common branch b.

The *principal subgraph* G_1 with respect to common chords is the union of the K-subgraphs with respect to all the common chords. The *principal subgraph* G_2 with respect to the common branches is the union of the K-subgraphs with respect to all the common branches.

For example, the principal subgraphs G_1 and G_2 of the graph in Fig. 11.4 with respect to the pair of spanning trees $T_1 = \{2,3,4,7\}$ and $T_2 = \{1,3,5,6\}$ are:

$$G_1 = \{6,7,8\},$$

$$G_2 = \{1,2,3\}.$$

Kishi and Kajitani [11.7] have shown that the principal subgraphs G_1 and G_2 have no common edges. For if they have a common edge, then we can construct a derivation sequence from a common chord to a common branch.

Thus any graph G consists of three subgraphs:

G_1— Principal subgraph with respect to common chords.
G_2— Principal subgraph with respect to common branches.
G_0— The subgraph $G-(G_1 \cup G_2)$.

This partition (G_0, G_1, G_2) of G is called the *principal partition* of G.

It is interesting to point out [11.7] that the principal partition of a graph G is unique and is independent of the maximally distant spanning trees used to construct the partition.

Some useful properties of the principal subgraphs G_0, G_1, and G_2 are stated below. They follow from the definitions of these subgraphs.

P1. G_1 has all the common chords but no common branches; G_2 has all the common branches but no common chords.

P2. Any fundamental circuit with respect to T_1 or T_2 defined by an edge in G_1 consists only of edges in G_1.

P3. Any fundamental cutset with respect to T_1 or T_2 defined by an edge in G_2 consists only of edges in G_2.

P4. $T_1 \cap G_1$ and $T_2 \cap G_1$ are spanning forests of G_1.

P5. $T_1 \cap G_2$ and $T_2 \cap G_2$ are spanning trees of the graph G_2' obtained by contracting all the edges not in G_2.

P6. Any edge whose end vertices are both in the same component of G_1 is also in G_1.

P7. Any edge whose end vertices are both in the same component of $G_0 \cup G_1$ is also in $G_0 \cup G_1$.

P8. $T_1 \cap G_0$ and $T_2 \cap G_0$ are both spanning forests of the graph G_0' obtained by contracting all the edges of G_1 and removing all the edges of G_2. (This property can be proved using property P3.)

Note that $G_1 = \varnothing$ if G has no common chords, and $G_2 = \varnothing$ if G has no common branches.

The partition of the edge set E of G arising as a result of property P1 is shown in Fig. 11.6. It follows from this property that

$$E_1 = A \cup D \cup J,$$

$$E_2 = B \cup K \cup H,$$

$$E_0 = C \cup F,$$

where E_1, E_2, and E_0 denote, respectively, the edge sets of G_1, G_2, and G_0.

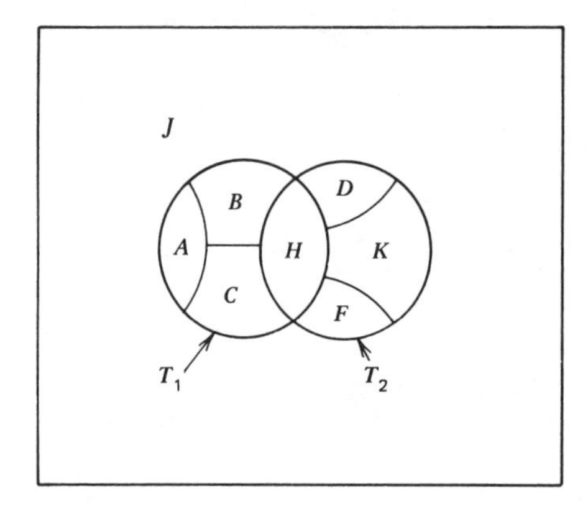

Figure 11.6.

Note that as a consequence of properties P4 and P5 we get $|A|=|D|$ and $|B|=|K|$.

The results presented in the foregoing discussions lead to the following theorem.

THEOREM 11.6. For a graph G with (G_0, G_1, G_2) as the principal partition:

1. If $G_1 \neq \varnothing$, then $\rho(G_1) < \mu(G_1)$.
2. If $G_2 \neq \varnothing$, then $\rho(G_2') > \mu(G_2')$, where G_2' is the graph obtained by contracting all the edges in $(G_0 \cup G_1)$.
3. $\rho(G_0') = \mu(G_0')$, where G_0' is obtained by contracting all the edges in G_1 and removing all the edges in G_2.
4. Maximum distance between any two spanning trees is $\rho(G_1) + \mu(G_2') + \rho(G_0')$.

Proof

1. From property P4 and the fact that $|A|=|D|$, we get

$$\begin{aligned} \rho(G_1) &= |A| \\ &< |D| + |J|, \qquad \text{if } G_1 \neq \varnothing \\ &= \mu(G_1). \end{aligned}$$

2. From property P5 and the fact that $|B|=|K|$, we get

$$\begin{aligned} \rho(G_2') &= |B| + |H|, \\ &> |K|, \qquad \text{if } G_2 \neq \varnothing \\ &= \mu(G_2'). \end{aligned}$$

3. From property P8 and the fact that $|C|=|F|$, we get

$$\begin{aligned} \rho(G_0') &= |C| \\ &= |F| \\ &= \mu(G_0'). \end{aligned}$$

4. Maximum distance between any two spanning trees of G is

$$|A| + |B| + |C| = \rho(G_1) + \mu(G_2') + \rho(G_0'). \quad \blacksquare$$

Let d_m denote the maximum distance between two spanning trees of a graph G. Let (G_0, G_1, G_2) be the principal partition of G. Then from the above theorem we have

$$d_m = \rho(G_1) + \mu(G_2') + \rho(G_0').$$

But $\mu(G_2') + \rho(G_0')$ is equal to the nullity of the graph $(G_0 \cup G_2)'$ obtained by contracting all the edges in G_1 (see Fig. 11.6). So

$$d_m = \rho(G_1) + \mu((G_0 \cup G_2)').$$

Let E_a and E_b form any partition of the edge set E of G. Let G_a be the subgraph on the edge set E_a, and let G_b' be the graph obtained by contracting all the edges in G_a. Ohtsuki, Ishizaki, and Watanabe [11.8] have shown that

$$d_m \leqslant \rho(G_a) + \mu(G_b').$$

Recall that $\rho(G_a) + \mu(G_b')$ is equal to the number of independent variables in the mixed-variable analysis if we use the partition (E_a, E_b) of E. So we can conclude that d_m is the minimum number of independent variables required in the mixed-variable analysis. This number may be less than both the rank and the nullity of G. Thus the number of independent variables in the mixed-variable analysis may be less than that required for the loop or cutset analysis. For example, the graph G in Fig. 11.4 has the principal partition

$$G_0 = \{4,5\},$$

$$G_1 = \{6,7,8\},$$

$$G_2 = \{1,2,3\}.$$

It can be verified that

$$\rho(G_1) = 1,$$

$$\mu(G_0 \cup G_2)' = 2.$$

So

$$d_m = 3.$$

Thus only three independent variables are required in the mixed-variable analysis of this network, whereas the loop and cutset methods both require four independent variables.

Ohtsuki, Ishizaki, and Watanabe [11.8] call d_m the *topological degrees of freedom* of a network. Several interesting properties of those decompositions (E_a, E_b) for which $\rho(G_a) + \mu(G_b') = d_m$ are discussed in [11.8].

Kishi and Kajitani [11.7] and Ohtsuki, Ishizaki, and Watanabe [11.8] contain several deep contributions to graph theory. Lin [11.9] discusses an algorithm for computing the principal partition of a graph. Bruno and Weinberg [11.10] extend the concept of principal partition to matroids.

11.5 STATE EQUATIONS

In Sections 11.2 through 11.4 we developed the loop, cutset, and mixed-variable systems of equations for describing a network. In the time domain these equations are integrodifferential equations. In this section we develop a description of networks in terms of first-order differential equations, without integrals. One of the reasons for seeking such a description is that in mathematics literature there is a wealth of information on solving such equations and on the properties of their solutions which can be readily applied to the case under consideration. Further, the state representation is more general in that it is applicable to time-varying and nonlinear networks. Again we restrict our attention to the special class of RLC networks with mutual inductances and independent voltage and current sources.

State equations of an electrical network N are formulated with the derivatives of the capacitance voltages and inductor currents as variables. Clearly, not all the capacitance voltages can be chosen as independent variables because in the network there may be a circuit consisting of only capacitances. Similarly not all the inductance currents can be chosen as independent variables because there may be a cutset consisting of only inductances. As in the case of the loop and cutset methods of analysis, the starting point for the development of state equations is the choice of an appropriate spanning tree. In particular we select a spanning tree such that it has:

1. All the voltage sources but no current sources;
2. The largest possible number of capacitances; and
3. The least possible number of inductances.

A spanning tree selected as above is called a *normal tree*.

Given a network N, let N_1 denote the subgraph of N which contains all the voltage sources and capacitances, and let N_2 be the subgraph of N which contains all the voltage sources, capacitances, and resistances. A normal tree of N can be selected as follows:

1. First select a spanning forest T_1 of N_1 such that it contains all the voltage sources.
2. Then select a spanning forest T_2 of N_2 such that $T_1 \subseteq T_2$.
3. To T_2 add as many inductances as necessary to obtain a spanning tree T of N. The tree so obtained is a normal tree of N.

As an example, a normal tree of the network of Fig. 11.7*a* is shown in Fig. 11.7*b*.

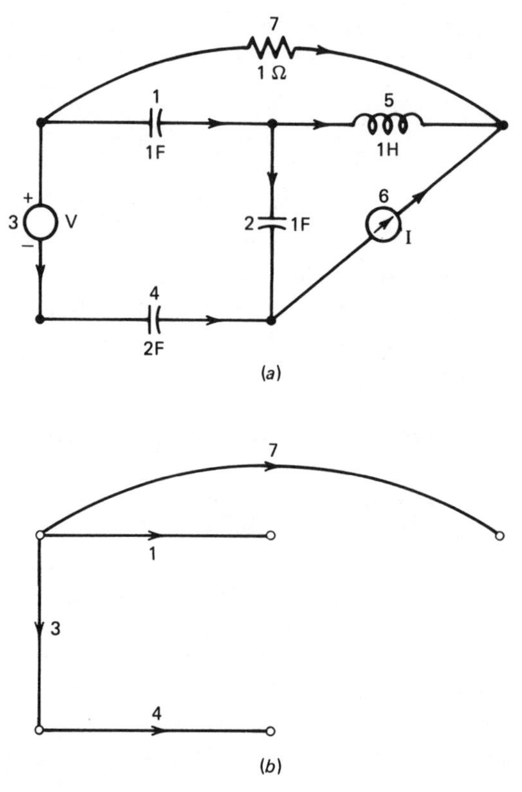

Figure 11.7. (*a*) A network. (*b*) A normal tree of network in (*a*).

After selecting a normal tree T, let us partition the element voltage vector V_e and the element current vector I_e as follows:

$$V_e = \begin{bmatrix} V_t \\ V_l \end{bmatrix} \quad \text{and} \quad I_e = \begin{bmatrix} I_t \\ I_l \end{bmatrix},$$

with

$$V_t = \begin{bmatrix} V_E \\ V_{Ct} \\ V_{Rt} \\ V_{Lt} \end{bmatrix}, \quad V_l = \begin{bmatrix} V_{Cl} \\ V_{Rl} \\ V_{Ll} \\ V_J \end{bmatrix},$$

$$I_t = \begin{bmatrix} I_E \\ I_{Ct} \\ I_{Rt} \\ I_{Lt} \end{bmatrix}, \quad I_l = \begin{bmatrix} I_{Cl} \\ I_{Rl} \\ I_{Ll} \\ I_J \end{bmatrix},$$

where the subscripts E, J, C, R, and L refer, respectively, to the voltage sources, current sources, capacitances, resistances, and inductances, and the subscripts t and l refer, respectively, to the branches and links of T.

The submatrix Q_{fl} of the fundamental cutset matrix $Q_f=[U \quad Q_{fl}]$ is then partitioned as:

$$Q_{fl} = \underset{\text{branches}}{\downarrow} \begin{matrix} \\ E \\ C \\ R \\ L \end{matrix} \overset{\text{links}\rightarrow}{\begin{matrix} C & R & L & J \\ \end{matrix}} \begin{bmatrix} Q_{EC} & Q_{ER} & Q_{EL} & Q_{EJ} \\ Q_{CC} & Q_{CR} & Q_{CL} & Q_{CJ} \\ 0 & Q_{RR} & Q_{RL} & Q_{RJ} \\ 0 & 0 & Q_{LL} & Q_{LJ} \end{bmatrix}. \quad (11.33)$$

Note that $Q_{RC}=0$. For if there were a capacitance in the fundamental cutset with respect to a resistance, then removing this resistance from T and adding this capacitance will result in a spanning tree having one more capacitance than T. This would then contradict the choice of the normal tree T. In a similar way we can show that $Q_{LC}=0$ and $Q_{LR}=0$.

From the KCL equations

$$Q_f I_e = 0$$

and hence we get the following:

$$I_E = -Q_{EC}I_{Cl} - Q_{ER}I_{Rl} - Q_{EL}I_{Ll} - Q_{EJ}I_J, \quad (11.34a)$$

$$I_{Ct} = -Q_{CC}I_{Cl} - Q_{CR}I_{Rl} - Q_{CL}I_{Ll} - Q_{CJ}I_J, \quad (11.34b)$$

$$I_{Rt} = -Q_{RR}I_{Rl} - Q_{RL}I_{Ll} - Q_{RJ}I_J, \quad (11.34c)$$

$$I_{Lt} = -Q_{LL}I_{Ll} - Q_{LJ}I_J. \quad (11.34d)$$

Further from the cutset transformation we get

$$V_{Cl} = Q_{EC}^t V_E + Q_{CC}^t V_{Ct}, \quad (11.35a)$$

$$V_{Rl} = Q_{ER}^t V_E + Q_{CR}^t V_{Ct} + Q_{RR}^t V_{Rt}, \quad (11.35b)$$

$$V_{Ll} = Q_{EL}^t V_E + Q_{CL}^t V_{Ct} + Q_{RL}^t V_{Rt} + Q_{LL}^t V_{Lt}, \quad (11.35c)$$

$$V_J = Q_{EJ}^t V_E + Q_{CJ}^t V_{Ct} + Q_{RJ}^t V_{Rt} + Q_{LJ}^t V_{Lt}. \quad (11.35d)$$

Using the element v–i relations and the above equations, we have to get the state equations in terms of the derivatives of V_{Ct} and I_{Ll}. All other nonsource variables must be eliminated.

For the capacitances the unwanted variables are I_{Ct}, V_{Cl}, and I_{Cl}, and for the inductances they are V_{Lt}, I_{Lt}, and V_{Ll}. We eliminate these variables as follows.

First we rewrite (11.34b) as

$$I_{Ct}+Q_{CC}I_{Cl}=\begin{bmatrix} U & Q_{CC}\end{bmatrix}\begin{bmatrix} I_{Ct} \\ I_{Cl}\end{bmatrix}$$

$$=-Q_{CR}I_{Rl}-Q_{CL}I_{Ll}-Q_{CJ}I_{J}.$$

After using the capacitance v–i relation in the above,

$$\begin{bmatrix} I_{Ct} \\ I_{Cl}\end{bmatrix}=\begin{bmatrix} C_t & 0 \\ 0 & C_l\end{bmatrix}\frac{d}{dt}\begin{bmatrix} V_{Ct} \\ V_{Cl}\end{bmatrix},$$

and substituting for V_{Cl} from (11.35a), we get

$$\mathcal{C}\frac{d}{dt}(V_{Ct})=-Q_{CR}I_{Rl}-Q_{CL}I_{Ll}-Q_{CJ}I_{J}+\mathcal{C}^*\frac{d}{dt}(V_E), \tag{11.36}$$

where

$$\mathcal{C}=\begin{bmatrix} U & Q_{CC}\end{bmatrix}\begin{bmatrix} C_t & 0 \\ 0 & C_l\end{bmatrix}\begin{bmatrix} U \\ Q^t_{CC}\end{bmatrix},$$

$$\mathcal{C}^*=-\begin{bmatrix} U & Q_{CC}\end{bmatrix}\begin{bmatrix} C_t & 0 \\ 0 & C_l\end{bmatrix}\begin{bmatrix} 0 \\ Q^t_{EC}\end{bmatrix}.$$

Note that $\mathcal{C}$ can be shown to be the cutset admittance matrix of the network obtained from N after removing all the link elements except the capacitances and contracting all the tree elements except the capacitances.

In a similar way, starting from (11.35c) and using the v–i relations,

$$\begin{bmatrix} V_{Ll} \\ V_{Lt}\end{bmatrix}=\begin{bmatrix} L_{ll} & L_{lt} \\ L_{tl} & L_{tt}\end{bmatrix}\frac{d}{dt}\begin{bmatrix} I_{Ll} \\ I_{Lt}\end{bmatrix},$$

and then substituting for I_{Lt} from (11.34d), we can get the following relation for the inductances:

$$\mathcal{L}\frac{d}{dt}(I_{Ll})=Q^t_{CL}V_{Ct}+Q^t_{RL}V_{Rt}+Q^t_{EL}V_E+\mathcal{L}^*\frac{d}{dt}(I_J), \tag{11.37}$$

where

$$\mathcal{L}=\begin{bmatrix} U & -Q^t_{LL}\end{bmatrix}\begin{bmatrix} L_{ll} & L_{lt} \\ L_{tl} & L_{tt}\end{bmatrix}\begin{bmatrix} U \\ -Q_{LL}\end{bmatrix},$$

$$\mathcal{L}^*=-\begin{bmatrix} U & -Q^t_{LL}\end{bmatrix}\begin{bmatrix} L_{ll} & L_{lt} \\ L_{tl} & L_{tt}\end{bmatrix}\begin{bmatrix} 0 \\ -Q_{LJ}\end{bmatrix}.$$

Note that $\mathcal{L}$ can be shown to be the loop impedance matrix of the network obtained after removing all the link elements except inductances and contracting all the tree elements except inductances.

Now we have to eliminate I_{Rl} and V_{Rt} from (11.36) and (11.37). For this purpose consider the following v–i relations:

$$I_{Rl}=G_l V_{Rl},$$

$$V_{Rt}=R_t I_{Rt}.$$

After substituting for V_{Rl} and I_{Rt} from (11.35b) and (11.34c), the above equations become

$$I_{Rl}=G_l Q_{RR}^t V_{Rt}+G_l Q_{CR}^t V_{Ct}+G_l Q_{ER}^t V_E, \tag{11.38}$$

$$V_{Rt}=-R_t Q_{RR} I_{Rl}-R_t Q_{RL} I_{Ll}-R_t Q_{RJ} I_J. \tag{11.39}$$

Substituting for I_{Rl} from (11.38), (11.39) reduces to

$$\begin{aligned}(U+R_t Q_{RR} G_l Q_{RR}^t)V_{Rt}=&-R_t Q_{RR} G_l Q_{CR}^t V_{Ct}-R_t Q_{RR} G_l Q_{ER}^t V_E\\&-R_t Q_{RL} I_{Ll}-R_t Q_{RJ} I_J.\end{aligned} \tag{11.40}$$

The above equation for V_{Rt} can be solved if and only if the inverse of the matrix $(U+R_t Q_{RR} G_l Q_{RR}^t)$ exists. This matrix can be written as $R_t G$, where

$$G=G_t+Q_{RR} G_l Q_{RR}^t.$$

The matrix $G_t+Q_{RR}G_l Q_{RR}^t$ can be shown to be the cutset admittance matrix of the network obtained from N after removing all the link elements except the resistances and contracting all the tree elements except the resistances. Thus the inverse of this matrix exists, and so (11.40) can be solved for V_{Rt}.

We next substitute in (11.36) and (11.37) for I_{Rl} and V_{Rt} as obtained above and get the following state equations:

$$\begin{aligned}\frac{d}{dt}\begin{bmatrix}\mathcal{C} & 0\\ 0 & \mathcal{L}\end{bmatrix}\begin{bmatrix}V_{Ct}\\ I_{Ll}\end{bmatrix}=&\begin{bmatrix}-\mathcal{Y} & \mathcal{H}\\ \mathcal{G} & -\mathcal{P}\end{bmatrix}\begin{bmatrix}V_{Ct}\\ I_{Ll}\end{bmatrix}\\&+\begin{bmatrix}-\mathcal{Y}^* & \mathcal{H}^*\\ \mathcal{G}^* & -\mathcal{P}^*\end{bmatrix}\begin{bmatrix}V_E\\ I_J\end{bmatrix}+\frac{d}{dt}\begin{bmatrix}\mathcal{C}^* & 0\\ 0 & \mathcal{L}^*\end{bmatrix}\begin{bmatrix}V_E\\ I_J\end{bmatrix},\end{aligned} \tag{11.41}$$

where

$$R=R_l+Q_{RR}^t R_t Q_{RR},$$

$$\mathcal{Y}=Q_{CR}R^{-1}Q_{CR}^t, \qquad \mathcal{H}=-Q_{CL}+Q_{CR}R^{-1}Q_{RR}^t R_t Q_{RL},$$

$$\mathcal{P}=Q_{RL}^t G^{-1}Q_{RL}, \qquad \mathcal{G}=Q_{CL}^t-Q_{RL}^t G^{-1}Q_{RR}G_l Q_{CR}^t=-\mathcal{H}^t, \tag{11.42}$$

and

$$\mathcal{Y}^* = Q_{CR}R^{-1}Q_{ER}, \qquad \mathcal{H}^* = -Q_{CJ} + Q_{CR}R^{-1}Q_{RR}^t R_t Q_{RJ},$$

$$\mathcal{P}^* = Q_{RL}^t G^{-1} Q_{RJ}, \qquad \mathcal{G}^* = Q_{EL}^t - Q_{RL}^t G^{-1} Q_{RR} G_l Q_{ER}^t. \tag{11.43}$$

The matrices $\mathcal{C}$ and $\mathcal{L}$ are nonsingular in the case of RLC networks, and so in such cases we can rewrite (11.41) as

$$\frac{d}{dt}\begin{bmatrix} V_{Ct} \\ I_{Ll} \end{bmatrix} = \mathcal{A}\begin{bmatrix} V_{Ct} \\ I_{Ll} \end{bmatrix} + \mathcal{B}\begin{bmatrix} V_E \\ I_J \end{bmatrix} + \mathcal{D}\frac{d}{dt}\begin{bmatrix} V_E \\ I_J \end{bmatrix}. \tag{11.44}$$

Once we have determined V_{Ct} and I_{Ll} after solving (11.44), we can obtain all the other voltages and currents in terms of V_{Ct}, I_{Ll}, V_E, and I_J. Balabanian and Bickart [11.11] and Kuh and Rohrer [11.12] may be consulted for further details.

The state equations as given in (11.44) are not in the normal form because of the presence of derivatives of V_E and I_J. By appropriately defining a new set of variables, (11.44) can be reduced to the normal form. See Balabanian and Bickart [11.11].

As an example consider the network shown in Fig. 11.7*a*. A normal tree of this network is shown in Fig. 11.7*b*. The Q_{fl} matrix in partitioned form is given below:

$$Q_{fl} = \begin{array}{c} \\ 3 \\ 1 \\ 4 \\ 7 \end{array}\begin{array}{c} \begin{array}{ccc} 2 & 5 & 6 \end{array} \\ \left[\begin{array}{c|c|c} 1 & 0 & -1 \\ \hline -1 & -1 & 0 \\ 1 & 0 & -1 \\ \hline 0 & 1 & 1 \end{array}\right] \end{array}.$$

From this we get

$$Q_{EC} = [1], \qquad Q_{RC} = [0],$$

$$Q_{EL} = [0], \qquad Q_{RL} = [1],$$

$$Q_{EJ} = [-1], \qquad Q_{RJ} = [1],$$

$$Q_{CC} = \begin{bmatrix} -1 \\ 1 \end{bmatrix}, \qquad Q_{CL} = \begin{bmatrix} -1 \\ 0 \end{bmatrix}, \qquad Q_{CJ} = \begin{bmatrix} 0 \\ -1 \end{bmatrix}.$$

Using the above matrices we get

$$\mathcal{C} = \begin{bmatrix} 2 & -1 \\ -1 & 3 \end{bmatrix}, \qquad \mathcal{C}^* = \begin{bmatrix} 1 \\ -1 \end{bmatrix}, \qquad \mathcal{L} = [1], \qquad \mathcal{L}^* = [0],$$

$$G = [1], \qquad R = [0],$$

$$\mathcal{Y} = \begin{bmatrix} 0 & 0 \\ 0 & 0 \end{bmatrix}, \qquad \mathcal{Y}^* = \begin{bmatrix} 0 \\ 0 \end{bmatrix}, \qquad \mathcal{H} = \begin{bmatrix} 1 \\ 0 \end{bmatrix}, \qquad \mathcal{H}^* = \begin{bmatrix} 0 \\ 1 \end{bmatrix},$$

$$\mathcal{P} = [1], \qquad \mathcal{P}^* = [1], \qquad \mathcal{G} = [-1 \quad 0], \qquad \mathcal{G}^* = [0].$$

The state equations of the network in Fig. 11.7 are then obtained as given below, using (11.41):

$$\frac{d}{dt}\left[\begin{array}{cc|c} 2 & -1 & 0 \\ -1 & 3 & 0 \\ \hline 0 & 0 & 1 \end{array}\right]\left[\begin{array}{c} v_1 \\ v_4 \\ \hline i_5 \end{array}\right]=\left[\begin{array}{cc|c} 0 & 0 & 1 \\ 0 & 0 & 0 \\ \hline -1 & 0 & -1 \end{array}\right]\left[\begin{array}{c} v_1 \\ v_4 \\ \hline i_5 \end{array}\right]+\left[\begin{array}{c|c} 0 & 0 \\ 0 & 1 \\ \hline 0 & -1 \end{array}\right]\left[\begin{array}{c} v_3 \\ \hline i_6 \end{array}\right]$$

$$+\left[\begin{array}{c|c} 1 & 0 \\ -1 & 0 \\ \hline 0 & 0 \end{array}\right]\left[\begin{array}{c} \dfrac{dv_3}{dt} \\ \hline \dfrac{di_6}{dt} \end{array}\right].$$

11.6 NO-GAIN PROPERTY OF RESISTANCE NETWORKS

We conclude this chapter with an interesting application of graph theory in the study of electrical networks.

It is well known in network theory that, given a network of resistances and sources, the magnitude of the voltage across any resistance is not greater than the sum of the magnitudes of the voltages across the sources, and similarly the magnitude of the current through any resistance is not greater than the magnitudes of the currents through the sources. This property of resistance networks is known as the *no-gain property*. In this section we give a proof of the no-gain property. This proof, due to Wolaver [11.13], is purely graph-theoretic.

Recall that a circuit in which all the edges are oriented in the same way with respect to the circuit orientation is called a directed circuit. A cutset in which all the edges are oriented in the same way with respect to the cutset orientation is called a *directed cutset*.

Wolaver's proof of the no-gain property is based on a special case of Minty's arc coloring lemma (Theorem 10.31), namely that in a directed graph each edge lies in either a directed circuit or a directed cutset, but not both.

THEOREM 11.7. Given a network of sources and (linear/nonlinear) positive resistances, the magnitude of the current through any resistance with nonzero voltage is not greater than the sum of the magnitudes of the currents through the sources.

Proof

Let all the elements with zero voltage be eliminated by considering them as short circuits. Then let the element reference directions be chosen so that all the element voltages are positive. Then consider any resistance with nonzero voltage. There can be no directed circuit that contains that resistance. For if such a directed circuit were present, then the sum of all the voltages around the circuit would be nonzero, violating Kirchhoff's voltage law. So, as we observed earlier, there exists a directed cutset that contains the resistance under consideration.

Let the current through the considered resistance be i_0. Select a directed cutset that contains the resistance. Let R be the set of all other resistances in this cutset, and let S be the set of all the sources in the cutset. Applying Kirchhoff's law to the cutset, we get

$$i_0+\sum_{k\in R} i_k+\sum_{s\in S} \pm i_s=0.$$

Since the resistances are positive, $v_i \geqslant 0$ for each resistance. Since all the voltages are positive, the current in each resistance is not negative, and we can write

$$|i_0|+\sum_{k\in R} |i_k|+\sum_{s\in S} \pm i_s=0.$$

Therefore

$$|i_0|\leqslant\sum_{s\in S} \mp i_s\leqslant\sum_{s\in S} |i_s|.$$

Thus follows the theorem. ■

Following is the dual of the above theorem. Proof follows in an exactly dual manner.

THEOREM 11.8. Given a network of sources and (linear/nonlinear) positive resistances, the magnitude of the voltage across any resistance is not greater than the sum of the magnitudes of the voltages across all the sources. ■

Talbot [11.14] and Schwartz [11.15] are among some of the earlier papers which discuss the no-gain property.

11.7 FURTHER READING

Seshu and Reed [11.16] is highly recommended for further reading on the graph-theoretic study of electrical networks. We also recommend Kim and Chien [11.17], Chen [11.18], and Mayeda [11.19]. Balabanian and Bickart [11.11] may be consulted for a more detailed discussion of some of the topics presented in this chapter, and in particular for their extension to more general classes of networks.

A method of network analysis, called *diakoptics*, based upon the concept of tearing, was introduced by Kron [11.20], [11.21]. He applied the concept to solve a certain class of networks. His derivation of the approach is based on concepts from tensor analysis. Happ [11.22], [11.23] and Branin [11.24], [11.25] have clarified the concepts of diakoptics and have contributed much to make this approach more known to electrical engineers. Other descriptions of diakoptics may be found in Onodera [11.26], Roth [11.27], Amari [11.28], Wang and Chao [11.29], and Wu [11.30]. See also Chua and Chen [11.31].

Recently Chua and Chen [11.32] introduced a generalized form of mixed-variable analysis which includes all existing forms (including diakoptics) as special cases. Chua and Chen [11.33] have also discussed the computational efficiency of the loop and cutset methods of analysis. They have shown that the choice of an optimum mode of analysis will give rise to the sparsest loop impedance matrix and the sparsest cutset admittance matrix.

Chua and Green [11.34] have used graph-theoretic ideas, in particular Tellegen's theorem, and a special case of the arc coloring lemma to establish several properties of nonlinear networks and nonlinear multiport resistive networks.

State equations for RLC networks were first derived by Bashkow [11.35] and Bryant [11.36]. A number of papers have subsequently appeared discussing the formulation of state equations for general classes of active networks. For example, see Purslow [11.37], Tosun and Dervisoglu [11.38], and Mark and Swamy [11.39].

Seshu and Reed [11.16] established necessary and sufficient conditions for the solvability of an RLC network containing independent sources. Using their basic approach, several others subsequently discussed the network solvability problem for more general classes of networks. For example, see Purslow [11.37] and Chen and Chan [11.40].

Matroid-theoretic study of electrical networks is fast emerging as a fertile area of research. The definition of an electrical network is based on graphs. As a generalization of this, Bruno and Weinberg [11.41], [11.42], [11.43] have introduced the concept of generalized networks—networks based on matroids—and have established several properties of such networks. It is expected that a closer study of generalized networks will provide much insight into some of the classical unsolved problems of network theory, in particular, the multiport resistive network synthesis problem.

Duffin and Morley [11.44] have recently discussed matroids using the Wang algebra [11.45], [11.46] as a tool. Petersen [11.47] has investigated the network solvability problem using the concept of the union of matroids. Iri and Tomizawa [11.48] have formulated a unified approach to three fundamental problems in network theory by showing that they all reduce to the problem of determining a basis of the union of two appropriately chosen matroids.

Network theory continues to be an abundant source of mathematical problems. Duffin [11.49] discusses several such problems.

11.8 EXERCISES

11.1 The *path matrix* $P=[p_{ij}]$ of a tree T with reference vertex v_r is defined as follows: If branch j is in the unique path in T from vertex v_i to v_r then $p_{ij}=+1$ or -1 depending on whether the branch orientation agrees or disagrees with that of the path; otherwise $p_{ij}=0$. Using node transformation, show that $P=(A^{-1})^t$, where A is the incidence matrix of T with v_r as reference, and deduce Theorem 6.12.

11.2 Derive the loop system of equations of a planar network using the meshes as independent circuits.

11.3 Let N be a planar network. Let a planar network $\hat{N}$ be constructed as follows:

(a) The graph of $\hat{N}$ is the dual of the graph of N. Let e and e' be the corresponding elements of N and $\hat{N}$, respectively.

(b) If e is a resistance of R ohms, then e' is a resistance of $1/R$ ohms.

(c) If e is a capacitance (inductance) of K farads (henrys), then e' is an inductance (capacitance) of K henrys (farads).

(d) If e is a current (voltage) source of value $g(t)$, then e' is a voltage (current) source of value $g(t)$.

Determine the orientations of the current and voltage sources of $\hat{N}$ so that the loop equations of N (with meshes chosen as independent circuits) become the node equations of $\hat{N}$, where the loop current variables are replaced by node voltage variables. It may be assumed that all the meshes of N are oriented clockwise. (The networks N and $\hat{N}$ defined as above are called *dual networks*.)

11.4 Construct the dual of the network shown in Fig. 11.8.

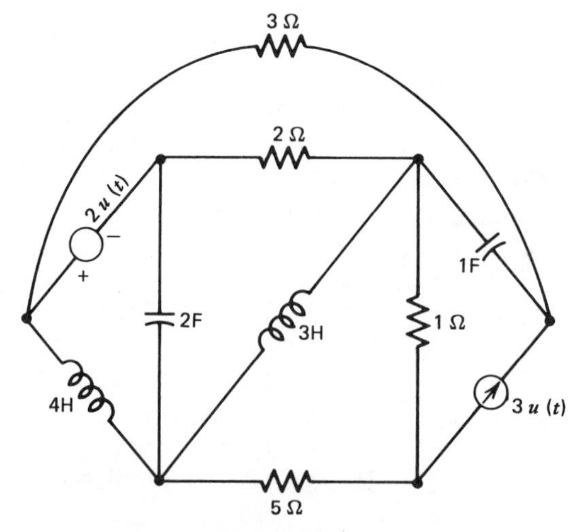

Figure 11.8.

11.5 Determine all the voltages and currents in the network shown in Fig. 11.9, using the loop method of analysis.

11.6 Repeat Exercise 11.5 using the cutset and mixed-variable methods.

11.7 Find the principal partition of the graph shown in Fig. 11.10.

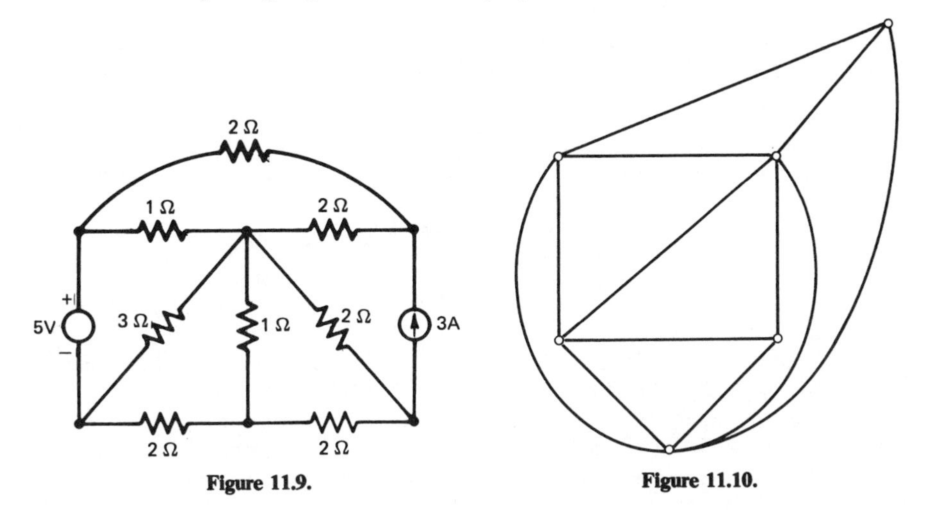

Figure 11.9. **Figure 11.10.**

11.8 (a) Derive the state equations of the network shown in Fig. 11.11.
(b) Express the voltages and currents through the resistances in terms of the state variables and the source variables.

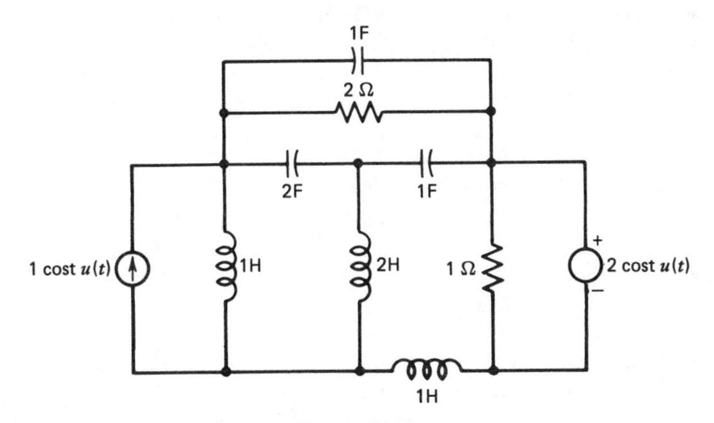

Figure 11.11.

11.9 The *order of complexity* of an electrical network N is the maximum number of initial conditions that can be specified for N. It is the same as the maximum number of dynamically independent element currents and voltages whose instantaneous values are sufficient to determine the instantaneous state of the network. Furthermore, it is also equal to the number of natural frequencies of N.

For an RLC network prove the following:

(a) The order of complexity of N is equal to the number of reactive elements less the sum of the number of linearly independent all-inductance cutsets and the number of linearly independent all-capacitor circuits.

(b) The number of nonzero natural frequencies of N is equal to the order of complexity of N less the sum of the number of linearly independent all-capacitor cutsets and the number of linearly independent all-inductance circuits.

11.9 REFERENCES

11.1 B. D. H. Tellegen, "A General Network Theorem with Applications," *Philips Res. Rept.*, Vol. 7, 259–269 (1952).

11.2 P. Penfield, Jr., R. Spence, and S. Duinker, "A Generalized Form of Tellegen's Theorem," *IEEE Trans. Circuit Theory*, Vol. CT-17, 302–305 (1970).

11.3 P. Penfield, Jr., R. Spence, and S. Duinker, *Tellegen's Theorem and Electrical Networks*, M.I.T. Press, Cambridge, Mass., 1970.

11.4 J. L. Bordewijk, "Inter-Reciprocity Applied to Electrical Networks," *Appl. Sci. Res.*, Vol. B6, 1–74 (1956).

11.5 S. W. Director and R. A. Rohrer, "Automated Network Design—The Frequency Domain Case," *IEEE Trans. Circuit Theory*, Vol. CT-16, 330–337 (1969).

11.6 G. Kishi and T. Kida, "Edge-Port Conservation in Networks," *IEEE Trans. Circuit Theory*, Vol. CT-15, 274–276 (1968).

11.7 G. Kishi and Y. Kajitani, "Maximally Distant Trees and Principal Partition of a Linear Graph," *IEEE Trans. Circuit Theory*, Vol CT-16, 323–330 (1969).

11.8 T. Ohtsuki, Y. Ishizaki, and H. Watanabe, "Topological Degrees of Freedom and Mixed Analysis of Electrical Networks," *IEEE Trans. Circuit Theory*, Vol. CT-17, 491–499 (1970).

11.9 P. M. Lin, "An Improved Algorithm for Principal Partition of Graphs," *Proc. IEEE Intl. Symp. Circuits and Systems*, 1976, pp. 145–148.

11.10 J. Bruno and L. Weinberg, "The Principal Minors of a Matroid," *Linear Algebra and Its Appl.*, Vol. 4, 17–54 (1971).

11.11 N. Balabanian and T. A. Bickart, *Electrical Network Theory*, Wiley, New York, 1969.

11.12 E. S. Kuh and R. A. Rohrer, "The State Variable Approach to Network Analysis," *Proc. IEEE*, Vol. 53, 672–686 (1965).

11.13 D. H. Wolaver, "Proof in Graph Theory of the 'No-Gain' Property of Resistor Networks," *IEEE Trans. Circuit Theory*, Vol. CT-17, 436–437 (1970).

11.14 A. Talbot, "Some Fundamental Properties of Networks without Mutual Inductance," *Proc. IEE (London)*, Vol. 102, 168–175 (1955).

11.15 R. J. Schwartz, "A Note on the Transfer Ratio of Resistive Networks with Positive Elements," *Proc. IRE*, Vol. 43, 1670 (1955).

11.16 S. Seshu and M. B. Reed, *Linear Graphs and Electrical Networks*, Addison-Wesley, Reading, Mass., 1961.

11.17 W. H. Kim and R. T. Chien, *Topological Analysis and Synthesis of Communication Networks*, Columbia Univ. Press, New York, 1962.

11.18 W. K. Chen, *Applied Graph Theory*, North-Holland, Amsterdam, 1971.

11.19 W. Mayeda, *Graph Theory*, Wiley-Interscience, New York, 1970.

11.20 G. Kron, "A Set of Principles to Interconnect the Solution of Physical Systems," *J. Appl. Phys.*, Vol. 24, 965–980 (1953).

11.21 G. Kron, *Diakoptics: The Piecewise Solution of Large Scale Systems*, MacDonald, London 1963.

11.22 H. H. Happ, *Diakoptics and Networks*, Academic Press, New York, 1971.

11.23 H. H. Happ, "Diakoptics—The Solution of System Problems by Tearing," *Proc. IEEE*, Vol. 62, 930–940 (1974).

11.24 F. H. Branin, "The Relation Between Kron's Method and the Classical Methods of Network Analysis," *Matrix and Tensor Quart*, Vol. 12, 69–115 (1962).

11.25 F. H. Branin, "A Sparse Matrix Modification of Kron's Method of Piecewise Analysis," *Proc. IEEE Intl. Symp. Circuits and Systems*, 1975, pp. 21–23.

11.26 R. Onodera, "Diakoptics and Codiakoptics of Electric Network," *RAAG Memoirs*, Vol. 2, 369–388 (1958).

11.27 J. Roth, "An Application of Algebraic Topology: Kron's Method of Tearing," *Quart. Appl. Math.*, Vol. 17, 1–24 (1959).

11.28 S. Amari, "Topological Foundations of Kron's Tearing of Electrical Networks," *RAAG Memoirs*, Vol. 3, 322–350 (1962).

11.29 K. U. Wang and T. Chao, "An Algebraic Theory of Network Topology," *Proc. IEEE Intl. Symp. Circuits and Systems*, 1974, pp. 324–328.

11.30 F. F. Wu, "Solutions of Large-Scale Networks by Tearing," *IEEE Trans. Circuits and Sys.*, Vol. CAS-23, 706–713 (1976).

11.31 L. O. Chua and L. K. Chen, "Nonlinear Diakoptics," *Proc. IEEE Intl. Symp. Circuits and Systems*, 1975, pp. 373–376.

11.32 L. O. Chua and L. K. Chen, "Diakoptic and Generalized Hybrid Analysis," *IEEE Trans. Circuits and Sys.*, Vol. CAS-23, 694–705 (1976).

11.33 L. O. Chua and L. K. Chen, "On Optimally Sparse Cycle and Coboundary Basis for a Linear Graph," *IEEE Trans. Circuit Theory*, Vol. CT-20, 495–503 (1973).

11.34 L. O. Chua and D. N. Green, "Graph-Theoretic Properties of Dynamic Nonlinear Networks," *IEEE Trans. Circuits and Sys.*, Vol. CAS-23, 292–312 (1976).

11.35 T. R. Bashkow, "The *A*-Matrix—A New Network Description," *IRE Trans. Circuit Theory*, Vol. CT-4, 117–119 (1957).

11.36 P. R. Bryant, "The Explicit Form of Bashkow's *A*-Matrix," *IRE Trans. Circuit Theory*, Vol. CT-9, 303–306 (1962).

11.37 E. J. Purslow, "Solvability and Analysis of Linear Active Networks by Use of the State Equations," *IEEE Trans. Circuit Theory*, Vol. CT-17, 469–475 (1970).

11.38 O. Tosun and A. Dervisoglu, "Formulation of State Equations in Active RLC Networks," *IEEE Trans. Circuits and Sys.*, Vol. CAS-21, 36–38 (1974).

11.39 S. K. Mark and M. N. S. Swamy, "The Generalized Tree for State Variables in Linear Networks," *Int. J. Circuit Theory and Appl.*, Vol. 4, 87–92 (1976).

11.40 W. K. Chen and F. N. Chan, "On the Unique Solvability of Linear Active Networks," *IEEE Trans. Circuits and Sys.*, Vol. CAS-21, 26–35 (1974).

11.41 J. Bruno and L. Weinberg, "Generalized Networks: Networks Embedded on a Matroid —Part I," *Networks*, Vol. 6, 53–94 (1976).

11.42 J. Bruno and L. Weinberg, "Generalized Networks: Networks Embedded on a Matroid —Part II," *Networks*, Vol. 6, 231–272 (1976).

11.43 L. Weinberg, "Matroids, Generalized Networks, and Electric Network Synthesis," *J. Combinat. Theory B*, Vol. 23, 106–126 (1977).

11.44 R. J. Duffin and T. D. Morley, "Wang Algebra and Matroids," *IEEE Trans. Circuits and Sys.*, Vol. CAS-25, 755–762 (1978).

11.45 R. J. Duffin, "An Analysis of the Wang Algebra of Networks," *Trans. Am. Math. Soc.*, Vol. 93, 114–131 (1959).

11.46 K. T. Wang, "On a New Method of Analysis of Electrical Networks," *Memoirs* 2, Nat. Res. Inst. Eng. Academia Sinica (1934).

11.47 B. Petersen, "Investigating Solvability and Complexity of Linear Active Networks by means of Matroids," *IEEE Trans. Circuits and Sys.*, Vol. CAS-26, 330–342 (1979).

11.48 M. Iri and N. Tomizawa, "A Unifying Approach to Fundamental Problems in Network Theory by means of Matroids," *Electron. and Commun. in Japan*, Vol. 58-A, 28–35 (1975).

11.49 R. J. Duffin, "Electrical Network Models," in *Studies in Graph Theory, Part II*, The Mathematical Association of America, 1975, pp. 94–138.

Chapter 12

Resistance N-Port Networks

12.1 INTRODUCTION

A network is called an *n-port network* if it has n pairs of accessible terminals for connection to external devices like current or voltage sources. Each such pair of accessible terminals is called a *port*. Clearly, an n-port network can have a maximum of $2n$ distinct port terminals, in which case each port has a pair of terminals of its own, and a minimum of $n+1$ distinct port terminals. An example of the latter situation arises when one terminal is common to all the ports. Vertices of an n-port network other than the port terminals are called *internal vertices* of the network. Each port of an n-port network is associated with two variables, the voltage across the port and the current through the port. Figure 12.1 shows a 3-port network with 6 terminals. The references for port voltages and port currents will be chosen as in Fig. 12.1.

In the directed graph representation, each port of an n-port network will be shown as an edge, called the *port edge* connecting the corresponding port terminals. A port edge may be regarded as representing the external device connected across the port. The subgraph of the port edges along with an indication of the positive and negative reference terminals of each port will be called the *port configuration* of the n-port network.

A port edge may be oriented either from the positive reference terminal to the negative reference terminal of the port or in the opposite direction. In view of the convention shown in Fig. 12.1 for the port voltage and current references, it should be clear that in the former choice of port orientation the port voltage is the same as the voltage associated with the port edge, and the port current is the negative of the current associated with the port edge. The corresponding relations in the latter choice of port orientation are obvious.

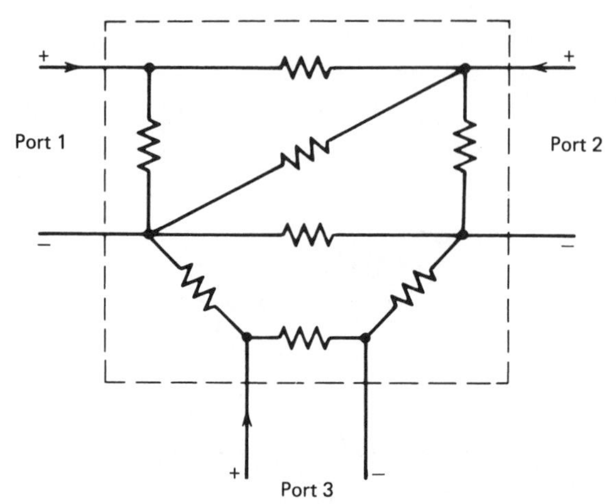

Figure 12.1. A 3-port network with 6 terminals.

An n-port network with the external devices connected across the ports will be referred to as the *augmented n-port network.*

We assume that the augmented n-port network of an n-port network is connected and that there are no internal vertices in the network. If the augmented n-port network is not connected, then each component of the augmented network can be treated separately as a multiport network. If there are some internal vertices in the given network, we can suppress them by star-delta transformation and obtain a network which is equivalent to the original network as far as the relationship between the port variables is concerned. Thus the two assumptions we have just made involve no loss of generality. We also do not allow the presence of parallel conductances. If such conductances are present in a network, they may be combined into a single conductance.

The properties of an n-port network can be described by any set of n independent equations relating the $2n$ port variables. Of special interest to us are the short-circuit admittance and the open-circuit impedance matrix descriptions.

In the short-circuit admittance matrix description, the port voltages are chosen as the independent variables. The *short-circuit admittance matrix* $Y=[y_{ij}]$ is an $n\times n$ matrix which transforms the port voltage vector V_p into the port current vector I_p. Thus we have

$$YV_p=I_p. \tag{12.1}$$

If $V_p=[v_1, v_2,\ldots, v_n]^t$ and $I_p=[i_1, i_2,\ldots, i_n]^t$, where v_j and i_j are, respectively, the voltage and current variables associated with the jth port, then it is clear that

$$y_{kj}=\left.\frac{i_k}{v_j}\right|_{v_i=0,\text{ for all } i\neq j}.$$

Physically this means that y_{kj} is the current through the kth port when the jth port is excited with a source of unit voltage keeping all the other ports short-circuited.

A necessary condition for the admittance matrix to exist is that the port configuration has no circuits. For otherwise it will not be possible to choose the port voltages as independent variables.

In the open-circuit impedance matrix description of an n-port network, the port currents are chosen as the independent variables. The *open-circuit impedance matrix* $Z=[z_{ij}]$ is an $n\times n$ matrix which transforms the port current vector I_p into the port voltage vector V_p. Thus we have

$$ZI_p=V_p. \tag{12.2}$$

If V_p and I_p are as defined earlier, then we have

$$z_{kj}=\left.\frac{v_k}{i_j}\right|_{i_m=0,\text{ for all } m\neq j}.$$

Physically this means that z_{kj} is the voltage across port k when port j is excited with a source of unit current keeping all the other ports open-circuited.

For the impedance matrix description to exist, it is necessary that in the augmented n-port network (which is assumed to be connected), there is no cutset consisting of only port edges, because in such a description the port currents are chosen as independent variables. In other words, the existence of the impedance matrix requires that the n-port network be connected.

We now proceed to derive expressions for the Y- and Z-matrices of an RLC n-port network N. We denote the corresponding augmented network as $\hat{N}$.

Let us first consider the derivation of the Y-matrix of N. Let T denote the port configuration of N. As we have seen earlier, we will require T to be acyclic so that each port edge may now be regarded as representing an independent voltage source. We assume that each port edge is oriented from the positive reference terminal to the negative reference terminal of the corresponding port.

Let T_0 denote a spanning tree of $\hat{N}$ such that T is a subgraph of T_0. Edges of T will be called the *port branches* and the remaining edges of T_0 the *nonport branches*. The fundamental cutset matrix of the augmented network $\hat{N}$ with respect to T_0 can now be written as

$$\left[\begin{array}{c|c} Q_1 & U \\ \hline Q_2 & 0 \end{array}\right],$$

where the submatrix

$$Q_f=\left[\begin{array}{c} Q_1 \\ \hline Q_2 \end{array}\right]$$

corresponds to the network N, with the rows of Q_1 corresponding to the port branches and those of Q_2 corresponding to the nonport branches.

Let the column vectors V_p, V_{nb}, and V_e be defined as follows:

V_p = Vector of port voltages.
V_{nb} = Vector of nonport branch voltages.
V_e = Vector of voltages across the edges in N.

Let the current vectors I_p and I_e be defined in a similar manner. Let Y_e denote the diagonal matrix of edge admittances in N so that

$$I_e = Y_e V_e. \tag{12.3}$$

We can now write a maximal set of independent KCL equations for the augmented network $\hat{N}$ as follows:

$$\left[\begin{array}{c|c} Q_1 & U \\ \hline Q_2 & 0 \end{array}\right]\left[\begin{array}{c} I_e \\ \hline -I_p \end{array}\right]=\left[\begin{array}{c} 0 \\ \hline 0 \end{array}\right],$$

that is,

$$\left[\begin{array}{c} Q_1 \\ \hline Q_2 \end{array}\right][I_e]=\left[\begin{array}{c} I_p \\ \hline 0 \end{array}\right]. \tag{12.4}$$

We also have the relation (cutset transformation, Theorem 11.2)

$$V_e=\left[\begin{array}{c} Q_1 \\ \hline Q_2 \end{array}\right]^t\left[\begin{array}{c} V_p \\ \hline V_{nb} \end{array}\right]. \tag{12.5}$$

Using (12.3) and (12.5) in (12.4), we get:

$$\left[\begin{array}{c|c} Y_{11} & Y_{12} \\ \hline Y_{21} & Y_{22} \end{array}\right]\left[\begin{array}{c} V_p \\ \hline V_{nb} \end{array}\right]=\left[\begin{array}{c} I_p \\ \hline 0 \end{array}\right], \tag{12.6}$$

where

$$Y_{11}=Q_1 Y_e Q_1^t,$$

$$Y_{12}=Q_1 Y_e Q_2^t = Y_{21}^t,$$

$$Y_{22}=Q_2 Y_e Q_2^t.$$

The matrix

$$Y_0 = Q_f Y_e Q_f^t = \left[\begin{array}{c|c} Y_{11} & Y_{12} \\ \hline Y_{21} & Y_{22} \end{array}\right]$$

is called the *cutset admittance matrix* of the n-port network N with respect to T_0.

Solving the second set of equations in (12.6), we get

$$V_{nb} = -Y_{22}^{-1}Y_{21}V_p. \tag{12.7}$$

Note that Y_{22} is nonsingular since Q_2 is of maximum rank, and Y_e is diagonal with no zero diagonal element.

Using (12.7) in the first set of equations in (12.6), namely,

$$Y_{11}V_p + Y_{12}V_{nb} = I_p,$$

we get

$$\left(Y_{11} - Y_{12}Y_{22}^{-1}Y_{21}\right)V_p = I_p. \tag{12.8}$$

Thus the short-circuit admittance matrix Y of N is given by

$$Y = Y_{11} - Y_{12}Y_{22}^{-1}Y_{21}. \tag{12.9}$$

Note that if the port configuration T has $p > 1$ components, then the network N will have $n+p$ port terminals. If $p=1$, then T will be a spanning tree of $\hat{N}$. In such a case, $T = T_0$, that is, there will be no nonport branches and the network N will be called an *n-port network of rank n*. It can be seen that the short-circuit admittance matrix Y of an n-port network of rank n is the same as its cutset admittance matrix Y_0, and so for such a network

$$Y = Q_f Y_e Q_f^t. \tag{12.10}$$

Note that in general Q_f is not a fundamental cutset matrix of $\hat{N}$, though it is in fact a submatrix of the fundamental cutset matrix of N with respect to T_0. However, we shall find it convenient to refer to Q_f as the fundamental cutset matrix of N with respect to T_0. If we admit the presence of edges with zero admittances, then the matrix Q_f will in fact be a fundamental cutset matrix of N.

Let us next consider the derivation of the Z-matrix of an RLC n-port network N. As we have seen earlier, we require the network N to be connected so that the port edges can be made part of a cospanning tree of some spanning tree T_0 of $\hat{N}$. The port edges will now be called *port chords* and the remaining edges of the cospanning tree will be called *nonport chords*. In this case we may regard each port edge as representing an independent current source and orient it from the negative reference terminal to the positive reference terminal of the port. The fundamental circuit matrix of $\hat{N}$ with respect to T_0 can now be written as

$$\left[\begin{array}{c|c} B_1 & U \\ \hline B_2 & 0 \end{array}\right]$$

where the submatrix

$$B_f = \begin{bmatrix} B_1 \\ \hline B_2 \end{bmatrix}$$

corresponds to the n-port network N, with the rows of B_1 corresponding to the port chords and those of B_2 corresponding to the nonport chords.

Let V_p, V_e, I_p, and I_e be defined as before. In addition, let I_{nc} denote the vector of nonport chord currents. We can now write a maximal set of independent KVL equations as follows:

$$\left[\begin{array}{c|c} B_1 & U \\ \hline B_2 & 0 \end{array}\right] \begin{bmatrix} V_e \\ \hline -V_p \end{bmatrix} = \begin{bmatrix} 0 \\ \hline 0 \end{bmatrix}. \tag{12.11}$$

If Z_e is the diagonal matrix of edge impedances in N, then

$$V_e = Z_e I_e. \tag{12.12}$$

We also have the relation (loop transformation, Theorem 11.2)

$$I_e = \begin{bmatrix} B_1 \\ \hline B_2 \end{bmatrix}^t \begin{bmatrix} I_p \\ \hline I_{nc} \end{bmatrix}. \tag{12.13}$$

Using (12.11), (12.12), and (12.13), we get

$$\left[\begin{array}{c|c} Z_{11} & Z_{12} \\ \hline Z_{21} & Z_{22} \end{array}\right] \begin{bmatrix} I_p \\ \hline I_{nc} \end{bmatrix} = \begin{bmatrix} V_p \\ \hline 0 \end{bmatrix}, \tag{12.14}$$

where

$$Z_{11} = B_1 Z_e B_1^t,$$

$$Z_{12} = B_1 Z_e B_2^t = Z_{21}^t,$$

$$Z_{22} = B_2 Z_e B_2^t.$$

Solving the second set of equations in (12.14), we get

$$I_{nc} = -Z_{22}^{-1} Z_{21} I_p. \tag{12.15}$$

We may note that Z_{22} is nonsingular because B_2 is of maximum rank and Z_e is a diagonal matrix with no zero element along the diagonal.

Finally using (12.15) in the first set of equations in (12.14), we get

$$(Z_{11} - Z_{12} Z_{22}^{-1} Z_{21}) I_p = V_p. \tag{12.16}$$

Thus the open-circuit impedance matrix Z of N is given by

$$Z = Z_{11} - Z_{12} Z_{22}^{-1} Z_{21}. \tag{12.17}$$

If the network N has no circuits, then it will be a spanning tree of $\hat{N}$, and hence the port edges will form the corresponding cospanning tree. In such a case there will be no nonport chords, and N will then be referred to as an *n-port network of nullity n*. So the open-circuit impedance matrix Z of an n-port network of nullity n is given by

$$Z = B_f Z_e B_f^t. \tag{12.18}$$

As in the case of the Q_f matrix, we refer to B_f as a fundamental circuit matrix of the network N.

We next illustrate with examples, derivation of the Y- and Z-matrices of n-port networks.

First let us calculate the Y-matrix of the resistance 4-port network of Fig. 12.2a. The port configuration of this network is shown in Fig. 12.2b. Choosing the edges (3,4) and (5,6) as nonport branches, we get the matrix Q_f as:

$$Q_f = \left[\begin{array}{c} Q_1 \\ \hline Q_2 \end{array}\right]$$

$$\begin{array}{r} \\ \text{port 1} \\ \text{port 2} \\ \text{port 3} \\ \text{port 4} \\ \text{nonport branch (3,4)} \\ \text{nonport branch (5,6)} \end{array} \begin{array}{c} (1,2)(3,4)(5,6)(1,4)(2,5)(3,7)(1,7) \\ \left[\begin{array}{ccccccc} 1 & 0 & 0 & 1 & 0 & 0 & 1 \\ 0 & 0 & 0 & 1 & 1 & 0 & 1 \\ 0 & 0 & 0 & 0 & 1 & 1 & 1 \\ 0 & 0 & 0 & 0 & 0 & 1 & 1 \\ \hline 0 & 1 & 0 & 1 & 1 & 1 & 1 \\ 0 & 0 & 1 & 0 & 0 & 1 & 1 \end{array}\right] \end{array}.$$

For this network

$$\begin{aligned} Y_e &= \text{diag}\ [g_{12}\ \ g_{34}\ \ g_{56}\ \ g_{14}\ \ g_{25}\ \ g_{37}\ \ g_{17}] \\ &= \text{diag}\ [\,1\ \ 1\ \ 1\ \ 2\ \ 1\ \ 2\ \ 1\,], \end{aligned}$$

where g_{ij} refers to the admittance of the edge (i, j) connecting the vertices i and j. We now obtain

$$Q_f Y_e Q_f^t = \begin{bmatrix} Y_{11} & Y_{12} \\ Y_{21} & Y_{22} \end{bmatrix}$$

$$= \left[\begin{array}{cccc|cc} 4 & 3 & 1 & 1 & 3 & 1 \\ 3 & 4 & 2 & 1 & 4 & 1 \\ 1 & 2 & 4 & 3 & 4 & 3 \\ 1 & 1 & 3 & 3 & 3 & 3 \\ \hline 3 & 4 & 4 & 3 & 7 & 3 \\ 1 & 1 & 3 & 3 & 3 & 4 \end{array}\right].$$

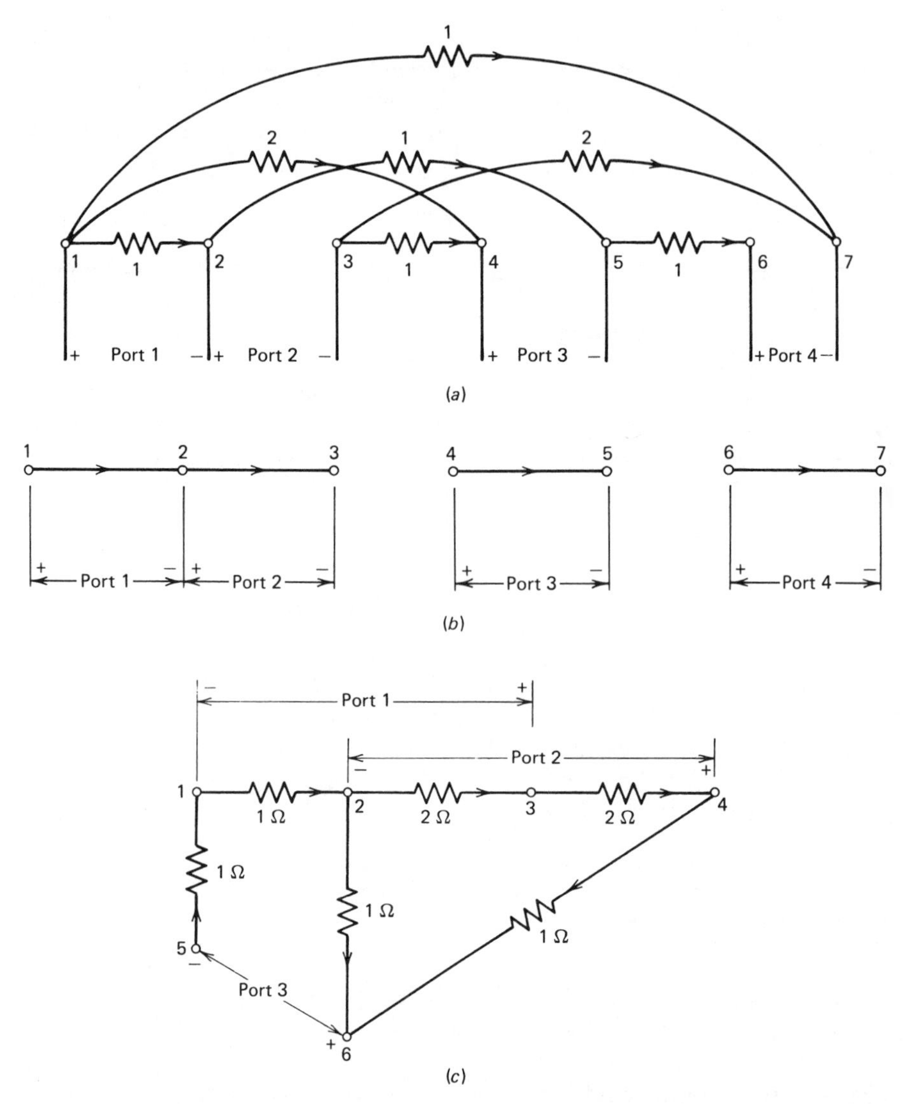

Figure 12.2. (*a*) A resistance 4-port network (all the conductances are in Siemens). (*b*) Port configuration of the network in (*a*). (*c*) A resistance 3-port network. (*d*) Port configuration of the network in (*c*).

The short-circuit admittance matrix of the 4-port network of Fig. 12.2*a* is given by

$$Y = Y_{11} - Y_{12}Y_{22}^{-1}Y_{21}$$

$$= \frac{1}{19}\begin{bmatrix} 51 & 23 & -11 & -2 \\ 23 & 29 & 1 & -5 \\ -11 & 1 & 21 & 9 \\ -2 & -5 & 9 & 12 \end{bmatrix}.$$

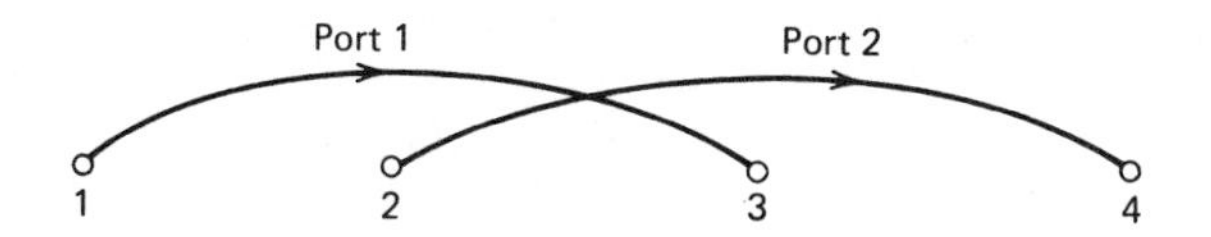

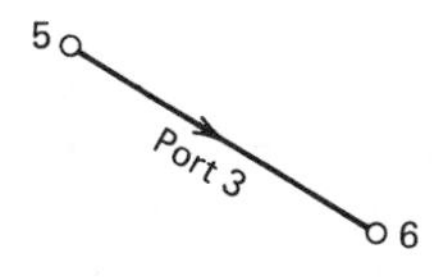

(d)

Figure 12.2. (*continued*)

Consider next the resistance 3-port network shown in Fig. 12.2c. The port configuration of this network is shown in Fig. 12.2d. Choosing the edge (4, 6) as the nonport chord, we obtain the matrix B_f as

$$B_f = \left[\begin{array}{c} B_1 \\ \hline B_2 \end{array}\right]$$

$$= \begin{array}{r} \\ \text{port 1} \\ \text{port 2} \\ \text{port 3} \\ \text{nonport chord } (4,6) \end{array} \begin{array}{c} \begin{array}{cccccc} (1,2) & (2,3) & (3,4) & (1,5) & (2,6) & (4,6) \end{array} \\ \left[\begin{array}{cccccc} -1 & -1 & 0 & 0 & 0 & 0 \\ 0 & -1 & -1 & 0 & 0 & 0 \\ -1 & 0 & 0 & -1 & -1 & 0 \\ \hline 0 & 1 & 1 & 0 & -1 & 1 \end{array}\right] \end{array}.$$

The matrix Z_e for this network is

$$\begin{aligned} Z_e &= \text{diag } [r_{12} \quad r_{23} \quad r_{34} \quad r_{15} \quad r_{26} \quad r_{46}] \\ &= \text{diag } [1 \quad 2 \quad 2 \quad 1 \quad 1 \quad 1]. \end{aligned}$$

We now obtain

$$B_f Z_e B_f^t = \begin{bmatrix} Z_{11} & Z_{12} \\ Z_{21} & Z_{22} \end{bmatrix}$$

$$= \left[\begin{array}{ccc|c} 3 & 2 & 1 & -2 \\ 2 & 4 & 0 & -4 \\ 1 & 0 & 3 & 1 \\ \hline -2 & -4 & 1 & 6 \end{array}\right].$$

The open-circuit impedance matrix of the network is

$$Z = Z_{11} - Z_{12} Z_{22}^{-1} Z_{21}$$

$$= \frac{1}{6}\begin{bmatrix} 14 & 4 & 8 \\ 4 & 8 & 4 \\ 8 & 4 & 17 \end{bmatrix}.$$

As we can see from (12.10) and (12.18), the properties of the Y-matrix of an n-port network of rank n and the properties of the Z-matrix of an n-port network of nullity n are closely related to those of the corresponding fundamental cutset and fundamental circuit matrices. In the following sections of this chapter, we establish several properties of these matrices and discuss procedures for their realization.

12.2 Y-MATRICES OF RESISTANCE n-PORT NETWORKS OF RANK n

In this section, we discuss several important properties of the Y-matrices of resistance n-port networks of rank n. Clearly, these n-port networks will have $n+1$ nodes. So we refer to them as $(n+1)$-node n-port networks. In all our discussions, we assume that there are no negative conductances in the networks under consideration.

12.2.1 Basic Properties

Consider an $(n+1)$-node resistance n-port network N with a port configuration T. We know from (12.10) that the short-circuit conductance matrix Y of N is given by

$$Y = Q_f G_e Q_f^t, \tag{12.19}$$

where Q_f is the fundamental cutset matrix of N with respect to T, and G_e is the diagonal matrix of edge conductances of N.

For example for the 6-port network shown in Fig. 12.3,

$$\begin{array}{cccccccccc} (1,2) & (1,4) & (1,6) & (1,7) & (2,5) & (2,6) & (3,4) & (3,5) & (4,6) & (5,6) \end{array}$$

$$Q_f = \begin{bmatrix} 1 & 1 & 1 & 0 & -1 & 0 & 1 & 0 & 0 & 1 \\ 0 & 0 & 0 & 0 & 0 & 0 & -1 & -1 & 0 & 0 \\ 0 & 1 & 0 & 0 & 0 & 0 & 1 & 0 & -1 & 0 \\ 0 & 0 & 0 & 1 & 1 & 0 & 0 & 1 & 0 & -1 \\ 0 & 0 & 1 & 0 & 0 & 1 & 0 & 0 & 1 & 1 \\ 0 & 0 & 0 & 0 & 1 & 0 & 0 & 1 & 0 & -1 \end{bmatrix}$$

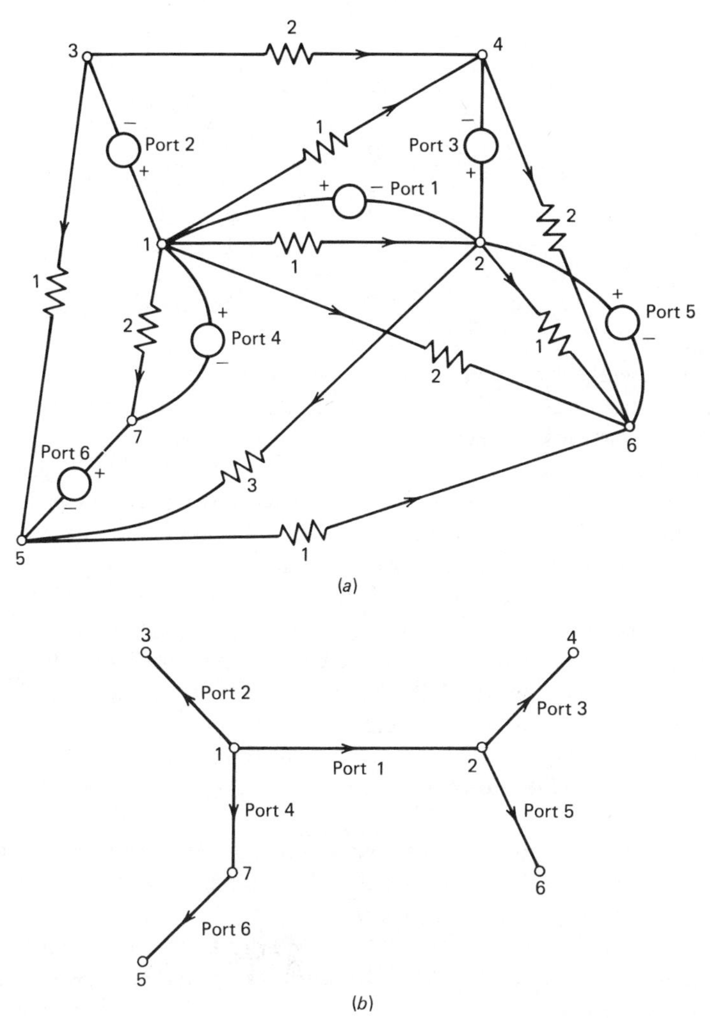

Figure 12.3. A 6-port network and its port configuration. (All the conductances are in Siemens.)

and

$$G_e = \begin{bmatrix} 1 & 0 & 0 & 0 & 0 & 0 & 0 & 0 & 0 & 0 \\ 0 & 1 & 0 & 0 & 0 & 0 & 0 & 0 & 0 & 0 \\ 0 & 0 & 2 & 0 & 0 & 0 & 0 & 0 & 0 & 0 \\ 0 & 0 & 0 & 2 & 0 & 0 & 0 & 0 & 0 & 0 \\ 0 & 0 & 0 & 0 & 3 & 0 & 0 & 0 & 0 & 0 \\ 0 & 0 & 0 & 0 & 0 & 1 & 0 & 0 & 0 & 0 \\ 0 & 0 & 0 & 0 & 0 & 0 & 2 & 0 & 0 & 0 \\ 0 & 0 & 0 & 0 & 0 & 0 & 0 & 1 & 0 & 0 \\ 0 & 0 & 0 & 0 & 0 & 0 & 0 & 0 & 2 & 0 \\ 0 & 0 & 0 & 0 & 0 & 0 & 0 & 0 & 0 & 1 \end{bmatrix}.$$

So, for this network

$$Y = Q_f G_e Q_f^t$$

$$= \begin{bmatrix} 10 & -2 & 3 & -4 & 3 & -4 \\ -2 & 3 & -2 & -1 & 0 & -1 \\ 3 & -2 & 5 & 0 & -2 & 0 \\ -4 & -1 & 0 & 7 & -1 & 5 \\ 3 & 0 & -2 & -1 & 6 & -1 \\ -4 & -1 & 0 & 5 & -1 & 5 \end{bmatrix}. \tag{12.20}$$

In the following, we denote by g_{ij} the conductance of the edge (i, j) connecting vertices i and j. We assume that row i in Q_f corresponds to port i. For further discussions we need a few definitions.

A conductance g_{ij} is said to *span* a port of N if the port lies in the unique path in T between the vertices i and j. For example, in the network of Fig. 12.3, the conductance g_{35} spans the ports 2, 4, and 6. Note that a conductance spans all those ports with which it forms a fundamental circuit.

In network theory literature, a pendant vertex in T is usually referred to as a *tip vertex*. Similarly, a pendant port in T is referred to as a *tip port*. So in all our discussions in this chapter, we use the terms "tip vertex" and "tip port" in place of "pendant vertex" and "pendant port." Further, we use the term *short-circuiting* in place of "contracting".

A tree is a *linear tree* if it has exactly two tip vertices. In other words, a linear tree is a path. For example, the ports 1, 2, and 5 in Fig. 12.3 form a linear tree.

A tree is a *star tree* if all its edges have a common end vertex called the *star vertex*. In other words, a star tree has exactly one nontip vertex. For example the ports 1, 2, and 4 in Fig. 12.3 form a star tree.

Two ports i and j are said to be *similarly oriented* if they are oriented in the same direction in any path in T containing these ports. Otherwise they are *dissimilarly oriented*.

Consider now any diagonal entry y_{ii} of the matrix Y. By Theorem 2.15 the nonzero entries in row i of Q_f correspond to all those conductances which span port i. So we get from (12.19)

$$y_{ii} = \text{sum of all the conductances which span port } i. \tag{12.21}$$

Consider next any off-diagonal entry y_{ij} of Y. Let q_{ir} denote the (i, r) entry of Q_f. Then we know (Exercise 6.1) that if, for any i and j and for any r and s, the products $q_{ir}q_{jr}$ and $q_{is}q_{js}$ are nonzero, then they will have the same sign. So we get from (12.19)

$$y_{ij} = \pm \left\{ \begin{array}{c} \text{sum of all the conductances which span both} \\ \text{port } i \text{ and port } j. \end{array} \right\} \tag{12.22}$$

Now we need to fix the sign of y_{ij}. Suppose for some r the product $q_{ir}q_{jr} \neq 0$. Then the sign of y_{ij} is the same as the sign of $q_{ir}q_{jr}$.

Since q_{ir} and q_{jr} are both nonzero, it is clear that ports i and j both lie in the fundamental circuit C produced by the conductance which corresponds to column r of Q_f. Now two cases arise.

Assume first that ports i and j are similarly oriented. Then the orientations of ports i and j will both agree or both disagree with the orientation of circuit C. In other words, both q_{ir} and q_{jr} will have the same sign. Hence in this case y_{ij} is positive.

On the other hand, if ports i and j are dissimilarly oriented, then ports i and j will have opposite relative orientations in the circuit C so that q_{ir} and q_{jr} will have opposite signs. Hence in this case y_{ij} is negative. Thus we have the following theorem.

THEOREM 12.1. If ports i and j are similarly oriented, then y_{ij} is positive; otherwise y_{ij} is negative. ■

The next result follows easily from Theorem 12.1.

THEOREM 12.2.

1. If ports i, j, and k form a linear tree when all the other ports are short-circuited, then

$$y_{ij}y_{ik}y_{jk} \geqslant 0.$$

2. If ports i, j, and k form a star tree when all the other ports are short-circuited, then

$$y_{ij}y_{ik}y_{jk} \leqslant 0. \quad ■$$

We may verify Theorems 12.1 and 12.2 with the Y-matrix of the 6-port network of Fig. 12.3.

THEOREM 12.3. Suppose that ports i, j, and k form a linear tree when all the other ports are short-circuited. Further, assume that in this linear tree the ports appear in the order i, j, and k. Then

$$|y_{ij}| \geqslant |y_{ik}|. \quad ■$$

The above result follows from the fact that every conductance which spans ports i and k also necessarily spans port j. An important corollary of Theorem 12.3 is stated next.

Corollary 12.3.1 Let y_{ij} be a nonzero entry of Y with minimum magnitude. Then there is exactly one conductance which spans both ports i and j. The value of this conductance is $|y_{ij}|$. In other words, rows i and j of Q_f have nonzero entries simultaneously in only one column. (Note that it is assumed that parallel conductances are combined into one.) ■

For example, in the matrix Y of (12.20), y_{24} is a nonzero entry with minimum magnitude and g_{35} (Fig. 12.3) is the only conductance spanning both ports 2 and 4.

THEOREM 12.4. Let the maximum magnitude of the entries in the ith row of Y be M_i, and let entries having magnitude M_i occur simultaneously in this row at the column positions $j, k, l, \ldots$, and only there. Then the subgraph formed by ports $i, j, k, l, \ldots$ is connected.

Proof

If the subgraph formed by the set $\{i, j, k, l, \ldots\}$ of ports is not connected, then there exists some port d in this set which is not present in the component containing port i. Since T is connected, there exists a port p of T, p being not in the set $\{i, j, k, l, \ldots\}$, such that the ports d, p, i form a linear tree with the three ports appearing in that order when all the other ports of T are short-circuited. From Theorem 12.3 we note that $|y_{ip}| \geqslant |y_{id}|$. Since y_{id} is a maximum magnitude entry in the ith row of Y, it follows that $|y_{ip}| = |y_{id}|$. So p also should be an element of the set $\{i, j, k, l, \ldots\}$, which is a contradiction. Therefore the subgraph consisting of the ports $i, j, k, l, \ldots$ is connected. ■

THEOREM 12.5. Let the smallest entry in row i of Y be m_i, where m_i may possibly be zero. Let in this row the smallest magnitude entry occur simultaneously in column positions $p, q, r, \ldots$, and only there. Then the subgraph formed by all the ports of T other than $p, q, r, \ldots$ is connected.

Proof

If the subgraph as specified is not connected, then there exists a port j in the set $\{p, q, r, \ldots\}$ and a port k not in $\{p, q, r, \ldots\}$ such that the ports i, j, and k form a linear tree when all the other ports of T are short-circuited. We then have by Theorem 12.3 that $|y_{ij}| \geqslant |y_{ik}|$. Since $|y_{ij}| = m_i$, a minimum magnitude entry in the ith row of Y, it follows that $|y_{ik}| = m_i$ and that k is in $\{p, q, r, \ldots\}$, which is a contradiction. ■

We may verify the above two theorems with the matrix Y of (12.20).

We see from (12.21) and (12.22) that $y_{ii} \geqslant |y_{ij}|$ for all i and j. This is only a special case of a more general property, known as *paramountcy*, exhibited by the Y-matrix.

A real symmetric matrix is said to be *paramount* if every principal minor of the matrix is not smaller than the magnitude of any other minor built from the same rows (or columns).

Cederbaum [12.1] has shown that any matrix $P = KDK^t$ where K is a unimodular matrix and D is a diagonal matrix with real positive entries is paramount. Since the matrix Q_f is unimodular, we can conclude from (12.19) that the short-circuit conductance matrix of an $(n+1)$-node resistance n-port network is paramount. In fact, we can also show that the short-circuit conductance and the open-circuit resistance matrices of any n-port network are paramount.

For example, consider the minors

$$M_1 = \begin{vmatrix} 10 & -2 & 3 \\ -2 & 3 & -2 \\ 3 & -2 & 5 \end{vmatrix} \quad \text{and} \quad M_2 = \begin{vmatrix} 10 & 3 & -4 \\ -2 & -2 & -1 \\ 3 & 5 & 0 \end{vmatrix}$$

of the matrix Y of (12.20). Both these minors are built from the first three rows of the matrix, the former being a principal minor. We may now check that $M_1 = 87$ and $M_2 = 57$, thereby verifying that $M_1 \geqslant |M_2|$.

For a detailed discussion of the properties of paramount matrices see Weinberg [12.2].

Next we characterize the Y-matrices of $(n+1)$-node resistance n-port networks having a star-tree or linear-tree port configuration. In the following, we say that a matrix Y is presentable in a particular form if Y can be brought to that form by one or more applications of the following operations:

1. Interchange any two rows and the corresponding columns.
2. Change the signs of all the entries in any row and the corresponding column.

Note that these two operations correspond, respectively, to renumbering the ports and changing the orientations of some of the ports.

12.2.2 Star-Tree Port Configuration

Consider the Y-matrix of an $(n+1)$-node n-port network N having a star-tree port configuration T. Assume that all the ports in T are oriented toward the star vertex (see Fig. 12.4). Let the star vertex be labeled 0. Let the other vertices be labeled $1, 2, \ldots, n$ so that the vertices i and 0 form the terminals of port i. We then have from (12.21)

$$y_{ii} = \sum_{\substack{j=0 \\ j \neq i}}^{n} g_{ij}. \tag{12.23}$$

Further by Theorem 12.1,

$$y_{ij} \leqslant 0. \tag{12.24}$$

So we get from (12.22)

$$y_{ij} = -g_{ij}. \tag{12.25}$$

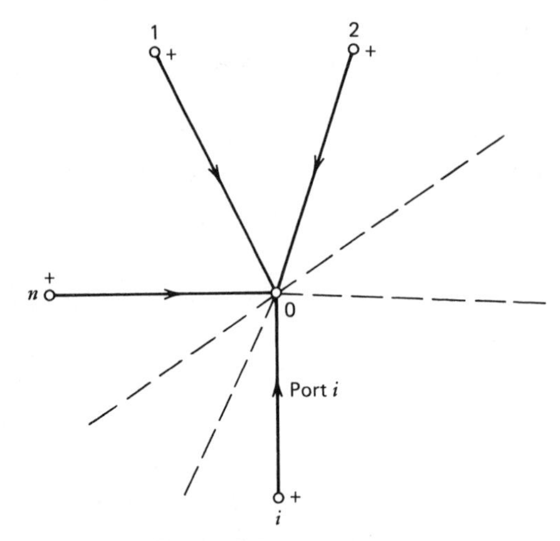

Figure 12.4. A star tree.

It can now be seen from (12.23) and (12.25) that for all i

$$y_{ii} \geqslant \sum_{\substack{j=1\\j\neq i}}^{n} |y_{ij}|. \tag{12.26}$$

A real symmetric matrix $Y=[y_{ij}]$ is *hyperdominant* if it satisfies (12.24) and (12.26), that is,

1. y_{ij} is nonpositive for all i and $j\neq i$; and
2. $y_{ii} \geqslant \sum_{\substack{j=1\\j\neq i}}^{n} |y_{ij}|.$

We can easily obtain from (12.23) and (12.25) the following expressions for the conductances of N:

$$g_{ij} = -y_{ij}, \qquad j\neq i, \quad i\neq 0, \quad j\neq 0,$$

$$g_{i0} = y_{ii} - \sum_{\substack{j=1\\j\neq i}}^{n} |y_{ij}|, \qquad i\neq 0.$$

The conductances g_{ij} calculated as above will be nonnegative if the matrix Y satisfies (12.24) and (12.26). Thus we have proved the following.

THEOREM 12.6. A real symmetric matrix Y is realizable as the short-circuit conductance matrix of an $(n+1)$-node resistance n-port network having a star-tree port configuration and containing no negative conductances if and only if it is presentable in the hyperdominant form. ■

12.2.3 Linear-Tree Port Configuration

Consider next the Y-matrix of an $(n+1)$-node resistance n-port network having a linear-tree port configuration T. Let the vertices of T be labeled consecutively starting from a tip vertex, and let the vertices i and $i+1$ form the positive and negative reference terminals of port i (Fig. 12.5). Since all the ports in T are similarly oriented, by Theorem 12.1

$$y_{ij} \geqslant 0, \qquad \text{for all } i \text{ and } j. \tag{12.27}$$

Further by Theorem 12.3, we have, for all i,

$$y_{ii} \geqslant y_{i,i+1} \geqslant \cdots \geqslant y_{in}, \tag{12.28}$$

$$y_{ii} \geqslant y_{i-1,i} \geqslant \cdots \geqslant y_{1i}. \tag{12.29}$$

Next we derive a simple expression for the conductance g_{ij} in terms of the entries of the Y-matrix.

Suppose we short-circuit all the ports except the ports $i-1$, i, $j-1$, and j, where $i>1$ and $i+1<j\leqslant n$. Then the resulting network will be as in Fig. 12.6, where only the conductances of interest to us are marked. Now we can easily show that

$$y_{i-1,j-1} = e+c,$$

$$y_{i-1,j} = e,$$

$$y_{i,j-1} = g_{ij}+c+d+e,$$

$$y_{i,j} = e+d.$$

From the above we get

$$g_{ij} = (y_{i,j-1}+y_{i-1,j}) - (y_{ij}+y_{i-1,j-1}), \qquad i>1, \quad i+1<j\leqslant n. \tag{12.30}$$

We can also prove the following:

$$g_{i\,i+1} = (y_{ii}+y_{i-1,i+1}) - (y_{i,i+1}+y_{i-1,i}), \qquad 1<i<n,$$

$$g_{1j} = y_{1,j-1} - y_{1j}, \qquad 2\leqslant j\leqslant n,$$

$$g_{1,n+1} = y_{1n},$$

$$g_{i,n+1} = y_{in} - y_{i-1,n}, \qquad 1<i\leqslant n. \tag{12.31}$$

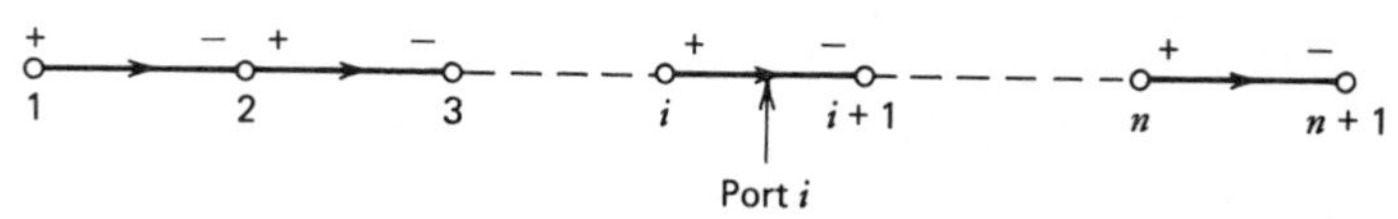

Figure 12.5. A linear tree.

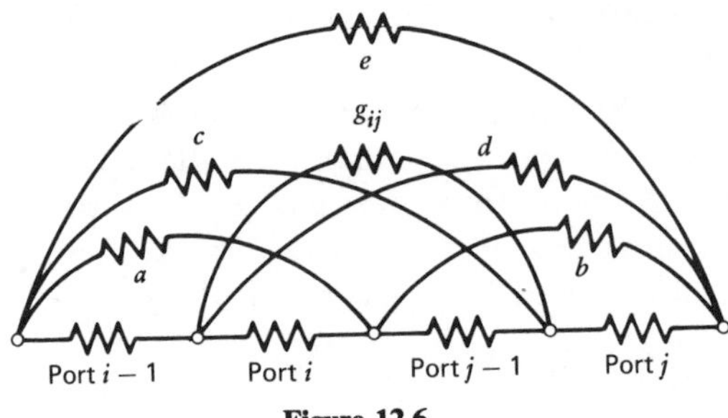

Figure 12.6.

A real symmetric matrix $Y=[y_{ij}]$ is *uniformly tapered* if

1. $y_{ij} \geqslant 0$, for all i and j.
2. $y_{ij} \geqslant y_{i,j+1}$, for $j \geqslant i$.
3. $y_{ij} \geqslant y_{i-1,j}$, for $j \geqslant i$.
4. $y_{i,j-1} + y_{i-1,j} \geqslant y_{ij} + y_{i-1,j-1}$, for $i > 1$, $i+1 \leqslant j \leqslant n$.

From (12.27) through (12.31), we get the following.

THEOREM 12.7. A real symmetric matrix $Y=[y_{ij}]$ is realizable as the short-circuit conductance matrix of an $(n+1)$-node resistance n-port network having a linear-tree port configuration and containing no negative conductances if and only if Y is presentable in the uniformly tapered form. ■

Given a uniformly tapered matrix Y, Guillemin [12.3] has given a simple procedure to obtain the conductances of an n-port network realizing Y. This procedure is as follows:

1. First construct from the given uniformly tapered $n \times n$ matrix Y a new $n \times n$ matrix Y' such that
   - **a.** The first row of Y' is the same as the first row of Y; and
   - **b.** For all $i > 1$, the ith row of Y' is the difference of the ith and the $(i-1)$th rows of Y.
2. Construct then another $n \times n$ matrix Y'' such that
   - **a.** The last column of Y'' is the same as the last column of Y'; and
   - **b.** For all $i < n$, the ith column of Y'' is the difference of the ith and the $(i+1)$th columns of Y'.

Then we can see that $g_{ij} = y''_{i,j-1}$ for all $j > i$.

For example, for the uniformly tapered matrix

$$Y=\begin{bmatrix} 5 & 4 & 3 & 1 & 1 \\ 4 & 5 & 3 & 1 & 1 \\ 3 & 3 & 11 & 8 & 5 \\ 1 & 1 & 8 & 9 & 5 \\ 1 & 1 & 5 & 5 & 6 \end{bmatrix}$$

the corresponding Y' and Y'' matrices are

$$Y'=\begin{bmatrix} 5 & 4 & 3 & 1 & 1 \\ & 1 & 0 & 0 & 0 \\ & & 8 & 7 & 4 \\ & & & 1 & 0 \\ & & & & 1 \end{bmatrix},$$

$$Y''=\begin{bmatrix} 1 & 1 & 2 & 0 & 1 \\ & 1 & 0 & 0 & 0 \\ & & 1 & 3 & 4 \\ & & & 1 & 0 \\ & & & & 1 \end{bmatrix}$$

$$=\begin{bmatrix} g_{12} & g_{13} & g_{14} & g_{15} & g_{16} \\ & g_{23} & g_{24} & g_{25} & g_{26} \\ & & g_{34} & g_{35} & g_{36} \\ & & & g_{45} & g_{46} \\ & & & & g_{56} \end{bmatrix}.$$

In the above we have not shown the entries below the diagonals of the matrices Y' and Y'' because these entries are not of interest to us in calculating the conductances g_{ij}.

12.2.4 Port Transformation

Consider next two different $(n+1)$-node n-port networks N and N^* constructed on the same resistance network. Let T and T^* be the port configurations of N and N^*, respectively. We now derive an expression relating the short-circuit conductance matrices Y and Y^* of these n-port networks.

Let Q_f and Q_f^* be the fundamental cutset matrices of N and N^* with respect to T and T^*, respectively. Then

$$Y=Q_f G_e Q_f^t$$

and

$$Y^*=Q_f^* G_e (Q_f^*)^t,$$

where G_e is the diagonal edge conductance matrix of the resistance network under consideration.

Consider now the graph $T' = T \cup T^*$. We can write the fundamental cutset matrix of T' with respect to T as

$$\begin{matrix} T-T^* & T^* \\ [P & | \; M]. \end{matrix}$$

We can then express Q_f^* as

$$Q_f^* = M^{-1} Q_f. \tag{12.32}$$

Therefore,

$$\begin{aligned} Y^* &= Q_f^* G_e (Q_f^*)^t \\ &= M^{-1} Q_f G_e Q_f^t (M^{-1})^t \\ &= M^{-1} Y (M^{-1})^t. \end{aligned} \tag{12.33}$$

Now we have from (11.8)

$$V_p^* = M^t V_p,$$

where V_p and V_p^* are the vectors of port voltages of N and N^*, respectively. So,

$$V_p = (M^{-1})^t V_p^*. \tag{12.34}$$

Combining (12.33) and (12.34), we get the following theorem.

THEOREM 12.8. Let N and N^* be two n-port networks constructed on the same resistance network. If $V_p = K V_p^*$, then

$$Y^* = K^t Y K. \quad \blacksquare$$

Note that the matrix K in the above theorem can be obtained easily by inspection of T and T^*. For example, consider the two n-port networks N and N^* shown in Fig. 12.7. These n-port networks are constructed from the same parent network. The short-circuit conductance matrix of N is

$$Y = \begin{bmatrix} 3 & 2 & 0 & 1 & -1 & 2 \\ 2 & 5 & -2 & 0 & 2 & 2 \\ 0 & -2 & 9 & 0 & -3 & 0 \\ 1 & 0 & 0 & 9 & -1 & -1 \\ -1 & 2 & -3 & -1 & 8 & 0 \\ 2 & 2 & 0 & -1 & 0 & 8 \end{bmatrix}.$$

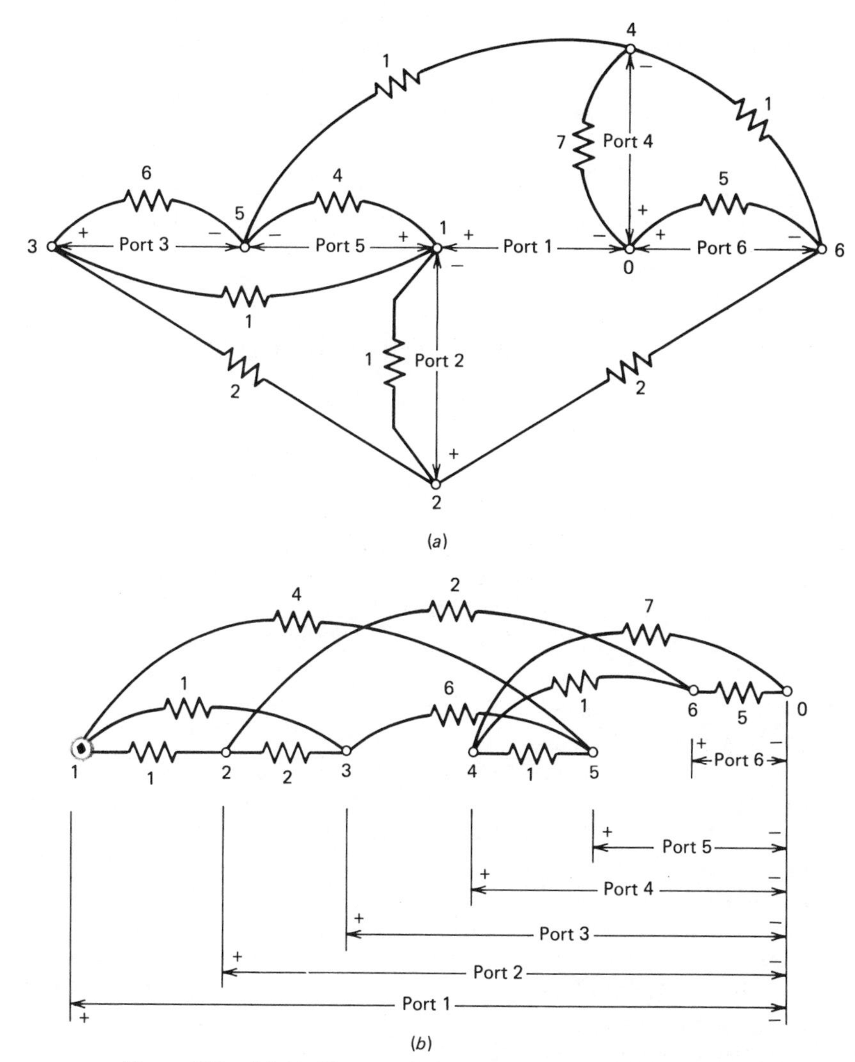

Figure 12.7. (*a*) N. (*b*) N^*. (All the conductances are in Siemens.)

The vectors V_p and V_p^* are related as

$$V_p = \begin{bmatrix} v_1 \\ v_2 \\ v_3 \\ v_4 \\ v_5 \\ v_6 \end{bmatrix} = \begin{bmatrix} 1 & 0 & 0 & 0 & 0 & 0 \\ -1 & 1 & 0 & 0 & 0 & 0 \\ 0 & 0 & 1 & 0 & -1 & 0 \\ 0 & 0 & 0 & -1 & 0 & 0 \\ 1 & 0 & 0 & 0 & -1 & 0 \\ 0 & 0 & 0 & 0 & 0 & -1 \end{bmatrix} \begin{bmatrix} v_1^* \\ v_2^* \\ v_3^* \\ v_4^* \\ v_5^* \\ v_6^* \end{bmatrix} = KV_p^*.$$

We may now verify that $Y^* = K^t Y K$:

$$Y^* = K^t Y K = \begin{bmatrix} 6 & -1 & -1 & 0 & -4 & 0 \\ -1 & 5 & -2 & 0 & 0 & -2 \\ -1 & -2 & 9 & 0 & -6 & 0 \\ 0 & 0 & 0 & 9 & -1 & -1 \\ -4 & 0 & -6 & -1 & 11 & 0 \\ 0 & -2 & 0 & -1 & 0 & 8 \end{bmatrix}.$$

12.3 REALIZATION OF $(n+1)$-NODE RESISTANCE n-PORT NETWORKS—I

In this section, we consider the problem of realizing a real symmetric $n \times n$ matrix Y as the short-circuit conductance matrix of an $(n+1)$-node resistance n-port network containing no negative conductances. For this problem, two basic approaches due to Cederbaum and Guillemin are available.

Cederbaum's [12.1] approach requires the decomposition of the matrix Y into the triple product $Y = Q_f G_e Q_f^t$, where Q_f is a matrix whose elements are -1, 1, or 0, and G_e is a diagonal matrix with nonnegative entries representing the edge conductances of the required n-port network N. Realization of Y is accomplished if $[U \quad Q_f]$ can be realized as the fundamental cutset matrix of a connected graph. This connected graph will then be the graph of the augmented n-port network $\hat{N}$ with the columns of Q_f corresponding to the edges representing the conductances of the required network N. In fact as we see later in this section, Cederbaum's approach is also applicable to the realization of open-circuit resistance matrices of n-port networks of nullity n.

Guillemin's approach [12.3] requires the determination of the port configuration (if one exists) and the transformation of the matrix Y to a matrix Y' which corresponds to a star-tree port configuration. If the matrix Y' is hyperdominant, then the realization (as we have seen in the previous section) can be accomplished by inspection. Through an appropriate redefinition of ports, the same network also realizes the given Y.

We now describe Cederbaum's algorithm to decompose Y into the triple product $Y = Q_f G_e Q_f^t$. In the next section, we consider the realization of $[U \quad Q_f]$ as a fundamental cutset matrix. An algorithm, based on Guillemin's approach to realize the Y-matrix of an $(n+1)$-node n-port network, is discussed in Section 12.5.

Let $\hat{q}_1, \hat{q}_2, \ldots$ denote the columns of the matrix Q_f. Let $g_1, g_2, \ldots$ denote the diagonal entries of G_e. $\hat{q}_i$ is the ith column of Q_f, and g_i is the ith diagonal entry of G_e. Note that $\hat{q}_i$ corresponds to g_i. Further the entries of $\hat{q}_i$ will be denoted as $q_{1i}, q_{2i}, \ldots, q_{ni}$.

Starting with the matrix $Y^{(0)} = Y$, Cederbaum's algorithm constructs a sequence of matrices $Y^{(0)}, Y^{(1)}, \ldots$ corresponding at each step to the removal from the network of an appropriately identified conductance. This process is continued until a diagonal matrix is obtained or it is recognized that the decomposition is not possible.

At each step of the algorithm, we need to consider two cases. Using the matrix Y, we describe below the procedure to be followed in each of these two cases.

Case 1 Y is diagonal with k zero diagonal entries.

Let Y' be the matrix obtained by deleting from Y all the rows and the columns which contain the zero diagonal entries. We can easily see that the decomposition in this case is

$$Y=\left[\frac{U}{0}\right]Y'\left[\frac{U}{0}\right]^t,$$

where it is taken that the zero diagonal entries of Y occur in the last k positions. For example, the matrix

$$\begin{bmatrix} 3 & 0 & 0 & 0 & 0 \\ 0 & 1 & 0 & 0 & 0 \\ 0 & 0 & 2 & 0 & 0 \\ 0 & 0 & 0 & 0 & 0 \\ 0 & 0 & 0 & 0 & 0 \end{bmatrix}$$

can be decomposed as

$$Y=\begin{bmatrix} 1 & 0 & 0 \\ 0 & 1 & 0 \\ 0 & 0 & 1 \\ \hdashline 0 & 0 & 0 \\ 0 & 0 & 0 \end{bmatrix}\begin{bmatrix} 3 & 0 & 0 \\ 0 & 1 & 0 \\ 0 & 0 & 2 \end{bmatrix}\left[\begin{array}{ccc:cc} 1 & 0 & 0 & 0 & 0 \\ 0 & 1 & 0 & 0 & 0 \\ 0 & 0 & 1 & 0 & 0 \end{array}\right].$$

Case 2 Y has some nonzero off-diagonal entries.

In this case, we first select a nonzero off-diagonal entry of Y with the smallest magnitude. Let us assume that y_{12} is such an entry of Y. If Y is realizable, then we know from Corollary 12.3.1 that in the required network N, there must be an edge with conductance value equal to $|y_{12}|$. In other words, G_e must have a diagonal entry equal to $|y_{12}|$. Let us assume without loss of generality that

$$g_1=|y_{12}|.$$

We can determine the column $\hat{q}_1$ corresponding to g_1 as follows:

Clearly the edge with conductance value $|y_{12}|$ spans both ports 1 and 2 of N. So the first two entries of $\hat{q}_1$ are nonzero. By Corollary 12.3.1, the first two rows of Q_f cannot have nonzero entries simultaneously in any but the first column. So we arbitrarily assign $+1$ to the first entry of $\hat{q}_1$, thereby fixing the orientation of the edge having conductance g_1, in relation to the orientation of port 1. Then, by Theorem 12.1, the sign of the second entry in $\hat{q}_1$ is the same as the sign of y_{12}. Thus

$$\hat{q}_1 = \begin{bmatrix} 1 \\ \pm 1 \\ \cdot \\ \cdot \\ \cdot \\ \cdot \end{bmatrix}.$$

Note that if y_{rs} is the nonzero off-diagonal entry with the smallest magnitude, then the rth and sth entries of $\hat{q}_1$ will be determined in a similar manner.

It now remains to obtain the other entries of $\hat{q}_1$. This can be achieved by using the following rule which is based on Theorem 12.2:

$$q_{j1} = \begin{cases} 1, & \text{if } y_{12}y_{1j}y_{2j} > 0 \quad \text{and} \quad y_{1j} > 0. \\ -1, & \text{if } y_{12}y_{1j}y_{2j} > 0 \quad \text{and} \quad y_{1j} < 0. \\ 0, & \text{otherwise.} \end{cases}$$

This completes the discussion of the two cases.

In the next step, we have to find the second column of Q_f and the corresponding diagonal entry g_2 of G_e. This can be done by considering the short-circuit conductance matrix $Y^{(1)}$ of the network $N^{(1)}$ which is obtained by removing from N the conductance g_1, that is, by setting $g_1 = 0$ in N.

Let us delete from Q_f its first column, and from G_e its first row and first column. Let the resulting matrices be denoted as $Q_f^{(1)}$ and $G_e^{(1)}$. Then it can be seen that

$$\begin{aligned} Y^{(1)} &= Q_f^{(1)} G_e^{(1)} \left(Q_f^{(1)}\right)^t \\ &= Y - \hat{q}_1 [g_1] (\hat{q}_1)^t. \end{aligned} \tag{12.35}$$

Hence

$$y_{ij}^{(1)} = y_{ij} - q_{i1} q_{j1} g_1. \tag{12.36}$$

Note that if the matrix is realizable, then all the zero entries of Y remain unchanged in $Y^{(1)}$, and no nonzero entry y_{ij} can change its sign or increase in magnitude. Therefore, in such a case, the matrix $Y^{(1)}$ will have more zero off-diagonal entries than Y.

Suppose $Y^{(1)}$ is diagonal. Then the decomposition described under case 1 will give all the remaining columns of Q_f and the remaining diagonal entries of G_e.

If $Y^{(1)}$ is not diagonal, then we repeat the procedure described under case 2 to obtain the conductance g_2 and the second column $\hat{q}_2$ of Q_f. The matrix $Y^{(2)}$ is then given by

$$Y^{(2)} = Y^{(1)} - \hat{q}_2[g_2](\hat{q}_2)^t.$$

If Y is realizable, then clearly each matrix in the sequence $Y^{(0)}, Y^{(1)}, Y^{(2)}, \ldots$ will fall into either of the two cases described earlier, and may therefore be decomposed.

If at any step the off-diagonal entries of the corresponding matrix start increasing in magnitude or changing signs, then the decomposition is not possible, and so we may terminate the algorithm at that step. In some cases the decomposition may be possible, but Q_f may not be a unimodular matrix or G_e may have negative diagonal elements, or both. Y is not realizable in this case also.

We now illustrate Cederbaum's algorithm with the matrix

$$Y = Y^{(0)} = \begin{bmatrix} 3 & 2 & 1^* & -2 & 0 \\ 2 & 5 & -2 & -2 & 1 \\ 1 & -2 & 3 & 0 & 0 \\ -2 & -2 & 0 & 5 & 3 \\ 0 & 1 & 0 & 3 & 4 \end{bmatrix}.$$

Since Y is not diagonal, case 1 does not arise. We therefore proceed as described under case 2.

y_{13} is a minimum magnitude nonzero entry in the matrix of Y. So we set

$$g_1 = |y_{13}| = 1,$$

$$q_{11} = 1,$$

$$q_{31} = 1 \qquad \text{(since } y_{13} \text{ is positive)}.$$

We then determine the remaining entries of the first column $\hat{q}_1$ of Q_f using the following rules:

$$q_{j1} = \begin{cases} 1, & \text{if } y_{1j}y_{3j}y_{13} > 0 \quad \text{and} \quad y_{1j} > 0. \\ -1, & \text{if } y_{1j}y_{3j}y_{13} > 0 \quad \text{and} \quad y_{1j} < 0. \\ 0, & \text{otherwise.} \end{cases}$$

For example, $y_{12}y_{32}y_{13} < 0$. So $q_{21} = 0$. Thus we get

$$\hat{q}_1 = \begin{bmatrix} 1 \\ 0 \\ 1 \\ 0 \\ 0 \end{bmatrix}.$$

We then obtain $Y^{(1)}$ as follows:

$$Y^{(1)} = Y^{(0)} - \hat{q}_1[g_1](\hat{q}_1)^t$$

$$= \begin{bmatrix} 2 & 2 & 0 & -2 & 0 \\ 2 & 5 & -2 & -2 & 1^* \\ 0 & -2 & 2 & 0 & 0 \\ -2 & -2 & 0 & 5 & 3 \\ 0 & 1 & 0 & 3 & 4 \end{bmatrix}.$$

In $Y^{(1)}$, the (2,5) entry has minimum nonzero magnitude. So we get

$$g_2 = 1,$$

$$q_{22} = 1,$$

$$q_{52} = 1.$$

We then obtain the remaining entries of $\hat{q}_2$ as before and

$$\hat{q}_2 = \begin{bmatrix} 0 \\ 1 \\ 0 \\ 0 \\ 1 \end{bmatrix}.$$

The matrix $Y^{(2)}$ is obtained as

$$Y^{(2)} = Y^{(1)} - \hat{q}_2[g_2](\hat{q}_2)^t$$

$$= \begin{bmatrix} 2 & 2 & 0 & -2^* & 0 \\ 2 & 4 & -2 & -2 & 0 \\ 0 & -2 & 2 & 0 & 0 \\ -2 & -2 & 0 & 5 & 3 \\ 0 & 0 & 0 & 3 & 3 \end{bmatrix}.$$

Continuing in this way, we get the following:

$$g_3 = 2, \qquad \hat{q}_3 = \begin{bmatrix} 1 \\ 1 \\ 0 \\ -1 \\ 0 \end{bmatrix};$$

and

$$Y^{(3)}=\begin{bmatrix} 0 & 0 & 0 & 0 & 0 \\ 0 & 2 & -2^* & 0 & 0 \\ 0 & -2 & 2 & 0 & 0 \\ 0 & 0 & 0 & 3 & 3 \\ 0 & 0 & 0 & 3 & 3 \end{bmatrix}, \qquad g_4=2, \qquad \hat{q}_4=\begin{bmatrix} 0 \\ 1 \\ -1 \\ 0 \\ 0 \end{bmatrix};$$

$$Y^{(4)}=\begin{bmatrix} 0 & 0 & 0 & 0 & 0 \\ 0 & 0 & 0 & 0 & 0 \\ 0 & 0 & 0 & 0 & 0 \\ 0 & 0 & 0 & 3 & 3^* \\ 0 & 0 & 0 & 3 & 3 \end{bmatrix}, \qquad g_5=3, \qquad \hat{q}_5=\begin{bmatrix} 0 \\ 0 \\ 0 \\ 1 \\ 1 \end{bmatrix}.$$

In the above, the entries chosen for consideration at each stage are starred. Following is a decomposition of $Y^{(0)}$:

$$Y^{(0)}=\begin{bmatrix} 3 & 2 & 1 & -2 & 0 \\ 2 & 5 & -2 & -2 & 1 \\ 1 & -2 & 3 & 0 & 0 \\ -2 & -2 & 0 & 5 & 3 \\ 0 & 1 & 0 & 3 & 4 \end{bmatrix}=\begin{bmatrix} 1 & 0 & 1 & 0 & 0 \\ 0 & 1 & 1 & 1 & 0 \\ 1 & 0 & 0 & -1 & 0 \\ 0 & 0 & -1 & 0 & 1 \\ 0 & 1 & 0 & 0 & 1 \end{bmatrix}$$

$$\times\begin{bmatrix} 1 & 0 & 0 & 0 & 0 \\ 0 & 1 & 0 & 0 & 0 \\ 0 & 0 & 2 & 0 & 0 \\ 0 & 0 & 0 & 2 & 0 \\ 0 & 0 & 0 & 0 & 3 \end{bmatrix}\begin{bmatrix} 1 & 0 & 1 & 0 & 0 \\ 0 & 1 & 0 & 0 & 1 \\ 1 & 1 & 0 & -1 & 0 \\ 0 & 1 & -1 & 0 & 0 \\ 0 & 0 & 0 & 1 & 1 \end{bmatrix}.$$

We would like to point out that Cederbaum's decomposition algorithm which we have just discussed is applicable to any paramount matrix; that is, if a paramount matrix K can be decomposed as PDP^t, where P is unimodular and D is diagonal with positive entries, then one such decomposition can be found using this algorithm. So we can see that this algorithm can also be used to decompose the open-circuit resistance matrix of an n-port network of nullity n as the triple product $B_f R B_f^t$. Thus Cederbaum's approach for realizing the Y-matrix of an $(n+1)$-node n-port network is also valid for realizing the Z-matrix of an n-port network of nullity n.

After determining B_f or Q_f, the next step in Cederbaum's approach is to determine the directed graph corresponding to these matrices. Suppose we have determined the undirected graph corresponding to B_f or Q_f. Then we can easily determine the appropriate directed graph (if one exists) by inspection of B_f or Q_f, as the case may be (Exercise 6.1).

Thus arises the need for an algorithm to realize a given matrix as the circuit or cutset matrix of an undirected graph. Such an algorithm is discussed in the next section.

12.4 REALIZATION OF CUTSET AND CIRCUIT MATRICES

We discuss in this section the problem of constructing an undirected graph having a given set of circuits or cutsets. We may assume, without any loss of generality, that we are given the fundamental cutset matrix Q_f or the fundamental circuit matrix B_f of the graph to be constructed.

Suppose that B_f is the fundamental circuit matrix with respect to a spanning tree T. Then we can write it as $B_f=[U \quad F]$, where F and U correspond, respectively, to T and the corresponding cospanning tree. It follows from the definition of B_f that the nonzero entries in each row of F correspond to the branches which form a path in T between the end vertices of an appropriate chord. In other words, each row of F corresponds to a path in T. For this reason F is called the *tree-path matrix* with respect to T.

Given the matrix F, we know from (6.13) that $Q_f=[F^t \quad U]$. So we can see that realizations of B_f and Q_f are not two independent problems. They are both equivalent to realizing F as a tree-path matrix.

Realizing F essentially involves determining the structure of a tree having branches corresponding to the columns of F such that the tree path prescribed by each row of F is present in it. It is clear that the task of constructing the required graph from the tree so determined entails merely the formality of adding to the latter a chord between the appropriate vertices for every row of F.

We now discuss an algorithm, due to Rao [12.4], for realizing a matrix F as the tree-path matrix of an undirected graph G.

12.4.1 Preamble

Let us first consider some rules which we may use to reduce the complexity of F before proceeding with its realization.

If the given F matrix is a direct sum matrix, that is, it is partitionable in the form

$$\begin{bmatrix} F_1 & 0 & \dots & 0 \\ 0 & F_2 & \dots & 0 \\ \cdot & \cdot & \cdots & \cdot \\ \cdot & \cdot & \cdots & \cdot \\ \cdot & \cdot & \cdots & \cdot \\ 0 & 0 & \dots & F_n \end{bmatrix},$$

then each of the submatrices $F_1, F_2, \dots, F_n$ can be realized separately and the realized subgraphs (trees) can be interconnected arbitrarily to obtain the required tree T.

If a set of identical rows is present in F, it indicates the existence of parallel chords in the graph G. All but one of these rows may be deleted from F for the purpose of realization of the associated tree T. A row with only one nonzero entry can also be suppressed, since it places no constraint on the interconnection of the branches of T.

In the event of a particular branch being present in only one of the prescribed paths, the realization of the tree can be attempted after deleting from F the associated column. The branch can be added later at either end of the appropriate path. Further, if some k columns of F are identical, it is sufficient to realize the matrix obtained by deleting $k-1$ of these columns. The branch corresponding to the retained column is later replaced by a series-connected set of branches ordered in an arbitrary manner.

Using the foregoing rules, the given matrix F may be reduced in complexity by deleting the redundant rows and columns successively. This procedure is called the *preamble* to the realization of F.

12.4.2 Identification of a Tip Branch

The first main step in Rao's algorithm, for realizing an F matrix is the identification of a tip branch of the tree realizing F. We now discuss a procedure to identify such a tip branch. This procedure is applicable to any general F matrix, and if F admits alternate realizations, then there exists at least one realization with the identified branch as a tip branch.

Let $W=[w_{ij}]$, a symmetric matrix, be defined as

$$W=F^tF, \tag{12.37}$$

where multiplications and additions are in the field of real numbers. Clearly, the entry w_{ij} equals the number of paths prescribed by F, in which both the branches i and j are present. Note that we have assumed that the ith row or column of W corresponds to the ith branch of the tree T.

Suppose G is the graph realizing the given matrix F and T is the corresponding spanning tree. If we regard each edge of G as a conductance of 1 Siemens, and if $Y=[y_{ij}]$ is the short-circuit conductance matrix of the resulting n-port network with respect to the port configuration T, then it can be seen that for $i\neq j, w_{ij}=|y_{ij}|$. Therefore we have the following properties which are merely reiterations of Theorems 12.3, 12.4, and 12.5.

Property 12.1 If the branches i, j, and k form a linear tree in that order when all the other branches of T are short-circuited, then $w_{ij}\geqslant w_{ik}$.

Property 12.2 Let the maximum value of the entries in the ith row of W be M_i, and let this entry M_i occur simultaneously in this row at the column positions $j, k, l, \ldots,$ and only there. Then the branches corresponding to $i, j, k, l, \ldots$ form a subtree of T.

Property 12.3 Let the smallest entry in row i of W be m_i, where m_i may possibly be zero. Let in this row the smallest entry occur simultaneously in column positions $p, q, r, \dots$, and only there. Then the branches of T other than $p, q, r, \dots$ form a subtree of T.

For further discussions, we need the following definitions.

An *L-set* is a set of branches which constitute a subtree of T having only one vertex in common with its complement in T, and having only one branch of the set, called the leading branch of the L-set, incident at that vertex.

To illustrate the above definition, we note that in Fig. 12.8 the branches l, k, m, and n constitute an L-set with branch l as its leading branch, while the branches k, m, and n do not constitute an L-set. A tip branch of a tree may be regarded as an L-set in its simplest form.

It is evident that the complement of an L-set with respect to T is a subtree of T. For any branch r other than a tip branch of T, there exist two L-sets, each having r as its leading branch.

We now examine the conditions under which a tip branch of a subtree of T can also be a tip branch of T. Let the branches of T be partitioned into two sets S and R such that the branches in R form a subtree of T. Further let the following relation be satisfied:

$$w_{r_i s_i} = w_{r_i s_j}, \qquad \text{for all } r_i \in R \quad \text{and} \quad s_i, s_j \in S. \tag{12.38}$$

We now show that a tip branch of the tree that would be obtained by short-circuiting all the branches in R is a tip branch of T or of a tree T', formed by the edges corresponding to the branches of T in a graph that is 2-isomorphic with the given graph.

If the relation

$$w_{rs} = 0, \qquad \text{for every } r \in R \quad \text{and} \quad s \in S, \tag{12.39}$$

is satisfied, then the matrix F is a direct sum matrix, and the submatrices of F corresponding to the branches contained in S and R can be realized separately. In such a case, a tip branch of the subtree constituted by the branches of S can evidently be made a tip branch of T in the overall realization.

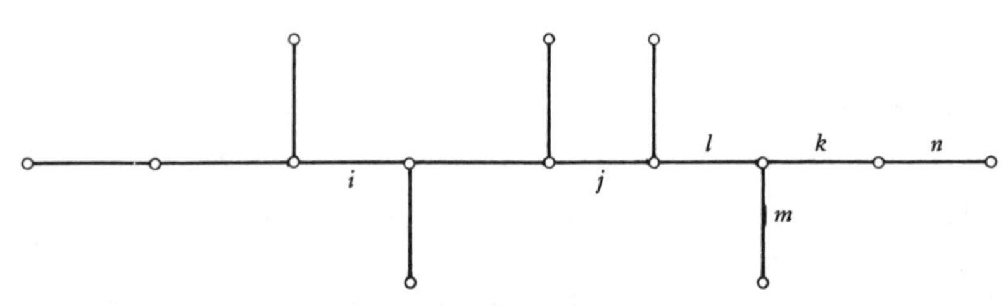

Figure 12.8. Illustration of L-set.

Let, on the other hand, there exist some $r_i \in R$ for which $w_{r_i s}$, $s \in S$, is not equal to zero. The set S can be partitioned into disjoint L-sets, since the branches in R form a subtree of T. We first prove that each of these L-sets is a linear subtree of T.

Let, in one of the L-sets having s_i as its leading branch, there exist branches s_j and s_k such that s_i, s_j, and s_k form a star tree when the other branches of T are short-circuited. Since the paths in which both s_k and r_i are present are distinct from those in which both s_j and r_i are present, and since s_i is present in both types of paths, we get

$$w_{r_i s_i} \geqslant w_{r_i s_j} + w_{r_i s_k}. \tag{12.40}$$

But by (12.38), $w_{r_i s_i} = w_{r_i s_j} = w_{r_i s_k}$. Further, they are all nonzero. So (12.40) cannot be satisfied. Hence s_i, s_j, and s_k should be contained in a linear subtree of T. Extending this argument, it can be shown that all the branches in each of the L-sets into which S is partitioned constitute a linear subtree of T.

Thus when the branches of R are short-circuited, one obtains from G a new graph G_1 which in general is separable with each nonseparable component containing one or more L-sets into which S is partitioned.

Let a particular nonseparable component of G_1 contain only one L-set, its branches constituting a linear subtree L_1 with x and y as its tip branches. Consider the specifications of F regarding the paths which include the branches in L_1. Some of these paths may contain branches from L_1 only. The others would contain branches from L_1 as well as some of those not in L_1, but every one of the latter paths would include all the branches of L_1, because all the relevant entries of W are equal. Therefore the subgraph of G_1 consisting of the branches of L_1 and the chords whose paths contain branches only from L_1 can be "turned around" to obtain a graph that is 2-isomorphic to G_1. If in one graph the tip branch is x, in the other it is y.

For example, for the graph G shown in Fig. 12.9*a*, the corresponding graph G_1 contains three nonseparable components as shown in Fig. 12.9*b*. We notice that s_6 is a tip branch of the spanning tree of this graph. The graph G_1' shown in Fig. 12.9*c* is 2-isomorphic with G_1, and s_4 is a tip branch of the corresponding tree of G_1'. Thus an alternate realization G' (Fig. 12.9*d*) of the tree-path matrix F of G exists with s_4 as a tip. By making use of the algorithm to be presented later, one can identify a tip branch of some tree realizing the given F matrix. Using this algorithm, any one of the branches s_1, s_4, s_6, s_7, s_9, and s_{12} may be identified as a tip branch of a tree realizing the F_1 matrix. For each one of these choices, there exist tree realizations for F, having the particular branch as a tip.

The point that emerges from the above discussion is the following:

When the branches of T can be partitioned into two sets R and S having the property specified by (12.38), then in order to find a tip branch of T, which definitely exists in the set S, one can consider the graph obtained by

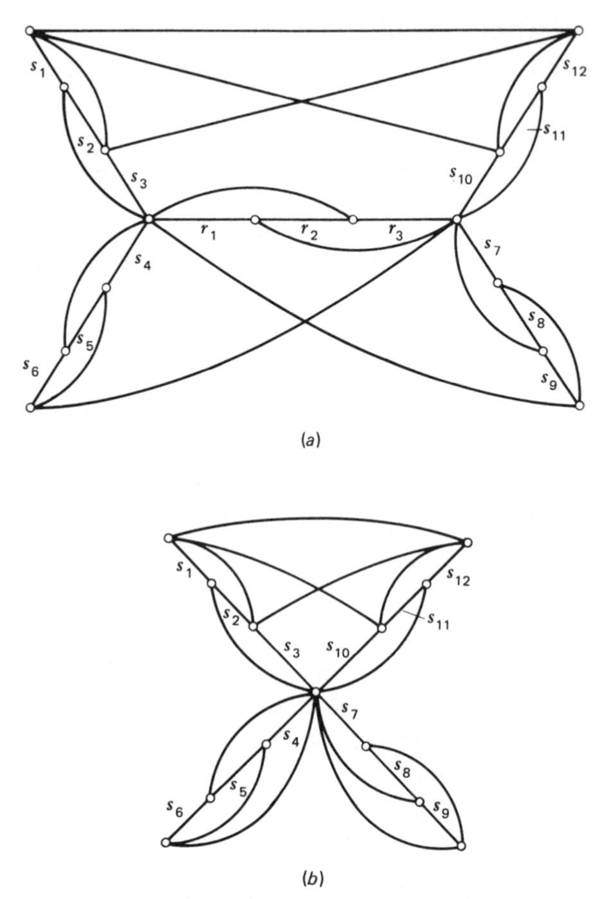

Figure 12.9. Example for tip identification from the derived graph. (*a*) G. (*b*) G_1. (*c*) G_1'. (*d*) G'.

short-circuiting the branches in R, and find a tip branch of the tree constituted by the branches of S. The tip branch so determined can be regarded as a tip branch of the tree of the graph required.

Following is an algorithm to find a tip branch of a tree T realizing a given F matrix.

Step 1. Obtain the matrix $W = F^t F$.

Step 2. Consider any row i of W. Let m_i (which may possibly be zero) be the minimum entry in that row. If there is only one entry in the ith row, say the entry w_{ij}, having this minimum value, then j is a tip branch. Otherwise let the columns at which the entries in row i equal m_i form the set $S = \{s_1, s_2, \ldots, s_n\}$. Evidently the set S contains a tip branch of T. Let the subset of branches of T which are not in S be designated as the set R.

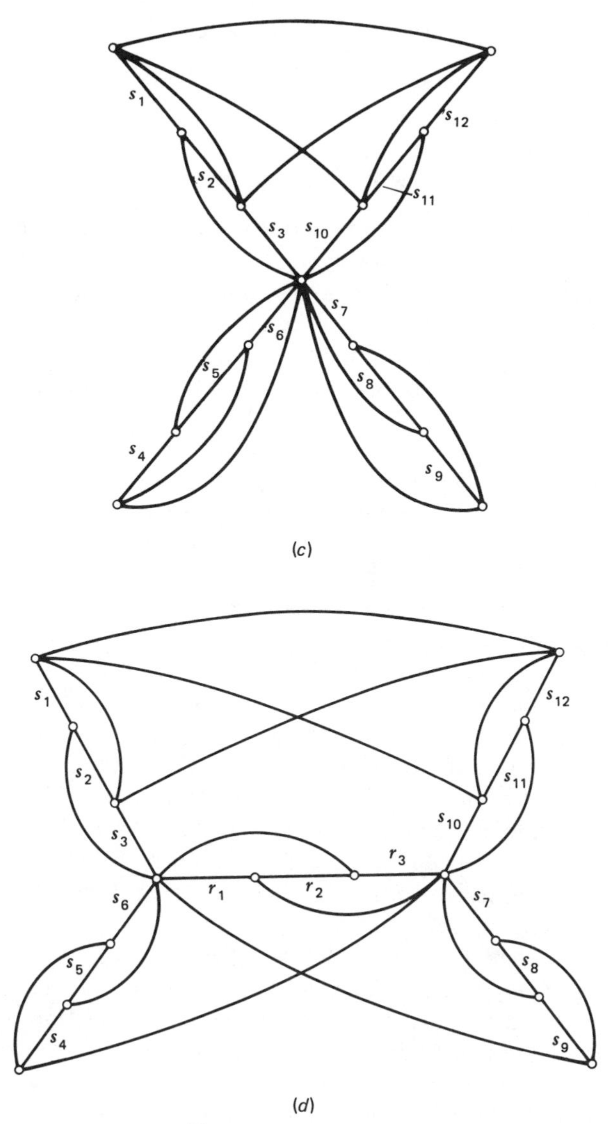

Figure 12.9. (*continued*)

Step 3. Consider any row $r_i \in R$. If $w_{r_i s_i} > w_{r_i s_j}$, where $s_i, s_j \in S$, we delete s_i from S. Applying this criterion to all relevant pairs of entries in row r_i, delete all possible elements from S, and designate the reduced set as S_1, and its complementary set as R_1.

Step 4. Repeat step 3 with other rows of W, choosing each time a row r_j such that r_j is an element of the R-set, say R_i, applicable at that stage. This procedure is continued till either

a. S_k contains only one element which then corresponds to a tip branch, or

b. S_k is not further reducible, that is, $w_{r_i s_i} = w_{r_i s_j}$ for all $r_i \in R_k$ and every pair of elements s_i and s_j of the set S_k.

Step 5. If a set S_k contains more than one element (case b of step 4), choose arbitrarily an element of S_k to take the place of r_i in step 3, and proceed through steps 3, 4, and 5, that is, delete an arbitrary element from S_k, add it to R_k, and continue with the process of further reduction of S_k. Repeated applications of these steps eventually lead to a set S_p containing only one element, which can be regarded as a tip branch of T.

The proof of correctness of the above algorithm now follows. We first note that by Property 12.3 the elements of the set R in step 2 form a subtree of T. So the complementary set S necessarily contains at least one tip branch. Further note that if s_k is an element of S, it follows from Property 12.1 that all elements of the particular L-set which does not contain i and for which s_k is the leading branch are present in S.

In step 3 we obtain a reduced set S_1 by deleting elements like s_k, such that $w_{r_i s_k} > w_{r_i s_j}$, $r_i \in R$, and $s_k, s_j \in S$. We proceed to show that S_1 contains at least one tip branch of T. This fact is evident (1) when s_k and s_j are in two different L-sets out of those referred to in the above, or (2) when s_k and s_j are in the same L-set, and the branches r_i, s_k, and s_j are contained in a linear subtree of T. On the other hand, if s_k and s_j are in the same L-set and the branches r_i, s_k, and s_j form a star tree when all the other branches of T are short-circuited, this L-set contains at least two tip branches. Hence in this case, even if s_k is a tip branch, the reduced set S_1 (obtained by omitting such s_k) contains a tip branch of T. Incidentally, we obtain that every branch present in the tree path having r_i and s_k as its end branches would also be deleted when s_k is deleted (Property 12.1). Thus once again after the completion of step 3, the set of branches contained in R constitute a subtree of T.

The above recursive procedure is repeated in step 4 by taking another pertinent row of W. At every stage the reduced set S_i definitely contains a tip branch and the branches in R_i continue to form a subtree of T.

In many cases it is possible to identify a tip branch by carrying out this procedure (application of step 4), until an S_k containing only one element is obtained. However, when this is not the case, and it is not possible to reduce the number of elements in S_k any further by the application of step 4, it means that $w_{r_i s_i} = w_{r_i s_j}$ for all $r_i \in R_k$ and all $s_i, s_j \in S_k$. Thus at this stage the elements of R_k form a subtree of T, and the elements of R_k and S_k satisfy (12.38).

If G_1 is the graph obtained from G by short-circuiting the branches of R_k, then, as mentioned earlier, G_1 is a separable graph with, say, $G_{11}, G_{12}, \ldots, G_{1q}$

as its nonseparable components. The set S_k now comprises the branches of a spanning tree T_1 of G_1.

In step 5 we choose arbitrarily an element of S_k, say a branch from G_{1i}, transfer this element to the set R_k, and carry out steps 3 and 4 with the row corresponding to this element. As can be readily seen, the surviving elements of the set S_k at the end of these steps are the branches in all the nonseparable components of G_1 except G_{1i}. Repetition of these steps eventually leads to a set containing branches present in one component of G_1 only and therefrom to a tip branch of a required T itself.

The above arguments prove the correctness of the algorithm for identifying a tip branch of the tree realizing the given F matrix.

12.4.3 Augmentation Paths

Let b be a tip branch determined as above. Let F_1 be the matrix obtained by deleting from F the column corresponding to b. If a tree T_1 realizing F_1 is obtained, then the tree T of F can be grown by attaching the tip branch b to T_1 only if the paths in T containing b have a common tip vertex. However, this may not be the case always. For example, consider the F matrix given below:

$$F = \begin{array}{c} \\ 6 \\ 7 \\ 8 \\ 9 \\ 10 \end{array} \begin{array}{c} \begin{array}{ccccc} 1 & 2 & 3 & 4 & 5 \end{array} \\ \begin{bmatrix} 1 & 1 & 1 & 0 & 0 \\ 1 & 0 & 1 & 0 & 1 \\ 0 & 1 & 1 & 0 & 0 \\ 0 & 0 & 1 & 1 & 0 \\ 0 & 0 & 0 & 1 & 1 \end{bmatrix} \end{array}.$$

We can readily identify branch 1 as a tip branch. Then the matrix F_1 obtained by deleting column 1 from F is

$$F_1 = \begin{bmatrix} 1 & 1 & 0 & 0 \\ 0 & 1 & 0 & 1 \\ 1 & 1 & 0 & 0 \\ 0 & 1 & 1 & 0 \\ 0 & 0 & 1 & 1 \end{bmatrix}.$$

It can be seen that F_1 is realizable by the tree T_1 shown in Fig. 12.10*a*. But this tree cannot be grown to obtain the tree realizing F, because the paths corresponding to chords 6 and 7 cannot be fitted in. This does not, however, mean that F is nonrealizable, as one of the alternate realizations of F_1 may lead to a realization of F. In fact the tree T_1' shown in Fig. 12.10*b* also realizes F_1 and can be grown to obtain the tree T (Fig. 12.10*c*) realizing the matrix F.

To overcome the difficulty mentioned above, we prescribe additional paths in T_1, that is, augment F_1 by additional rows such that the tree realizing the

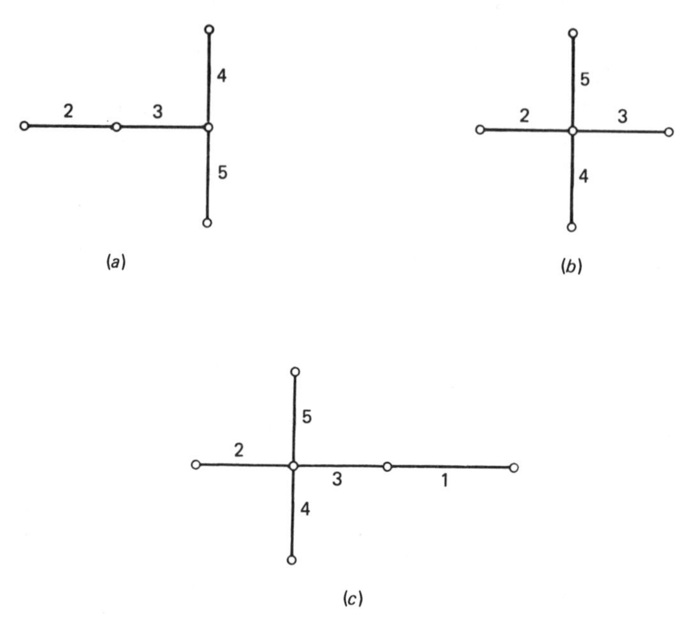

Figure 12.10. (*a*) Tree T_1. (*b*) Tree T_1'. (*c*) Tree T.

augmented matrix definitely turns out to be the complement of the tip branch b in the required realization T.

Let the paths containing the tip branch b as prescribed by F be denoted as $p_1', p_2', \ldots, p_r'$. All these paths are assumed to be distinct, and each is assumed to contain more than one branch. For purposes of our discussion, we define a path to be trivial if there exists another path identical to it, or when it contains only one branch. Thus under the foregoing assumptions, there is no trivial path among $p_1', p_2', \ldots, p_r'$. The complement of the tip branch in each of these paths is also a path in T. Let these paths be $p_1, p_2, \ldots, p_r$. Note that all these paths have a common end vertex in T.

Further any tree realizing F_1 also contains these paths. We now describe a procedure to augment F_1 such that in every tree realizing F_1 the paths $p_1, p_2, \ldots, p_r$ have a common end vertex. From such a tree we can easily grow the required tree T by attaching the branch b at the common end vertex of these paths.

Consider any three paths p_i, p_j, and p_k from the set $p_1, p_2, \ldots, p_r$. If $p_i \oplus p_j = p_k$, then no realization exists in which the paths p_i, p_j, and p_k have a common end vertex. This follows from the fact that p_i, p_j, and $p_i \oplus p_j$ can all be paths in a tree only in two particular ways illustrated in Fig. 12.11. Thus the matrix F is not realizable if the ring sum of any three of the paths in $p_1, p_2, \ldots, p_r$ is an empty set. To check the presence of such a situation, we need to examine separately ${}_rC_3$ sets of paths. Our aim, however, being the realization of F, we examine only $r-2$ of the above sets for consistency, and

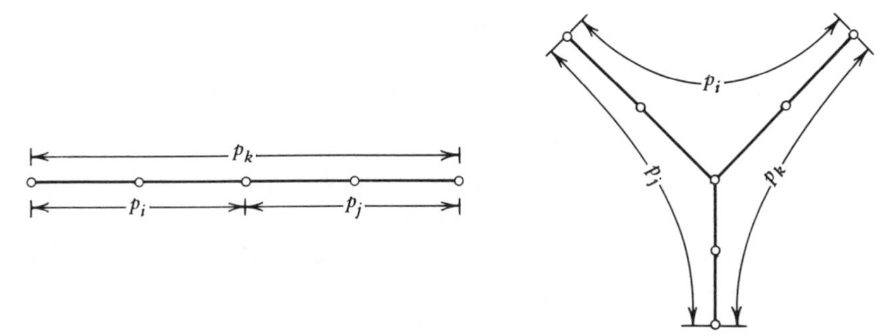

Figure 12.11.

then augment F_1 by $2r-3$ additional rows to ensure that the augmented F_1 turns out to be nonrealizable if F is.

We first make sure that $p_1 \oplus p_2 \neq p_k$, $k=3,4,\ldots,r$. If $p_1 \oplus p_2 = p_k$ for some k, then the matrix F is of course not realizable. After this test, the matrix F_1 is augmented by rows corresponding to the following paths:

1. $p_1 \oplus p_2, p_1 \oplus p_3, \ldots, p_1 \oplus p_r$.
2. $p_2 \oplus p_3, p_2 \oplus p_4, \ldots, p_2 \oplus p_r$.

The additional paths specified in 1 impose the condition that each one of the paths $p_2, p_3, \ldots, p_r$ shares a common end vertex with p_1 in the tree realizing the augmented F_1 matrix because $p_1 \oplus p_k$ is a path only if p_1 and p_k have a common end vertex. Similarly, the stipulation of the existence of the second set of paths forces each one of the paths $p_3, p_4, \ldots, p_r$ to have a common end vertex with p_2 as well. As no circuit can exist in T_1 and as the preliminary test ensuring $p_1 \oplus p_2 \neq p_k$, $k=3,4,\ldots,r$, rules out the existence in T_1 of paths like those shown in Fig. 12.11, it follows that the common end vertex of p_1 and p_2 is also an end vertex of every path p_k, $k=3,4,\ldots,r$. Thus if the augmented F_1 is realizable, then in any realization T_1 of this matrix the paths $p_1, p_2, \ldots, p_r$ have a common end vertex. In such a case T_1 can be grown to the tree realizing F. On the other hand, if the augmented F_1 matrix is nonrealizable, then it means either of the following:

1. F_1 is nonrealizable, or
2. There exists no realization of F_1 in which the paths $p_1, p_2, \ldots, p_r$ have a common end vertex.

In both these cases, F is also not realizable.

The procedure for the determination of a tree T realizing a given matrix F may now be formulated.

12.4.4 Construction of the Tree

We first apply on F the preamble operations repeatedly until we obtain a matrix F' which is not reducible any further. When a tree T' realizing F' is obtained, the tree realizing F can be grown from it.

We next identify a tip branch of T'. The preliminary check on the consistency of paths is next carried out. The column corresponding to the identified tip branch is deleted from F', and the resultant matrix is then augmented by additional rows as described earlier. The preamble operations on rows is now effected on this augmented matrix, and the resulting matrix is designated as F_1'.

We now identify a tip branch of the tree T_1 that would realize F_1' and proceed on similar lines as above to find F_2'. This procedure is repeated until a matrix F_k' containing only two columns is obtained. In the event of F', and hence F, being not realizable, this fact would have been detected at some stage in the reduction process.

If F' is realizable, then starting with T_k, the tree realizing F_k', the tips are attached in the reverse order of the sequence in which they are removed. Application of the preamble ensures that there exist two paths containing the tip branch, and hence the position of the tip can be uniquely fixed at each stage. Thus we can construct T' and hence T. By adding to the latter, chords corresponding to the rows of F, the graph G is realized.

It may be mentioned that it is not essential to reduce the given F matrix to the matrix F' by the application of the preamble. The reduction process can be directly applied to F itself, provided that the paths $p_1', p_2', \ldots, p_r'$ considered at each stage are distinct, and each one of them contains more than one branch. If during the process of growing T from T_k two different paths containing the tip branch are not available at any stage, the tip branch can be attached at either end of the relevant path.

We illustrate with two examples the discussions of this section.

First, let it be required to realize the matrix

$$F = \begin{array}{c} \begin{matrix} 1 & 2 & 3 & 4 \end{matrix} \\ \begin{bmatrix} 1 & 0 & 1 & 1 \\ 0 & 1 & 1 & 1 \\ 0 & 1 & 1 & 0 \\ 1 & 0 & 0 & 1 \end{bmatrix} \end{array}.$$

It is seen that the given matrix cannot be reduced by the application of the preamble. We then obtain

$$F^tF = \begin{array}{c} \\ 1 \\ 2 \\ 3 \\ 4 \end{array} \begin{array}{c} \begin{matrix} 1 & 2 & 3 & 4 \end{matrix} \\ \begin{bmatrix} 2 & 0 & 1 & 2 \\ 0 & 2 & 2 & 1 \\ 1 & 2 & 3 & 2 \\ 2 & 1 & 2 & 3 \end{bmatrix} \end{array}.$$

Considering the first row, we conclude that branch 2 is a tip branch of T. As there are only two paths which include this branch, there is no necessity to check the consistency of paths as envisaged in Section 12.4.3. Deleting column 2, and augmenting the matrix by a row corresponding to $p_2 \oplus p_3$, we obtain

$$F_1\,(\text{augmented}) = \begin{matrix} 1 & 3 & 4 \end{matrix} \\ \begin{bmatrix} 1 & 1 & 1 \\ 0 & 1 & 1 \\ 0 & 1 & 0 \\ 1 & 0 & 1 \\ 0 & 0 & 1 \end{bmatrix}.$$

After deletion of the rows corresponding to trivial paths we get

$$F_1' = \begin{matrix} 1 & 3 & 4 \end{matrix} \\ \begin{bmatrix} 1 & 1 & 1 \\ 0 & 1 & 1 \\ 1 & 0 & 1 \end{bmatrix},$$

for which the W matrix is

$$\begin{matrix} & 1 & 3 & 4 \\ 1 & 2 & 1 & 2 \\ 3 & 1 & 2 & 2 \\ 4 & 2 & 2 & 3 \end{matrix}.$$

Examining the first row, we conclude that branch 3 is a tip of T_1. The tree T_2 realizing F_2, which is obtained by deleting from F_1' the column corresponding to 3, has two branches only, and is shown in Fig. 12.12*a*. To this tree, we attach branch 3, such that $\{1,3,4\}$ and $\{3,4\}$ are paths, and obtain T_1 as shown in Fig. 12.12*b*. Branch 2 is now attached to T_1 such that $\{2,3,4\}$ and $\{2,3\}$ are paths. Thus the tree T shown in Fig. 12.12*c* is obtained.

As a second example, consider next the matrix

$$F = \begin{matrix} 1 & 2 & 3 & 4 \end{matrix} \\ \begin{bmatrix} 1 & 0 & 1 & 0 \\ 1 & 0 & 1 & 1 \\ 1 & 1 & 1 & 1 \\ 1 & 1 & 0 & 1 \\ 1 & 1 & 0 & 0 \end{bmatrix}.$$

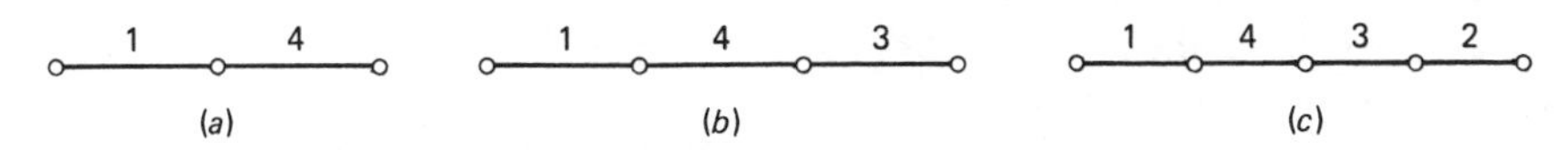

Figure 12.12. (*a*) Tree T_2. (*b*) Tree T_1. (*c*) Tree T.

The corresponding matrix W is now computed as

$$W=\begin{array}{c} \\ 1\\ 2\\ 3\\ 4 \end{array}\begin{array}{c} \begin{array}{cccc} 1 & 2 & 3 & 4 \end{array} \\ \begin{bmatrix} 5 & 3 & 3 & 3\\ 3 & 3 & 1 & 2\\ 3 & 1 & 3 & 2\\ 3 & 2 & 2 & 3 \end{bmatrix} \end{array}.$$

With respect to row 1, the set $S=\{2,3,4\}$. Since this set S is not further reducible, we delete an arbitrary element, say 2, from it. Now we find $w_{2,4}>w_{2,3}$, and hence branch 3 is a tip. Thus $p_1=\{1\}$; $p_2=\{1,4\}$, and $p_3=\{1,2,4\}$. Further $p_1\oplus p_2\neq p_3$. We now delete column 3 from F and augment it with the rows $p_1\oplus p_2$, $p_1\oplus p_3$, and $p_2\oplus p_3$. After removing the rows corresponding to trivial paths, we obtain F_1' as

$$F_1'=\begin{array}{c} \begin{array}{ccc} 1 & 2 & 4 \end{array} \\ \begin{bmatrix} 1 & 0 & 1\\ 1 & 1 & 1\\ 1 & 1 & 0\\ 0 & 1 & 1 \end{bmatrix} \end{array}.$$

The corresponding W matrix is

$$\begin{array}{c} \\ 1\\ 2\\ 4 \end{array}\begin{array}{c} \begin{array}{ccc} 1 & 2 & 4 \end{array} \\ \begin{bmatrix} 3 & 2 & 2\\ 2 & 3 & 2\\ 2 & 2 & 3 \end{bmatrix} \end{array}.$$

The set S with respect to the first row of W consists of 2 and 4. Since this set S is not further reducible, we delete an arbitrary element, say 2, from it. Now 4 is a tip branch since there is only one element in S at this stage. Then we have $p_1=\{1\}$, $p_2=\{1,2\}$, and $p_3=\{2\}$. The ring sum of p_1 and p_2 being equal to p_3, the matrix F is not realizable.

12.5 REALIZATION OF $(n+1)$-NODE RESISTANCE n-PORT NETWORKS—II

We mentioned in Section 12.3 that there are two basic approaches— one due to Cederbaum and the other due to Guillemin—to the problem of realizing a given real symmetric matrix $Y=[y_{ij}]$ as the short-circuit conductance matrix of an $(n+1)$-node resistance n-port network containing no negative conductances. We have discussed in Sections 12.3 and 12.4 the two main steps in Cederbaum's approach.

We now discuss an algorithm due to Rao [12.4] which uses Guillemin's approach. Guillemin's approach, as outlined earlier, consists of the following main steps:

1. Obtain the port configuration T of the n-port network N realizing the matrix Y.
2. Obtain from Y and T the short-circuit conductance matrix Y' of an n-port network described on N, which has a star-tree port configuration in which the star vertex is the negative reference terminal for all the ports (see Section 12.2.4).
3. If the matrix Y' is hyperdominant, then we can easily obtain the required network (see Section 12.2.2); otherwise Y is not realizable.

12.5.1 Identification of Tip Port

The first main step in Rao's algorithm for realizing the Y matrix of an $(n+1)$-node resistance n-port network is the identification of a tip port of the port configuration of the required network. We can easily see that by applying to the matrix $[|y_{ij}|]$ the algorithm described in Section 12.4.2, we can determine such a tip port. For the sake of completeness we give below the steps of this algorithm.

Step 1. Consider any row i of Y. Let m_i be the minimum of the absolute values of the entries in that row, and let S be the set of columns in which the entry at row i is $\pm m_i$. If S contains only one element, then this element corresponds to a tip port. Otherwise let $S=\{s_1, s_2, \ldots, s_k\}$.

Step 2. Consider any other row j of Y, where $j \notin S$. If $|y_{js_p}|>|y_{js_q}|$, for some s_p and s_q contained in S, delete s_p from S. Delete all such s_p with respect to row j.

Step 3. Repeat step 2 with some other row k, where k is not contained in the reduced set S. If the repeated application of this step does not result in a set S containing only one element, then delete any element arbitrarily from the set S at that stage, and repeat steps 2 and 3. Finally S would contain only one element, which can be identified as a tip port.

12.5.2 Realization of a Y-Matrix with No Zero Entries

We now consider the realization of Y-matrices which contain no zero entries. The general case where zero entries are also permitted is treated in [12.4].

A network N realizing a Y-matrix with all nonzero entries is characterized by the fact that there exists a chord spanning every pair of ports of N. If we designate a path in T from one tip vertex to another as a *maximal path*, then for each such maximal path, there exists a chord in N. Further, the conductance of this chord is equal to the magnitude of the transfer conductance between the two tip ports of the maximal path. In the following, each maximal path will be identified and indicated by a sequence of the constituent ports in the order in which they occur with a horizontal line below. If the exact order of all or some of the ports contained in the path is not known, we enclose the designations of the corresponding ports in parentheses. Thus, for example, $\underline{i,(j,k),p,q,(r,s,t)}$ is a maximal path containing i, j, k, p, q, r, s, and t, in which the exact locations of j and k in the second and third positions, and the relative positions of r, s, and t in the last three positions are yet to be determined.

Let port i be identified as a tip port by the procedure outlined earlier. Our aim is to determine the set of maximal paths from the tip vertex of T. Evidently, these ordered maximal paths are sufficient to establish the port tree configuration.

Let m_i be the minimum of the absolute values of the entries in the ith row of Y, and let the transfer conductance y_{ij}, $j \neq i$, among others, if any, be equal to $\pm m_i$. Port j may or may not be a tip port. In the event of its not being a tip port, the ports of the L-set (Section 12.4), which does not contain port i and which has j as its leading port, form a linear subtree of T, and there exists only one chord in N which spans i and the ports of the L-set. Therefore, whether or not j is a tip, there exists in T only one maximal path which contains i and j.

Let y_{ij} be positive (negative); for any k, if the signs of y_{ik} and y_{jk} agree (differ), then k is in the maximal path containing i and j (Theorem 12.2). In this way, all ports that are contained in this maximal path can be determined. Let this path be $\underline{i,(j,k,l,\ldots)}$, where the order of the ports will be determined later. Evidently, none of the ports within the parentheses in the foregoing representation of the maximal path can be a tip port in another maximal path containing i. To aid us in keeping these ports aside from consideration in the immediately following steps, the entries in the ith row of Y against the columns $j, k, l, \ldots$ are enclosed by circles. We now select another entry y_{ip}, $p \neq i$, having the minimum absolute value among the uncircled entries in the ith row and obtain another maximal path, that is, determine all x's for which $y_{ip}y_{xi}y_{xp}$ is positive. Let this path be $\underline{i,(p,q,\ldots)}$. We now encircle the entries $y_{ip}, y_{iq}, \ldots$. The procedure is repeated until all entries of Y in the ith row, except y_{ii}, are encircled. We note that each of the ports $j, p, \ldots$ is present in only one maximal path, while the others may be present in more than one maximal path.

The data needed for the construction of the tree are complete, if we arrange the ports of each of these maximal paths in the order in which they are present in it. To this end, we order the ports in each maximal path in the descending order of the magnitude of their transfer conductances with port i. When more than one port has the same magnitude of transfer conductance with respect to port i, we include those ports in parentheses. Thus when the first stage is completed, we have a set of partially ordered sequences like

$$\underline{i,m,(h,k,n,o),s,(t,u,v),(w,x,j)}\,.$$

Now the ports within each set of parentheses have to be ordered. For this purpose we consider the row corresponding to port m, the port just preceding the parentheses containing h, k, n, and o, and arrange the ports within the parentheses in the descending order of the magnitude of their transfer conductances with respect to port m. Again if some transfer conductances are found to have equal magnitude, we include the respective ports in parentheses and repeat the procedure. The ordering of ports can also be carried out by considering port s, the port immediately after the parentheses, and arranging the ports within the parentheses in increasing order of the magnitude of their transfer conductances with respect to port s. In this way, the port sequence in each maximal path can be determined.

However, in certain cases we may have a situation where the transfer conductances of all ports within parentheses to any port outside have equal magnitude. For example, for a sequence like $\underline{\ldots, s, (p, q, r), t, \ldots}$ we may have

$$|y_{sp}| = |y_{sq}| = |y_{sr}| \qquad \text{and} \qquad |y_{pt}| = |y_{qt}| = |y_{rt}|.$$

In such a case, we extract the square submatrix of Y corresponding to the ports within parentheses and determine its linear-tree port structure separately. Let q, r, p be the order so determined. Then either $\underline{\ldots, s, q, r, p, t, \ldots}$ or $\underline{\ldots, s, p, r, q, t, \ldots}$ may be taken as the ordered sequence, and the two corresponding graphs are 2-isomorphic.

Once the ordering of paths in one maximal path is completed, the ordering of the common branches in other maximal paths is the same as the earlier one for realizable matrices. Advantage can be taken of this fact when ordering the branches in a new path.

We illustrate the above procedure by determining the port-tree configuration of the 10th-order conductance matrix Y given below:

$$Y = \begin{array}{c} \\ 1 \\ 2 \\ 3 \\ 4 \\ 5 \\ 6 \\ 7 \\ 8 \\ 9 \\ 10 \end{array} \begin{array}{c} \begin{array}{cccccccccc} 1 & 2 & 3 & 4 & 5 & 6 & 7 & 8 & 9 & 10 \end{array} \\ \left[\begin{array}{cccccccccc} 67 & 20 & 10 & 15 & 45 & 17 & 23 & 33 & 15 & 26 \\ 20 & 30 & 3 & -6 & 23 & 5 & 12 & 8 & -3 & 29 \\ 10 & 3 & 30 & 2 & 5 & -4 & -1 & 11 & 5 & 3 \\ 15 & -6 & 2 & 45 & 30 & 9 & 4 & 11 & -15 & -12 \\ 45 & 23 & 5 & 30 & 79 & 14 & 16 & 25 & -18 & 29 \\ 17 & 5 & -4 & 9 & 14 & 30 & -7 & 24 & 3 & 5 \\ 23 & 12 & -1 & 4 & 16 & -7 & 49 & -8 & 7 & 12 \\ 33 & 8 & 11 & 11 & 25 & 24 & -8 & 42 & 8 & 14 \\ 15 & -3 & 5 & -15 & -18 & 3 & 7 & 8 & 36 & -3 \\ 26 & 29 & 3 & -12 & 29 & 5 & 12 & 14 & -3 & 41 \end{array}\right] \end{array}.$$

1. Considering the first row, we determine port 3 to be a tip port.
2. In row 3, the minimum magnitude entry is in the 7th column position, and y_{37} is negative. Scanning down columns 3 and 7, we note that for $k=3$, 7, and 8, only y_{k3} and y_{k7} are of opposite signs.

Thus $\underline{3, (7,8)}$ is a maximal path. We encircle the entries at columns 7 and 8 in the third row, which now appears as

$$\begin{bmatrix} 10 & 3 & 30 & 2 & 5 & -4 & \textcircled{-1} & \textcircled{11} & 5 & 3 \end{bmatrix}.$$

3. Among the uncircled entries in row 3, y_{34} has the minimum magnitude, and it is positive. Scanning through columns 3 and 4, we note that for $k=1$, 3, 4, 5, and 8, the entries y_{k3} and y_{k4} are of like sign, and $\underline{3, (1,4,5,8)}$ is another maximal path. The entries y_{31}, y_{34}, and y_{35} are now encircled. Similarly, we obtain the other maximal paths as $\underline{3, (2,1,5,8,10)}$; $\underline{3,6}$; and $\underline{3, (9,1,8)}$.
4. Consider the path $\underline{3, (2,1,5,8,10)}$. Arranging the ports within parentheses in decreasing order of the magnitude of their transfer conductances, with respect to port 3, we obtain

$$\underline{3,8,1,5,(2,10)}\,.$$

Now considering the 5th row of Y, we find $|y_{5,10}|>|y_{5,2}|$. Thus $\underline{3,8,1,5,10,2}$ is an ordered maximal path.
5. Since the other paths have common branches with this path, they can be easily arranged. Thus the other ordered paths are $\underline{3,8,7}$; $\underline{3,8,1,5,4}$; and $\underline{3,8,1,9}$.
6. The port tree T can be easily determined now and is shown in Fig. 12.13*a*.
7. The orientation of any one of the ports, say port 1, can be fixed arbitrarily. The orientations of the remaining ports are then obtained using Theorem 12.1. The resulting port orientations are shown in Fig. 12.13*a*.
8. Let a star tree T^* having the same vertex set as T be chosen as in Fig. 12.13*b*.

The matrix K relating the port voltage vectors V_p and V_p^* is given below. (Note that $V_p=KV_p^*$.)

$$K=\begin{bmatrix} 0 & 0 & 1 & -1 & 0 & 0 & 0 & 0 & 0 & 0 \\ 0 & 0 & 0 & 0 & 0 & 1 & -1 & 0 & 0 & 0 \\ 1 & -1 & 0 & 0 & 0 & 0 & 0 & 0 & 0 & 0 \\ 0 & 0 & 0 & 0 & 1 & 0 & 0 & 0 & 0 & -1 \\ 0 & 0 & 0 & 1 & -1 & 0 & 0 & 0 & 0 & 0 \\ 0 & -1 & 0 & 0 & 0 & 0 & 0 & 1 & 0 & 0 \\ 0 & 0 & -1 & 0 & 0 & 0 & 0 & 0 & 0 & 0 \\ 0 & 1 & -1 & 0 & 0 & 0 & 0 & 0 & 0 & 0 \\ 0 & 0 & 0 & 1 & 0 & 0 & 0 & 0 & -1 & 0 \\ 0 & 0 & 0 & 0 & 1 & -1 & 0 & 0 & 0 & 0 \end{bmatrix}.$$

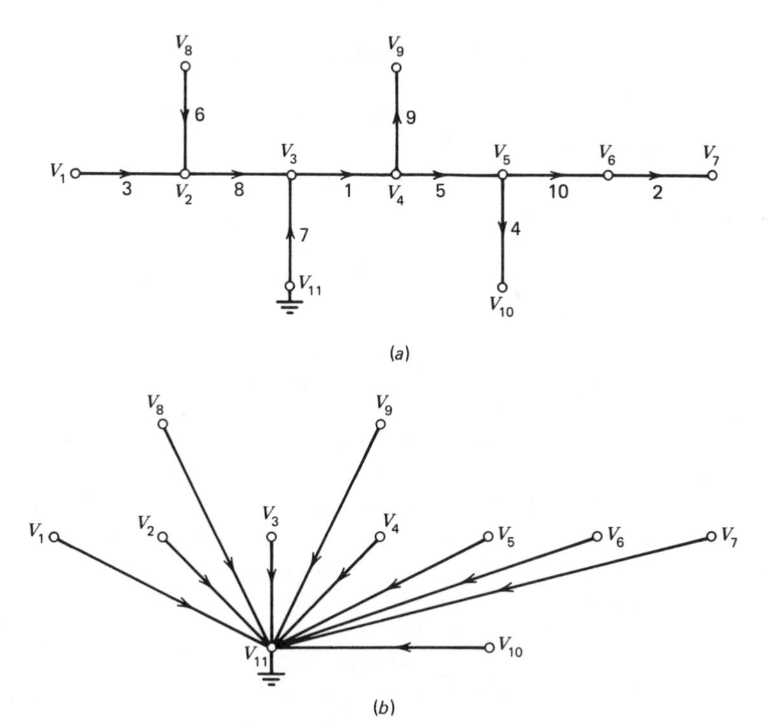

Figure 12.13. (*a*) Tree T. (*b*) Tree T^*.

The short-circuit conductance matrix Y^* with respect to T^* is given by

$$Y^* = K^t Y K$$

$$= \begin{bmatrix} 30 & -15 & 0 & 0 & 0 & 0 & -3 & -4 & -5 & -2 \\ -15 & 24 & -1 & 0 & 0 & -6 & 0 & -2 & 0 & 0 \\ 0 & -1 & 30 & -7 & -4 & 0 & 0 & 0 & 0 & 0 \\ 0 & 0 & -7 & 26 & -16 & 0 & 0 & 0 & -3 & 0 \\ 0 & 0 & -4 & -16 & 23 & 0 & 0 & 0 & 0 & -3 \\ 0 & -6 & 0 & 0 & 0 & 13 & -1 & 0 & 0 & -6 \\ -3 & 0 & 0 & 0 & 0 & -1 & 30 & -5 & -3 & -6 \\ -4 & -2 & 0 & 0 & 0 & 0 & -5 & 30 & -3 & -9 \\ -5 & 0 & 0 & -3 & 0 & 0 & -3 & -3 & 36 & -15 \\ -2 & 0 & 0 & 0 & -3 & -6 & -6 & -9 & -15 & 45 \end{bmatrix}.$$

The matrix Y^* is hyperdominant. The conductances of the 10-port network realizing Y can be calculated from Y^*, and they are given by the following relation:

$$g_{ij} = -y_{ij}, \qquad 1 \leqslant i, j \leqslant 10,$$

$$g_{i,11} = y_{ii} - \sum_{\substack{j=1 \\ j \neq i}}^{10} |y_{ij}|, \qquad 1 \leqslant i \leqslant 10.$$

12.6 FURTHER READING

Studies in the area of resistance n-port networks have been mainly concerned with the following topics:

1. Y-matrices of n-port networks on $n+1$ nodes, that is, n-port networks of rank n.
2. Y-matrices of n-port networks on more than $n+1$ nodes.
3. Z-matrices of n-port networks of nullity n.

Cederbaum's decomposition algorithm given in Section 12.3 and Rao's algorithm given in Section 12.4 for realizing a cutset or a circuit matrix together completely solve the problem of constructing an n-port network on $n+1$ nodes having a specified Y matrix. Rao [12.4] also has given an alternative algorithm for this problem. This algorithm, as we mentioned in Section 12.5, is based on Guillemin's approach which involves constructing the port configuration from a given Y-matrix. For an algorithm to realize a Y-matrix having no nonzero entries see Boesch and Youla [12.5]. Some of the early papers on $(n+1)$-node resistance n-port networks include Biorci and Civalleri [12.6], and Halkias, Cederbaum, and Kim [12.7].

Guillemin [12.8] has given a general approach to study the problem of realizing a Y-matrix by an n-port network having more than $n+1$ nodes. Using this approach, many attempts have subsequently been made to discover the properties of the Y-matrices of such networks. Not many significant results have been obtained, except in the special case of $(n+2)$-node networks. See Swaminathan and Frisch [12.9], Biorci and Civalleri [12.10], and Reddy and Thulasiraman [12.11].

As we observed in Section 12.3, Cederbaum's algorithm can be applied to decompose the Z-matrix of an n-port network of nullity n. This and the algorithm in Section 12.4 together completely solve the problem of realizing the open-circuit resistance matrix of an n-port network of nullity n. Alternative approaches to this problem are discussed in Brown [12.12] and Eswaran and Murti [12.13]. See also Eswaran and Murti [12.14] and Frisch and Kim [12.15], where a relationship between resistance networks and communication nets is discussed.

Cederbaum [12.16] introduced the concepts of modified cutset and circuit matrices. For some applications of these matrices in the study of the n-port problem see Reddy, Murti, and Thulasiraman [12.17] and Lempel and Cederbaum [12.18].

Many other interesting questions relating to the n-port resistance network problem have been discussed in the literature. For example, see Lempel and Cederbaum [12.19], Prigozy and Weinberg [12.20], and Naidu, Reddy, and Thulasiraman [12.21].

See Tutte [12.22], [12.23] and Mayeda [12.24], [12.25] for the problem of realizing cutset and circuit matrices.

Kim and Chien [12.26] discuss several results in the area of resistance n-port networks, and also an application of the problem of circuit matrix realization in the synthesis of switching networks. The concepts from the area of resistance n-port networks can be used to advantage to realize state matrices of RLC networks; see Swamy and Thulasiraman [12.27].

12.7 EXERCISES

12.1 Consider an n-port network N. Assume that there is no circuit consisting only of ports. If I_p is the port current vector and I is the current vector associated with the nonport elements of N, then show that there exists a matrix M whose entries are 1 or -1 or 0 such that $I_p = MI$.

Hint Use Theorem 10.31 (Chua and Green [12.28]).

12.2 Repeat Exercise 12.1 replacing circuits by cutsets, I_p by V_p, and I by V, where V_p and V are the vectors associated with the ports and nonport elements of N, respectively.

12.3 Using Cederbaum's approach obtain the resistance 6-port network having the following short-circuit conductance matrix:

$$\begin{bmatrix} 3 & 2 & 0 & 1 & -1 & 2 \\ 2 & 5 & -2 & 0 & 2 & 2 \\ 0 & -2 & 9 & 0 & -3 & 0 \\ 1 & 0 & 0 & 9 & -1 & -1 \\ -1 & 2 & -3 & -1 & 8 & 0 \\ 2 & 2 & 0 & -1 & 0 & 8 \end{bmatrix}.$$

12.4 Realize the following as a tree-path matrix:

$$\begin{bmatrix} 1 & 1 & 0 & 0 & 0 & 1 & 0 & 0 & 1 & 1 & 0 & 0 & 0 & 1 \\ 0 & 0 & 1 & 1 & 0 & 1 & 0 & 0 & 1 & 0 & 1 & 0 & 0 & 0 \\ 0 & 0 & 1 & 0 & 1 & 1 & 0 & 1 & 0 & 0 & 0 & 0 & 0 & 0 \\ 0 & 1 & 0 & 0 & 0 & 1 & 1 & 1 & 0 & 0 & 0 & 1 & 0 & 0 \\ 0 & 0 & 1 & 0 & 1 & 1 & 0 & 0 & 1 & 0 & 1 & 0 & 0 & 0 \\ 1 & 1 & 1 & 0 & 1 & 0 & 0 & 0 & 0 & 0 & 0 & 0 & 0 & 0 \\ 0 & 0 & 1 & 1 & 0 & 1 & 0 & 1 & 0 & 0 & 0 & 0 & 0 & 0 \\ 0 & 1 & 0 & 0 & 0 & 0 & 1 & 0 & 0 & 0 & 0 & 0 & 1 & 0 \end{bmatrix}.$$

12.5 Let $Z = [z_{ij}]$ be the open-circuit resistance matrix of an n-port network N of nullity n. Show that Z satisfies the following relationship if the ports of N form a spanning star tree of $\hat{N}$, the augmented network: For every three indices i, j, and k, either all the three entries z_{ij}, z_{jk}, and z_{ki} are equal, or the smaller two are equal (Eswaran and Murti [12.14]).

12.6 Let the matrices Q_1, Q_2, Y_{12}, Y_{22}, etc., associated with an n-port network N, be defined as in Section 12.1. The matrix $Q_1 - Y_{12}Y_{22}^{-1}Q_2$ is called the *modified cutset matrix* of N. Prove the following:

(a) $Y = Q_m Y_e Q_1^t$, where Q_m is the modified cutset matrix of N.

(b) A matrix Q_m is the modified cutset matrix of an n-port network N if and only if $Q_m Y_e Q_2^t = 0$.

12.7 Let N be the parallel combination of two n-port networks N_1 and N_2 having Y_1 and Y_2 for their short-circuit admittance matrices. Show that the short-circuit admittance matrix of N is equal to $Y_1 + Y_2$ if and only if both N_1 and N_2 have the same modified cutset matrix (Lempel and Cederbaum [12.18]).

12.8 Using the results of Exercises 12.6 and 12.7 construct a 4-port network whose short-circuit admittance matrix is equal to the short-circuit admittance matrix of the 4-port network shown in Fig. 12.14, where all the conductances are in Siemens.

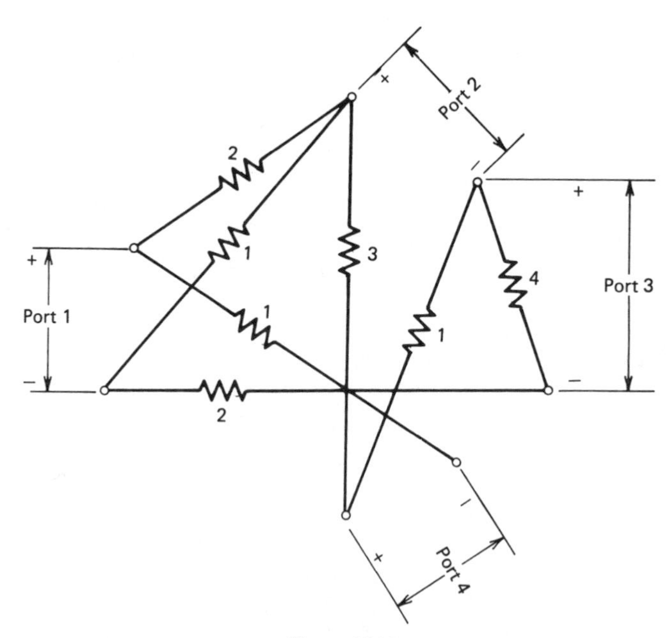

Figure 12.14.

12.9 Let the matrices B_1, B_2, Z_{12}, Z_{22}, etc., associated with an n-port network N, be defined as in Section 12.1. Then the matrix $(B_1 - Z_{12}Z_{22}^{-1}B_2)$ is called the *modified circuit matrix* of N. Prove the following:

(a) $Z = B_m Z_e B_1^t$, where B_m is the modified circuit matrix of N.

(b) A matrix B_m is the modified circuit matrix of an n-port network N if and only if $B_m Z_e B_2^t = 0$.

12.10 Consider two n-port networks N_1 and N_2 having identical port configurations and edge configurations and orientations. Let Z_1 and Z_2 be the open-circuit impedance matrices of N_1 and N_2. Let us construct a third n-port network N_3 having the same edge and port configurations and orientations as N_1 and N_2, but having the impedance of each edge as the sum of the impedances of the corresponding edges of N_1 and N_2. The network N_3 is referred to as the *pseudo-series combination* of N_1 and N_2. (Two networks N_1 and N_2 and their pseudo-series combination N_3 are shown in Fig. 12.15.) Show that the open-circuit impedance matrix of N_3 is equal to $Z_1 + Z_2$ if and only if both N_1 and N_2

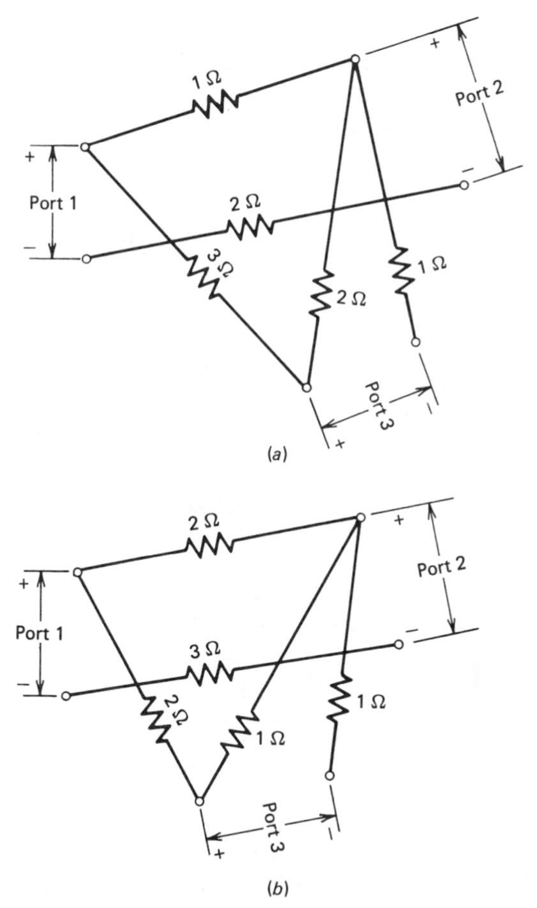

Figure 12.15. (*a*) N_1. (*b*) N_2. (*c*) Pseudo-series combination of N_1 and N_2.

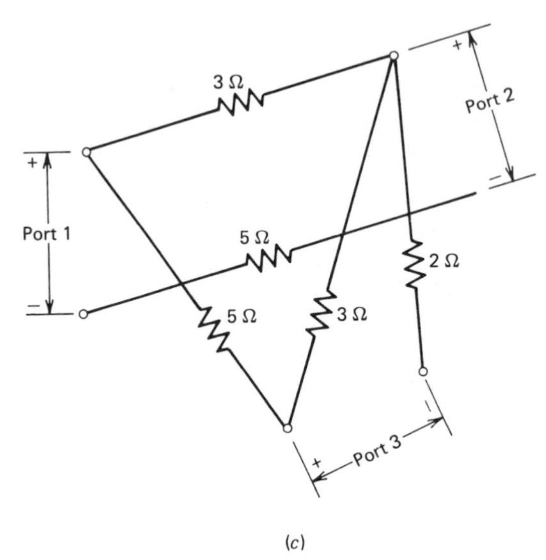

(c)

Figure 12.15. *(continued)*

have the same modified circuit matrix (Thulasiraman and Murti [12.29]).

12.11 Using the result of Exercise 12.9b, construct a 4-port network whose modified circuit matrix is the same as that of the network shown in Fig. 12.14.

12.12 Given an n-port network N, using the results of Exercises 12.9 and 12.10, try to construct a new 4-port network whose open-circuit impedance matrix is equal to that of the network shown in Fig. 12.14. Discuss any difficulties you may encounter.

12.8 REFERENCES

12.1 I. Cederbaum, "Applications of Matrix Algebra to Network Theory," *IRE Trans. Circuit Theory* (special supplement), Vol. CT-6, 127–137 (1959).

12.2 L. Weinberg, *Network Analysis and Synthesis*, McGraw-Hill, New York, 1962.

12.3 E. A. Guillemin, "On the Analysis and Synthesis of Single-Element Kind Networks," *IRE Trans. Circuit Theory*, Vol. CT-7, 303–312 (1960).

12.4 V. V. Bapeswara Rao, "The Tree-Path Matrix of a Network and Its Applications," *Ph.D. Thesis*, Dept. of Electrical Engineering, Indian Institute of Technology, Madras, India, 1970.

12.5 F. T. Boesch and D. C. Youla, "Synthesis of $(n+1)$-Node Resistor n-Ports," *IEEE Trans. Circuit Theory*, Vol. CT-12, 515–520 (1965).

12.6 G. Biorci and P. P. Civalleri, "On the Synthesis of Resistive N-Port Networks," *IRE Trans. Circuit Theory*, Vol. CT-8, 22–28 (1961).

12.7 C. C. Halkias, I. Cederbaum, and W. H. Kim, "Synthesis of Resistive n-Ports with $(n+1)$ Nodes," *IRE Trans. Circuit Theory*, Vol. CT-9, 69–73, (1962).

12.8 E. A. Guillemin, "On the Realization of an n-th Order G-Matrix," *IRE Trans. Circuit Theory*, Vol. CT-8, 318–323 (1961).

12.9 K. R. Swaminathan and I. T. Frisch, "Necessary Conditions for the Realizability of n-Port Resistive Networks with more than $(n+1)$ Nodes," *IEEE Trans. Circuit Theory*, Vol. CT-12, 520–527 (1965).

12.10 G. Biorci and P. P. Civalleri, "Analysis of Resistive N-Port Networks Based on $(n+2)$ Nodes," in *Aspects of Network and System Theory* (Eds. R. E. Kalman and N. DeClaris), Holt, Rinehart, Winston, New York, 1971.

12.11 P. Subbarami Reddy and K. Thulasiraman, "Synthesis of $(n+2)$-Node Resistive n-Port Networks," *IEEE Trans. Circuit Theory*, Vol. CT-19, 20–25 (1972).

12.12 D. P. Brown, "N-Port Synthesis of N-order Positive Entry Resistance Matrices," *J. Franklin Inst.*, Vol. 284, 26–38 (1967).

12.13 C. Eswaran and V. G. K. Murti, "Realization of Positive Entry Resistance Matrices," *AEÜ*, Vol. 29, 212–216 (1975).

12.14 C. Eswaran and V. G. K. Murti, "On a Relationship between Terminal Capacity and Impedance Matrices," *IEEE Trans. Circuits and Sys.*, Vol. CAS-21, 732–734 (1974).

12.15 I. T. Frisch and W. H. Kim, "n-Port Resistive Networks and Communication Nets," *IRE Trans. Circuit Theory*, Vol. CT-8, 493–496 (1961).

12.16 I. Cederbaum, "On Equivalence of Resistive N-Port Networks," *IEEE Trans. Circuit Theory*, Vol. CT-12, 338–344 (1965).

12.17 P. Subbarami Reddy, V. G. K. Murti, and K. Thulasiraman, "Realization of Modified Cutset Matrix and Applications," *IEEE Trans. Circuit Theory*, Vol. CT-17, 475–486 (1970).

12.18 A. Lempel and I. Cederbaum, "Parallel Interconnection of n-port Networks," *IEEE Trans. Circuit Theory*, Vol. CT-14, 274–279 (1967).

12.19 A. Lempel and I. Cederbaum, "Terminal Configurations of n-Port Networks," *IEEE Trans. Circuit Theory*, Vol. CT-15, 51–53 (1968).

12.20 S. Prigozy and L. Weinberg, "Realization of Fourth-Order Singular and Quasi-Singular Resistance and Conductance Matrices," *IEEE Trans. Circuits and Sys.*, Vol. CAS-23, 245–253 (1976).

12.21 M. G. Govindarajulu Naidu, P. Subbarami Reddy, and K. Thulasiraman, "$(n+2)$-Node Resistive n-Port Realizability of Y-Matrices," *IEEE Trans. Circuits and Sys.*, Vol. CAS-23, 254–261 (1976).

12.22 W. T. Tutte, "An Algorithm for Determining whether a Given Binary Matroid is Graphic," *Proc. Am. Math. Soc.*, Vol. 11, 905–917 (1960).

12.23 W. T. Tutte, "From Matrices to Graphs," *Can. J. Math.*, Vol. 56, 108–127 (1964).

12.24 W. Mayeda, "Necessary and Sufficient Conditions for the Realizability of Cutset and Circuit Matrices," *IRE Trans. Circuit Theory*, Vol. CT-7, 79–81 (1960).

12.25 W. Mayeda, "A Proof of Tutte's Realizability Condition," *IEEE Trans. Circuit Theory*, Vol. CT-17, 506–511 (1970).

12.26 W. H. Kim and R. T. Chien, *Topological Analysis and Synthesis of Communication Networks*, Columbia Univ. Press, New York, 1962.

12.27 M. N. S. Swamy and K. Thulasiraman, "Realization of the A–Matrix of RLC Networks," *IEEE Trans. Circuit Theory*, Vol. CT-19, 515–518 (1972).

12.28 L. O. Chua and D. N. Green, "Graph-Theoretic Properties of Dynamic Nonlinear Networks," *IEEE Trans. Circuits and Sys.*, Vol. CAS-23, 292–302 (1976).

12.29 K. Thulasiraman and V. G. K. Murti, "Pseudo-Series Combination of n-Port Networks," *Proc. IEEE*, Vol. 56, 1143–1144 (1968).

Chapter 13

Network Functions and Network Sensitivity

In this chapter we first derive formulas for network functions in terms of admittances associated with certain subgraphs of a given network. Such formulas, called topological formulas, were first given in 1847 by Kirchhoff in terms of resistances and later in 1892 by Maxwell in terms of admittances. Our development in this chapter is based on the indefinite admittance matrix, and we obtain topological formulas by considering network functions as ratios of appropriate cofactors of the indefinite admittance matrix. Thus many of the results to be presented here would follow easily from those in Chapter 6.

In the latter part of the chapter we describe a method for computing sensitivities of network functions. This method is based on the concept of the adjoint of a network and Tellegen's theorem (Theorem 11.3).

13.1 TOPOLOGICAL FORMULAS FOR RLC NETWORKS WITHOUT MUTUAL INDUCTANCES

In this section we derive topological formulas for RLC networks without mutual inductances. The node-admittance matrix is the starting point for the development of these formulas.

Consider first a 1-port RLC network N without mutual inductances. Let the network N have $n+1$ nodes denoted by $0, 1, 2, \ldots, n$, and let the nodes 1 and 0 be, respectively, the positive and negative reference terminals of the port (Fig. 13.1). Assume that all the initial conditions (capacitor voltages and

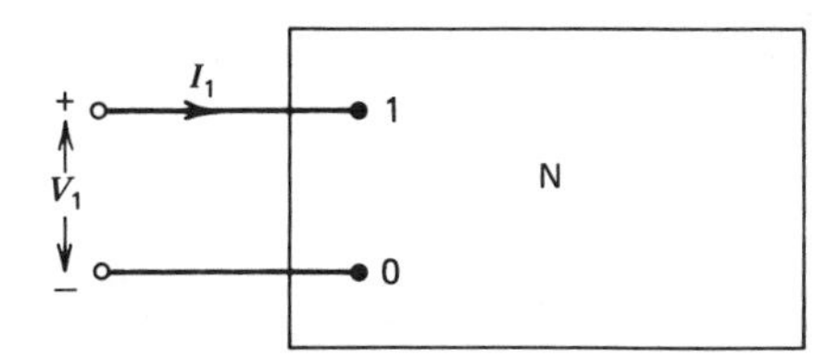

Figure 13.1. A 1-port network.

inductor currents) in N are equal to zero. Further, the voltage and current variables are all Laplace transforms of the complex frequency variable. Let us now excite the network by connecting a current source of value I_1 across the port. If $V_1, V_2, \ldots, V_n$ denote the voltages of the nodes $1,2,\ldots, n$ with respect to node 0, and Y denotes the node-admittance matrix of N with node 0 as reference, then the node equations for N are

$$YV=I, \tag{13.1}$$

where

$$V=\begin{bmatrix} V_1 \\ V_2 \\ \vdots \\ V_n \end{bmatrix}$$

and

$$I=\begin{bmatrix} I_1 \\ 0 \\ 0 \\ \vdots \\ 0 \end{bmatrix}.$$

Solving (13.1) for V_1, we get

$$V_1=\frac{\Delta_{11}}{\Delta}I_1,$$

where

$$\Delta=\det Y$$

and

$$\Delta_{11}=(1,1) \text{ cofactor of } Y.$$

So the driving-point impedance z of N is given by

$$z=\frac{V_1}{I_1}=\frac{\Delta_{11}}{\Delta}, \tag{13.2}$$

and the driving-point admittance y of N is given by

$$y=\frac{1}{z}=\frac{\Delta}{\Delta_{11}}. \tag{13.3}$$

To derive the topological formulas for z and y we need to express Δ_{11} and Δ in terms of appropriate quantities associated with certain subgraphs of N. This can be done easily as shown below.

In the following, the *admittance product* of a subgraph of N will refer to the product of the admittances associated with the edges of the subgraph. If a subgraph has no edges, then its admittance product is defined as equal to 1. In a similar manner the *impedance product* of a subgraph of N is defined.

With these definitions, let

W = Sum of the admittance products of all spanning trees of N.
$W_{i,j}$ = Sum of the admittance products of all spanning 2-trees $T_{i,j}$ of N.
$C(W)$ = Sum of the impedance products of all cospanning trees of N.
$C(W_{i,j})$ = Sum of the impedance products of the complements of all the spanning 2-trees $T_{i,j}$ of N. (13.4)

If $z_1, z_2, \ldots, z_m$ are the impedances of the edges of N, then clearly,

$$C(W)=W\prod_{i=1}^{m} z_i \tag{13.5}$$

and

$$C(W_{i,j})=W_{i,j}\prod_{i=1}^{m} z_i. \tag{13.6}$$

Consider now the node-admittance matrix of N. If we regard N as a weighted graph with the admittances representing the weights of the corresponding edges, then we can see (Exercise 6.16) that

$$\Delta=\det Y=W \tag{13.7}$$

and

$$\Delta_{ij}=W_{ij,0}. \tag{13.8}$$

Using (13.5) through (13.8) in (13.2) and (13.3), we get the following theorem.

THEOREM 13.1. Let z and y denote, respectively, the driving-point impedance and the driving-point admittance of a 1-port RLC network without mutual inductances. If 1 and 0 are the terminals of the port of the network, then

1. $z=\dfrac{W_{1,0}}{W}=\dfrac{C(W_{1,0})}{C(W)}$.

2. $y=\dfrac{W}{W_{1,0}}=\dfrac{C(W)}{C(W_{1,0})}$. ■

We next derive topological formulas for the open-circuit impedance and short-circuit admittance functions of a 2-port RLC network N (Fig. 13.2) without mutual inductances. Again assume that all the initial conditions in N are equal to zero. If the ports of N are excited by current sources of values I_1 and I_2, then the node equations of N can be written as

$$YV=I,$$

where

$$I=\begin{bmatrix} I_1 \\ I_2 \\ -I_2 \\ 0 \\ \vdots \\ 0 \end{bmatrix}.$$

Solving for the node voltages V_1, V_2, and V_3, we get

$$V_1=\frac{1}{\Delta}(\Delta_{11}I_1+\Delta_{21}I_2-\Delta_{31}I_2),$$

$$V_2=\frac{1}{\Delta}(\Delta_{12}I_1+\Delta_{22}I_2-\Delta_{32}I_2),$$

$$V_3=\frac{1}{\Delta}(\Delta_{13}I_1+\Delta_{23}I_2-\Delta_{33}I_2).$$

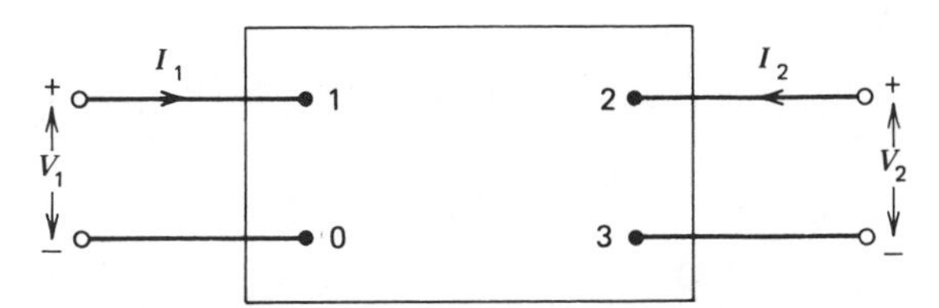

Figure 13.2. A 2-port network.

From the above relations we get

$$\begin{bmatrix} V_1 \\ V_2 - V_3 \end{bmatrix} = \frac{1}{\Delta}\begin{bmatrix} \Delta_{11} & \Delta_{21}-\Delta_{31} \\ \Delta_{12}-\Delta_{13} & \Delta_{22}+\Delta_{33}-\Delta_{32}-\Delta_{23} \end{bmatrix}\begin{bmatrix} I_1 \\ I_2 \end{bmatrix}. \tag{13.9}$$

Note that since Y is symmetric, we have

$$\Delta_{ij} = \Delta_{ji}.$$

Since V_1 is the voltage across port 1 and $V_2 - V_3$ is the voltage across port 2, the coefficient matrix in (13.9) is equal to the open-circuit impedance matrix Z_{oc} of N. Thus,

$$Z_{oc} = \frac{1}{\Delta}\begin{bmatrix} \Delta_{11} & \Delta_{12}-\Delta_{13} \\ \Delta_{12}-\Delta_{13} & \Delta_{22}+\Delta_{33}-2\Delta_{23} \end{bmatrix}. \tag{13.10}$$

To express the elements of Z_{oc} in terms of appropriate admittance products of N, first observe that

$$\Delta_{11} = W_{1,0}. \tag{13.11}$$

We also have from (13.8)

$$\Delta_{12} - \Delta_{13} = W_{12,0} - W_{13,0}. \tag{13.12}$$

Since each spanning 2-tree $T_{12,0}$ is either a spanning 2-tree $T_{12,30}$ or a spanning 2-tree $T_{123,0}$, we get

$$W_{12,0} = W_{12,30} + W_{123,0}. \tag{13.13}$$

Similarly,

$$W_{13,0} = W_{13,20} + W_{123,0}. \tag{13.14}$$

Using (13.13) and (13.14) in (13.12), we get

$$\Delta_{12} - \Delta_{13} = W_{12,30} - W_{13,20}. \tag{13.15}$$

By a similar reasoning,

$$\begin{aligned} \Delta_{22}+\Delta_{33}-2\Delta_{23} &= W_{2,0}+W_{3,0}-2W_{23,0} \\ &= W_{23,0}+W_{2,30}+W_{23,0}+W_{3,20}-2W_{23,0} \\ &= W_{2,30}+W_{3,20} \\ &= W_{2,3}. \end{aligned} \tag{13.16}$$

Using (13.15) and (13.16) in (13.10), we get the following theorem.

THEOREM 13.2. Let N be an RLC 2-port network without mutual inductances. Let the positive and negative reference terminals of the ports of N be as shown in Fig. 13.2. Then the open-circuit impedance matrix Z_{oc} of N is given by

$$Z_{oc}=\frac{1}{W}\begin{bmatrix} W_{1,0} & W_{12,30}-W_{13,20} \\ W_{12,30}-W_{13,20} & W_{2,3} \end{bmatrix}. \quad \blacksquare$$

It is clear from (13.5) and (13.6) that the elements of Z_{oc} can also be expressed in terms of impedance products of appropriate cospanning trees and cospanning 2-trees.

The short-circuit admittance matrix Y_{sc} can be obtained by inverting Z_{oc}. Thus

$$Y_{sc}=\frac{1}{\Delta_{11,22}+\Delta_{11,33}-2\Delta_{11,23}}\begin{bmatrix} \Delta_{22}+\Delta_{33}-2\Delta_{23} & \Delta_{13}-\Delta_{12} \\ \Delta_{13}-\Delta_{12} & \Delta_{11} \end{bmatrix}, \quad (13.17)$$

which is obtained using the identity

$$\Delta_{ii,jk}=\frac{1}{\Delta}(\Delta_{ii}\Delta_{jk}-\Delta_{ji}\Delta_{ik}). \quad (13.18)$$

Note that $\Delta_{ii,jk}$ is the second-order cofactor of Y with respect to its (i,i) and (j,k) elements, and it is given by

$$\Delta_{ii,jk}=\operatorname{sgn}(i-j)\operatorname{sgn}(i-k)(-1)^{j+k}\det(Y_{ii,jk}) \quad (13.19)$$

where

1. $\operatorname{sgn}(r-s)=\begin{cases} 1, & \text{if } (r-s)>0 \\ -1, & \text{if } (r-s)<0; \text{ and} \end{cases}$

2. $Y_{ii,jk}$ is the matrix obtained by deleting from Y the rows i and j and the columns i and k.

In fact, $\Delta_{ii,jk}$ is the cofactor of Y_{ii} with respect to the (j,k) element of Y, and (13.18) is known as Jacobi's identity [13.1].

We now express $\Delta_{ii,jk}$ in terms of appropriate admittance products.

Consider the network N' which is obtained from N by short-circuiting nodes i and 0. Then the node-admittance matrix Y' of N' with node 0 as the reference is equal to Y_{ii}. If Δ'_{jk} denotes the cofactor of Y' with respect to the (j,k) element of Y, then, as we have just mentioned,

$$\Delta'_{jk}=\Delta_{ii,jk}.$$

If $T'_{jk,0}$ denotes a spanning 2-tree of N' with nodes j and k in one component and node 0 in the other component, and $W'_{jk,0}$ denotes the sum of admittance products of all spanning 2-trees $T'_{jk,0}$ of N', then from (13.8) we get

$$\Delta'_{jk} = W'_{jk,0}.$$

But a spanning 2-tree $T'_{jk,0}$ of N' is a spanning 3-tree of N of the type $T_{i,jk,0}$ because in N' the nodes 0 and i are represented by a single node. Thus

$$\Delta_{ii,jk} = \Delta'_{jk} = W'_{jk,0} = W_{i,jk,0} \tag{13.20}$$

where $W_{i,jk,0}$ is the sum of the admittance products of all spanning 3-trees of the type $T_{i,jk,0}$.

Now, consider the term $\Delta_{11,22} + \Delta_{11,33} - 2\Delta_{11,23}$ in (13.17). It is clear from (13.20) that

$$\Delta_{11,22} + \Delta_{11,33} - 2\Delta_{11,23} = W_{1,2,0} + W_{1,3,0} - 2W_{1,23,0}.$$

Since

$$W_{1,2,0} = W_{13,2,0} + W_{1,23,0} + W_{1,2,30}$$

and

$$W_{1,3,0} = W_{12,3,0} + W_{1,23,0} + W_{1,3,20},$$

we get

$$\Delta_{11,22} + \Delta_{11,33} - 2\Delta_{11,23} = W_{13,2,0} + W_{1,2,30} + W_{12,3,0} + W_{1,3,20}. \tag{13.21}$$

Using (13.8), (13.15), (13.16), and (13.21) we obtain the topological formulas for all the elements of Y_{sc}.

THEOREM 13.3. Let N be an RLC 2-port network without mutual inductances. Let the positive and negative reference terminals of the ports of N be as in Fig. 13.2. Then the short-circuit admittance matrix Y_{sc} of N is given by

$$Y_{sc} = \frac{1}{W_{13,2,0} + W_{1,2,30} + W_{12,3,0} + W_{1,3,20}} \begin{bmatrix} W_{2,3} & W_{13,20} - W_{12,30} \\ W_{13,20} - W_{12,30} & W_{1,0} \end{bmatrix}.$$

■

We now illustrate the topological evaluation of the open-circuit impedance and short-circuit admittance matrices of a 2-port RLC network. The following example is from Balabanian and Bickart [13.2].

Consider the network N shown in Fig. 13.3. The elements of this network are denoted by the symbols a, b, c, d, and e. Element values are as indicated in the figure.

Note that 3 and 0 refer to the same vertex. So N has no spanning 2-trees of the type $T_{13,20}$. For the same reason it has no spanning 3-trees of the types $T_{13,2,0}$, $T_{12,3,0}$, and $T_{1,3,20}$.

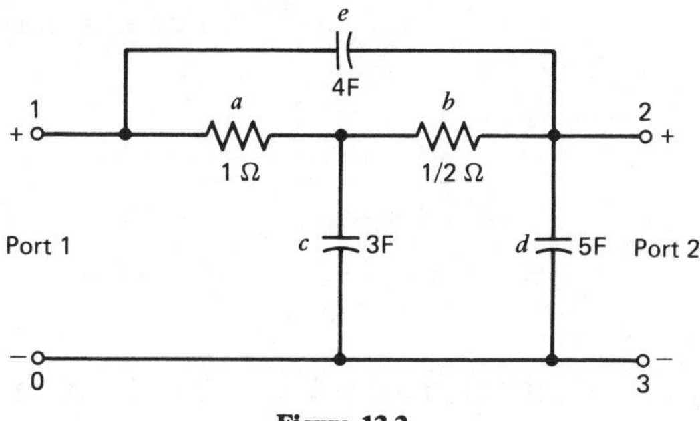

Figure 13.3.

The spanning trees, 2-trees, and 3-trees required to determine Z_{oc} and Y_{sc} are obtained as follows:

spanning trees:

$$(a,c,e),\quad (c,b,e),\quad (b,d,e),\quad (a,b,c),$$
$$(a,b,d),\quad (a,c,d),\quad (a,d,e),\quad \text{and}\quad (c,d,e).$$

$T_{1,0}$:

$$(b,c),\quad (b,d),\quad (c,d),\quad (a,d),$$
$$(a,b),\quad (b,e),\quad (c,e),\quad \text{and}\quad (a,e).$$

$T_{2,3}$:

$$(a,b),\quad (a,e),\quad (a,c),\quad (c,e),\quad \text{and}\quad (b,e).$$

$T_{12,30}$:

$$(a,b),\quad (a,e),\quad (b,e),\quad \text{and}\quad (c,e).$$

$T_{1,2,30}$:

$$(a),\quad (b),\quad \text{and}\quad (c).$$

From the above we can obtain W, $W_{1,0}$, $W_{2,3}$, $W_{13,,30}$, and $W_{1,2,30}$ as follows:

$$W = 60s^3 + 111s^2 + 16s,$$
$$W_{1,0} = 27s^2 + 33s + 2,$$
$$W_{2,3} = 12s^2 + 15s + 2,$$
$$W_{12,30} = 12s^2 + 12s + 2,$$
$$W_{1,2,30} = 3s + 3.$$

Using the above and Theorems 13.2 and 13.3, we can obtain Z_{oc} and Y_{sc}:

$$Z_{oc}=\frac{1}{60s^3+111s^2+16s}\begin{bmatrix} 27s^2+33s+2 & 12s^2+12s+2 \\ 12s^2+12s+2 & 12s^2+15s+2 \end{bmatrix},$$

$$Y_{sc}=\frac{1}{3s+3}\begin{bmatrix} 12s^2+15s+2 & -(12s^2+12s+2) \\ -(12s^2+12s+2) & 27s^2+33s+2 \end{bmatrix}.$$

13.2 TOPOLOGICAL FORMULAS FOR GENERAL LINEAR NETWORKS

In this section we develop topological formulas for general lumped linear time-invariant networks. These formulas are expressed in terms of the admittance products of directed trees and 2-trees of an appropriate directed graph associated with a given network. The starting point for the development of these formulas is the indefinite admittance matrix which we now define.

Consider a network N with n terminals $1,2,\ldots,n$. Let $V_1, V_2,\ldots,V_n$ denote the voltages of these terminals with respect to an arbitrary terminal 0 external to the network. Let $I_1, I_2,\ldots, I_n$ denote the currents entering these terminals (Fig. 13.4) when external connections are made.

The matrix $\hat{Y}$ relating these currents and the terminal voltages of N is known as the *indefinite admittance matrix* of N. Thus

$$\hat{Y}V=I,$$

that is,

$$\begin{bmatrix} \hat{y}_{11} & \hat{y}_{12} & \cdots & \hat{y}_{1n} \\ \hat{y}_{21} & \hat{y}_{22} & \cdots & \hat{y}_{2n} \\ \vdots & \vdots & \vdots & \vdots \\ \hat{y}_{n1} & \hat{y}_{n2} & \cdots & \hat{y}_{nn} \end{bmatrix}\begin{bmatrix} V_1 \\ V_2 \\ \vdots \\ V_n \end{bmatrix}=\begin{bmatrix} I_1 \\ I_2 \\ \vdots \\ I_n \end{bmatrix}. \tag{13.22}$$

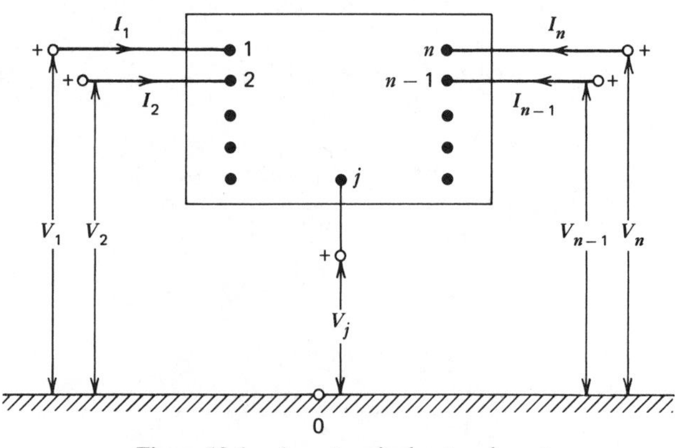

Figure 13.4. An n-terminal network.

If we regard the network N as an n-port network in which the ith terminal and the external terminal 0 constitute the ith port, then we can see that the indefinite admittance matrix $\hat{Y}$ is the short-circuit admittance matrix of such an n-port network. Thus

$$\hat{y}_{jk} = \left.\frac{I_j}{V_k}\right|_{\text{all } V_i=0,\, i\neq k}.$$

In other words, if we connect a voltage source of unit value between the terminal k and the external terminal 0, and short-circuit all other terminals to the external terminal, then the value of current I_j entering terminal j is equal to $\hat{y}_{jk}$. We can use this interpretation to compute all the elements of $\hat{Y}$. For example, the indefinite admittance matrix of the network in Fig. 13.5 can be shown to be equal to

$$\begin{bmatrix} 1+y & -y & -1 & 0 \\ -y & 2+y & -1 & -1 \\ -1 & -1-g_m & 2+g_p+g_m & -g_p \\ 0 & -1+g_m & -g_p-g_m & 1+g_p \end{bmatrix}. \tag{13.23}$$

This example is from Mitra [13.3].

Two useful properties of an indefinite admittance matrix will now be established.

By Kirchhoff's current law,

$$I_1 + I_2 + \cdots + I_n = 0.$$

Hence

$$(\hat{y}_{11}+\hat{y}_{21}+\cdots+\hat{y}_{n1})V_1 + (\hat{y}_{12}+\hat{y}_{22}+\cdots+\hat{y}_{n2})V_2 + \cdots + (\hat{y}_{1n}+\hat{y}_{2n}+\cdots+\hat{y}_{nn})V_n = 0.$$

Note that the voltages $V_1, V_2, \ldots, V_n$ are all independent. Suppose we now set $V_i=0$ for all $i\neq k$. Then the above expression reduces to

$$(\hat{y}_{1k}+\hat{y}_{2k}+\cdots+\hat{y}_{nk})V_k = 0.$$

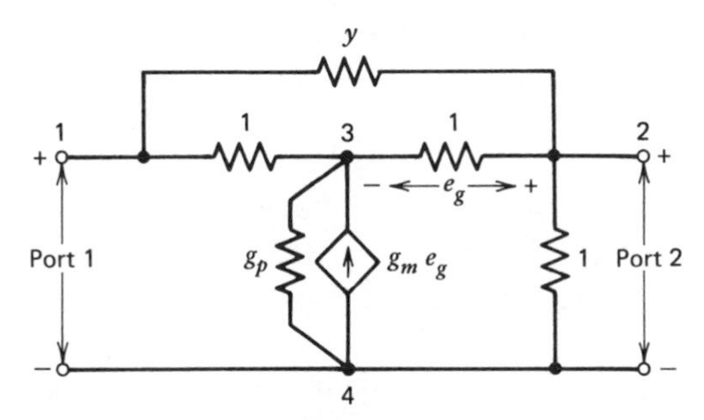

Figure 13.5.

Since

$$V_k \neq 0,$$

it follows that

$$(\hat{y}_{1k}+\hat{y}_{2k}+\cdots+\hat{y}_{nk})=0.$$

Thus the sum of all the elements in each column of an indefinite admittance matrix equals zero.

We can prove a similar result for the rows of $\hat{Y}$, namely, the sum of all the elements in any row of $\hat{Y}$ is equal to zero.

Thus we have the following result.

THEOREM 13.4. The indefinite admittance matrix of an electrical network is an equi-cofactor matrix. ■

Another attractive feature of an indefinite admittance matrix is that it is a very basic description of a network in the sense that several descriptions of a network can be obtained starting from this matrix. For example, the node-admittance matrix with the ith terminal as the reference can be obtained by deleting the ith row and the ith column from the indefinite admittance matrix. For a detailed discussion of the application of the indefinite admittance matrix in the analysis of active networks, we recommend Mitra [13.3].

We next consider the topological evaluation of the cofactors of an indefinite admittance matrix.

Given the indefinite admittance matrix $\hat{Y}$ of a network N having n nodes, first we construct a weighted directed graph $G(\hat{Y})$ as follows:

1. $G(\hat{Y})$ has n nodes labeled by the integers $1,2,3,\ldots,n$.
2. If $\hat{y}_{ij} \neq 0$, for $i, j=1,2,\ldots,n$, $i \neq j$, then there is an edge in $G(\hat{Y})$ directed from node i to node j with the associated weight $-y_{ij}$.

For example, the directed graph associated with the indefinite admittance matrix in (13.23) is shown in Fig. 13.6*a*.

In the following we denote by T_i a spanning directed tree in $G(\hat{Y})$ with node i as the root. A spanning directed 2-tree in $G(\hat{Y})$ is a spanning 2-tree in the underlying undirected graph such that each component of the directed 2-tree is a directed tree in $G(\hat{Y})$. Also, we denote by $T_{ij,kl}$ a spanning directed 2-tree such that:

1. Nodes i and j are in one component and nodes k and l are in the other component; and
2. Nodes i and k are the roots of the two components of $T_{ij,kl}$.

A directed spanning 3-tree $T_{ij,kl,rs}$ is defined in a similar manner. The nodes

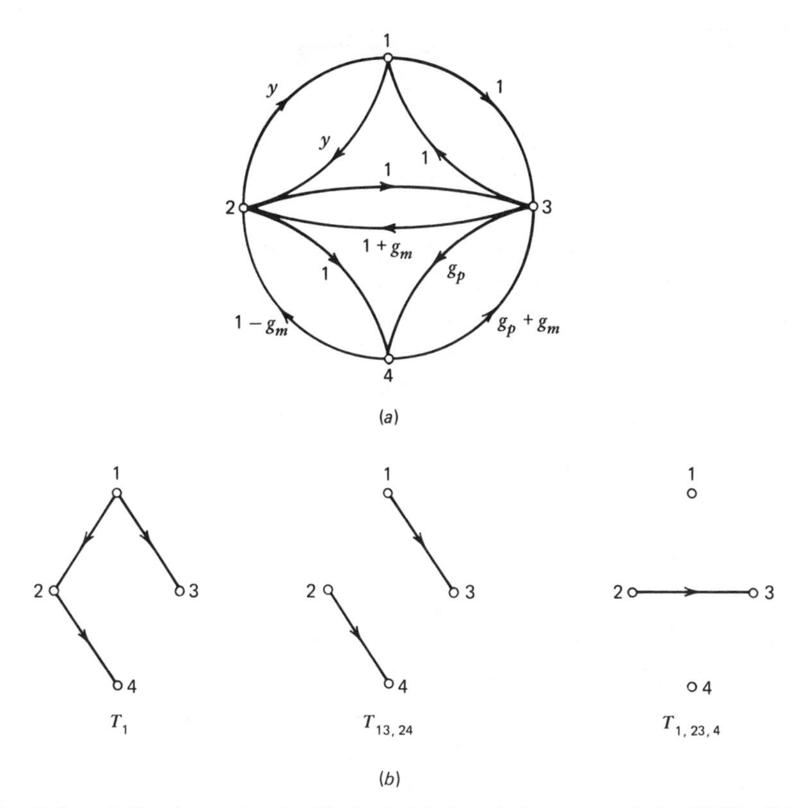

Figure 13.6. (*a*) Graph associated with the indefinite admittance matrix in (13.23). (*b*) A directed tree, a directed 2-tree, and a directed 3-tree.

i, k, and r are the roots of the three components of $T_{ij,kl,rs}$. Examples of a directed spanning tree, a 2-tree, and a 3-tree are shown in Fig. 13.6*b*.

Let us also define:

W_i = Sum of admittance products of all the spanning directed trees T_i of $G(\hat{Y})$.

$W_{ij,kl}$ = Sum of admittance products of all the spanning directed 2-trees $T_{ij,kl}$ of $G(\hat{Y})$.

$W_{ij,kl,rs}$ = Sum of admittance products of all the spanning directed 3-trees $T_{ij,kl,rs}$ of $G(\hat{Y})$.

It can be seen (Exercise 6.18) that

$$\hat{\Delta}_{kk} = W_k. \tag{13.24}$$

In the following theorem we relate second-order cofactors of $\hat{Y}$ to admittance products of appropriate directed 2-trees.

THEOREM 13.5. Let $\hat{Y}$ denote the indefinite admittance matrix of a network N. Then

$$\hat{\Delta}_{ij,kk} = W_{ji,k}.$$

Proof

Let G^* be the directed graph obtained from $G(\hat{Y})$ by altering the weights of the edges (i, j), (k, j), and (i, k) from $-y_{ij}$, $-y_{kj}$, and $-y_{ik}$ to $-y_{ij}+K$, $-y_{kj}-K$, and $-y_{ik}-K$, respectively. Let $\hat{Y}^*$ denote the associated equicofactor matrix. We shall regard the edge (i, j) in G^* as consisting of two parallel edges $(i, j)_1$ and $(i, j)_2$ with weights $-y_{ij}$ and K, respectively. Similarly we shall regard the edge (k, j) as consisting of two parallel edges $(k, j)_1$ and $(k, j)_2$ with weights $-y_{kj}$ and $-K$, respectively.

Clearly,

$$\hat{\Delta}^*_{kk} = W^*_k, \tag{13.25}$$

where $*$ is used to denote quantities relating to $\hat{Y}^*$ and G^*.

Let us now partition the set S of all directed spanning trees T^*_k of G^* into three sets S_1, S_2, and S_3 defined as follows:

$S_1 = \{T^*_k \mid T^*_k \in S$ and neither $(i, j)_2$ nor $(k, j)_2$ is in $T^*_k\}$,
$S_2 = \{T^*_k \mid T^*_k \in S$, $(k, j)_2$ is in T^*_k and there is a directed path from j to i in $T^*_k\}$,
$S_3 = S - (S_1 \cup S_2)$.

Since G can be obtained from G^* by setting $K=0$, it follows that

$$\hat{\Delta}_{kk} = \sum_{T^*_k \in S_1} \text{admittance product of } T^*_k. \tag{13.26}$$

For any directed spanning tree $T^*_k \in S_2$, $T^*_k - (k, j)_2$ is a directed spanning 2-tree $T_{ji,k}$ of $G(\hat{Y})$ and vice versa. Thus

$$\sum_{T^*_k \in S_2} \text{admittance product of } T^*_k = -K \sum_{\text{all } T_{ji,k}} \text{admittance product of } T_{ji,k}$$

$$= -K\, W_{ji,k}. \tag{13.27}$$

Now note that any directed tree T^*_k in S_3 is one of the following types:

1. T^*_k contains $(i, j)_2$.
2. T^*_k contains $(k, j)_2$, and there is no directed path in T^*_k from j to i.

It is easy to see that to each $T^*_k \in S_3$ which is of the former type there exists a unique directed tree $\{T_k^* - (i, j)_2 \cup (k, j)_2\} \in S_3$, which is of the latter type and vice versa. The admittance product of one is the negative of the admittance product of the other. So

$$\sum_{T_k^* \in S_3} \text{admittance product of directed tree } T_k^* = 0. \qquad (13.28)$$

From (13.26) through (13.28) we get

$$\hat{\Delta}_{kk}^* = \hat{\Delta}_{kk} - K\, W_{ji,k}. \qquad (13.29)$$

Finally, we also have the following relation:

$$\hat{\Delta}_{kk}^* = \hat{\Delta}_{kk} - K\, \hat{\Delta}_{ij,kk}. \qquad (13.30)$$

From (13.29) and (13.30) we get

$$\hat{\Delta}_{ij,kk} = W_{ji,k}. \quad \blacksquare$$

It is easy to show, using Theorem 13.5, that

$$\hat{\Delta}_{ii,jj,kl} = W_{i,j,lk}. \qquad (13.31)$$

We now proceed to derive topological formulas for the open-circuit and short-circuit functions of a general lumped linear network.

Consider the 2-port network N shown in Fig. 13.7. Let Y denote its node-admittance matrix with node n as the reference. This matrix can be obtained by deleting the nth row and the nth column from the indefinite admittance matrix $\hat{Y}$ of N. Thus

$$Y = \hat{Y}_{nn}. \qquad (13.32)$$

Starting from the node equations

$$YV = I,$$

we can obtain the open-circuit impedance and short-circuit admittance

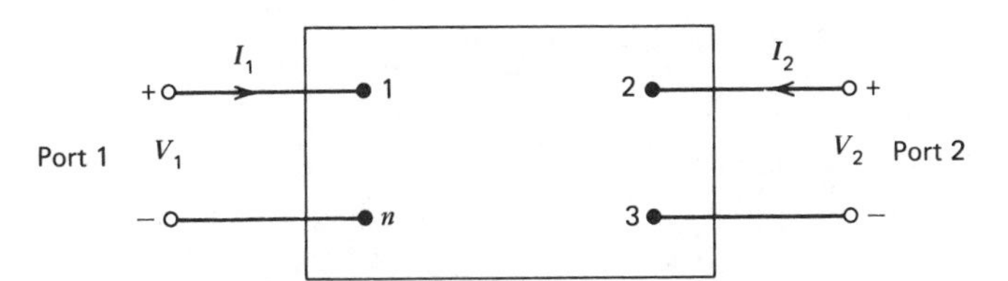

Figure 13.7. A 2-port network.

matrices of N as follows:

$$Z_{oc}=\frac{1}{\Delta}\begin{bmatrix}\Delta_{11} & \Delta_{21}-\Delta_{31}\\ \Delta_{12}-\Delta_{13} & \Delta_{22}+\Delta_{33}-\Delta_{23}-\Delta_{32}\end{bmatrix}, \tag{13.33}$$

$$Y_{sc}=\frac{1}{\Delta_{11,22}+\Delta_{11,33}-\Delta_{11,23}-\Delta_{11,32}}\begin{bmatrix}\Delta_{22}+\Delta_{33}-\Delta_{23}-\Delta_{32} & \Delta_{31}-\Delta_{21}\\ \Delta_{13}-\Delta_{12} & \Delta_{11}\end{bmatrix}. \tag{13.34}$$

To obtain the topological formulas for the elements of Z_{oc} and Y_{sc} first observe, using (13.24),

$$\Delta=\hat{\Delta}_{nn}=W_n.$$

Also from Theorem 13.5 we have

$$\Delta_{11}=\hat{\Delta}_{11,nn}=W_{1,n}, \tag{13.35}$$

$$\begin{aligned}\Delta_{21}-\Delta_{31}&=\hat{\Delta}_{21,nn}-\hat{\Delta}_{31,nn}\\ &=W_{12,n}-W_{13,n}\\ &=W_{123,n}+W_{12,n3}-W_{132,n}-W_{13,n2}\\ &=W_{12,n3}-W_{13,n2},\end{aligned} \tag{13.36}$$

$$\Delta_{12}-\Delta_{13}=W_{21,3n}-W_{31,2n}. \tag{13.37}$$

Now

$$\begin{aligned}\Delta_{22}+\Delta_{33}-\Delta_{32}-\Delta_{23}&=\hat{\Delta}_{22,nn}+\hat{\Delta}_{33,nn}-\hat{\Delta}_{32,nn}-\hat{\Delta}_{23,nn}\\ &=\hat{\Delta}_{22,33}.\end{aligned} \tag{13.38}$$

The last step in the above equation follows from the following identity (see Sharpe and Spain [13.4]):

$$\hat{\Delta}_{pq,rs}=\hat{\Delta}_{pq,uv}+\hat{\Delta}_{rs,uv}-\hat{\Delta}_{ps,uv}-\hat{\Delta}_{rq,uv}.$$

So we get from Theorem 13.5

$$\Delta_{22}+\Delta_{33}-\Delta_{32}-\Delta_{23}=\hat{\Delta}_{22,33}=W_{2,3}. \tag{13.39}$$

Further

$$\begin{aligned}\Delta_{11,22}+\Delta_{11,33}-\Delta_{11,23}-\Delta_{11,32}&=\hat{\Delta}_{11,22,nn}+\hat{\Delta}_{11,33,nn}-\hat{\Delta}_{11,23,nn}-\hat{\Delta}_{11,32,nn}\\&=W_{1,2,n}+W_{1,3,n}-W_{1,32,n}-W_{1,23,n}\\&=W_{13,2,n}+W_{1,2,n3}+W_{12,3,n}+W_{1,3,n2}.\end{aligned}\tag{13.40}$$

From the foregoing equations we get the following.

THEOREM 13.6. Let N be a 2-port network with the positive and negative reference terminals of the ports as shown in Fig. 13.7. Then the open-circuit impedance matrix and the short-circuit admittance matrix of N are given by

$$Z_{oc}=\frac{1}{W_n}\begin{bmatrix} W_{1,n} & W_{12,n3}-W_{13,n2}\\ W_{21,3n}-W_{31,2n} & W_{2,3}\end{bmatrix}$$

$$Y_{sc}=\frac{1}{W_{13,2,n}+W_{1,2,n3}+W_{12,3,n}+W_{1,3,n2}}\begin{bmatrix} W_{2,3} & W_{13,n2}-W_{12,n3}\\ W_{31,2n}-W_{21,3n} & W_{1,n}\end{bmatrix}.$$

■

We now illustrate the topological evaluation of Z_{oc} and Y_{sc} using Theorem 13.6.

Consider again the network N in Fig. 13.5 with nodes 1 and 4 forming port 1, and nodes 2 and 4 forming port 2. The indefinite admittance matrix $\hat{Y}$ of N is given in (13.23). The graph $G(\hat{Y})$ corresponding to $\hat{Y}$ is shown in Fig. 13.6*a*.

To calculate Z_{oc} and Y_{sc} using Theorem 13.6, we need to compute for $G(\hat{Y})$, W_4, $W_{1,4}$, $W_{12,4}$, $W_{21,4}$, $W_{2,4}$, and $W_{1,2,4}$. First we obtain the required spanning trees, 2-trees, and 3-trees of $G(\hat{Y})$:

T_4:

$$\{(2,1),(4,2),(4,3)\},\quad \{(4,2),(4,3),(3,1)\},$$

$$\{(1,2),(3,1),(4,3)\},\quad \{(2,1),(4,2),(1,3)\},$$

$$\{(2,1),(2,3),(4,2)\},\quad \{(3,1),(3,2),(4,3)\},$$

$$\{(2,1),(3,2),(4,3)\},\quad \{(3,1),(2,3),(4,2)\}.$$

(Note that (i, j) is the edge directed from i to j.)

$T_{1,4}$:

$$\{(1,2),(4,3)\},\ \{(1,2),(1,3)\},\ \{(1,2),(2,3),\}\ \{(1,3),(3,2)\},$$
$$\{(1,3),(4,2)\},\ \{(3,2),(4,3)\},\ \{(4,2),(2,3)\},\ \{(4,2),(4,3)\}.$$

$T_{2,4}$:

$$\{(2,1),(4,3)\},\ \{(2,1),(1,3)\},\ \{(2,1),(2,3)\},$$
$$\{(3,1),(2,3)\},\ \{(3,1),(4,3)\}.$$

$T_{12,4}$:

$$\{(1,2),(1,3)\},\ \{(1,2),(2,3)\},\ \{(1,3),(3,2)\},\ \{(1,2),(4,3)\}.$$

$T_{21,4}$:

$$\{(2,1),(1,3)\},\ \{(2,1),(2,3)\},\ \{(3,1),(2,3)\},\ \{(2,1),(4,3)\}.$$

$T_{1,2,4}$:

$$\{(1,3)\},\ \{(2,3)\},\ \{(4,3)\}.$$

Using the above we can obtain the following admittance products:

$$W_4 = g_m(y+1) + g_p(3y+2) + (2y+1),$$
$$W_{1,4} = (2+y)(2+g_p+g_m) - (1+g_m),$$
$$W_{2,4} = (1+y)(2+g_p+g_m) - 1,$$
$$W_{12,4} = y(2+g_p+g_m) + (1+g_m),$$
$$W_{21,4} = y(2+g_p+g_m) + 1,$$
$$W_{1,2,4} = 2+g_p+g_m.$$

So we obtain

$$Z_{oc} = \frac{1}{W_4}\begin{bmatrix} W_{1,4} & W_{12,4} \\ W_{21,4} & W_{2,4} \end{bmatrix}$$
$$= \frac{1}{g_m(y+1)+g_p(3y+2)+(2y+1)}$$
$$\times \begin{bmatrix} (2+y)(2+g_p+g_m)-(1+g_m) & y(2+g_p+g_m)+(1+g_m) \\ y(2+g_p+g_m)+1 & (1+y)(2+g_p+g_m)-1 \end{bmatrix}$$

and

$$Y_{sc}=\frac{1}{W_{1,2,4}}\begin{bmatrix} W_{2,4} & -W_{12,4} \\ -W_{21,4} & W_{1,4} \end{bmatrix}$$

$$=\frac{1}{2+g_p+g_m}\begin{bmatrix} (1+y)(2+g_p+g_m)-1 & -y(2+g_p+g_m)-(1+g_m) \\ -y(2+g_p+g_m)-1 & (2+y)(2+g_p+g_m)-(1+g_m) \end{bmatrix}.$$

13.3 ADJOINT NETWORK AND NETWORK SENSITIVITY COMPUTATION

In this section we develop a method for the computation of network sensitivities. This method is based on Tellegen's theorem and the concept of the adjoint of a network (see Bordewijk [13.5] and Director and Rohrer [13.6]). Out treatment here is based on [13.6].

Let N be a lumped linear time-invariant network consisting of resistances, capacitances, inductances, transformers, gyrators, voltage controlled voltage and current sources, and current controlled voltage and current sources. We assume that N is a 2-port network, although the discussion that follows is valid even when N has more than two-ports.

Let $\tilde{N}$ be a 2-port network which is topologically equivalent to N. In other words, the graph of $\tilde{N}$ is identical to that of N. The corresponding elements of N and $\tilde{N}$ will be denoted by the same symbol. Note that we have not yet defined the elements of $\tilde{N}$ and their values.

Let V_e and I_e denote, respectively, the voltage and the current associated with an element e in N, and ψ_e and λ_e denote, respectively, the voltage and the current associated with the corresponding element e in $\tilde{N}$. V_i and I_i, $i=1,2$, will denote the voltage and current variables associated with the ports of N, and ψ_i and λ_i, $i=1,2$, will denote the corresponding variables for the ports of $\tilde{N}$.

Applying Tellegen's theorem to N and the network $\tilde{N}$, yields

$$V_1\lambda_1+V_2\lambda_2=\sum_e V_e\lambda_e \tag{13.41a}$$

and

$$I_1\psi_1+I_2\psi_2=\sum_e I_e\psi_e, \tag{13.41b}$$

where the summation is over all the elements of N and $\tilde{N}$. Suppose the element values of the network N are changed. Then Tellegen's theorem applied to the perturbed network N and the network $\tilde{N}$ yields the following:

$$(V_1+\Delta V_1)\lambda_1+(V_2+\Delta V_2)\lambda_2=\sum_e (V_e+\Delta V_e)\lambda_e \tag{13.42a}$$

and

$$(I_1+\Delta I_1)\psi_1+(I_2+\Delta I_2)\psi_2=\sum_e (I_e+\Delta I_e)\psi_e, \qquad (13.42b)$$

where ΔV and ΔI represent the changes in the voltage and the current which occur as a result of the perturbation of the element values in N.

Subtracting (13.41) from (13.42), we get

$$\Delta V_1\lambda_1+\Delta V_2\lambda_2=\sum_e \Delta V_e\lambda_e \qquad (13.43)$$

and

$$\Delta I_1\psi_1+\Delta I_2\psi_2=\sum_e \Delta I_e\psi_e. \qquad (13.44)$$

Subtracting (13.43) from (13.44), yields

$$(\Delta V_1\lambda_1-\Delta I_1\psi_1)+(\Delta V_2\lambda_2-\Delta I_2\psi_2)=\sum_e (\Delta V_e\lambda_e-\Delta I_e\psi_e). \qquad (13.45)$$

We now wish to define the elements in $\tilde{N}$ so that (13.45) becomes independent of all ΔV_e and ΔI_e terms.

Consider first the resistance elements. We have

$$V_R=RI_R, \qquad (13.46)$$

where V_R and I_R are the voltage and current associated with a resistance element R. Let R be changed to $R+\Delta R$. Then

$$V_R+\Delta V_R=(R+\Delta R)(I_R+\Delta I_R),$$

which, on neglecting second-order terms, simplifies to

$$V_R+\Delta V_R=RI_R+R\Delta I_R+I_R\Delta R. \qquad (13.47)$$

Subtracting (13.46) from (13.47), yields

$$\Delta V_R=R\Delta I_R+I_R\Delta R. \qquad (13.48)$$

Therefore the terms in (13.45) associated with the resistance elements of N can be written as

$$\sum_R \left[(R\lambda_R-\psi_R)\Delta I_R+I_R\lambda_R\Delta R\right], \qquad (13.49)$$

where the summation is over all the resistance elements in N. Note that the subscripts on λ and ψ do not necessarily indicate resistance types on $\tilde{N}$, but merely the correspondence between the elements of N and $\tilde{N}$.

If we now choose

$$\psi_R = R\lambda_R, \tag{13.50}$$

then (13.49) reduces to

$$\sum_R I_R \lambda_R \Delta R, \tag{13.51}$$

which is independent of ΔV_R and ΔI_R.

Equation (13.50) is the relation for a resistance of value R. So the element in $\tilde{N}$ which corresponds to a resistance element of value R in N is also a resistance of value R.

Consider next a voltage controlled voltage source defined by the relation:

$$V_{VV2} = \mu V_{VV1}$$

and

$$I_{VV1} = 0.$$

Then, neglecting second-order terms, we have

$$\Delta V_{VV2} = \mu \Delta V_{VV1} + V_{VV1} \Delta\mu$$

and

$$\Delta I_{VV1} = 0.$$

Terms of (13.45) corresponding to voltage controlled voltage sources can now be written as

$$\sum \left[(\lambda_{VV1} + \mu\lambda_{VV2}) \Delta V_{VV1} - \psi_{VV2} \Delta I_{VV2} + V_{VV1} \lambda_{VV2} \Delta\mu \right]. \tag{13.52}$$

If we now choose

$$\lambda_{VV1} = -\mu\lambda_{VV2}$$

$$\psi_{VV2} = 0, \tag{13.53}$$

then (13.52) reduces to

$$\sum V_{VV1} \lambda_{VV2} \Delta\mu.$$

Note that (13.53) represents a current controlled current source of amplification factor $-\mu$. Also note the changes in the roles between controlling and dependent elements in the adjoint network.

Table 13.1

Element Type	Branch Relation	Branch Relation in Adjoint	Sensitivity (Component of $\mathcal{G}$)	Component of Δp
Resistance	$V_R = RI_R$	$\psi_R = R\lambda_R$	$+I_R\lambda_R$	ΔR
Conductance	$I_G = GV_G$	$\lambda_G = G\psi_G$	$-V_G\psi_G$	ΔG
Capacitance	$I_C = j\omega CV_C$	$\lambda_C = j\omega C\psi_C$	$-j\omega V_C\psi_C$	ΔC
Elastance (Recriprocal capacitance)	$SI_S = j\omega V_S$	$S\lambda_S = j\omega\psi_S$	$+\frac{1}{j\omega}I_S\lambda_S$	ΔS
Inductance	$V_L = j\omega LI_L$	$\psi_L = j\omega L\lambda_L$	$+j\omega I_L\lambda_L$	ΔL
Reciprocal inductance	$\Gamma V_\Gamma = j\omega I_\Gamma$	$\Gamma\psi_\Gamma = j\omega\lambda_\Gamma$	$-\frac{1}{j\omega}V_\Gamma\psi_\Gamma$	$\Delta\Gamma$
Transformer	$V_{T2} = nV_{T1}$ $I_{T1} = -nI_{T2}$	$\psi_{T2} = n\psi_{T1}$ $\lambda_{T1} = -n\lambda_{T2}$	$+(V_{T1}\lambda_{T2} + I_{T2}\psi_{T1})$	Δn
Gyrator	$V_{GY1} = \alpha I_{GY2}$ $V_{GY2} = -\alpha I_{GY1}$	$\psi_{GY1} = -\alpha\lambda_{GY2}$ $\psi_{GY2} = \alpha\lambda_{GY1}$	$+(I_{GY2}\lambda_{GY1} - I_{GY1}\lambda_{GY2})$	$\Delta\alpha$
Voltage controlled voltage source	$V_{VV2} = \mu V_{VV1}$ $I_{VV1} = 0$	$\lambda_{VV1} = -\mu\lambda_{VV2}$ $\psi_{VV2} = 0$	$+V_{VV1}\lambda_{VV2}$	$\Delta\mu$
Voltage controlled current source	$I_{VI2} = g_m V_{VI1}$ $I_{VI1} = 0$	$\lambda_{VI1} = g_m\psi_{VI2}$ $\lambda_{VI2} = 0$	$-V_{VI1}\psi_{VI2}$	Δg_m
Current controlled voltage source	$V_{IV2} = r_m I_{IV1}$ $V_{IV1} = 0$	$\psi_{IV1} = r_m\lambda_{IV2}$ $\psi_{IV2} = 0$	$+I_{IV1}\lambda_{IV2}$	Δr_m
Current controlled current source	$I_{II2} = \beta I_{II1}$ $V_{II1} = 0$	$\psi_{II1} = -\beta\psi_{II2}$ $\lambda_{II2} = 0$	$-I_{II1}\psi_{II2}$	$\Delta\beta$

The terms of (13.45) for the remaining element types of the network N can be obtained in a similar manner. The results are given in Table 13.1 with the appropriate branch relations to be chosen for the related network $\tilde{N}$.

If the elements of $\tilde{N}$ are chosen as given in Table 13.1, then (13.45) reduces to

$$\sum(\Delta V_1\lambda_1 - \Delta I_1\psi_1) + (\Delta V_2\lambda_2 - \Delta I_2\psi_2) = \mathcal{G}^t\Delta p, \tag{13.54}$$

where $\mathcal{G}$ and Δp are vectors whose components are given in Table 13.1.

The network $\tilde{N}$ whose elements have beeen defined as in Table 13.1 is called the *adjoint* of N. The representations of the different network elements of N and their corresponding elements in the adjoint network $\tilde{N}$ are given in Fig. 13.8.

Note that the adjoint network $\tilde{N}$ is related to the original network N in the following way:

1. The graph of $\tilde{N}$ is identical to that of N.
2. All resistance, capacitance, and inductance elements in N are associated with resistance, capacitance, and inductance elements, respectively, in $\tilde{N}$ of like values.
3. All transformers of turns ratio $1:n$ in N are associated with transformers of turns ratio $1:n$ in $\tilde{N}$.
4. All gyrators of gyration ratio α in N are associated with gyrators of gyration ratio $-\alpha$ in $\tilde{N}$ (or the two gyrator ports are reversed).
5. Voltage controlled voltage sources with voltage amplification factor μ in N are associated with current controlled current sources with current amplification factor $-\mu$ in $\tilde{N}$, and the roles of controlling and dependent elements in N are reversed in $\tilde{N}$.
6. Current controlled current sources with current amplification factor β in N are associated with voltage controlled voltage sources with voltage amplification factor $-\beta$ in $\tilde{N}$, and the roles of controlling and dependent elements in N are reversed in $\tilde{N}$.
7. Voltage controlled current sources, and current controlled voltage sources in N are associated with voltage controlled current sources and current controlled voltage sources respectively, in $\tilde{N}$, and the roles of controlling and dependent elements in N are reversed in $\tilde{N}$.

We now illustrate the application of the adjoint network in the computation of the sensitivities of a network function.

The sensitivity S_x^F of a network function $F(s)$ with respect to a parameter x of a network is defined as

$$S_x^F = \frac{x}{F}\frac{\partial F}{\partial x}.$$

Clearly, the sensitivity S_x^F is a measure of the effect on $F(s)$ of an incremental variation in x. Computing S_x^F essentially involves determining $\partial F/\partial x$. This can be done as follows:

1. Select the port variables so that the left-hand side of (13.54) reduces to ΔF. (In other words, excite the network N and the adjoint network $\tilde{N}$ at their ports suitably.) For example, suppose $F(s)$ is the open-circuit

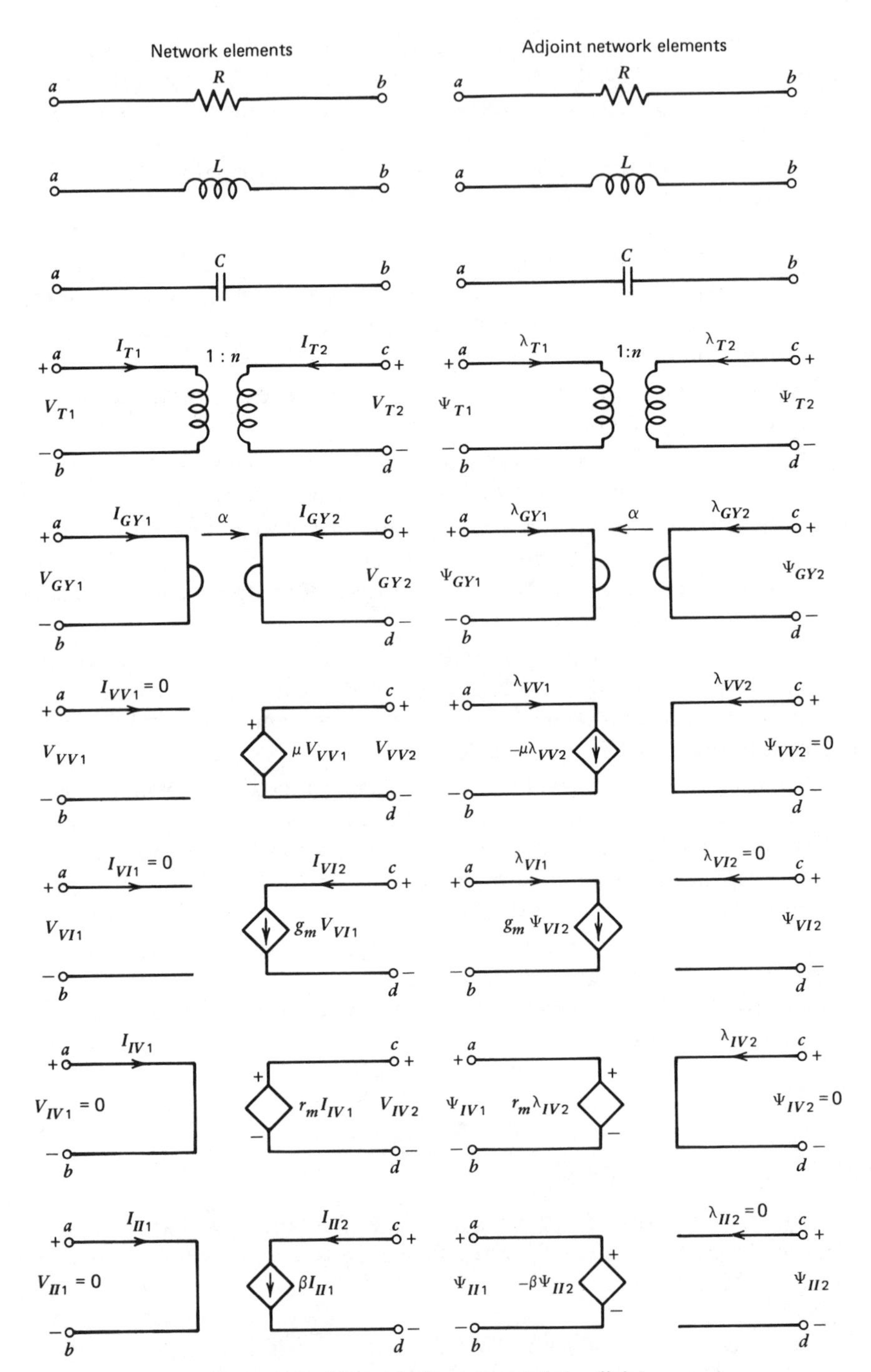

Figure 13.8. Network elements and their adjoints.

voltage ratio, that is

$$F(s)=\frac{V_2(s)}{V_1(s)}.$$

Then excite the networks N and $\tilde{N}$ as follows:

a. Connect across port 1 of N an independent voltage source of constant value 1. Thus

$$V_1=1 \qquad \text{and} \qquad \Delta V_1=0.$$

b. Open-circuit port 2 of N. Thus

$$I_2=0.$$

c. Short-circuit port 1 of $\tilde{N}$ so that

$$\psi_1=0.$$

d. Connect across port 2 of $\tilde{N}$ an independent current source of value 1. Thus

$$\lambda_2=1.$$

We may easily verify that the above choice of port excitations reduces the left-hand side of (13.54) to

$$\Delta V_2=\Delta F.$$

2. Analyze N and $\tilde{N}$ after exciting as described above, and determine all the element voltages and currents.

3. The term on the right-hand side of (13.54) corresponding to a parameter x is a product of Δx and the voltages and/or currents associated with this parameter (see Table 13.1). So $\partial F/\partial x$ for each parameter x can be determined once the voltages and currents in N and $\tilde{N}$ are determined as in 2.

Thus sensitivity computation using the adjoint network concept requires the analysis of the given network N and its adjoint $\tilde{N}$.

We now illustrate with an example the above method for sensitivity computation. This example is from Mitra [13.3].

Consider the network N shown in Fig. 13.9*a*. The adjoint $\tilde{N}$ is shown in Fig. 13.9*b*. The open-circuit voltage ratio

$$F(s)=\frac{V_2(s)}{V_1(s)}$$

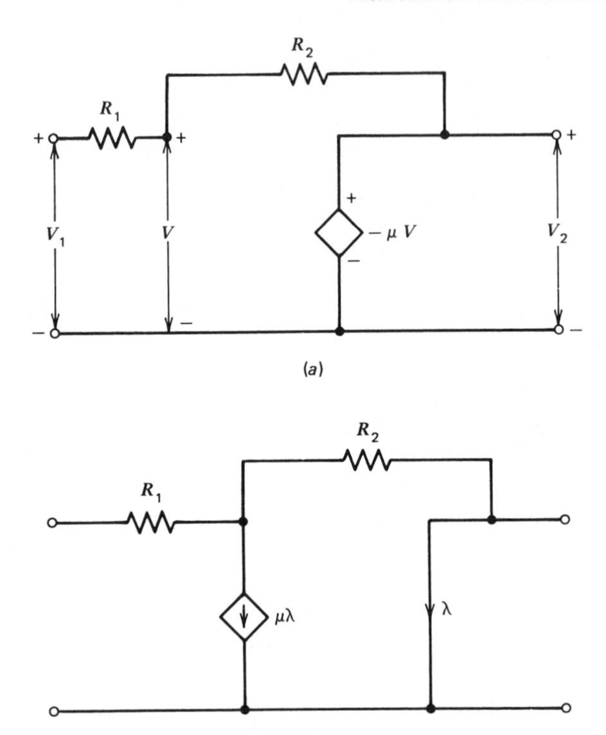

Figure 13.9. A network and its adjoint.

of N is given by

$$F(s)=\frac{-\mu R_2}{R_1+R_2+\mu R_1}.$$

As described before, excite N and $\tilde{N}$ as shown in Fig. 13.10. From Table 13.1 we have

$$\frac{\partial F}{\partial(-\mu)}=V\lambda,$$

$$\frac{\partial F}{\partial R_1}=I_1\lambda_1,$$

$$\frac{\partial F}{\partial R_2}=I_2\lambda_2.$$

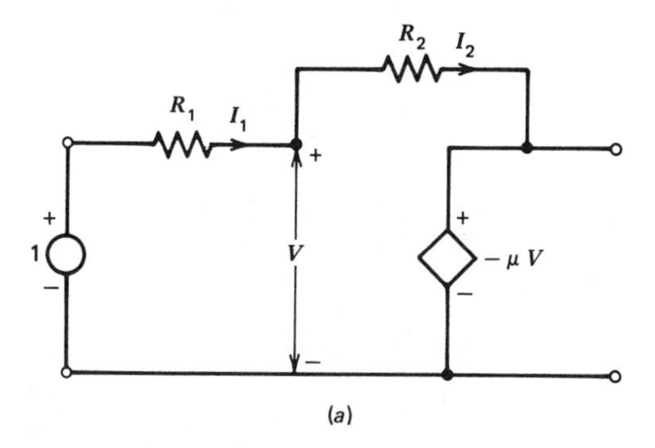

(a)

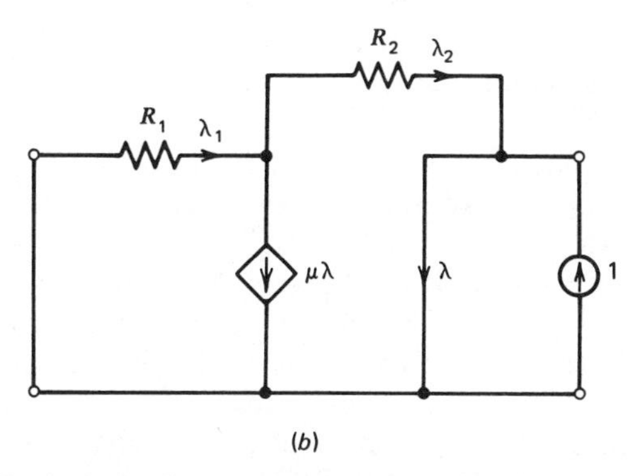

(b)

Figure 13.10. Port excitations of networks in Fig. 13.9.

It is easy to verify that

$$I_1 = I_2 = \frac{1+\mu}{R_1 + R_2 + \mu R_1},$$

$$V = \frac{R_2}{R_1 + R_2 + \mu R_1},$$

$$\lambda = \frac{R_1 + R_2}{\mu R_1 + R_1 + R_2},$$

$$\lambda_1 = \lambda + \mu\lambda - 1 = \frac{\mu R_2}{\mu R_1 + R_1 + R_2},$$

$$\lambda_2 = \lambda - 1 = \frac{-\mu R_1}{\mu R_1 + R_1 + R_2}.$$

Therefore we obtain

$$S_{\mu}^{F}=\frac{R_1+R_2}{R_1+R_2+\mu R_1},$$

$$S_{R_1}^{F}=\frac{-R_1(\mu+1)}{R_1+R_2+\mu R_1},$$

$$S_{R_2}^{F}=\frac{R_1(1+\mu)}{R_1+R_2+\mu R_1}.$$

We saw in Chapter 11 (Exercise 11.3) that the loop impedance matrix of a planar network is equal to the cutset admittance matrix of its dual. We may now ask the question whether for a given network N there exists another network N_T such that the indefinite admittance matrix of N_T is equal to the transpose of the indefinite admittance matrix of N. The answer is yes. Bhattacharyya and Swamy [13.7] have shown how to construct such a network N_T, called the *transpose* of N. It so happens that N_T is the same as the adjoint of N. See also Swamy, Bhushan and Bhattacharyya [13.8].

Swamy, Bhushan, and Bhattacharyya [13.9] have obtained a theorem, similar to Tellegen's theorem, but applicable to any two planar networks having dual topologies. Using this theorem they develop from a given planar network a new network called *generalized dual transpose*. If N_D is the dual of N and N_{DT} is the transpose of N_D, then it can be shown that N_{DT} is a special case of the generalized dual transpose of N. The concepts of transpose and generalized dual transpose have helped to see the unity between the various network realizations (reported in the literature) which look seemingly different but in fact are interrelated through the transpose/generalized dual transpose operations.

For example, Yanagisawa (see [13.3]) proposed two network structures to realize a voltage transfer function. It can be shown that one of them is a generalized dual transpose of the other. As another example, Dagget and Vlach [13.10] proposed two structures, one to realize a voltage transfer function and the other to realize a current transfer function. Again one of these is the transpose of the other. More examples which show how to obtain, using the concepts of transpose and generalized dual transpose, different equivalent structures realizing a given network function may be found in [13.9].

13.4 FURTHER READING

Chen [13.11] is highly recommended for further reading on a detailed treatment of topological formulas for network functions. Seshu and Reed [13.12], Chan [13.13], and Mayeda [13.14] are some of the other books which discuss topological formulas.

The formulas for general networks given in this chapter are in terms of appropriate trees associated with the indefinite admittance matrix of a network. Several authors have presented formulas which can be written directly from a given network. For example, see Numata and Iri [13.15], where several related papers are included as references. See Seshu and Reed [13.12] for interesting properties of network functions of RLC networks which are derived using topological formulas.

For a general form of Tellegen's theorem and its applications see Penfield, Spence, and Duinker [13.16], [13.17]. See also Desoer [13.18]. For an application of the concept of the adjoint network in the comput- aided design of resistance n-port networks see Director and Rohrer [13.19]. The adjoint network concept has also been used to establish sensitivity invariants and bounds on the sum of element sensitivity magnitudes for network functions; see for example, Swamy, Bhushan, and Thulasiraman [13.20], [13.21].

13.5 EXERCISES

13.1 (a) Prove that the admittance matrix of an RLC network without mutual inductance will have a pole at $s=0$ if and only if there is an all-inductor path between the terminals of the port.

(b) Prove that the admittance matrix of an RLC network without mutual inductance will have a pole at infinity if and only if there is an all-capacitor path between the terminals of the port.

13.2 Let

$$z_{11}=\frac{a_0+a_1s+\cdots+a_ns^n}{D(s)},$$

$$z_{22}=\frac{b_0+b_1s+\cdots+b_ms^m}{D(s)},$$

$$z_{21}=\frac{c_0+c_1s+\cdots+c_rs^r}{D(s)}$$

be the open-circuit impedance functions of an RLC 2-port network without mutual inductance. Using topological formulas, show that

$$a_k \geqslant |c_k|$$

and

$$b_k \geqslant |c_k|.$$

13.3 Derive topological formulas for the open-circuit voltage transfer ratio and the short-circuit current transfer ratio of a general linear network.

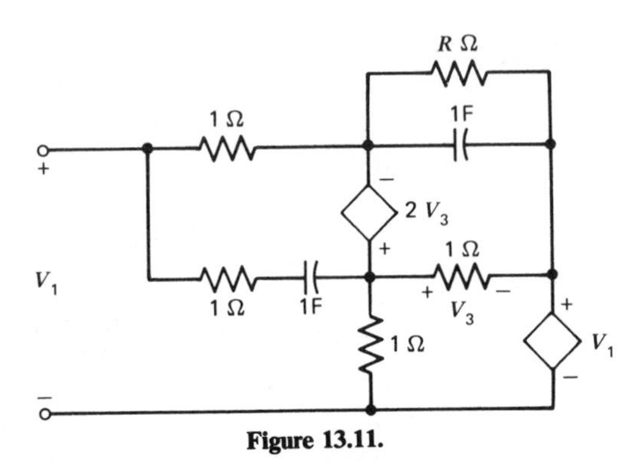

Figure 13.11.

13.4 Find the sensitivity, with respect to R, of the driving-point impedance of the network shown in Fig. 13.11.

13.5 Show that the indefinite admittance matrix of a network N is the transpose of the indefinite admittance matrix of its adjoint.

Hint Use Tellegen's theorem.

13.6 In the graph representation of a network containing multiport elements, each port is represented by an edge. Let N and N_D be two planar networks having dual graphs. Derive a theorem, similar to Tellegen's theorem, which involves only the current variables or only the voltage variables of N and N_D.

13.7 Let N be a planar network. Let N_D be the dual of N and let N_{DT} be the transpose or adjoint of N_D. Using the theorem in Exercise 13.6 and following a procedure similar to the one described in Section 13.3, determine directly from N (without constructing N_D) the element types and values for the network N_{DT} (Swamy, Bhushan, and Bhattacharyya [13.9]).

13.8 For the network N shown in Fig. 13.12 construct the network N_{DT} (the dual transpose of N) defined in Exercise 13.7.

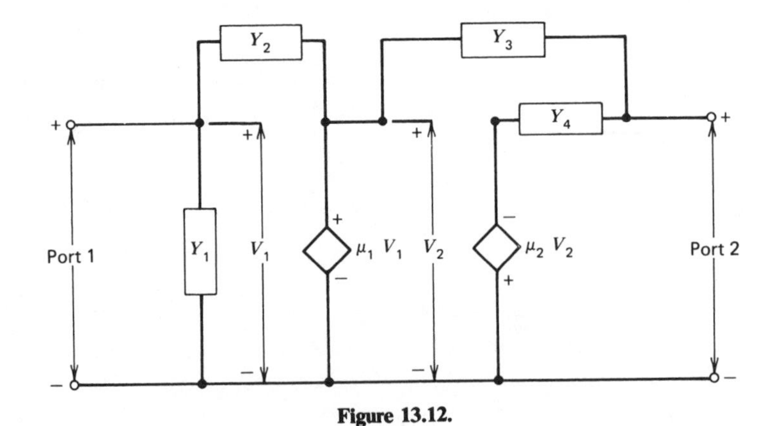

Figure 13.12.

13.6 REFERENCES

13.1 A. C. Aitken, *Determinants and Matrices*, Interscience, New York, 1956.

13.2 N. Balabanian and T. A. Bickart, *Electrical Network Theory*, Wiley, New York, 1969.

13.3 S. K. Mitra, *Analysis and Synthesis of Linear Active Networks*, Wiley, New York, 1969.

13.4 G. E. Sharpe and B. Spain, "On the Solution of Networks by means of the Equi-Cofactor Matrix," *IRE Trans. Circuit Theory*, Vol. CT-7, 230–239 (1960).

13.5 J. L. Bordewijk, "Inter-Reciprocity Applied to Electrical Networks," *Appl. Sci. Res.*, Vol. B6, 1–74 (1956).

13.6 S. W. Director and R. A. Rohrer, "Automated Network Design—The Frequency Domain Case," *IEEE Trans. Circuit Theory*, Vol. CT-16, 330–337 (1969).

13.7 B. B. Bhattacharyya and M. N. S. Swamy, "Network Transposition and Its Application in Synthesis," *IEEE Trans. Circuit Theory*, Vol. CT-18, 394–397 (1971).

13.8 M. N. S. Swamy, C. Bhushan, and B. B. Bhattacharyya, "Generalized Duals, Generalized Inverses, and Their Applications," *Radio and Electron. Eng.*, Vol. 44, 95–97 (1974).

13.9 M. N. S. Swamy, C. Bhushan, and B. B. Bhattacharyya, "Generalized Dual Transposition and Its Applications," *J. Franklin Inst.*, Vol. 301, 465–474 (1976).

13.10 K. Dagget and J. Vlach, "Sensitivity-Compensated Active Networks," *IEEE Trans. Circuit Theory*, Vol. CT-16, 416–422 (1969).

13.11 W. K. Chen, *Applied Graph Theory*, North-Holland, Amsterdam, 1971.

13.12 S. Seshu and M. B. Reed, *Linear Graphs and Electrical Networks*, Addison-Wesley, Reading, Mass., 1961.

13.13 S. P. Chan, *Introductory Topological Analysis of Electrical Networks*, Holt, Rinehart and Winston, New York, 1969.

13.14 W. Mayeda, *Graph Theory*, Wiley-Interscience, New York, 1972.

13.15 J. Numata and M. Iri, "Mixed-Type Topological Formulas for General Linear Networks," *IEEE Trans. Circuit Theory*, Vol. CT-20, 488–494 (1973).

13.16 P. Penfield, Jr., R. Spence, and S. Duinker, *Tellegen's Theorem and Electrical Networks*, M.I.T. Press, Cambridge, Mass., 1970.

13.17 P. Penfield, Jr., R. Spence, and S. Duinker, "A Generalized Form of Tellegen's Theorem," *IEEE Trans. Circuit Theory*, Vol. CT-17, 302–305 (1970).

13.18 C. A. Desoer, "On the Description of Adjoint Networks," *IEEE Trans. Circuits and Sys.*, Vol. CAS-22, 585–587 (1975), Erratum, *IEEE Trans. Circuits and Sys.*, Vol. CAS-23, 58 (1976).

13.19 S. W. Director and R. A. Rohrer, "On the Design of Resistance n-Port Networks by Digital Computer," *IEEE Trans. Circuit Theory*, Vol. CT-16, 337–346 (1969).

13.20 M. N. S. Swamy, C. Bhushan, and K. Thulasiraman, "Bounds on the Sum of Element Sensitivity Magnitudes for Network Functions," *IEEE Trans. Circuit Theory*, Vol. CT-19, 502–504 (1972).

13.21 M. N. S. Swamy, C. Bhushan, and K. Thulasiraman, "Sensitivity Invariants for Linear Time-Invariant Networks," *IEEE Trans. Circuit Theory*, Vol. CT-20, 21–24 (1973).

III

Algorithmic Graph Theory

Chapter 14

Algorithmic Analysis

Graphs arise in the study of several practical problems. The first step in such studies is to discover graph-theoretic properties of the problem under consideration which would help us in the formulation of a method of solution to the problem. For example, in Chapters 11 through 13, we established several useful graph-theoretic properties of electrical networks, and using these properties we developed, among other things, different methods for formulating network equations. As another example, in the study of transport networks we are interested in determining a maximum flow. As a first step, we develop in Section 15.7 several properties of these networks which would lead us to the maximum flow minimum cut theorem. This theorem forms the basis of Ford-Fulkerson's labeling algorithm for finding a maximum flow. Usually solving a problem involves analysis of a graph or testing a graph for some specified property. Graphs which arise in the study of real-life problems are very large and complicated. Analysis of such graphs in an efficient manner, therefore, involves the design of efficient computer algorithms.

In this part of the book we discuss several graph algorithms. Our main concern is to establish the theoretical foundation on which the design of the algorithms is based. We also develop results concerning the computational complexity of some of these algorithms. In certain cases, computational complexity depends crucially on the computational complexity of implementing certain basic operations such as finding the union of disjoint sets (see, for example, Section 14.5). In such cases we provide adequate references for the interested reader to pursue further.

The computational complexity of an algorithm is a measure of the running time of the algorithm. Thus it is a function of the size of the input. In the case of graph algorithms, complexity results will be in terms of the number of vertices and the number of edges in the graph. In the following, a function $g(n)$ is said to be $O(f(n))$ if and only if there exist constants c and n_0 such that $|g(n)| \leqslant c|f(n)|$ for all $n \geqslant n_0$. Furthermore all the complexity results will be with respect to the worst-case analysis.

There are different methods of representing a graph on a computer. Two of the most common methods use the adjacency matrix (Section 6.10) and the adjacency list. Adjacency matrix representation is not a very efficient one in the case of sparse graphs. In the adjacency list representation, we associate with each vertex a list which contains all the edges incident on it. A detailed discussion of data structures for representing a graph may be found in some of the references listed at the end of this chapter.

There are several algorithms which are hard to design but whose proof of correctness is trivial. There are several others which are easy to conceive but whose proof of correctness is very hard. Furthermore there are trivial algorithms whose analysis is not. Examples of these classes of algorithms are found in different sections of this and the next chapter.

For purposes of our discussions here we classify algorithms into two groups: algorithms which concern the analysis of graphs and those which concern optimization problems on graphs. In this chapter, we study algorithms which belong to the former group. Broadly speaking, this chapter discusses algorithms relating to the following main topics :

1. Transitive closure,
2. Transitive orientation,
3. Depth-first search,
4. Biconnectivity and strong connectivity,
5. Program graphs.

14.1 TRANSITIVE CLOSURE

We may recall (Section 5.2) that a binary relation on a set is a collection of ordered pairs of the elements of the set. The *transitive closure* of a binary relation R is a relation R^* defined as follows:

$x\,R^*\,y$ if and only if there exists a sequence

$$x_0 = x, x_1, x_2, \ldots, x_k = y$$

such that $k > 0$ and $x_0\,R\,x_1, x_1\,R\,x_2, \ldots, x_{k-1}\,R\,x_k$.

Clearly, if $x\,R\,y$, then $x\,R^*\,y$. Hence $R \subseteq R^*$. Further, it can be easily shown that R^* is transitive. In fact, it is the smallest transitive relation containing R. So if R is transitive, then $R^* = R$.

As we have already pointed out in Section 5.2, a binary relation can be represented by a directed graph. Suppose that G is the directed graph

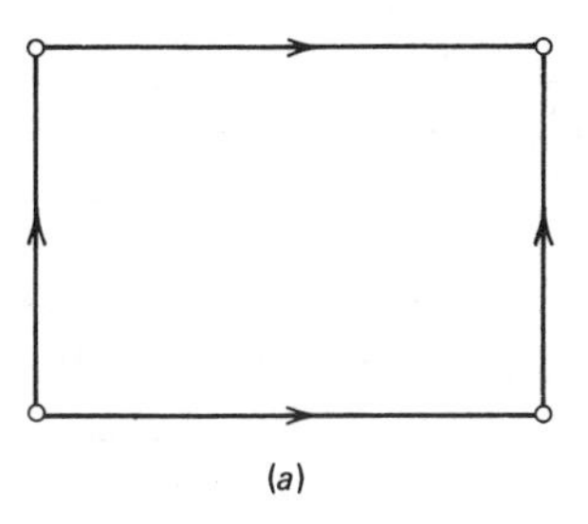

(a)

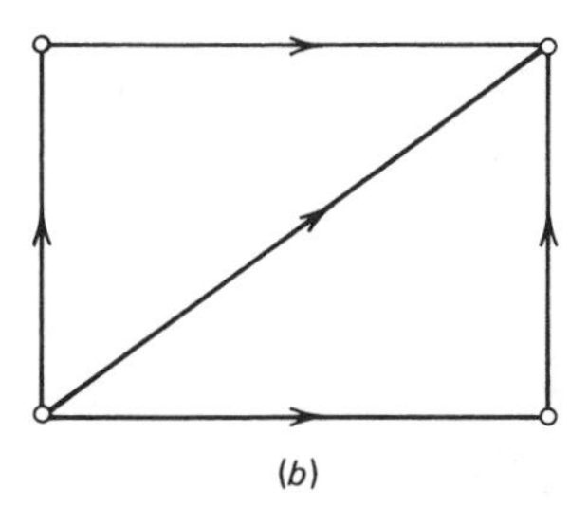

(b)

Figure 14.1. (*a*) Graph G. (*b*) G^*, transitive closure of G.

representing a relation R. The directed graph G^* representing the transitive closure R^* of R is called the *transitive closure* of G. It follows from the definition of R^* that the edge (x, y), $x \neq y$, is in G^* if and only if there exists in G a directed path from the vertex x to the vertex y. Similarly the self-loop (x, x) at vertex x is in G^* if and only if there exists in G a directed circuit containing x. For example, the graph shown in Fig. 14.1*b* is the transitive closure of the graph of Fig. 14.1*a*.

Suppose that we define the *reachability matrix* of an n-vertex directed graph G as an $n \times n$ (0, 1) matrix in which the (i, j) entry is equal to 1 if and only if there exists a directed path from vertex i to vertex j when $i \neq j$, or a directed circuit containing vertex i when $i = j$. In other words, the (i, j) entry of the reachability matrix is equal to 1 if and only if vertex j is reachable from vertex i through a sequence of directed edges. It is now easy to see that the adjacency matrix of G^* is the same as the reachability matrix of G.

The problem of constructing the transitive closure of a directed graph arises in several applications. For examples, see Gries [14.1]. In this section we discuss an elegant and computationally efficient algorithm due to Warshall [14.2] for computing the transitive closure. We also discuss a variation of Warshall's algorithm given by Warren [14.3].

Let G be an n-vertex directed graph with its vertices denoted by the integers $1, 2, \ldots, n$. Let $G^0 = G$. Warshall's algorithm constructs a sequence of graphs so that $G^i \subseteq G^{i+1}$, $0 \leqslant i \leqslant n-1$, and G^n is the transitive closure of G. The graph G^i, $i \geqslant 1$, is obtained from G^{i-1} by processing vertex i in G^{i-1}. Processing vertex i in G^{i-1} involves addition of new edges to G^{i-1} as described below.

Let, in G^{i-1}, the edges $(i, k), (i, l), (i, m), \ldots$ be incident out of vertex i. Then for each edge (j, i) incident into vertex i, add to G^{i-1} the edges $(j, k), (j, l), (j, m), \ldots$ if these edges are not already present in G^{i-1}. The graph that results after vertex i is processed is denoted as G^i.

Warshall's algorithm is illustrated in Fig. 14.2. It is clear that $G^i \subseteq G^{i+1}$, $i \geqslant 0$. To show that G^n is the transitive closure of G we need to prove the following result.

THEOREM 14.1.

1. Suppose that, for any two vertices s and t, there exists in G a directed path P from vertex s to vertex t such that all its vertices other than s and t are from the set $\{1, 2, \ldots, i\}$. Then G^i contains the edge (s, t).

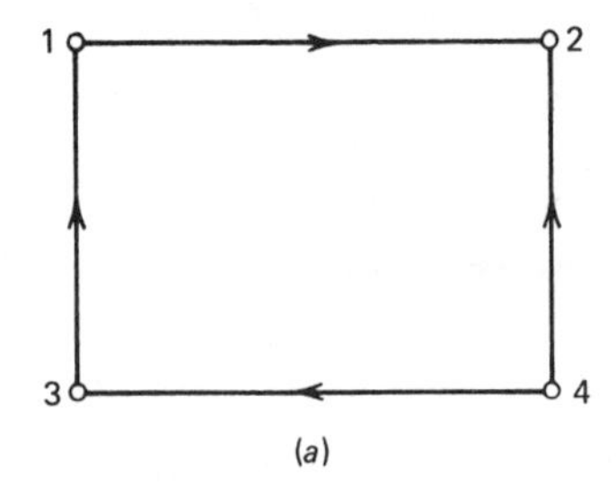

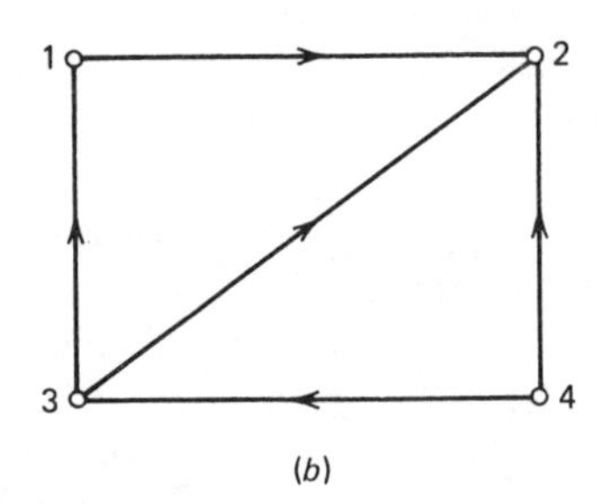

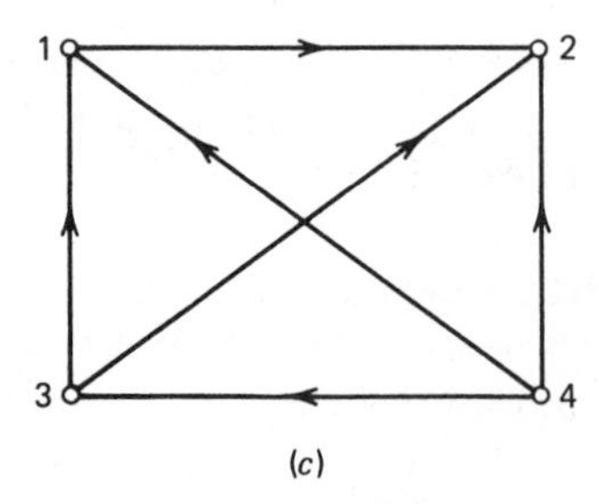

Figure 14.2. Illustration of Warshall's algorithm. (a) G^0. (b) $G^1 = G^2$. (c) $G^3 = G^4$.

2. Suppose that, for any vertex s, there exists in G a directed circuit C containing s such that all its vertices other than s are from the set $\{1,2,\ldots,i\}$. Then G^i contains the self-loop (s, s).

Proof

1. Proof is by induction on i.
   Clearly the result is true for G^1 since Warshall's construction, while processing vertex 1, introduces the edge (s, t) if G^0 $(=G)$ contains the edges $(s, 1)$ and $(1, t)$.
   Let the result be true for all G^k, $k<i$.
   Suppose that i is not an internal vertex of P. Then it follows from the induction hypothesis that G^{i-1} contains the edge (s, t). Hence G^i also contains (s, t) because $G^{i-1} \subseteq G^i$.
   Suppose that i is an internal vertex of P. Then again from the induction hypothesis it follows that G^{i-1} contains the edges (s, i) and (i, t). Therefore, while processing vertex i in G^{i-1}, the edge (s, t) is added to G^i.
2. Proof follows along the same lines as above. ■

As an immediate consequence of this theorem we get the following.

Corollary 14.1.1 G^n is the transitive closure of G. ■

We next give a formal description of Warshall's algorithm. In this description the graph G is represented by its adjacency matrix M and the symbol $\vee$ stands for Boolean addition.

Algorithm 14.1 Transitive Closure (Warshall)

S1. (Initialization) M is the adjacency matrix of G.
S2. Do S3 for $i=1,2,\ldots, n$.
S3. Do S4 for $j=1,2,\ldots, n$.
S4. If $M(j, i)=1$, do S5 for $k=1,2,\ldots, n$.
S5. $M(j, k)=M(j, k) \vee M(i, k)$.
S6. HALT. (M is the adjacency matrix of G^*.) ■

Note that the matrix M (when the algorithm begins to execute step S3 with $i=p$) is the adjacency matrix of G^{p-1}. Further, processing a diagonal entry does not result in adding new nonzero entries.
A few observations are now in order:

1. Warshall's algorithm transforms the adjacency matrix M of a graph G to the adjacency matrix of the transitive closure of G by suitably overwriting on M. It is for this reason that the algorithm is said to work "in place."

2. The algorithm processes all the edges incident into a vertex before it begins to process the next vertex. In other words, it processes the matrix M columnwise. Hence we describe Warshall's algorithm as *column-oriented.*
3. While processing a vertex no new edge (that is, an edge which is not present when the processing of that vertex begins) incident into the vertex is added to the graph. This means that while processing a vertex we can choose the edges incident into the vertex in any arbitrary order.
4. Suppose that the edge (j, i) incident into the vertex i is not present while vertex i is processed, but that it is added subsequently while processing some vertex k, $k>i$. Clearly this edge is not processed while processing vertex i. Neither will it be processed later since no vertex is processed more than once. In fact, such an edge will not result in adding any new edges.
5. Warshall's algorithm is said to work in one pass since each vertex is processed exactly once.

Suppose that we wish to modify Warshall's algorithm so that it becomes *row-oriented*. In a row-oriented algorithm, while processing a vertex, all the edges incident out of the vertex are to be processed. The processing of the edge (i, j) introduces the edges (i, k) for every edge (j, k) incident out of vertex j. Therefore new edges incident out of a vertex may be added while processing a vertex "rowwise." Some of these newly added edges may not be processed before the processing of the vertex under consideration is completed. If the processing of these edges is necessary for the computation of the transitive closure, then such a processing can be done only in a second pass. Thus, in general, a row-oriented algorithm may require more than one pass to compute the transitive closure.

For example, consider the graph G of Fig. 14.3*a*. After processing rowwise the vertices of G we obtain the graph G' shown in Fig. 14.3*b*. Clearly, G' is not the transitive closure of G since the edge $(1,2)$ is yet to be added. It may be noted that the edge $(1,3)$ is not processed in this pass because it is added only while processing the edge $(1,4)$. The same is the case with the edge $(4,2)$.

Suppose we next process the vertices of G'. In this second pass the edge $(1, 2)$ is added while processing vertex 1 and we get the transitive closure G^* shown in Fig. 14.3*c*. Thus in the case of the graph of Fig. 14.3*a* two passes of the row-oriented algorithm are required.

Now the question arises whether two passes always suffice. The answer is in the affirmative, and Warren [14.3] has demonstrated this by devising a clever two-pass row-oriented algorithm. In this algorithm, while processing a vertex, say vertex i, in the first pass only edges connected to vertices less than i are processed, and in the second pass only edges connected to vertices greater than i are processed. In other words, the algorithm transforms the adjacency matrix M of the graph G to the adjacency matrix of G^* by

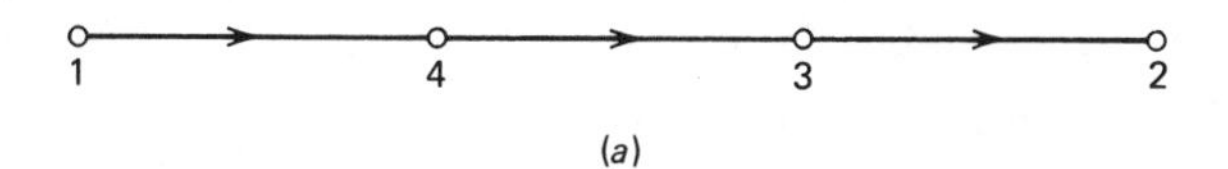

(a)

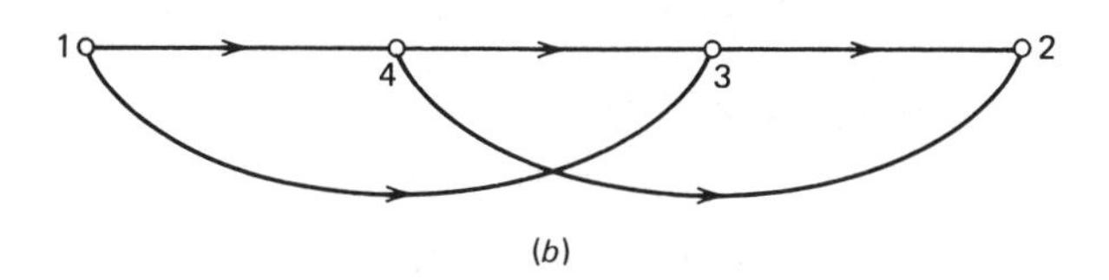

(b)

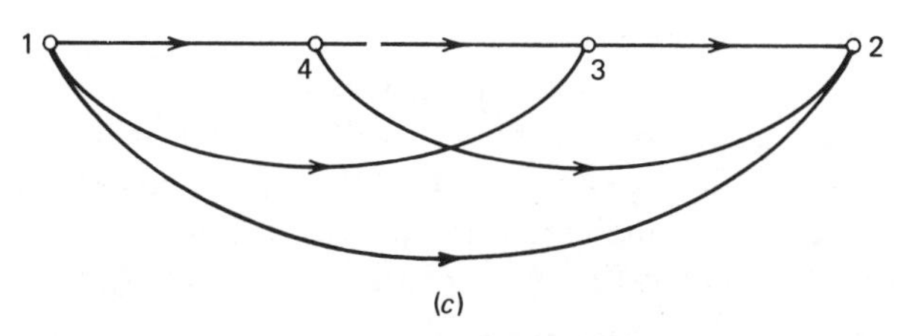

(c)

Figure 14.3. An example of row-oriented transitive closure algorithm. (*a*) G. (*b*) G'. (*c*) G^*.

processing in the first pass only entries below the main diagonal of M and in the second pass only entries above the main diagonal. Thus during each pass at most $n(n-1)/2$ edges are processed. A description of Warren's modification of Warshall's algorithm now follows.

Algorithm 14.2 Transitive Closure (Warren)

S1. M is the adjacency matrix of G.

S2. Do S3 for $i=2,3,\ldots,n$.

S3. Do S4 for $j=1,2,\ldots,i-1$.

S4. If $M(i,j)=1$, then do S5 for $k=1,2,\ldots,n$.

S5. $M(i,k)=M(i,k)\vee M(j,k)$.

S6. Do S7 for $i=1,2,\ldots,n-1$.

S7. Do S8 for $j=i+1,i+2,\ldots,n$.

S8. If $M(i, j)=1$, then do S9 for $k=1,2,\ldots, n$.

S9. $M(i, k)=M(i, k)\vee M(j, k)$.

S10. HALT. (M is the adjacency matrix of G^*.) ■

Note that in the above algorithm the steps S2 through S5 correspond to the first pass and steps S6 through S9 correspond to the second pass.

As an example, consider again the graph shown in Fig. 14.3*a*. At the end of the first pass of Warren's algorithm we obtain the graph shown in Fig. 14.4*a*, and at the end of the second pass we get the transitive closure G^* shown in Fig. 14.4*b*

The proof of correctness of Warren's algorithm is based on the following lemma.

LEMMA 14.1. Suppose that, for any two vertices s and t, there exists in G a directed path P from s to t. Then the graph that results after processing vertex s in the first pass (steps S2 through S5) of Warren's algorithm contains an edge (s, r), where r is a successor of s on P and either $r>s$ or $r=t$.

Proof

Proof is by induction on s.

If $s=1$, then the lemma is clearly true because all the successors of 1 on P are greater than 1. Assume that the lemma is true for all $s<k$ and let $s=k$. Suppose (s, i_1) is the first edge on P. If $i_1>s$, then clearly the lemma is true.

If $i_1<s$, then by the induction hypothesis the graph that results after processing vertex i_1 in the first pass contains an edge (i_1, i_2), where i_2 is a successor of i_1 on P and either $i_2>i_1$ or $i_2=t$.

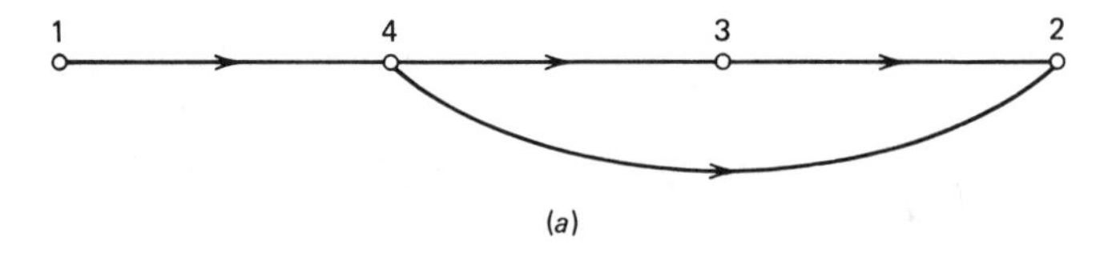

(*a*)

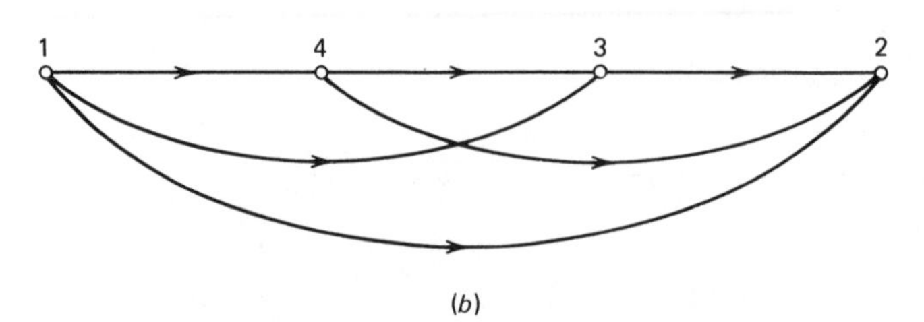

(*b*)

Figure 14.4. Illustration of Warren's algorithm.

If $i_2 \neq t$ and $i_2 < s$, then again by the induction hypothesis the graph that results after processing vertex i_2 in the first pass contains an edge (i_2, i_3) where i_3 is a successor of i_2 on P, and either $i_3 > i_2$ or $i_3 = t$.

If $i_3 \neq t$ and $i_3 < s$, we repeat the above arguments on i_3 until we locate an i_m such that either $i_m > s$ or $i_m = t$. Thus the graph that we have before the processing of vertex s begins contains the edges $(s, i_1), (i_1, i_2), \ldots, (i_{m-1}, i_m)$ such that the following conditions are satisfied:

1. i_p is a successor of i_{p-1} on P, $p \geqslant 2$.
2. $i_{m-1} > i_{m-2} > i_{m-3} > \cdots > i_1$, and $i_k < s$ for $k \neq m$.
3. $i_m = t$ or $i_m > s$.

We now begin to process vertex s. Processing of (s, i_1) introduces the edge (s, i_2) because of the presence of (i_1, i_2). Since $i_2 > i_1$, the edge (s, i_2) is subsequently processed. Processing of this edge introduces (s, i_3) because of the presence of (i_2, i_3), and so on. Thus when the processing of s is completed, the required edge (s, i_m) is present in the resulting graph. ■

THEOREM 14.2. Warren's algorithm computes the transitive closure of a graph G.

Proof

We need to consider two cases:

Case 1 For any two vertices s and t there exists in G a directed path P from s to t.

Let (i, j) be the first edge on P (as we proceed from s to t) such that $i > j$. Then it follows from the previous lemma that the graph that we have, before the second pass of Warren's algorithm begins, contains an edge (i, k), where k is a successor of i on P and either $k = t$ or $k > i$. Thus after the first pass is completed there exists a path P': $s, i_1, i_2, \ldots, i_m, t$ such that $s < i_1 < i_2 < \cdots < i_m$ and each i_{j+1} is a successor of i_j on P.

When in the second pass we process vertex s, the edge (s, i_1) is first encountered. The processing of this edge introduces the edge (s, i_2) because of the presence of the edge (i_1, i_2). Since $i_2 > i_1$, the edge (s, i_2) is processed subsequently. This, in turn, introduces the edge (s, i_3), and so on. Thus when the processing of s is completed, we have the edge (s, t) in the resulting graph.

Case 2 There exists in G a directed circuit containing a vertex s.

In this case we can prove along the same lines as above that when the processing of vertex s is completed in the second pass, the resulting graph contains the self-loop (s, s). ■

Clearly both Warshall's and Warren's algorithms have the worst-case complexity $O(n^3)$. However, Warren's algorithm will execute faster than Warshall's in the case of large sparse matrices, particularly in the case of paging environment. Warren [14.3] refers to other row-oriented algorithms.

Arlazarov, Dinic, Kronrod, and Faradžev [14.4] have presented a $O(n^3/\log n)$ algorithm. This is based on the "Four Russians" algorithm for multiplying Boolean matrices.

Strassen [14.5] has given a $O(n^{\log_2 7})$ algorithm for multiplying two $n \times n$ matrices. Using this algorithm, Fischer and Meyer [14.6] and Furman [14.7] have presented $O(n^{\log_2 7} \log n)$ algorithms for the transitive closure problem. Other algorithms, which are based on matrix multiplication, are those due to Munro [14.8] and O'Neil and O'Neil [14.9].

Eve and Kurki-Suonio [14.10] have presented an algorithm which is based on Tarjan's algorithm for finding the strongly connected components of a directed graph (see Section 14.4). Munro's algorithm also determines as a first step the strongly connected components of the given graph.

Purdom's algorithm [14.11] is $O(n^2)$ in many cases, though in the worst case it is $O(n^3)$. However, this algorithm is much more complicated than many of the other transitive closure algorithms.

Recently Schnorr [14.12] presented an algorithm with expected time $O(n + m^*)$, where m^* is the expected number of edges in the transitive closure.

Syslo and Dzikiewicz [14.13] discuss computational experiences with several of the transitive closure algorithms.

14.2 TRANSITIVE ORIENTATION

An undirected graph G is *transitively orientable* if we can assign orientations to the edges of G so that the resulting directed graph is transitive. If G is transitively orientable, then $\vec{G}$ will denote a *transitive orientation* of G.

For example, the graph shown in Fig. 14.5*a* is transitively orientable. A transitive orientation of this graph is shown in Fig. 14.5*b*.

In this section we discuss an algorithm due to Pnueli, Lempel, and Even [14.14] to test whether a simple undirected graph G is transitively orientable and obtain a transitive orientation $\vec{G}$ if one exists. To aid the development and presentation of this algorithm, we introduce some notations:

- $i \to j$ means that vertex i is connected to vertex j by an edge oriented from i to j.
- $i \leftarrow j$ is similarly defined.
- $i - j$ means that there is an edge connecting vertex i and vertex j.
- $i \not- j$ means that there is no edge connecting vertex i and vertex j.
- $i \not\to j$ means that either $i \not- j$ or $i \leftarrow j$ or the edge $i - j$ is not oriented.

$i \to j$, $i \leftarrow j$, $i - j$, etc., will also be used to denote the corresponding edges.

Consider now an undirected graph $G = (V, E)$ which is transitively orientable. Let $\vec{G} = (V, \vec{E})$ denote a transitive orientation of G.

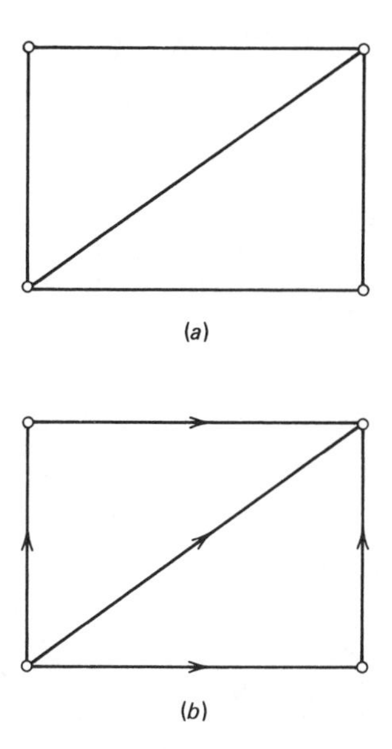

Figure 14.5. (*a*) Graph G. (*b*) A transitive orientation of G.

Suppose that there exist three vertices $i, j, k \in V$ such that $i \rightarrow j$, $j — k$, and $i \not— k$; then transitivity of $\vec{G}$ requires that $j \leftarrow k$. Similarly, if, for $i, j, k \in V$, $i \rightarrow j$, $i — k$, and $j \not— k$, then transitivity of $\vec{G}$ requires that $i \rightarrow k$.

These two observations lead to the following simple rules which form the basis of Pnueli, Lempel and Even's algorithm.

Rule $\mathbf{R_1}$ For $i, j, k \in V$, if $i \rightarrow j$, $j — k$, and $i \not— k$, then orient the edge $j — k$ as $j \leftarrow k$.

Rule $\mathbf{R_2}$ For $i, j, k \in V$, if $i \rightarrow j$, $i — k$, and $j \not— k$, then orient the edge $i — k$ as $i \rightarrow k$.

These two rules are illustrated in Fig. 14.6, where a dashed line indicates the absence of the corresponding edge.

A description of the transitive orientation algorithm now follows.

Algorithm 14.3 Transitive Orientation (Pnueli, Lempel, and Even)

S1. G is the given simple undirected graph. Set $i = 1$.

S2. (Phase i begins.) Select an edge e of the graph G and assign an arbitrary orientation to e. Assign, whenever possible, orientations to the edges in G adjacent to e, using Rule R_1 or Rule R_2. The directed edge e is now labeled "examined."

S3. Test if there exists in G a directed edge which has not been labeled "examined." If yes, go to step S4. Otherwise go to step S6.

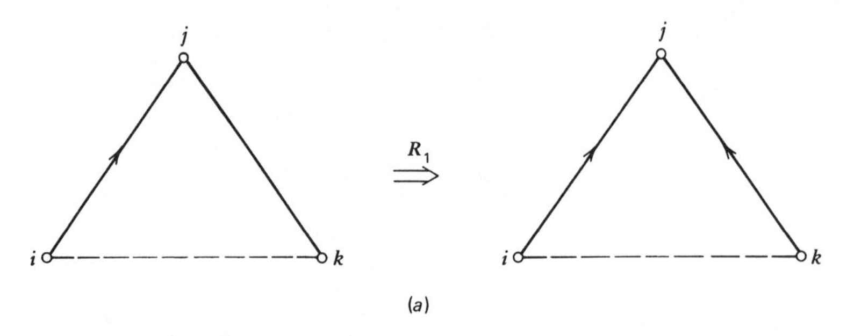

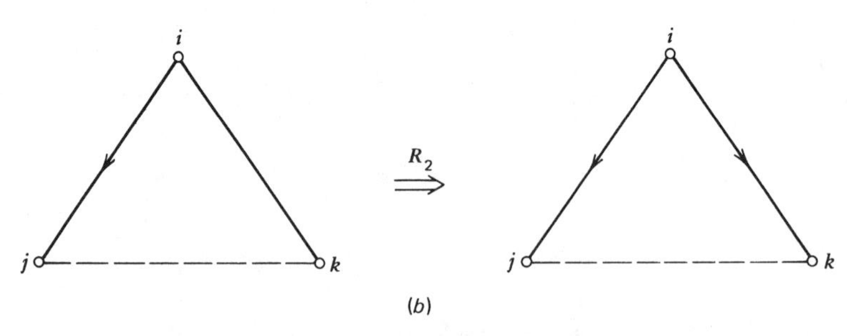

Figure 14.6. (*a*) Rule R_1. (*b*) Rule R_2.

S4. Let $i \to j$ be a directed edge in G which has not been labeled "examined." Now do the following, whenever applicable, for each edge in G incident on i or j, and then label $i \to j$ "examined":

Case 1 Let the edge under consideration be j—k.

a. (Application of rule R_1.) If $i \not\!\!- k$ and the edge j—k is not oriented, then orient j—k as $j \leftarrow k$.

b. (Contradiction of rule R_1.) If $i \not\!\!- k$ and the edge j—k is already oriented as $j \to k$, then a contradiction of rule R_1 has occurred. Go to step S9.

Case 2 Let the edge under consideration be i—k.

a. (Application of Rule R_2.) If $j \not\!\!- k$ and the edge i—k is not oriented, then orient i—k as $i \to k$.

b. (Contradiction of Rule R_2.) If $j \not\!\!- k$ and edge i—k is already oriented as $i \leftarrow k$, then a contradiction of Rule R_2 has occurred. Go to step S9.

S5. Go to step S3.

S6. (Phase i has ended successfully.) Test whether all the edges of G have been assigned orientations. If yes, go to step S8. Otherwise remove from G all its directed edges and let G' be the resulting graph.

S7. Set $G = G'$ and $i = i + 1$. Go to Step S2.

S8. (All the edges of the given graph have been assigned orientations which are consistent with Rules R_1 and R_2. These orientations define a transitive orientation of the given graph.) HALT.

S9. (The graph G is not transitively orientable.) HALT. ■

The main step in the above algorithm is S4. In this step we examine each edge adjacent to a directed edge, say, edge $i \rightarrow j$. If such an edge is already oriented, then we test whether its orientation and that of $i \rightarrow j$ are consistent with Rule R_1 or Rule R_2. If an edge under examination is not yet oriented, then we assign to it, if possible, an orientation using Rule R_1 or Rule R_2.

As we can see, the algorithm consists of different phases. Each phase involves execution of step S2 and repeated executions of step S4 as more and more edges get oriented. If a phase ends without detecting any contradiction of Rule R_1 or Rule R_2, then it means that no more edges can be assigned orientations in this phase by application of the two rules, and that all the orientations assigned in this phase are consistent with these rules.

The algorithm terminates either (1) by detecting a contradiction of Rule R_1 or Rule R_2, or (2) by assigning orientations to all the edges of the given graph such that these orientations are consistent with Rules R_1 and R_2. In the former case the graph is not transitively orientable, and in the latter case the graph is transitively orientable with the resulting directed graph defining a transitive orientation.

The complexity of the algorithm depends on the complexity of executing step S4. This step is executed at most m times, where m is the number of edges in the given graph. Each execution of step S4 involves examining all the edges adjacent to an oriented edge. So the number of operations required to execute step S4 is proportional to 2Δ, where Δ is the maximum degree in the given graph. Thus the overall complexity of the algorithm is $O(2m\Delta)$.

Next we illustrate the algorithm with two examples. Consider first the graph G shown in Fig. 14.7*a*.

Phase 1 We begin by orienting edge 7—2 as $7 \rightarrow 2$. By Rule R_2, $7 \rightarrow 2$ implies that $7 \rightarrow 4$ and $7 \rightarrow 5$, and $7 \rightarrow 4$ implies that $7 \rightarrow 6$. Rule R_1 is not applicable to any edge adjacent to $7 \rightarrow 2$.

By Rule R_2, $7 \rightarrow 5$ and $7 \rightarrow 6$ imply that $7 \rightarrow 1$ and $7 \rightarrow 3$, respectively. In this phase no more edges can be assigned orientations. We can also check that all the assigned orientations are consistent with Rules R_1 and R_2. Phase 1 now terminates, and the edges oriented in this phase are shown in Fig. 14.7*b*.

Now remove from the graph G of Fig. 14.7*a* all the edges which are oriented in phase 1. The resulting graph G' is shown in Fig. 14.7*c*. Phase 2 now begins, and G' is the graph under consideration.

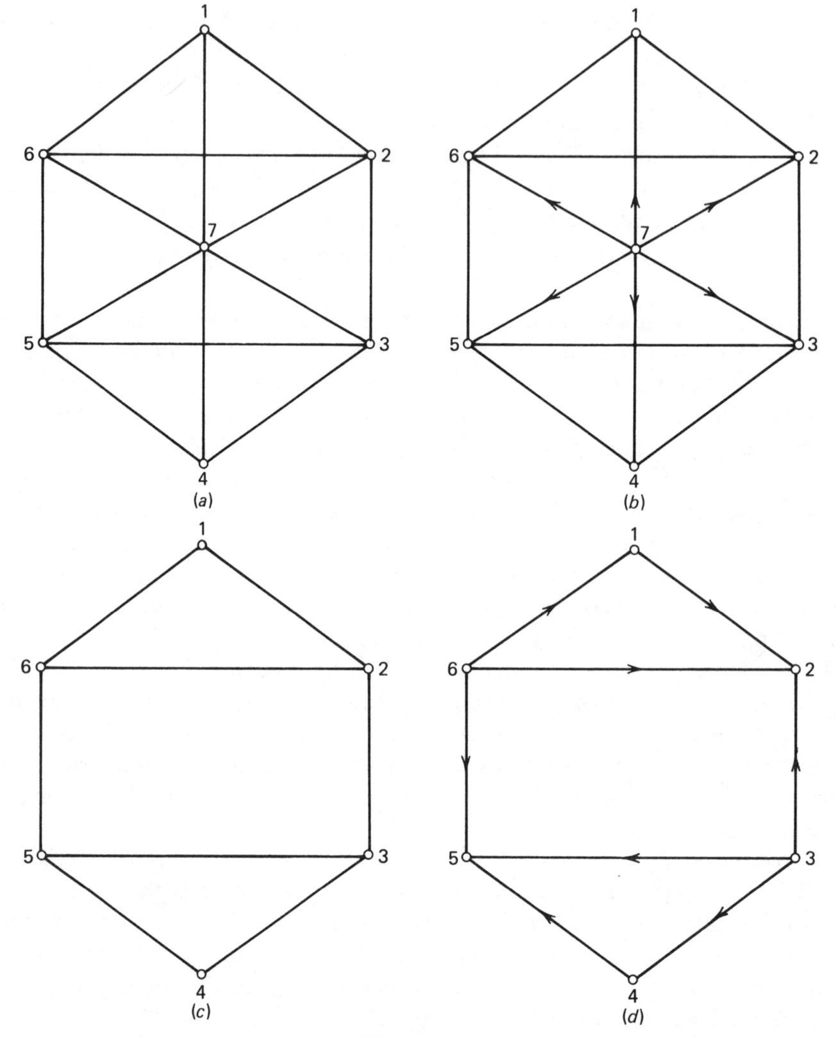

Figure 14.7.

Phase 2 We begin by orienting edge 1—2 as 1→2. This results in the following sequences of implications:

$$(1\rightarrow 2) \underset{R_1}{\Rightarrow} (2\leftarrow 3) \underset{R_2}{\Rightarrow} (3\rightarrow 5) \underset{R_1}{\Rightarrow} (6\rightarrow 5) \underset{R_1}{\Rightarrow} (4\rightarrow 5)$$

$$(2\leftarrow 3) \underset{R_2}{\Rightarrow} (3\rightarrow 4)$$

$$(6\rightarrow 5) \underset{R_2}{\Rightarrow} (6\rightarrow 1)$$

$$(6\rightarrow 5) \underset{R_2}{\Rightarrow} (6\rightarrow 2)$$

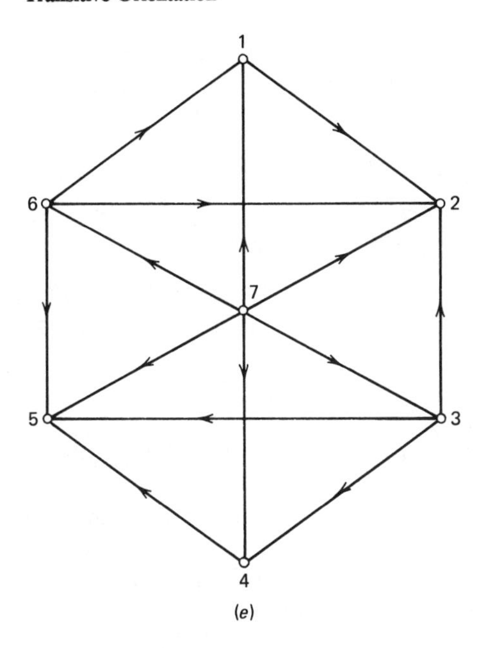

Figure 14.7. (*continued*)

Thus all the edges of G' have now been oriented as shown in Fig. 14.7*d*. These orientations of G' are also consistent with the Rules R_1 and R_2. Therefore phase 2 terminates successfully.

The resulting transitive orientation of G is shown in Fig. 14.7*e*.

Consider next the graph shown in Fig. 14.8. We begin by orienting edge 1—2 as 1→2. This leads to the following sequence of implications:

$$(1\rightarrow 2) \underset{R_1}{\Rightarrow} (2\leftarrow 3) \underset{R_2}{\Rightarrow} (3\rightarrow 4) \underset{R_1}{\Rightarrow} (4\leftarrow 5) \underset{R_2}{\Rightarrow} (5\rightarrow 1) \underset{R_1}{\Rightarrow} (1\leftarrow 2),$$

which requires that 1—2 be directed as 1←2, contrary to the orientation we

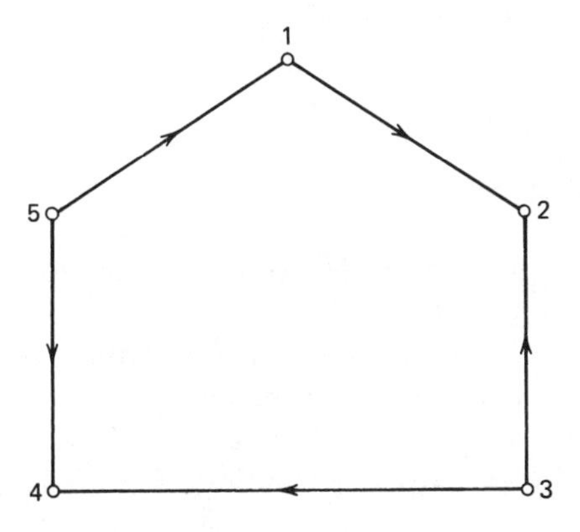

Figure 14.8.

have already assigned to the edge 1—2. Thus a contradiction of Rule R_1 is observed. Hence the graph of Fig. 14.8 is not transitively orientable.

We now proceed to prove the correctness of Algorithm 14.3. To do so we need to prove the following two assertions.

Assertion 1 If Algorithm 14.3 terminates successfully (step S8), then the resulting directed graph is a transitive orientation of the given graph.

Assertion 2 If the given graph is transitively orientable, then Algorithm 14.3 terminates successfully.

We first consider Assertion 1.

Given an undirected graph $G=(V, E)$. Suppose that the algorithm terminates successfully.

Consider now any two edges $i \rightarrow j$ and $k \rightarrow l$ which are assigned orientations in the same phase of the algorithm. Then we can construct a sequence of implications, which starts with the directed edge $i \rightarrow j$ and ends orienting the edge $k—l$ as $k \rightarrow l$. Such a sequence will be called a *derivation chain* from $i \rightarrow j$ to $k \rightarrow l$. Note that there may be more than one derivation chain from $i \rightarrow j$ to $k \rightarrow l$. For example, in the directed graph of Fig. 14.7*d* the following are two of the derivation chains from $2 \leftarrow 6$ to $3 \rightarrow 4$:

$$(2 \leftarrow 6) \Rightarrow (3 \rightarrow 2) \Rightarrow (3 \rightarrow 4),$$
$$(2 \leftarrow 6) \Rightarrow (5 \leftarrow 6) \Rightarrow (3 \rightarrow 5) \Rightarrow (3 \rightarrow 2) \Rightarrow (3 \rightarrow 4).$$

Thus it is clear that it is meaningful to talk about a shortest derivation chain between any pair of directed edges which are assigned orientations in the same phase of Algorithm 14.3. Proof of Assertion 1 is based on the following important Lemma.

LEMMA 14.2. After a successful completion of phase 1 it is impossible to have three vertices i, j, k such that $i \rightarrow j$ and $j \rightarrow k$ with $i \nrightarrow k$.

Proof

Note that "$i \nrightarrow k$" means that either $i \not- k$ or $i \leftarrow k$ or edge $i—k$ is not oriented.

It is clear that there is an edge connecting i and k. For otherwise we get a contradiction because by Rule R_1, $i \rightarrow j$ implies that $j \leftarrow k$.

Now assume that forbidden situations of the type $i \rightarrow j$ and $j \rightarrow k$ with $i \nrightarrow k$ exist after phase 1 of Algorithm 14.3. Then select, from among all the derivation chains which lead to a forbidden situation, a chain which is shortest with the minimum number of directed edges incident into k. Clearly, any such chain must be of length at least 3. Let one such chain be as follows:

$$(i \to j) \Rightarrow (\alpha_1) \Rightarrow (\alpha_2) \Rightarrow \cdots \Rightarrow (\alpha_{p-1}) \Rightarrow (j \to k).$$

Now α_{p-1} must be either $j \to j'$ for some j' or $k' \to k$ for some k'. Thus we need to consider two cases.

Case 1 Let α_{p-1} be $j \to j'$.

Then the derivation $(\alpha_{p-1}) \Rightarrow (j \to k)$ requires that $j' \not\to k$. Further $i \to j'$, for otherwise the derivation chain

$$(i \to j) \Rightarrow (\alpha_1) \Rightarrow \cdots \Rightarrow (\alpha_{p-2}) \Rightarrow (j \to j')$$

would lead to the forbidden situation $i \to j$ and $j \to j'$ with $i \not\to j'$. But this chain is shorter than our chain, contradicting its minimality.

The situation arising out of the above arguments is depicted in Fig. 14.9*a*, where a dashed line indicates the absence of the corresponding edge.

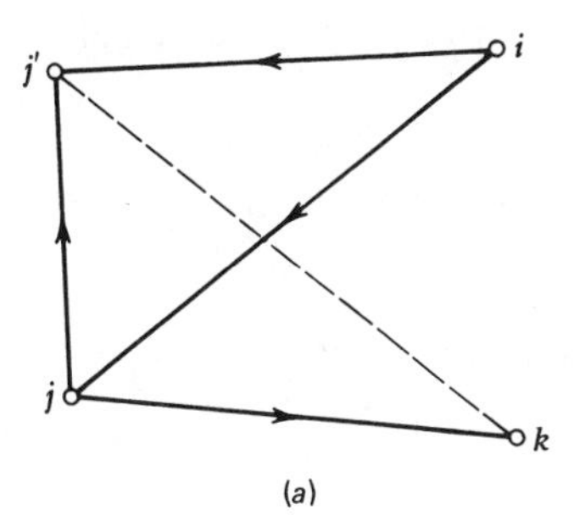

(*a*)

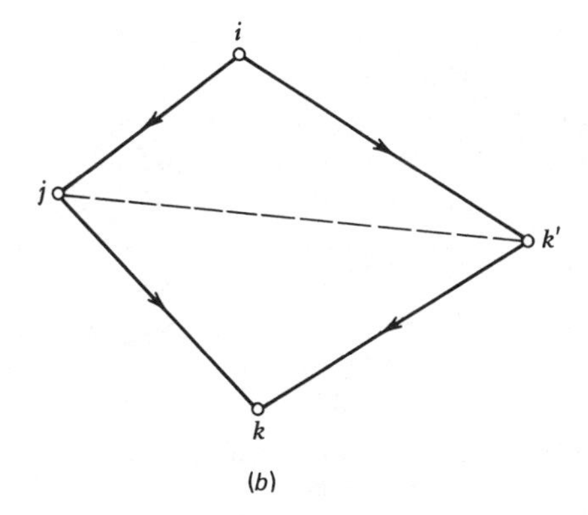

(*b*)

Figure 14.9.

Now $i \rightarrow j'$ and $j' \not\longrightarrow k$ imply that $i \rightarrow k$ by Rule R_2. But this contradicts our assumption that $i \not\rightarrow k$.

Case 2 Let α_{p-1} be $k' \rightarrow k$.

As in the previous case the derivation $(\alpha_{p-1}) \Rightarrow (j \rightarrow k)$ requires that $j \not\longrightarrow k'$. Further i—k', for otherwise $k' \rightarrow k$ would imply that $i \rightarrow k$, contrary to our assumption that $i \not\rightarrow k$. In addition, $i \rightarrow j$ and $j \not\longrightarrow k'$ imply that $i \rightarrow k'$.

The situation resulting from the above arguments is depicted in Fig. 14.9*b*. Now the derivation chain

$$(i \rightarrow k') \Rightarrow (i \rightarrow j) \Rightarrow (\alpha_1) \Rightarrow \cdots \Rightarrow (\alpha_{p-2}) \Rightarrow (k' \rightarrow k),$$

leading to the forbidden situation $i \rightarrow k'$ and $k' \rightarrow k$ with $i \not\rightarrow k$, is of the same length as our chain, but with one less edge entering k. This is again a contradiction of the assumption we have made about the choice of our chain. ■

Let E' be the set of edges which are assigned orientations in the first phase, and let $\vec{E}'$ be the corresponding set of directed edges. The following result is an immediate consequence of Lemma 14.2.

THEOREM 14.3. The subgraph $\vec{G}' = (V, \vec{E}')$ of the directed edges of the set $\vec{E}'$ is transitive. ■

THEOREM 14.4. If Algorithm 14.3 terminates successfully, then it gives a transitive orientation.

Proof

Proof is by induction on the number of phases in the algorithm. If the algorithm orients all the edges of the given graph in phase 1, then, by Theorem 14.3, the resulting orientation is transitive.

Let the given graph $G = (V, E)$ be oriented in p phases. Let E' be the set of edges which are assigned orientations in the first phase. Then, by Theorem 14.3, the subgraph $\vec{G}' = (V, \vec{E}')$ is transitive. Further, since the subgraph $G'' = (V, E - E')$ is orientable in $p - 1$ phases, it follows from the induction hypothesis that $\vec{G}'' = (V, \vec{E} - \vec{E}')$ is transitive. Now we prove that the directed graph $\vec{G} = (V, \vec{E})$ is transitive.

Suppose that $\vec{G}$ is not transitive, that is, there exist in $\vec{G}$ three vertices i, j, k such that $i \rightarrow j$ and $j \rightarrow k$, but $i \not\rightarrow k$. Then both $i \rightarrow j$ and $j \rightarrow k$ cannot belong to $\vec{E}'$ or to $\vec{E} - \vec{E}'$ because $\vec{G}'$ and $\vec{G}''$ are both transitive.

Without loss of generality, assume that $i \rightarrow j$ is in $\vec{E}'$ and $j \rightarrow k$ is in $\vec{E} - \vec{E}'$. Then there must be an edge i—k connecting i and k; for otherwise $i \rightarrow j$ would imply $j \leftarrow k$, by Rule R_1.

Suppose that the edge i—k is oriented as $i \leftarrow k$ in phase 1. Then $\vec{G}'$ is not transitive, resulting in a contradiction. On the other hand, if it is oriented as $i \leftarrow k$ in a latter phase, then $\vec{G}''$ is not transitive, again resulting in a contradiction.

Thus it is impossible to have in G three vertices i, j, k such that $i \rightarrow j$ and $j \rightarrow k$, but $i \not\rightarrow k$. Hence $\vec{G}$ is transitive. ■

Thus Assertion 1 is established.

We next proceed to establish Assertion 2.

Consider a graph $G=(V, E)$ which is transitively orientable. It is clear that if we reverse the orientations of all the edges in any transitive orientation of G, then the resulting directed graph is also a transitive orientation of G.

Suppose that we pick an edge of G and assign an arbitrary orientation to it. Let this edge be $i \rightarrow j$. If we now proceed to assign orientations to additional edges using Rules R_1 and R_2, then the edges so oriented will have the same orientations in all possible transitive orientations in which the edge i—j is oriented as $i \rightarrow j$. This is because once the orientation of i—j is specified, the orientations derived by Rules R_1 and R_2 are necessary for transitive orientability. It therefore follows that if we apply Algorithm 14.3 to the transitively orientable graph G, then phase 1 will terminate successfully without encountering any contradiction of Rule R_1 or Rule R_2. Further the edges oriented in the first phase will have the same orientations in some transitive orientation of G.

If we can prove that graph $G''=(V, E-E')$, where E' is the set of edges oriented in the first phase, is also transitively orientable, then it would follow that the second phase and also all other phases will terminate successfully giving a transitive orientation of G. Thus proving Assertion 2 is the same as establishing the transitive orientability of $G''=(V, E-E')$. Toward this end we proceed as follows.

Let the edges of the set E' be called *marked edges*, and the end vertices of these edges be called *marked vertices*. Let V' denote the set of marked vertices. Note that an unmarked edge may be incident on a marked vertex.

LEMMA 14.3. It is impossible to have three marked vertices i, j, k such that edges i—j and j—k are unmarked, and edge i—k is marked.

Proof

Assume that forbidden situations of the type mentioned in the lemma exist. In other words, assume that there exist triples of marked vertices i, j, k such that edges i—j and j—k are unmarked, and edge i—k is marked. For each such triple i, j, k there exists a marked edge j—l for some l, because j is a marked vertex. Therefore there exists a derivation chain from the marked edge i—k to the marked edge j—l.

Now select a forbidden situation with i, j, and k as the marked vertices such that there is a derivation chain P from i—k to j—l, which is a shortest one among all such chains that lead to a forbidden situation. This is shown in Fig. 14.10a, where a diamond on an edge indicates that the edge is marked, and a dashed line indicates the absence of the corresponding edge.

The next marked edge after i—k in the shortest chain P is either i—p or k—p, for some p. We assume, without loss of generality, that it is k—p. Hence $i \not- p$, for otherwise edge k—p would not have been marked from edge i—k. Further there exists the edge i—l connecting i and l, for otherwise edge i—j would have been marked. Thus p and l are distinct. Also there exists the edge j—p, for otherwise edge j—k would have been marked. The relations established so far are shown in Fig. 14.10b. Now j—p cannot be a marked edge, for marking it would result in marking i—j. Now we have a shorter derivation chain from edge k—p to edge j—l, leading to another forbidden situation where edges k—j and j—p are unmarked with edge k—p marked. A contradiction. ■

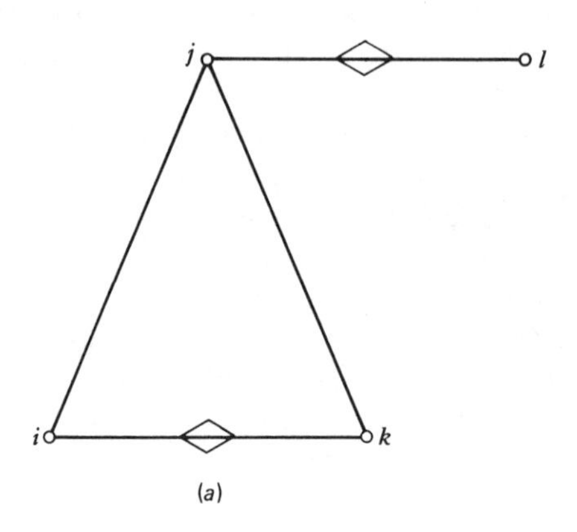

(a)

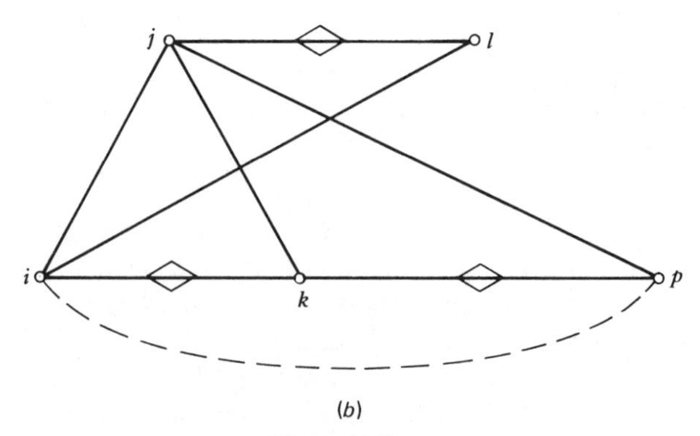

(b)

Figure 14.10.

THEOREM 14.5. If $G=(V, E)$ is transitively orientable, then $G''=(V, E-E')$ is also transitively orientable.

Proof

Since the Rules R_1 and R_2 mark only adjacent edges, it follows that the graph $G'=(V', E')$ is connected.

Consider now a vertex $v \in V-V'$. If v is connected to any vertex $v' \in V'$, then v should be connected to all the vertices of V' which are adjacent to v', for otherwise the edge $v-v'$ would have been marked. Since the graph G' is connected, it would then follow that v should be connected to all the vertices in V'.

Let $\vec{G}$ be a transitive orientation of G such that the orientations of the edges of $\vec{E}'$ agree in $\vec{G}$.

Next we partition the set $V-V'$ into four subsets as follows:

$$A=\{i \mid i \in V-V' \text{ and for all } j \in V',\ i \to j \text{ in } \vec{G}\},$$

$$B=\{i \mid i \in V-V' \text{ and for all } j \in V',\ j \to i \text{ in } \vec{G}\},$$

$$C=\{i \mid i \in V-V' \text{ and for all } j \in V',\ i \not\to j \text{ in } \vec{G}\},$$

$$D=V-(V' \cup A \cup B \cup C).$$

Note that D consists of all those vertices of $V-V'$ which are connected to all the vertices of V', but not all the edges connecting a vertex in D to the vertices in V' are oriented in the same direction.

Transitivity of $\vec{G}$ implies the following connections between the different subsets of V:

1. For every $i \in A, j \in D, k, \in B$, $i \to j$, $j \to k$, and $i \to k$.
2. For all $i \in C$ and $j \in D$, $i \not\to j$.
3. All edges connecting A and C are directed from A to C.
4. All edges connecting B and C are directed from C to B.

The situation so far is depicted in Fig. 14.11*a*.

Now reverse the orientations of all the edges directed from V' to D so that all the edges connecting V' and D are directed from D to V'. The resulting orientation is as shown in Fig. 14.11*b*. We now claim that this orientation is transitive.

To prove this claim we have to show that, in the graph of Fig. 14.11*b*, if $i \to j$ and $j \to k$, then $i \to k$ for all i, j, and k. Clearly, this is true if none of these three edges is among those edges whose directions have just been reversed. Thus we need to consider only the following four cases:

1. $i \in D, j \in V'$, and $k \in V'$.
2. $i \in D, j \in V'$, and $k \in B$.

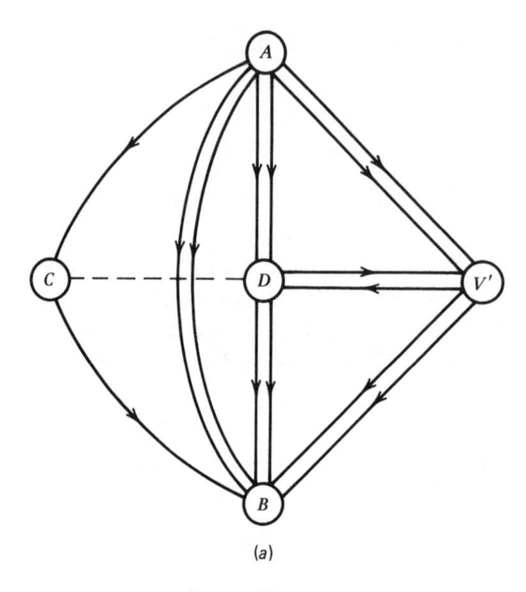

(a)

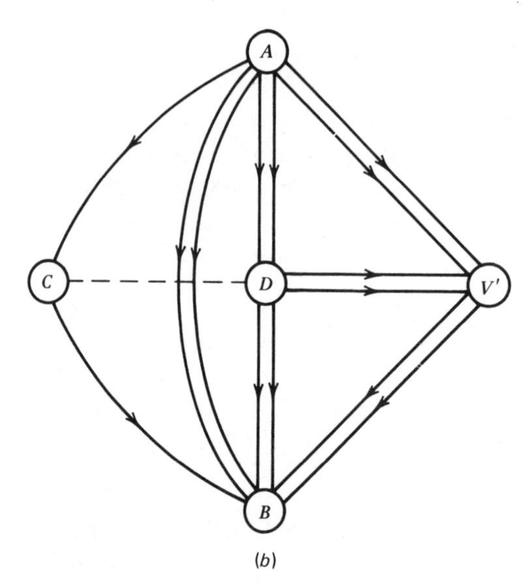

(b)

Figure 14.11.

3. $j \in D$, $k \in V'$, and $i \in A$.
4. $j \in D$, $k \in V'$, and $i \in D$.

In all these four cases $i \to k$ as shown in Fig. 14.11*b*. Thus the orientation of Fig. 14.11*b* is transitive.

Now remove from the graph of Fig. 14.11*b* all the edges of E', namely, all the marked edges.

Suppose that in the resulting graph there exist vertices i, j, and k such that $i \rightarrow j$ and $j \rightarrow k$, but $i \not\rightarrow k$. Here $i \not\rightarrow k$ means only that $i \not- k$, for $i \leftarrow k$ would imply that the orientation in Fig. 14.11*b* is not transitive. If edge i—k is not in $E-E'$, it should be in E', for otherwise the orientation in Fig. 14.11*b* is not transitive. Thus i and k are in V'.

Since there is no vertex outside V' which has both an edge into it from V' and an edge from it into V', it follows that j is also in V'.

Thus we have i, j, k marked, edges i—j and j—k unmarked, and edge i—k marked. This is not possible by Lemma 14.3.

Therefore the directed graph which results after removing the edges of $\vec{E}'$ from the graph of Fig. 14.11*b* is transitive. Hence $G''=(V, E-E')$ is transitively orientable. ■

Thus we have established Assertion 2 and hence the correctness of Algorithm 14.3.

It should now be clear that the transitive orientation algorithm discussed above is an example of an algorithm which is simple but whose proof if correctness is very involved. For an earlier algorithm for this problem see Gilmore and Hoffman [14.15].

Even, Pnueli, and Lempel [14.14], [14.16] introduce permutation graphs and establish a structural relationship between these graphs and transitively orientable graphs. They also discuss an algorithm to test whether a given graph is a permutation graph.

Certain graph problems which are in general very hard to solve become simple when the graph under consideration is transitively orientable. Problems of finding a maximum clique and a minimum coloration are examples of such problems. These problems arise in the study of memory relocation and circuit lay-out problems [14.16]. See also Even [14.17] and Liu [14.18].

14.3 DEPTH-FIRST SEARCH

In this section we describe a systematic method for exploring a graph. This method known as *depth-first search*, in short, *DFS*, has proved very useful in the design of several efficient algorithms. Some of these algorithms are discussed in the remaining sections of this chapter. Our discussion here is based on [14.19].

14.3.1 DFS of an Undirected Graph

We first describe DFS of an undirected graph. To start with, we assume that the graph under consideration is connected. If the graph is not connected, then DFS would be performed separately on each component of the graph. We also assume that there are no self-loops in the graph.

DFS of an undirected graph G proceeds as follows:

We choose any vertex, say v, in G and begin the search from v. The start vertex v, called the *root of the DFS*, is now said to be *visited*.

We then select an edge (v, w) incident on v and traverse this edge to visit w. We also orient this edge from v to w. The edge (v, w) is now said to be *examined* and is called a *tree edge*. The vertex v is called the *father* of w, denoted as FATHER(w).

In general, while we are at some vertex x, two possibilities arise:

1. If all the edges incident on x have already been examined, then we return to the father of x and continue the search from FATHER(x). The vertex x is now said to be *completely scanned*.
2. If there exist some unexamined edges incident on x, then we select one such edge (x, y) and orient it from x to y. The edge (x, y) is now said to be examined. Two cases need to be considered now:

   Case 1 If y has not been previously visited, then we traverse the edge (x, y), visit y, and continue the search from y. In this case (x, y) is a *tree edge* and $x = \text{FATHER}(y)$.

   Case 2 If y has been previously visited, then we proceed to select an other unexamined edge incident on x. In this case the edge (x, y) is called a *back edge*.

During the DFS, whenever a vertex x is visited for the first time, it is assigned a distinct integer DFN(x) such that DFN(x)$=i$, if x is the ith vertex to be visited during the search. DFN(x) is called the *depth-first number* of x. Clearly, depth-first numbers indicate the order in which the vertices are visited during DFS.

DFS terminates when the search returns to the root and all the vertices have been visited.

As we can see from the above description, DFS partitions the edges of G into tree edges and back edges. It is easy to show that the tree edges form a spanning tree of G. DFS also imposes directions on the edges of G. The resulting directed graph will be denoted by $\hat{G}$. The tree edges with their directions imposed by the DFS will form a directed spanning tree of $\hat{G}$. This directed spanning tree will be called the *DFS tree*.

Note that DFS of a graph is not unique, since the edges incident on a vertex may be chosen for examination in any arbitrary order.

As an example, we have shown in Fig. 14.12 DFS of an undirected graph. In this figure tree edges are shown as continuous lines, and back edges are shown as dashed lines. Next to each vertex we have shown its depth-first number. We have also shown in the figure the list of edges incident on each

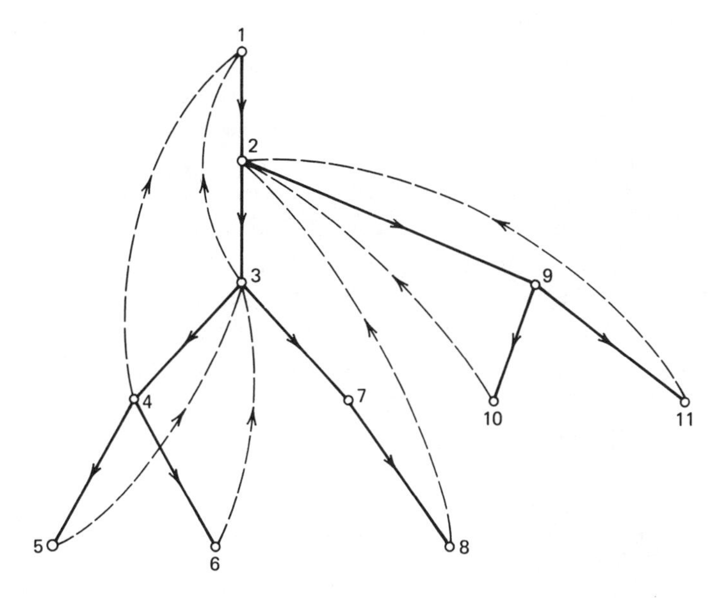

Vertex	Adjacency List
1	(1,2), (1,3), (1,4)
2	(2,1), (2,3), (2,8), (2,9), (2,10), (2,11)
3	(3,1), (3,2), (3,4), (3,5), (3,6), (3,7)
4	(4,1), (4,3), (4,5), (4,6)
5	(5,3), (5,4)
6	(6,3), (6,4)
7	(7,3), (7,8)
8	(8,2), (8,7)
9	(9,2), (9,10), (9,11)
10	(10,2), (10,9)
11	(11,2), (11,9)

Figure 14.12. DFS of an undirected graph.

vertex v. This list for a vertex v is called the *adjacency list* of v, and it gives the order in which the edges incident on v are chosen for examination.

We now present a formal description of the DFS algorithm. In this description the graph under consideration is not assumed to be connected. The array MARK used in the algorithm has one entry for each vertex. To begin with we set MARK(v)=0 for every vertex v in the graph, thereby indicating that no vertex has yet been visited. Whenever a vertex is visited, we set the corresponding entry in the MARK array equal to 1. The arrays DFN and FATHER are as defined before. TREE and BACK are two sets storing, respectively, the tree edges and the back edges as they are generated.

Algorithm 14.4 DFS of an Undirected Graph

S1. G is the given graph. Set TREE$=\varnothing$, BACK$=\varnothing$, and $i=1$. For every vertex v in G, set FATHER(v)$=0$ and MARK(v)$=0$.

S2. (DFS of a component of G begins.) Select any vertex, say vertex r, with MARK(r)$=0$. Set DFN(r)$=i$, MARK(r)$=1$, and $v=r$. (The vertex r is called the *root* of the component under consideration.)

S3. If all the edges incident on v have already been labeled "examined," then go to step S5. (v is now completely scanned.) Otherwise select an edge (v, w) which is not yet labeled "examined" and go to step S4.

S4. Orient the edge (v, w) from v to w and label it "examined." Do the following and then go to step S3.

1. If MARK(w)$=0$, set

$$i=i+1,$$

$$\text{DFN}(w)=i,$$

$$\text{TREE}=\text{TREE}\cup\{(v,w)\},$$

$$\text{MARK}(w)=1,$$

$$\text{FATHER}(w)=v,$$
$$v=w.$$

2. If MARK(w)$=1$, set

$$\text{BACK}=\text{BACK}\cup\{(v,w)\}.$$

S5. If FATHER(v)$\neq 0$ (that is, v is not the root of the component under consideration), then set $v=$FATHER(v) and go to step S3. Otherwise go to step S6.

S6. If, for every vertex x, MARK(x)$=1$, then go to step S7; otherwise set $i=i+1$ and go to step S2.

S7. (DFS is completed.) HALT. ■

Let T be a DFS tree of a connected undirected graph. As we mentioned before, T is a directed spanning tree of $\hat{G}$. For further discussions, we need to introduce some terminology.

If there is a directed path in T from a vertex v to a vertex w, then v is called an *ancestor* of w, and w is called a *descendant* of v. Furthermore, if $v\neq w$, v is called a *proper ancestor* of w, and w is called a *proper descendant* of v. If (v, w) is a directed edge in T, then v is called the *father* of w, and w is called a *son* of v. Note that a vertex may have more than one son. A vertex v and all its descendants form a subtree of T with vertex v as the root of this subtree.

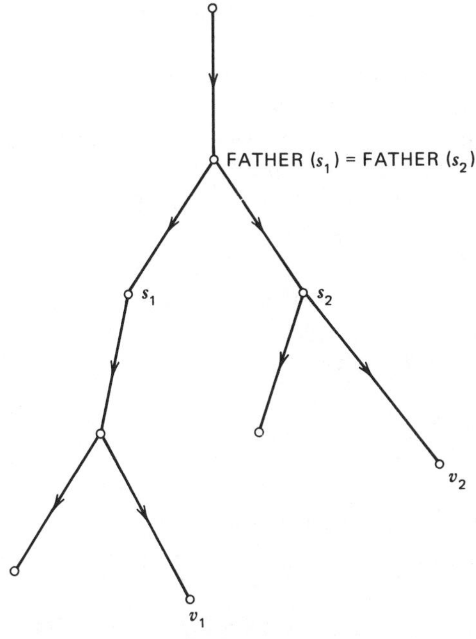

Figure 14.13.

Two vertices v and w are *related* if one of them is a descendant of the other. Otherwise v and w are *unrelated*. If v and w are unrelated and $\text{DFN}(v) < \text{DFN}(w)$, then v is said to be to the *left* of w; otherwise, v is to the *right* of w. Edges of G connecting unrelated vertices are called *cross edges*.

We now show that there are no cross edges in G.

Let v_1 and v_2 be any two unrelated vertices in T. Clearly then there are two distinct vertices s_1 and s_2 such that (1) $\text{FATHER}(s_1) = \text{FATHER}(s_2)$, and (2) v_1 and v_2 are descendants of s_1 and s_2, respectively (see Fig. 14.13).

Let T_1 and T_2 denote the subtrees of T rooted at s_1 and s_2, respectively. Assume without loss of generality that $\text{DFN}(s_1) < \text{DFN}(s_2)$. It is then clear from the DFS algorithm that vertices in T_2 are visited only after the vertex s_1 is completely scanned. Further, scanning of s_1 is completed only after all the vertices in T_1 are scanned completely. So there cannot exist an edge connecting v_1 and v_2. For if such an edge existed, it would have been visited before the scanning of s_1 is completed.

Thus we have the following.

THEOREM 14.6. If (v, w) is an edge in a connected undircted graph G, then in any DFS tree of G either v is a descendant of w or vice versa. ∎

The absence of cross edges in an undirected graph is an important property which forms the basis of an algorithm to be discussed in the next section for determining the biconnected components of a graph.

14.3.2 DFS of a Directed Graph

DFS of a directed graph is essentially similar to that of an undirected graph. The main difference is that in the case of a directed graph an edge is traversed only along its orientation. As a result of this constraint, edges in a directed graph G are partitioned into four categories (and not two as in the undirected case) by a DFS of G. An unexamined edge (v, w) encountered while at the vertex v would be classified as follows.

Case 1 w has not yet been visited.

In this case (v, w) is a *tree edge*.

Case 2 w has already been visited.

a. If w is a descendant of v in the DFS forest (that is, the subgraph of tree edges), then (v, w) is called a *forward edge*.

b. If w is an ancestor of v in the DFS forest, then (v, w) is called a *back edge*.

c. If v and w are not related in the DFS forest and $\text{DFN}(w) < \text{DFN}(v)$, then (v, w) is a *cross edge*. Note that there are no cross edges of the type (v, w) with $\text{DFN}(w) > \text{DFN}(v)$. The proof for this is along the same lines as that for Theorem 14.6.

A few useful observations are now in order:

1. An edge (v, w), with $\text{DFN}(w) > \text{DFN}(v)$, is either a tree edge or a forward edge. During the DFS it is easy to distinguish between a tree edge and a forward edge because a tree edge always leads to a new vertex.
2. An edge (v, w) with $\text{DFN}(w) < \text{DFN}(v)$ is either a back edge or a cross edge. Such an edge (v, w) is a back edge if w is not completely scanned when the edge is encountered while examining the edges incident out of v.
3. DFS forest, the subgraph of tree edges, may not be connected even if the directed graph under consideration is connected. The first vertex to be visited in each component of the DFS forest will be called the *root* of the corresponding component.

A description of the DFS algorithm for a directed graph is presented next. In this algorithm we use a new array SCAN which has one entry for each vertex in the graph. To begin with we set $\text{SCAN}(v) = 0$ for every vertex v, thereby indicating that none of the vertices is completely scanned. Whenever a vertex is completely scanned, the corresponding entry in the SCAN array is set to 1.

As we pointed out earlier, when we encounter an edge (v,w) with $\mathrm{DFN}(w) < \mathrm{DFN}(v)$, we shall classify it as a back edge if $\mathrm{SCAN}(w)=0$; otherwise (v,w) is a cross edge. We also use two arrays, FORWARD and CROSS, that store, respectively, forward and cross edges.

Algorithm 14.5 DFS of a Directed Graph

S1. G is the given directed graph with no self-loops. Set $\mathrm{TREE}=\varnothing$, $\mathrm{FORWARD}=\varnothing$, $\mathrm{BACK}=\varnothing$, $\mathrm{CROSS}=\varnothing$, and $i=1$. For every vertex v in G, set $\mathrm{MARK}(v)=0$, $\mathrm{FATHER}(v)=0$, and $\mathrm{SCAN}(v)=0$.

S2. (DFS with a new root begins.) Select any vertex, say vertex r, with $\mathrm{MARK}(r)=0$. Set $\mathrm{DFN}(r)=i$, $\mathrm{MARK}(r)=1$, and $v=r$.

S3. If all the edges incident out of v have already been labeled "examined," set $\mathrm{SCAN}(v)=1$, and then go to step S5. (v is now completely scanned.) Otherwise select an edge (v,w) which is not yet labeled "examined" and go to step S4.

S4. Label the edge (v,w) "examined", do the following, and then go to step S3.

1. If $\mathrm{MARK}(w)=0$, then set

$$i=i+1,$$

$$\mathrm{DFN}(w)=i,$$

$$\mathrm{TREE}=\mathrm{TREE}\cup\{(v,w)\},$$

$$\mathrm{MARK}(w)=1,$$

$$\mathrm{FATHER}(w)=v,$$

$$v=w.$$

2. Otherwise set

$$\mathrm{FORWARD}=\mathrm{FORWARD}\cup\{(v,w)\} \text{ if } \mathrm{DFN}(w)>\mathrm{DFN}(v),$$

$$\mathrm{BACK}=\mathrm{BACK}\cup\{(v,w)\} \text{ if } \mathrm{DFN}(w)<\mathrm{DFN}(v) \text{ and}$$

$$\mathrm{SCAN}(w)=0;$$

$$\mathrm{CROSS}=\mathrm{CROSS}\cup\{(\mathrm{v},\mathrm{w})\}, \text{ otherwise.}$$

S5. If $\mathrm{FATHER}(v)\neq 0$ (that is, v is not a root), then set $v=\mathrm{FATHER}(v)$ and go to step S3. Otherwise go to step S6.

S6. If for every vertex x, $\mathrm{MARK}(x)=1$, then go to step S7; otherwise set $i=i+1$ and go to step S2.

S7. (DFS is completed.) HALT. ■

As an example, DFS of a directed graph is shown in Fig. 14.14*a*. Next to each vertex we have shown its depth-first number. The tree edges are shown as continuous lines, and the other edges are shown as dashed lines. The DFS forest is shown separately in Fig. 14.14*b*.

We pointed out earlier that the DFS forest of a directed graph may not be connected, even if the graph is connected. This can also be seen from Fig. 14.14*b*. This leads us to the problem of discovering sufficient conditions for a DFS forest to be connected. In the following we prove that the DFS forest of

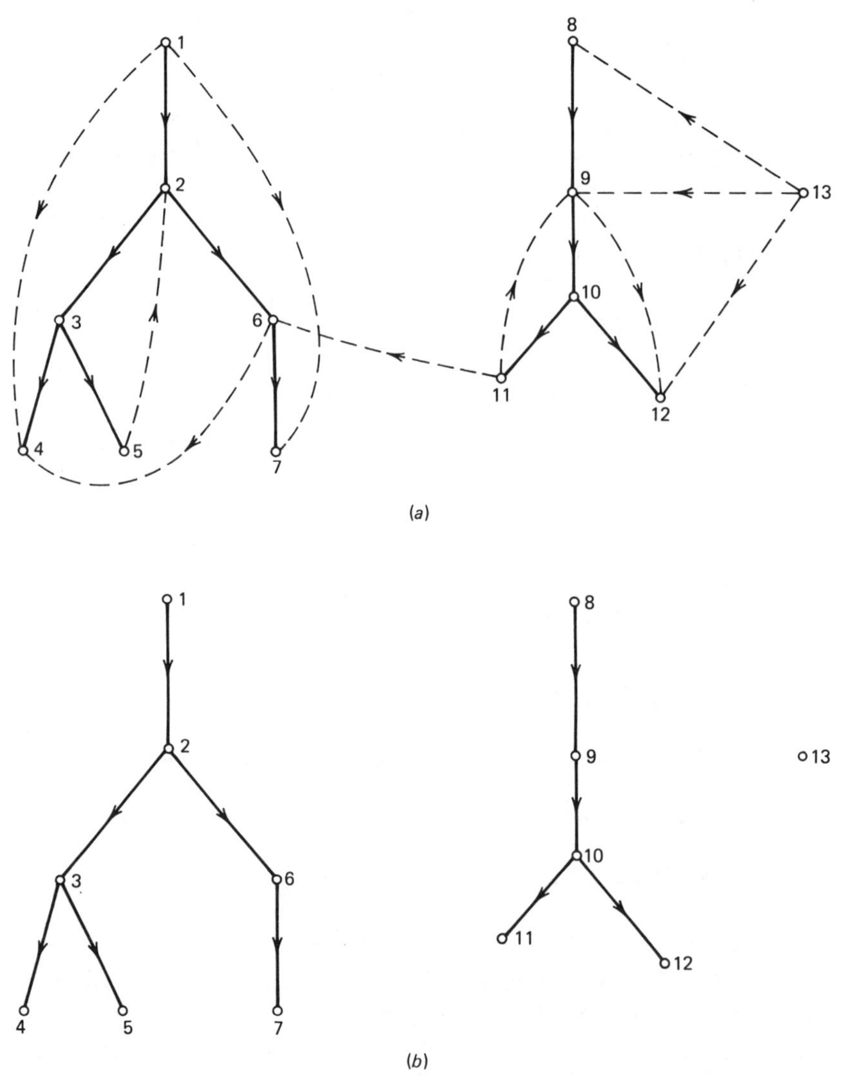

Figure 14.14. (*a*) DFS of a directed graph. (*b*) DFS forest of graph in (*a*).

a strongly connected graph is connected. In fact, we shall be establishing a more general result.

Let T denote a DFS forest of a directed graph $G=(V, E)$. Let $G_i=(V_i, E_i)$, with $|V_i|\geqslant 2$, be a strongly connected component of G. Consider any two vertices v and w in G_i. Assume without loss of generality that DFN$(v)<$ DFN(w). Since G_i is strongly connected, there exists a directed path P in G_i from v to w. Let x be the vertex on P with the lowest depth-first number and let T_x be the subtree of T rooted at x. Note that cross edges and back edges are the only edges which lead out of the subtree T_x. Since these edges lead to vertices having lower depth-first numbers than DFN(x), it follows that once path P reaches a vertex in T_x, then all the subsequent vertices on P will also be in T_x. In particular, w also lies in T_x. So it is a descendant of x. Since DFN$(x)\leqslant$ DFN$(v)<$ DFN(w), it follows from the DFS algorithm that v is also in T_x. Thus any two vertices v and w in G_i have a common ancestor which is also in G_i.

We may conclude from the above that all the vertices of G_i have a common ancestor r_i which is also in G_i. It may now be seen that among all the common ancestors in T of vertices in G_i, vertex r_i has the highest depth-first number. Further, it is easy to show that if v is a vertex in G_i, then any vertex on the tree path from r_i to v will also be in G_i. So the subgraph of T induced by V_i is connected. Thus we have the following.

THEOREM 14.7. Let $G_i=(V_i, E_i)$ be a strongly connected component of a directed graph $G=(V, E)$. If T is a DFS forest of G, then the subgraph of T induced by V_i is connected. ■

Following is an immediate corollary of the above theorem.

Corollary 14.7.1 The DFS forest of a strongly connected graph is connected. ■

It is easy to show that the DFS Algorithms 14.4 and 14.5 are both of complexity $O(n+m)$, where n is the number of vertices and m is the number of edges in a graph.

14.4 BICONNECTIVITY AND STRONG CONNECTIVITY

In this section we discuss algorithms due to Hopcroft and Tarjan [14.20] and Tarjan [14.19] for determining the biconnected components and the strongly connected components of a graph. These algorithms are based on DFS. We begin our discussion with the biconnectivity algorithm.

14.4.1 Biconnectivity

We may recall (Chapter 8) that a biconnected graph is a connected graph with no cut-vertices. A maximal biconnected subgraph of a graph is called a *biconnected component* of the graph.*

A crucial step in the development of the biconnectivity algorithm is the determination of a simple criterion which can be used to identify cut-vertices as we perform a DFS. Such a criterion is given in the following two lemmas.

Let $G=(V, E)$ be a connected undirected graph. Let T be a DFS tree of G with vertex r as the root. Then we have the following.

LEMMA 14.4. Vertex $v \neq r$ is a cut-vertex of G if and only if for some son s of v there is no back edge between any descendant in T of s (including itself) and a proper ancestor of v.

Proof

Let G' be the graph that results after removing vertex v from G. By definition, v is a cut-vertex of G if and only if G' is not connected.

Let $s_1, s_2, \ldots, s_k$ be the sons of v in T. For each i, $1 \leqslant i \leqslant k$, let V_i denote the set of descendants of s_i (including itself), and let G_i be the subgraph of G' induced on V_i. Further let $V'' = V' - \cup_{i=1}^{k} V_i$, where $V' = V - \{v\}$, and let G'' be the subgraph induced on V''. Note that all the proper ancestors of V are in V''.

Clearly, $G_1, G_2, \ldots, G_k$ and G'' are all subgraphs of G which together contain all the vertices of G'. We can easily show that all these subgraphs are connected. Further, by Theorem 14.6 there are no edges connecting vertices belonging to different G_i's. So it follows that G' will be connected if and only if for every i, $1 \leqslant i \leqslant k$, there exists an edge (a, b) between a vertex $a \in V_i$ and a vertex $b \in V''$. Such an edge (a, b) will necessarily be a back edge, and b will be a proper ancestor of v. We may therefore conclude that G' will be connected if and only if for every son s_i of v there exists a back edge between some descendant of s_i (including itself) and a proper ancestor of v. The proof of the lemma is now immediate. ■

LEMMA 14.5. The root vertex r is a cut-vertex of G if and only if it has more than one son.

Proof

Proof in this case follows along the same line as that for Lemma 14.4. ■

*Note that a biconnected component is the same as a block defined in Section 1.7.

In the following we refer to the vertices of G by their depth-first numbers. To embed into the DFS procedure the criterion given in Lemmas 14.4 and 14.5, we now define, for each vertex v of G,

$$\text{LOW}(v) = \min(\{v\} \cup \{w \mid \text{there exists a back edge } (x,w) \text{ such that } x \text{ is a descendant of } v, \text{ and } w \text{ is a proper ancestor of } v \text{ in } T\}) \tag{14.1}$$

Using the LOW values defined as above, we can restate the criterion given in Lemma 14.4 as in the following theorem.

THEOREM 14.8. Vertex $v \neq r$ is a cut-vertex of G if and only if v has a son s such that $\text{LOW}(s) \geqslant v$. ■

Noting that LOW(v) is equal to the lowest numbered vertex which can be reached from v by a directed path containing at most one back edge, we can rewrite (14.1) as

$$\text{LOW}(v) = \min(\{v\} \cup \{\text{LOW}(s) \mid s \text{ is a son of } v\} \cup \{w \mid (v,w) \text{ is a back edge}\}).$$

This equivalent definition of LOW(v) suggests the following steps for computing LOW(v):

1. When v is visited for the first time during DFS, set LOW(v) equal to the depth-first number of v.
2. When a back edge (v,w) incident on v is examined, set LOW(v) to the minimum of its current value and the depth-first number of w.
3. When the DFS returns to v after completely scanning a son s of v, set LOW(v) equal to the minimum of its current value and LOW(s).

Note that for any vertex v, computation of LOW(v) ends when the scanning of v is completed.

We next consider the question of identifying the edges belonging to a biconnected component. For this purpose we use an array STACK. To begin with STACK is empty. As edges are examined, they are added to the top of STACK.

Suppose DFS returns to a vertex v after completely scanning a son s of v. At this point, computation of LOW(s) will have been completed. Suppose it is now found that $\text{LOW}(s) \geqslant v$. Then, by Theorem 14.8, v is a cut-vertex. Further, if s is the first vertex with this property, then we can easily see that the edge (v,s) along with the edges incident on s and its descendants will form a biconnected component. These edges are exactly those which lie on

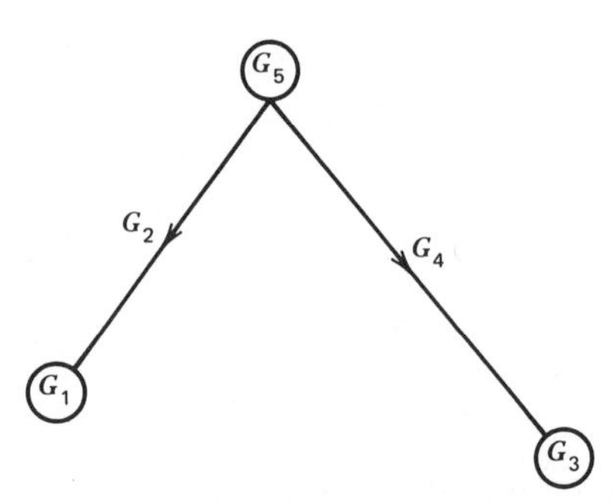

Figure 14.15. G_1, G_2, G_3, G_4, G_5—biconnected components of a graph.

top of STACK up to and including (v, s). They are now removed from STACK. From this point on the algorithm behaves in exactly the same way as it would on the graph G' which is obtained by removing from G the edges of the biconnected component which has just been identified.

For example, a DFS tree of a connected graph may be as in Fig. 14.15. $G_1, G_2, \ldots, G_5$ are the biconnected components in the order in which they are identified.

A description of the biconnectivity algorithm now follows. This algorithm is essentially the same as Algorithm 14.4, with the inclusion of appropriate steps for computing LOW(v) and identifying the cut-vertices and the edges belonging to the different biconnected components. Note that in this algorithm the root vertex r is treated as a cut-vertex, even if it is not one, for the purpose of identifying the biconnected component containing r.

Algorithm 14.6 Biconnectivity

S1. G is the given connected graph. For every vertex v in G, set FATHER(v)=0 and MARK(v)=0. Set $i=1$ and STACK=$\varnothing$.

S2. Select any vertex, say vertex r, with MARK(r)=0. Set

DFN(r)=i,

LOW(r)=i,

MARK(r)=1,

$v=r$.

S3. If all the edges incident on v have already been labeled "examined," then go to step S5. Otherwise select an edge (v, w) which is not yet labeled "examined." Label this edge "examined," add it to the top of STACK, and go to step S4.

S4. Do the following and then go to step S3:

1. If MARK(w)=0, set

i=i+1,

DFN(w)=i,

LOW(w)=i,

FATHER(w)=v,

MARK(w)=1,

$v=w$.

2. If MARK(w)=1, set

$$\text{LOW}(v)=\min\{\text{LOW}(v),\text{DFN}(w)\}.$$

S5. If FATHER(v)$\neq$0, go to step S6; otherwise go to step S8.

S6. If LOW(v)$\geqslant$DFN(FATHER(v)), then remove all the edges from the top of STACK up to and including the edge (FATHER(v), v). (A biconnected component has been found.)

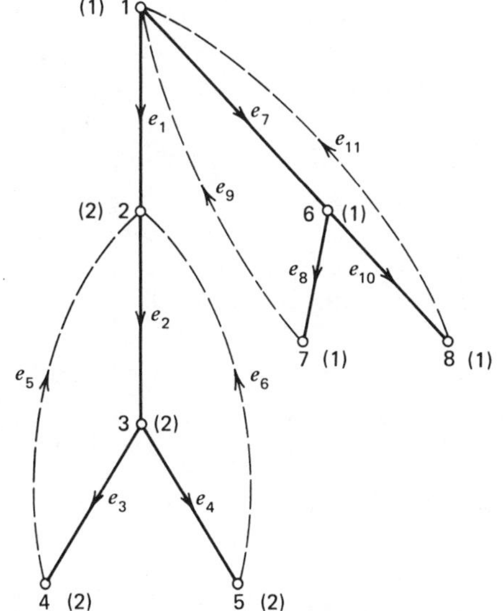

Figure 14.16. Illustration of Algorithm 14.6. LOW values are shown in parentheses. Biconnected components are $\{e_2, e_3, e_4, e_5, e_6\}$, $\{e_1\}$, and $\{e_7, e_8, e_9, e_{10}, e_{11}\}$.

S7. Set

$$\text{LOW}(\text{FATHER}(v)) = \min\{\text{LOW}(v), \text{LOW}(\text{FATHER}(v))\},$$

$$v = \text{FATHER}(v)$$

and go to step S3.

S8. (All biconnected components have been found.) HALT. ■

See Fig. 14.16 for an illustration of the above algorithm.

14.4.2 Strong Connectivity

Recall that a graph is strongly connected if for every two vertices v and w there exists in G a directed path from v to w and a directed path from w to v; further a maximal strongly connected subgraph of a graph G is called a strongly connected component of the graph.

Consider a directed graph $G=(V, E)$. Let $G_1=(V_1, E_1)$, $G_2=(V_2, E_2), \ldots,$ $G_k=(V_k, E_k)$ be the strongly connected components of G. Let T be a DFS forest of G and $T_1, T_2, \ldots, T_k$ be the induced subgraphs of T on the vertex sets $V_1, V_2, \ldots, V_k$, respectively. We know from Theorem 14.7 that $T_1, T_2, \ldots, T_k$ are connected.

Let r_i, $1 \leqslant i \leqslant k$, be the root of T_i. Assume that if $i<j$, then DFS terminates at vertex r_i earlier than at r_j. Then we can see that for each $i<j$, either r_i is to the left of r_j, or r_i is a descendant of r_j in T. Further G_i, $1 \leqslant i \leqslant k$, would consist of those vertices which are descendants of r_i but are in none of $G_1, G_2, \ldots, G_{i-1}$.

The first step in the development of the strong connectivity algorithm is the determination of a simple criterion which can be used to identify the roots as we perform a DFS. The following observations will be useful in deriving such a criterion. These observations are all direct consequences of the fact that there exist no directed circuits in the graph obtained by contracting all the edges in each one of the sets $E_1, E_2, \ldots, E_k$ (see Section 5.1).

1. There is no back edge of the type (v, w) with $v \in V_i$ and $w \in V_j$, $i \neq j$. In other words, all the back edges which leave vertices in V_i also end on vertices in V_i.
2. There is no cross edge of the type (v, w) with $v \in V_i$, $w \in V_j$, $i \neq j$, and r_j is an ancestor of r_i. Thus for each cross edge (v, w) one of the following two is true:
   - **a.** $v \in V_i$ and $w \in V_j$ for some i and j with $i \neq j$ and r_j to the left of r_i.
   - **b.** For some i, $v \in V_i$ and $w \in V_i$.

Assuming that the vertices of G are named by their DFS numbers, we define for each v in G,

$$\text{LOWLINK}(v) = \min(\{v\} \cup \{w \mid \text{there is a cross edge or a back edge from a descendant of } v \text{ to } w, \text{ and } w \text{ is in the same strongly connected component as } v\}).$$

Suppose $v \in V_i$. Then it follows from the above definition that LOWLINK(v) is the lowest numbered vertex in V_i which can be reached from v by a directed path which contains at most one back edge or one cross edge. From the observations that we have just made it follows that all the edges of such a directed path will necessarily be in G_i. As an immediate consequence we get

$$\text{LOWLINK}(r_i) = r_i, \qquad \text{for all } 1 \leqslant i \leqslant k. \tag{14.2}$$

Suppose $v \in V_i$ and $v \neq r_i$. Then there exists a directed path P in G_i from v to r_i. Such a directed path P should necessarily contain a back edge or a cross edge because $r_i < v$, and only cross edges and back edges lead to lower numbered vertices. In other words, P contains a vertex $w < v$. So for $v \neq r_i$, we get

$$\text{LOWLINK}(v) < v. \tag{14.3}$$

Combining (14.2) and (14.3) we get the following theorem which characterizes the roots of the strongly connected components of a directed graph.

THEOREM 14.9. A vertex v is the root of a strongly connected component of a directed graph G if and only if LOWLINK(v) $= v$. ∎

The following steps can be used to compute LOWLINK(v) as we perform a DFS.

1. On visiting v for the first time, set LOWLINK(v) equal to the DFS number of v.
2. If a back edge (v, w) is examined, then set LOWLINK(v) equal to the minimum of its current value and the DFS number of w.
3. If a cross edge (v, w) with w in the same strongly connected component as v is explored, set LOWLINK(v) equal to the minimum of its current value and the DFS number of w.
4. When the search returns to v after completely scanning a son s of v, set LOWLINK(v) to the minimum of its current value and LOWLINK(s).

To implement step 3 above we need a test to check whether w is in the same strongly connected component as v. For this purpose we use an array STACK1 to which vertices of G are added in the order in which they are visited during the DFS. STACK1 is also used to determine the vertices belonging to a strongly connected component.

Let v be the first vertex during DFS for which it is found that LOWLINK$(v)=v$. Then by Theorem 14.9, v is a root and in fact it is r_1. At this point, the vertices on top of STACK1 up to and including v are precisely those that belong to G_1. Thus G_1 can easily be identified. These vertices are now removed from STACK1. From this point on the algorithm behaves in exactly the same way as it would on the graph G' which is obtained by removing from G the vertices of G_1.

As regards the implementation of step 3 in LOWLINK computation, let $v \in V_i$, and let (v, w) be a cross edge encountered while examining the edges incident on v. Suppose w is not in the same strongly connected component as v. Then it would belong to a strongly connected component G_j whose root r_j is to the left of r_i (observation 2a p. 462). The vertices of such a component would already have been identified, and so they would no longer be on STACK1. Thus w will be in the same strongly connected component as v if and only if w is on STACK1.

A description of the strong connectivity algorithm now follows. This is the same as Algorithm 14.5 with the inclusion of appropriate steps for computing LOWLINK values and for identifying the vertices of the different strongly connected components. We use in this algorithm an array POINT. To begin with, POINT$(v)=0$ for every vertex v. This indicates that no vertex is on the array STACK1. POINT(v) is set to 1 when v is added to STACK1, and it is set to zero when v is removed from STACK1.

Algorithm 14.7 Strong Connectivity

S1. G is the given directed graph. For every vertex v in G, set MARK$(v)=0$, FATHER$(v)=0$, and POINT$(v)=0$. Set $i=1$ and STACK1 $=\emptyset$.

S2. Select any vertex, say vertex r, with MARK$(r)=0$. Set

DFN$(r)=i$,

LOWLINK$(r)=i$,

MARK$(r)=1$.

Add r to STACK1, set POINT$(r)=1$ and $v=r$.

S3. If all the edges incident on v have already been labeled "examined," then go to step S5. Otherwise select an edge (v, w) which is not yet labeled "examined". Label it "examined" and go to step S4.

S4. Do the following and then go to step S3:

1. If $\text{MARK}(w)=0$, set

$$i=i+1,$$

$$\text{DFN}(w)=i,$$

$$\text{LOWLINK}(w)=i,$$

$$\text{FATHER}(w)=v,$$

$$\text{MARK}(w)=1.$$

Add w to STACK1, and set $\text{POINT}(w)=1$ and $v=w$.

2. If $\text{MARK}(w)=1$, $\text{DFN}(w)<\text{DFN}(v)$, and $\text{POINT}(w)=1$, then set

$$\text{LOWLINK}(v)=\min\{\text{LOWLINK}(v),\text{DFN}(w)\}.$$

S5. If $\text{LOWLINK}(v)=\text{DFN}(v)$, then remove all the vertices from the top of STACK1 up to and including v. (These vertices form a strongly connected component.) Then set $\text{POINT}(x)=0$ for all such vertices x removed from STACK1.

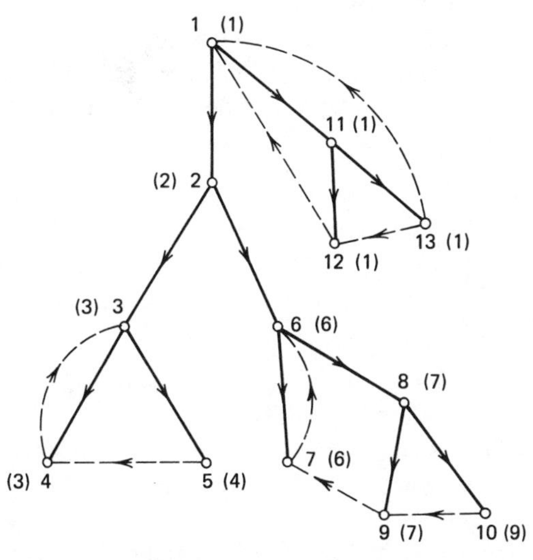

Figure 14.17. Illustration of Algorithm 14.7. LOWLINK values are shown in parentheses. Strongly connected components are $\{3,4,5\}$, $\{6,7,8,9,10\}$, $\{2\}$, and $\{1,11,12,13\}$.

S6. If FATHER(v)=0, go to step S7; otherwise set

1. LOWLINK(FATHER(v))= min{LOWLINK(FATHER(v)), LOWLINK(v)},
2. v=FATHER(v),

and then go to step S3.

S7. If for every vertex x, MARK(x)=1, then go to step S8; otherwise go to step S2.

S8. (All strongly connected components have been identified.) HALT. ∎

See Fig. 14.17 for an illustration of the above algorithm.

14.5 REDUCIBILITY OF A PROGRAM GRAPH

A *program graph* is a directed graph G with a distinguished vertex s such that there is a directed path from s to every other vertex of G. In other words, every vertex in G is reachable from s. The vertex s is called the *start vertex* of G. We assume that there are no parallel edges in a program graph. This assumption will involve no loss of generality as far as our discussions in this chapter are concerned.

The flow of control in a computer program can be modeled by a program graph in which each vertex represents a block of instructions which can be executed sequentially. Such a representation of computer programs has proved very useful in the study of several questions relating to what is known as the *code-optimization* problem.

For many of the code-optimization methods to work, the program graph must have a special property called *reducibility*. See [14.21] through [14.29].

Reducibility of a program graph G is defined in terms of the following two transformations on G:

S_1: Delete self-loop (v, v) in G.

S_2: If (v, w) is the only edge incident into w, and $w \neq s$, delete vertex w. For every edge (w, x) in G add a new edge (v, x) if (v, x) is not already in G. (This transformation is called *collapsing* vertex w into vertex v.)

For example collapsing vertex 5 into vertex 4 in the program graph of Fig. 14.18*a* results in the graph shown in Fig. 14.18*b*.

A program graph is *reducible* if it can be transformed into a graph consisting only of vertex s by repeated applications of the transformations S_1 and S_2.

For example, the graph in Fig. 14.18*a* is reducible. It can be verified that this graph can be reduced by collapsing the vertices in the order $5, 8, 4, 3, 10, 9, 7, 6, 2$.

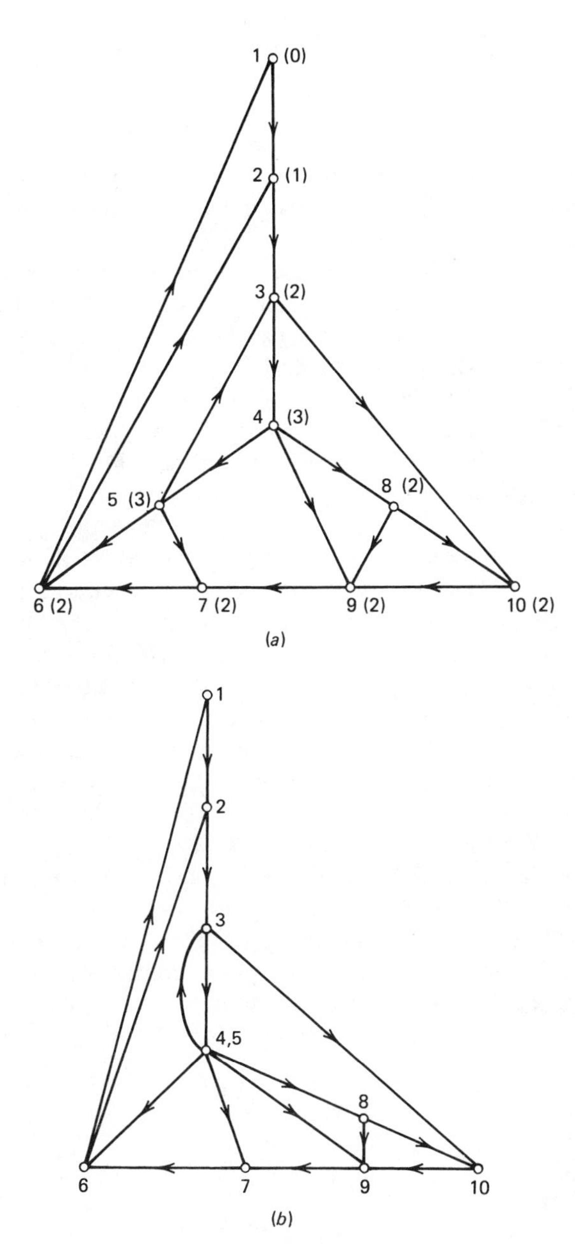

Figure 14.18. (*a*) A reducible program graph *G*. HIGHPT1 values are given in parentheses. (*b*) Graph obtained after collapsing in *G* vertex 5 into vertex 4.

Cocke [14.22] and Allen [14.27] were the original formulators of the notion of reducibility, and their definition is in terms of a technique called *interval analysis*. The definition given above is due to Hecht and Ullman [14.30], and it is equivalent to that of Cocke and Allen.

If a graph G is reducible, then it can be shown [14.30] that any graph G' obtained from G by one or more applications of the transformations S_1 and S_2 is also reducible. Thus the order of applying transformations does not matter in a test for reducibility. Further, some interesting classes of programs such as "go-to-less" programs give rise to graphs which are necessarily reducible [14.30], and most programs may be modeled by a reducible graph using a process of "node splitting" [14.31].

Suppose we wish to test the reducibility of a graph G. This may be done by first deleting self-loops using transformation S_1 and then counting the number of edges incident into each vertex. Next we may find a vertex w with only one edge (v, w) incident into it and apply transformation S_2, collapsing w into v. We may then repeat this process until we reduce the graph entirely or discover that it is not reducible. Clearly each application of S_2 requires $O(n)$ time, where n is the number of vertices in G and reduces the number of vertices by 1. Thus the complexity of this algorithm is $O(n^2)$. Hopcroft and Ullman [14.32] have improved this algorithm to $O(m \log m)$, where m is the number of edges in G. Tarjan [14.33], [14.34] has subsequently given an algorithm which compares favorably with that of Hopcroft and Ullman.

Hecht and Ullman [14.30], [14.35] have given several useful structural characterizations of program graphs. One of these is given in the following theorem.

THEOREM 14.10. Let G be a program graph with start vertex s. G is reducible if and only if there do not exist distinct vertices $v \neq s$ and $w \neq s$, directed paths P_1 from s to v and P_2 from s to w, and a directed circuit C containing v and w, such that C has no edges and only one vertex in common with each of P_1 and P_2 (see Fig. 14.19). ■

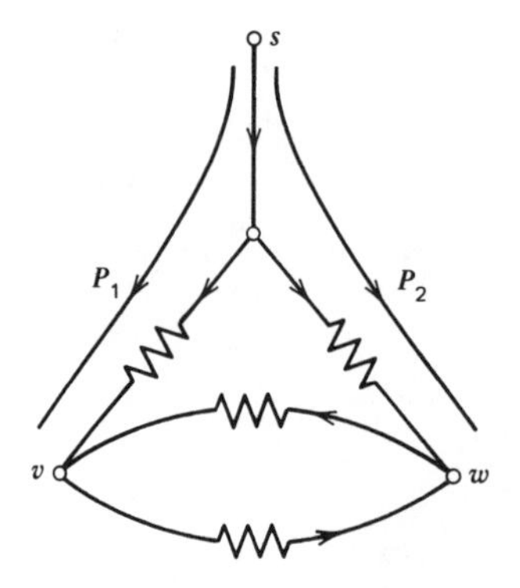

Figure 14.19. A basic nonreducible graph.

Proof of the above theorem may be found in [14.30] and [14.36].

We discuss in this section Tarjan's algorithm for testing the reducibility of a program graph. This algorithm uses DFS and is based on a characterization of reducible program graphs which we shall prove using Theorem 14.10. Our discussion here is based on [14.33].

Let G be a program graph with start vertex s. Let T be a DFS tree of G with s as the root. Henceforth we refer to vertices by their depth-first numbers.

THEOREM 14.11. G is reducible if and only if G contains no directed path P from s to some vertex v such that v is a proper ancestor in T of some other vertex on P.

Proof

Suppose G is not reducible. Then there exist vertices v and w, directed paths P_1 and P_2, and circuit C which satisfy the condition in Theorem 14.10. Assume that $v<w$. Let C_1 be the part of C from v to w. Then C_1 contains some common ancestor u of v and w. Then the directed path consisting of P_2 followed by the part of C from w to u satisfies the condition in the theorem.

Conversely, suppose there exists a directed path P which satisfies the condition in the theorem. Let then v be the first vertex on P which is a proper ancestor of some earlier vertex on P. Suppose w is the first vertex on P, which is a descendant of v. Let P_1 be the part of T from s to v, and let P_2 be the part of P from s to w. Also let C be the directed circuit consisting of the part of P from w to v followed by the path of tree edges from v to w. Then we can see that v, w, P_1, P_2, and C satisfy the condition in Theorem 14.10. So G is not reducible. ∎

For any vertex v, let HIGHPT1(v) be the highest numbered proper ancestor of v such that there is a directed path P from v to HIGHPT1(v) and P includes no proper ancestors of v except HIGHPT1(v). We define HIGHPT1(v)=0 if there is no directed path from v to a proper ancestor of v. As an example, in Fig. 14.18*a* we have indicated in parentheses the HIGHPT1 values of the corresponding vertices.

Note that in HIGHPT1(v) calculation we may ignore forward edges since if P is a directed path from v to w and P contains no ancestors of v except v and w, we may substitute for each forward edge in P a path of tree edges or a part of it and still have a directed path from v to w which contains no ancestors of v except v and w.

Tarjan's algorithm is based on the following characterization of program graphs.

THEOREM 14.12. G is reducible if and only if there is no vertex v with an edge (u, v) incident into v such that $w<$HIGHPT1(v), where w is the highest numbered common ancestor of u and v.

Proof

Suppose G is not reducible. Then by Theorem 14.11 there is a directed path P from s to v with v a proper ancestor of some other vertex on P. Choose P as short as possible. Let w be the first vertex on P which is a descendant of v. Then all the vertices except v which follow w on P are descendants of v in T. In other words, the part of P from w to v includes no proper ancestors of w except v. So $\text{HIGHPT1}(w) \geq v$. Thus w, $\text{HIGHPT1}(w)$, and the edge of P incident into w satisfy the condition in the theorem.

Conversely, suppose the condition in the theorem holds. Then the edge (u, v) is not a back edge because in such a case the highest numbered common ancestor w of u and v is equal to v, and so $w = v \geq \text{HIGHPT1}(v)$. Thus (u, v) is either a forward edge or a cross edge. Let P_1 be a directed path from v to $\text{HIGHPT1}(v)$ which passes through no proper ancestors of v except $\text{HIGHPT1}(v)$. Then the directed path consisting of the tree edges from s to u followed by the edge (u, v) followed by P_1 satisfies the condition in Theorem 14.11. So G is not reducible. ■

Now it should be clear that testing reducibility of a graph G using the above theorem involves the following main steps:

1. Perform a DFS of G with s as the root.
2. Calculate $\text{HIGHPT1}(v)$ for each vertex v in G.
3. For cross edges, check the condition in Theorem 14.12 during the HIGHPT1 calculation.
4. For forward edges, check the condition in Theorem 14.12 after the HIGHPT1 calculation.

Note that, as we observed earlier, forward edges may be ignored during the HIGHPT1 calculation. Further as we observed in the proof of Theorem 14.12, back edges need not be tested for the condition in Theorem 14.12.

To calculate HIGHPT1 values, we first order the back edges (u, v) by the number of v. Then we process the back edges in order, from highest to lowest v. Initially all vertices are unlabeled. To process the back edge (u, v), we proceed up the tree path from u to v, labeling each currently unlabeled vertex with v. (We do not label v itself.) If a vertex w gets labeled, we examine all cross edges incident into w. If (z, w) is such a cross edge (see Fig. 14.20), we proceed up the tree path from z to v, labeling each unlabeled vertex with v. If z is not a descendant of v, then G is not reducible by Theorem 14.12, and the calculation stops. We continue labeling until we run out of cross edges incident into just labeled vertices; then we process the next back edge. When all the back edges are processed, the labels give the HIGHPT1 values of the vertices. Each unlabeled vertex has HIGHPT1 equal to zero.

We now describe Tarjan's algorithm for testing reducibility. In this algorithm we use n queues, called buckets, one for each vertex. The bucket

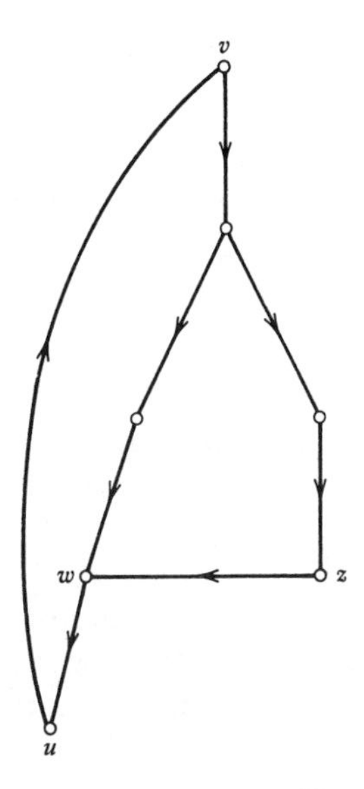

Figure 14.20.

BUCKET(w) corresponding to vertex w contains the list of back edges (u, w) incident into vertex w. While processing back edge (u, w) we need to keep track of certain vertices from which u can be reached by directed paths. The set CHECK is used for this purpose.

Algorithm 14.8 Program Graph Reducibility (Tarjan)

S1. Perform a DFS of the given n-vertex program graph G. Denote vertices by their DFS numbers. Order the back edges (u, v) of G by the number of v.

S2. For $i = 1, 2, \ldots, n$ set

HIGHPT1(i)$=0$,

BUCKET(i)$=$the empty list.

S3. Add each back edge (u, w) to BUCKET(w).

S4. Set $w = n - 1$.

S5. Test if BUCKET(w) is empty. If yes, go to step S6; otherwise go to step S7.

S6. Set $w = w - 1$. If $w < 1$, go to step S16; otherwise go to step S5.

S7. (Processing of a new back edge begins.) Delete a back edge (x, w) from BUCKET(w) and set CHECK$=\{x\}$.

S8. Test if CHECK is empty. If yes (processing of a back edge is over,) go to step S5; otherwise go to step S9.

S9. Delete u from CHECK.

S10. Test if u is a descendant of w. If yes, go to step S11; otherwise go to step S17.

S11. Test if $u=w$. If yes, go to step S8; otherwise go to step S12.

S12. Test if HIGHPT1$(u)=0$. If yes, go to step S13; otherwise go to step S15.

S13. Set HIGHPT1$(u)=w$.

S14. For each cross edge (v,u) add v to CHECK.

S15. Set $u=$FATHER(u). Go to step S11.

S16. If $u \geqslant$ HIGHPT1(v) for each forward edge (u,v) (the graph is reducible), then HALT. Otherwise go to step S17.

S17. HALT. The graph is not reducible. ■

Note that step S10 in Algorithm 14.8 requires that we be able to determine whether a vertex w is a descendant of another vertex u. Let ND(u) be the number of descendants of vertex u in T. Then we can show that w is a descendant of u if and only if $u \leqslant w < u+$ND(u). (See Exercise 14.3.) We can calculate ND(u) during the DFS in a straightforward fashion.

The efficiency of Algorithm 14.8 depends crucially on the efficiency of the HIGHPT1 calculation. To make the HIGHPT1 calculation efficient, in Algorithm 14.8 we need to avoid examining vertices which have already been labeled. Tarjan suggests a procedure to achieve this. The following observation forms the basis of this procedure:

Suppose at step S12 we are examining a vertex u for labeling. Let at this stage, u' be the highest unlabeled proper ancestor of u. This means that all the proper ancestors of u except u', which lie between u' and u in T, have already been labeled. Thus u' is the next vertex to be examined. Furthermore when u is labeled, then all the vertices for which u is the highest unlabeled proper ancestor will have u' as their highest unlabeled proper ancestor.

To implement the method which follows from the above observation, we shall use sets numbered 1 to n. A vertex $w \neq 1$ will be in the set numbered v, that is, SET(v), if v is the highest numbered unlabeled proper ancestor of w. Since vertex 1 never gets labeled, each vertex is always in a set. Initially a vertex is in the set whose number is its father in T. Thus initially add i to SET(FATHER(i)) for $i=2,3,\ldots,n$.

To carry out step S15, we find the number u' of the set containing u and let that be the new u. Further when u becomes labeled, we shall combine the sets numbered u and u' to form a new set numbered u'. Thus u' becomes the highest numbered unlabeled proper ancestor of all the vertices in the old SET(u).

It can now be seen that replacing steps S12 through S15 in Algorithm 14.8 by the following sequence of steps will implement the method described above. Of course, SETs are initialized as explained earlier.

S12′. Set $u'=$ the number of the set containing u.

S13′. Test if HIGHPT1(u)=0. If yes, go to step S14′; otherwise go to step S16′.

S14′. Set

1. HIGHPT1(u)=w; and
2. SET(u')=SET(u)∪SET(u').

S15′. For each cross edge (v, u) add v to CHECK.

S16′. Set $u=u'$. Go to step S11.

It can be verified that the HIGHPT1 calculations modified as above require $O(n)$ set unions, $O(m+n)$ executions of step S12′, and $O(m+n)$ time exclusive of set operations. If we use the algorithm described in Fischer [14.37] and Hopcroft and Ullman [14.38] for performing disjoint set unions and for performing step S12′, then it follows from the analysis given in Tarjan [14.39] that the reducibility algorithm has complexity $O(m\alpha(m,n))$, where $\alpha(m,n)$ is a very slowly growing function which is related to a functional inverse of Ackermann's function $A(p,q)$, and it is defined as follows:

$$\alpha(m,n)=\min\{z\geqslant 1 \mid A(z,4\lceil m/n\rceil)>\log_2 n\}.$$

The definition of *Ackermann's function* is

$$A(p,q)=\begin{cases} 2q, & p=0.\\ 0, & q=0 \text{ and } p\geqslant 1.\\ 2, & p\geqslant 1 \text{ and } q=1.\\ A(p-1,A(p,q-1)), & p\geqslant 1 \text{ and } q\geqslant 2.\end{cases} \tag{14.4}$$

Note that Ackermann's function is a very rapidly growing function. It is easy to see that $A(3,4)$ is a very large number, and it can be shown that $\alpha(m,n)\leqslant 3$ if $m\neq 0$ and $\log_2 n<A(3,4)$. The algorithm for set operations mentioned above is also described in Aho, Hopcroft, and Ullman [14.40] and Horowitz and Sahni [14.41].

Algorithm 14.8 is nonconstructive, that is, it does not give us the order in which vertices have to be collapsed to reduce a reducible graph. However, as we shall see now, this information can be easily obtained as this algorithm progresses.

During DFS let us assign to the vertices, numbers called SNUMBERs, from n to 1 in the order in which scanning at a vertex is completed. In fact this is the same as the order in which the corresponding entries of the array SCAN in Algorithm 14.5 are set to 1. It can be easily verified that

1. If (v,w) is a tree edge, then SNUMBER(v)<SNUMBER(w).
2. If (v,w) is a cross edge, then SNUMBER(v)<SNUMBER(w).
3. If (v,w) is a back edge, then SNUMBER(v)>SNUMBER(w).
4. If (v,w) is a forward edge, then SNUMBER(v)<SNUMBER(w).

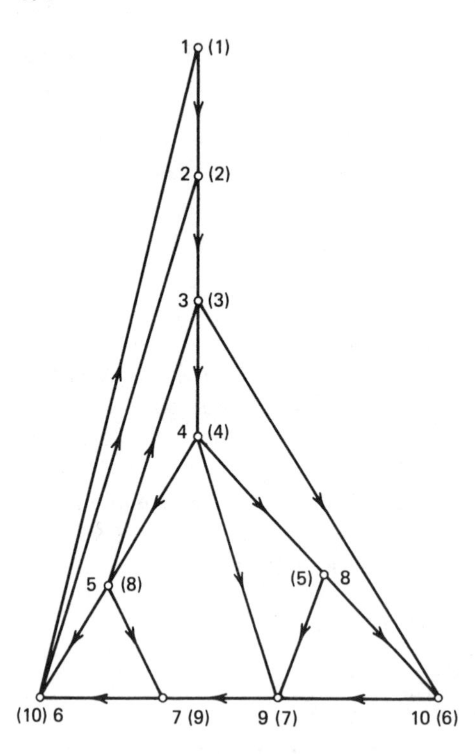

Figure 14.21. Graph of Fig. 14.18a with SNUMBER values in parentheses.

As an example, in Fig. 14.21 we have shown in parentheses the SNUMBERs of the corresponding vertices of the graph in Fig. 14.18a.

Suppose we apply the reducibility algorithm, and each time we label a vertex v let us associate with it a pair (HIGHPT1(v), SNUMBER(v)). When the algorithm is finished, let us order the vertices so that a vertex labeled (x_1, y_1) appears before a vertex labeled (x_2, y_2) if and only if $x_1 > x_2$ or $x_1 = x_2$ and $y_1 < y_2$. This order of vertices is called *reduction order*. Note that an unlabeled vertex v is associated with the pair (0, SNUMBER(v)).

Suppose $v_a, v_b, v_c, \ldots$ is a reduction order for a reducible program graph G. Let T be a DFS tree of G with vertex s as the root. Using Theorem 14.12 and the properties of SNUMBERs outlined earlier, it is easy to show that the tree edge (u, v_a) is the only edge incident into v_a. Suppose we collapse v_a into u and let G' be the resulting reducible graph. Also let T' be the tree obtained from T by contracting the edge (u, v_a). Then clearly T' is a DFS tree of G'. Further

1. A cross edge of G corresponds to nothing or to a cross edge or to a forward edge of G'.
2. A forward edge of G corresponds either to nothing or to a forward edge of G'.
3. A back edge of G corresponds either to nothing or to a back edge of G'.

It can now be verified that the relative HIGHPT1 values of the vertices in G' will be the same as those in G. This is also true for the SNUMBERs. Thus $v_b, v_c, \ldots$ will be a reduction order for G'. Repeating the above arguments, we get the following theorem.

THEOREM 14.13. If a program graph G is reducible, then we may collapse the vertices of G in the reduction order using the transformation S_2 (interspersed with applications of S_1). ■

For example, it can be verified from the HIGHPT1 values given in Fig. 14.18*a* and the SNUMBER values given in Fig. 14.21 that the sequence 4,5,3,8,10,9,7,6,2 is a reduction order for the graph in Fig. 14.18*a*.

14.6 DOMINATORS IN A PROGRAM GRAPH

Let G be a program graph with start vertex s. If in G, vertex v lies on every directed path from s to w, then v is called a *dominator* of w and is denoted by DOM(w). If v is a dominator of w and every other dominator of w also dominates v, then v is called the *immediate dominator* of w, and it is denoted by IDOM(w). For example, in the program graph G shown in Fig. 14.22*a* vertex 1 is the immediate dominator of vertex 9.

It can be shown that every vertex of a program graph $G=(V,E)$, except for the start vertex s, has a unique immediate dominator. The edges $\{(\text{IDOM}(w), w) | w \in V-\{s\}\}$ form a directed tree rooted at s, called the *dominator tree* of G, such that v dominates w if and only if v is a proper ancestor of w in the dominator tree. If G represents the flow of control in a computer program, then the dominator tree provides information about what kinds of code motion are safe. The dominator tree of the program graph in Fig. 14.22*a* is shown in Fig. 14.22*b*.

We now develop an algorithm due to Lengauer and Tarjan [14.42] for finding the dominator tree of a program graph. This algorithm is a simpler and faster version of an algorithm presented earlier by Tarjan [14.43].

Let G be a program graph with start vertex s. Let T be a DFS tree of G. In the following we shall identify the vertices of G by their DFS numbers. Furthermore, the notation $x \overset{*}{\to} y$ means that x is an ancestor of y in T, $x \overset{+}{\to} y$ means that $x \overset{*}{\to} y$ in T and $x \neq y$, and $x \to y$ means that x is father of y in T.

The following two lemmas are crucial in the development of the algorithm. Lemma 14.6 follows easily from our arguments leading to Theorem 14.7. Using this result, Lemma 14.7 can be proved.

LEMMA 14.6. If v and w are vertices of G such that $v<w$, then any directed path from v to w must contain a common ancestor of v and w in T. ■

LEMMA 14.7. Let $w \neq s$, $v \overset{*}{\to} w$, and P be a directed path from s to w. Let x be the last vertex on P such that $x<v$, and let y be the first vertex following x

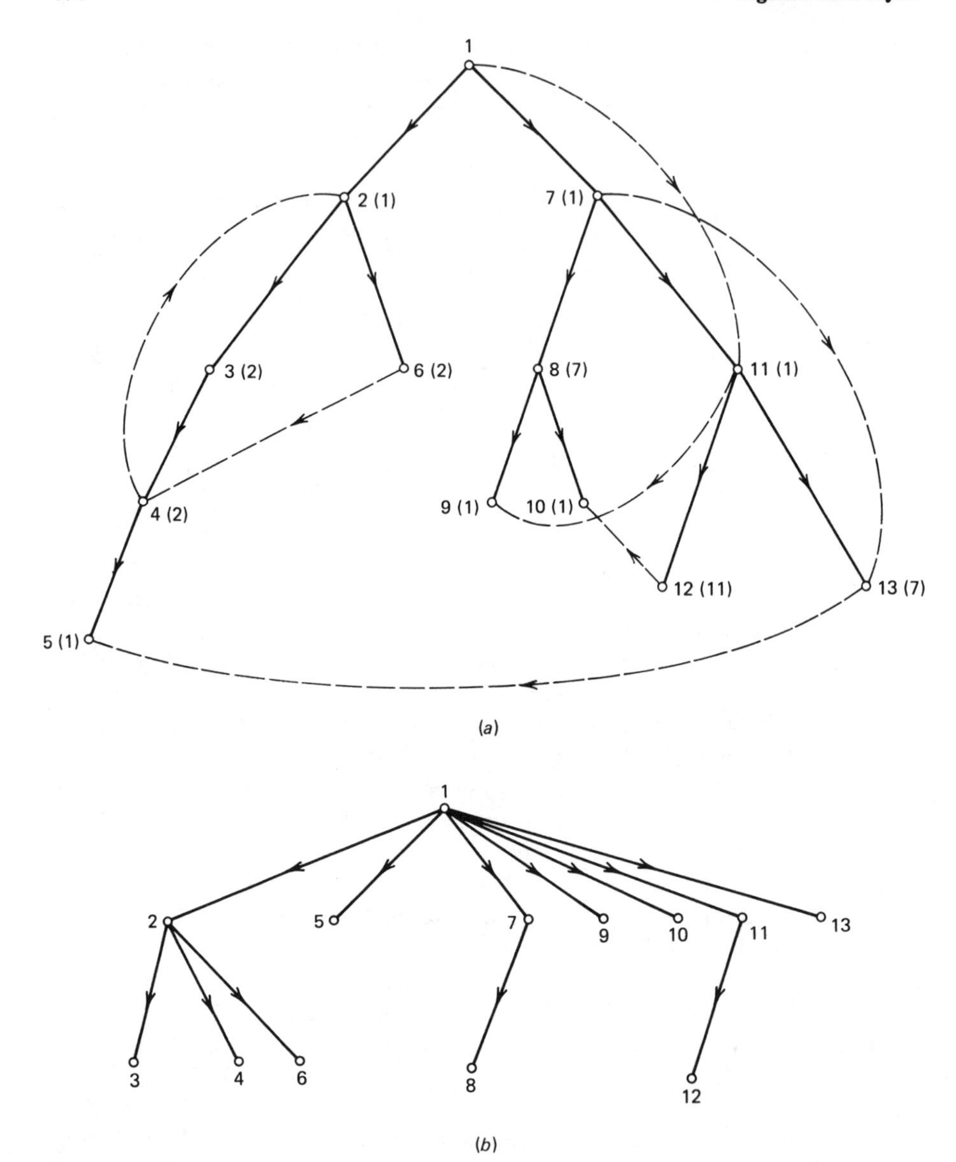

Figure 14.22. (a) A program graph. (b) Dominator tree of graph in (a).

on P and satisfying $v \overset{*}{\rightarrow} y \overset{*}{\rightarrow} w$. If Q: $x = v_0, v_1, v_2, \ldots, v_k = y$ is the part of P from x to y, then $v_i > y$ for $1 \leqslant i \leqslant k-1$. ■

Let us now define for each vertex $w \neq s$,

$$\text{SDOM}(w) = \min\{v \mid \text{there is a directed path } v = v_0, v_1, v_2, \ldots, v_k = w \text{ such that } v_i > w \text{ for } 1 \leqslant i \leqslant k-1\}.$$

SDOM(w) will be called the *semi-dominator* of w. It easily follows from the above definition that

$$\mathrm{SDOM}(w) < w. \tag{14.5}$$

In Fig. 14.22a the continuous edges are tree edges and the dashed edges are nontree edges with respect to a DFS. The semi-dominator of each vertex is shown in parentheses next to the vertex.

As a first step, Lengauer and Tarjan's algorithm computes semi-dominators of all the vertices. The semi-dominators are then used to compute the immediate dominators of the vertices.

The following theorem provides a way to compute semi-dominators.

THEOREM 14.14. For any vertex $w \neq s$,

$$\mathrm{SDOM}(w) = \min(\{v \mid (v,w) \in E \text{ and } v < w\} \cup \{\mathrm{SDOM}(u) \mid u > w \text{ and there is an edge } (v,w) \text{ such that } u \overset{*}{\rightarrow} v\}). \tag{14.6}$$

Proof

Let x equal the right-hand side of (14.6). It can be shown using the definition of semi-dominators that SDOM(w) $\leqslant x$.

To prove that SDOM(w) $\geqslant x$, let $y = \mathrm{SDOM}(w)$, and let $y = v_0, v_1, \ldots, v_k = w$ be a directed path such that $v_i > w$ for $1 \leqslant i \leqslant k-1$. If $k = 1$, then $(y,w) \in E$, and $y < w$ by (14.5). Thus $\mathrm{SDOM}(w) = y \geqslant x$. Suppose $k > 1$. Let j be minimum such that $j \geqslant 1$ and $v_j \overset{*}{\rightarrow} v_{k-1}$. Such a j exists since $k-1$ is a candidate for j. We now claim $v_i > v_j$ for $1 \leqslant i \leqslant j-1$.

Suppose to the contrary that $v_i < v_j$ for some $1 \leqslant i \leqslant j-1$. Then choose the i such that $1 \leqslant i \leqslant j-1$ and v_i is minimum. Then by Lemma 14.6 $v_i \overset{*}{\rightarrow} v_j$, contradicting the choice of j. This proves the claim.

The claim implies that $\mathrm{SDOM}(w) = y \geqslant \mathrm{SDOM}(v_j)$. Since $v_j > w$, $v_j \overset{*}{\rightarrow} v_{k-1}$, and $(v_{k-1}, w) \in E$ it follows from (14.6) that $\mathrm{SDOM}(v_j) \geqslant x$. So $\mathrm{SDOM}(w) \geqslant x$. Thus whether $k = 1$ or $k > 1$, we have $\mathrm{SDOM}(w) \geqslant x$, and the theorem is proved. ∎

We now need a way to compute immediate dominators from semi-dominators. Toward this end we proceed as follows.

The next three lemmas are easy to prove. Proof of Lemma 14.9 uses Lemma 14.6.

LEMMA 14.8. For any vertex $w \neq s$, $\mathrm{IDOM}(w) \overset{+}{\rightarrow} w$. ∎

LEMMA 14.9. For any vertex $w \neq s$, let $v = \mathrm{SDOM}(w)$. Then $v \overset{+}{\rightarrow} w$. ∎

LEMMA 14.10. For any vertex $w \neq s$, let $v = \mathrm{SDOM}(w)$. Then $\mathrm{IDOM}(w) \overset{*}{\rightarrow} v$. ∎

LEMMA 14.11. Let vertices v, w satisfy $v \xrightarrow{*} w$. Then either $v \xrightarrow{*} \text{IDOM}(w)$ or $\text{IDOM}(w) \xrightarrow{*} \text{IDOM}(v)$.

Proof

Let x be any proper descendant of IDOM(v) which is also a proper ancestor of v. Then there is a directed path from s to v which avoids x. By concatenating this path with the path in T from v to w, we obtain a directed path from s to w which avoids x. Thus IDOM(w) must be either a descendant of v or an ancestor of IDOM(v). ■

Using the foregoing lemmas we next prove two results which provide a way to compute immediate dominators from semi-dominators.

THEOREM 14.15. Let $w \neq s$ and let $v = \text{SDOM}(w)$. Suppose every u for which $v \xrightarrow{+} u \xrightarrow{*} w$ satisfies $\text{SDOM}(u) \geq \text{SDOM}(w)$. Then $\text{IDOM}(w) = v$.

Proof

By Lemma 14.10, $\text{IDOM}(w) \xrightarrow{*} v$. So to prove that $\text{IDOM}(w) = v$, it suffices to show that v dominates w.

Consider any directed path P from s to w. Let x be the last vertex on this path such that $x < v$. If there is no such x, then $v = s$ dominates w. Otherwise let y be the first vertex following x on the path and satisfying $v \xrightarrow{*} y \xrightarrow{*} w$. Let Q: $x = v_0, v_1, v_2, \ldots, v_k = y$ be the part of P from x to y. Then by Lemma 14.7, $v_i > y$ for $1 \leq i \leq k-1$. This together with the definition of semi-dominators implies that $\text{SDOM}(y) \leq x < v = \text{SDOM}(w)$. So $\text{SDOM}(y) < \text{SDOM}(w)$.

By the hypothesis of the theorem, $\text{SDOM}(u) \geq \text{SDOM}(w)$ for every u satisfying $v \xrightarrow{+} u \xrightarrow{*} w$. So y cannot be a proper descendant of v. Since y satisfies $v \xrightarrow{*} y \xrightarrow{*} w$, it follows that $y = v$, and v lies on P. Since the choice of P was arbitrary, v dominates w. ■

THEOREM 14.16. Let $w \neq s$, and let $v = \text{SDOM}(w)$. Let u be a vertex for which SDOM(u) is minimum among vertices u satisfying $v \xrightarrow{+} u \xrightarrow{*} w$. Then $\text{SDOM}(u) \leq \text{SDOM}(w)$ and $\text{IDOM}(u) = \text{IDOM}(w)$.

Proof

Let z be the vertex such that $v \rightarrow z \xrightarrow{*} w$. Then $\text{SDOM}(u) \leq \text{SDOM}(z) \leq v = \text{SDOM}(w)$.

By Lemma 14.10, IDOM(w) is an ancestor of v and hence a proper ancestor of u. Thus by Lemma 14.11, $\text{IDOM}(w) \xrightarrow{*} \text{IDOM}(u)$. To prove that $\text{IDOM}(u) = \text{IDOM}(w)$, it suffices to show that IDOM(u) dominates w.

Consider any directed path P from s to w. Let x be the last vertex on P satisfying $x < \text{IDOM}(u)$. If there is no such x, then $\text{IDOM}(u) = s$ dominates

w. Otherwise let y be the first vertex following x on P and satisfying $\text{IDOM}(u) \xrightarrow{*} y \xrightarrow{*} w$. As in the proof of Theorem 14.15, we can show using Lemma 14.7 that $\text{SDOM}(y) \leqslant x$. Since by Lemma 14.10 $\text{IDOM}(u) \leqslant \text{SDOM}(u)$, we have $\text{SDOM}(y) \leqslant x < \text{IDOM}(u) \leqslant \text{SDOM}(u)$. So $\text{SDOM}(y) < \text{SDOM}(u)$.

Since u has a minimum semi-dominator among the vertices on the tree path from z to w, y cannot be a proper descendant of v. Furthermore, y cannot be both a proper descendant of $\text{IDOM}(u)$ and an ancestor of u, for if this were the case, the directed path consisting of the tree path from s to $\text{SDOM}(y)$ followed by a path $\text{SDOM}(y) = v_0, v_1, v_2, \ldots, v_k = y$ such that $v_i > y$, for $1 \leqslant i \leqslant k-1$, followed by the tree path from y to u would avoid $\text{IDOM}(u)$, but no path from s to u avoids $\text{IDOM}(u)$.

The only remaining possibility is that $\text{IDOM}(u) = y$. Thus $\text{IDOM}(u)$ lies on the directed path P from s to w. Since the choice of P was arbitrary, $\text{IDOM}(u)$ dominates w. ■

The following main result is an immediate consequence of Theorems 14.15 and 14.16.

THEOREM 14.17. Let $w \neq s$ and let $v = \text{SDOM}(w)$. Let u be a vertex for which $\text{SDOM}(u)$ is minimum among vertices u satisfying $v \xrightarrow{+} u \xrightarrow{*} w$. Then

$$\text{IDOM}(w) = \begin{cases} v, & \text{if } \text{SDOM}(u) = \text{SDOM}(w) \\ \text{IDOM}(u), & \text{otherwise.} \end{cases}$$ ■

We are now ready to describe the dominator algorithm of Lengauer and Tarjan. Following are the main steps in this algorithm.

Algorithm 14.9 Dominators (Lengauer and Tarjan)

- **S1** Carry out a DFS of the given program graph $G = (V, E)$ with the start vertex as the root.
- **S2.** Compute the semi-dominators of all vertices by applying Theorem 14.14. Carry out the computation vertex by vertex in the decreasing order of their DFS numbers.
- **S3.** Implicitly define the immediate dominator of each vertex by applying Theorem 14.17.
- **S4.** Explicitly define the immediate dominator of each vertex, carrying out the computation vertex by vertex in the increasing order of their DFS numbers. ■

Implementation of Step S1 is straightforward. In the following we denote vertices by their DFS numbers assigned in step S1.

In our description of steps S2, S3 and S4 we use the arrays FATHER (defined as in Section 14.3), SEMI, BUCKET, and DOM, defined below.

SEMI(w)

1. Before the semi-dominator of w is computed,

$$\text{SEMI}(w)=w.$$

2. After the semi-dominator of w is computed,

$$\text{SEMI}(w)=\text{SDOM}(w).$$

BUCKET(w) It is a set of vertices whose semi-dominator is w.

DOM(w)

1. After step S3, if the semi-dominator of w is its immediate dominator, then DOM(w) is the immediate dominator of w. Otherwise DOM(w) is a vertex v such that $v<w$ and the immediate dominator of v is also the immediate dominator of w.
2. After step S4, DOM(w) is the immediate dominator of w.

After carrying out step S1, the algorithm carries out steps S2 and S3 simultaneously, processing the vertices $w\neq 1$ in the decreasing order of their DFS numbers. During this computation, the algorithm maintains a forest contained in the DFS tree of G. The forest consists of vertex set V and edge set $\{(\text{FATHER}(w), w)|\text{vertex } w \text{ has been processed}\}$. The algorithm uses one procedure to construct the forest and another to extract information from it. These procedures are:

LINK(v,w) Add edge (v,w) to the forest.

EVAL(v)

1. If v is the root of a tree in the forest, then EVAL(v)$=v$.
2. Otherwise let r be the root of the tree in the forest which contains v, and let u be a vertex for which SEMI(u) is minimum among vertices satisfying $r\overset{+}{\to}u\overset{*}{\to}v$. Then EVAL($v$)$=u$.

To process a vertex w, the algorithm computes the semi-dominator of w by applying Theorem 14.14. Thus the algorithm assigns

$$\text{SEMI}(w)=\min\{\text{SEMI}(\text{EVAL}(v))|(v,w)\in E\}.$$

After this computation, SEMI(w) is the semi-dominator of w. This follows from Theorem 14.14 and the definition of EVAL(v).

After computing SEMI(w), the algorithm adds w to BUCKET(SEMI(w)) and adds a new edge to the forest using LINK(FATHER(w), w). This completes step S2 for w.

The algorithm then carries out step S3 by considering each vertex in BUCKET(FATHER(w)).

Let v be such a vertex. The algorithm implicitly computes the immediate dominator of v by applying Theorem 14.17. Let $u = \text{EVAL}(v)$. Then u is the vertex satisfying $\text{FATHER}(w) \overset{+}{\to} u \overset{*}{\to} v$ whose semi-dominator is minimum. If $\text{SEMI}(u) = \text{SEMI}(v)$, then FATHER($w$) is the immediate dominator of v and the algorithm assigns $\text{DOM}(v) = \text{FATHER}(w)$. Otherwise u and v have the same immediate dominator, and the algorithm assigns $\text{DOM}(v) = u$. This completes step S3 for v.

In step S4 the algorithm examines vertices in the increasing order of their DFS numbers, filling in the immediate dominators not explicitly computed in step S3. Thus step S4 is as follows:

For each $i = 2, 3, \ldots, n$, if $\text{DOM}(i) \neq \text{SEMI}(i)$, then let

$$\text{DOM}(i) = \text{DOM}(\text{DOM}(i)).$$

For an illustration of the dominator algorithm consider the program graph shown in Fig. 14.22a. Just before vertex 11 is processed, the forest will be as in Fig. 14.23a. The entries of the SEMI array at this stage are shown in parentheses next to the corresponding vertices. Let us now process vertex 11.

The edges (1, 11) and (7, 11) are incident into 11. So

$$\text{SEMI}(11) = \min\{\text{SEMI}(\text{EVAL}(1)), \text{SEMI}(\text{EVAL}(7))\}.$$

Now EVAL(1) = 1 because vertex 1 is a tree root in the forest. For the same reason, EVAL(7) = 7. Thus

$$\begin{aligned}\text{SEMI}(11) &= \min\{\text{SEMI}(1), \text{SEMI}(7)\}\\ &= \min\{1, 7\}\\ &= 1.\end{aligned}$$

The algorithm now adds the edge (7, 11) to the forest and the vertex 11 to BUCKET(SEMI(11)) = BUCKET(1). The new forest with the SEMI array entries is shown in Fig. 14.23b. This completes step S2 for vertex 11.

The algorithm now considers BUCKET(FATHER(11)) = BUCKET(7). Vertex 13 is the only one whose semi-dominator is equal to 7. So BUCKET(7) = {13}. Now EVAL(13) = 11, since SEMI(11) is minimum among vertices u satisfying $7 \overset{+}{\to} u \overset{*}{\to} 13$ (see Fig. 14.23b). Since $\text{SEMI}(13) \neq \text{SEMI}(11)$, the algorithm sets DOM(13) = 11. This completes step S3 for vertex 11.

After steps S2 and S3 have been carried out for every vertex $w \neq 1$, the semi-dominators of all the vertices will be available. At this stage, the entries

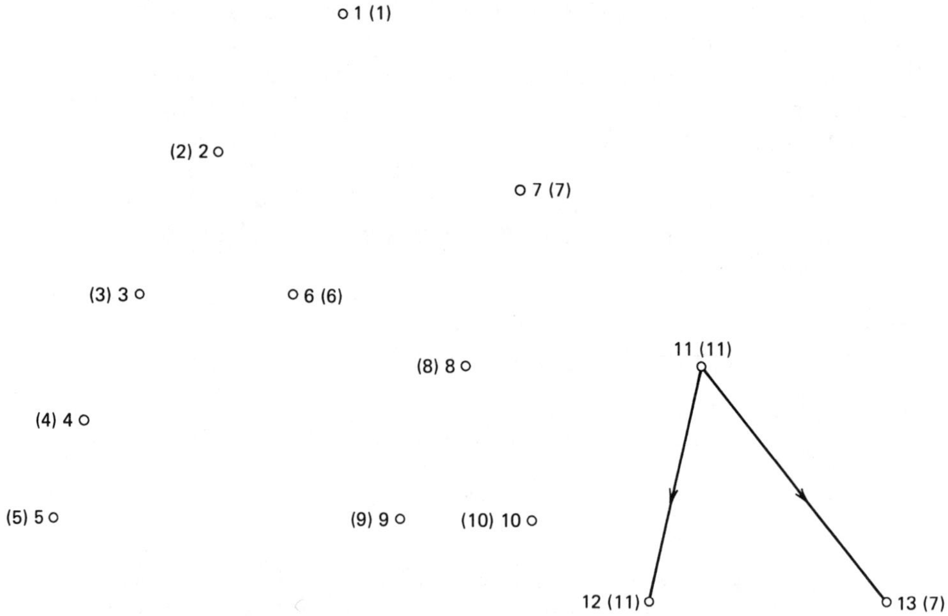

(a)

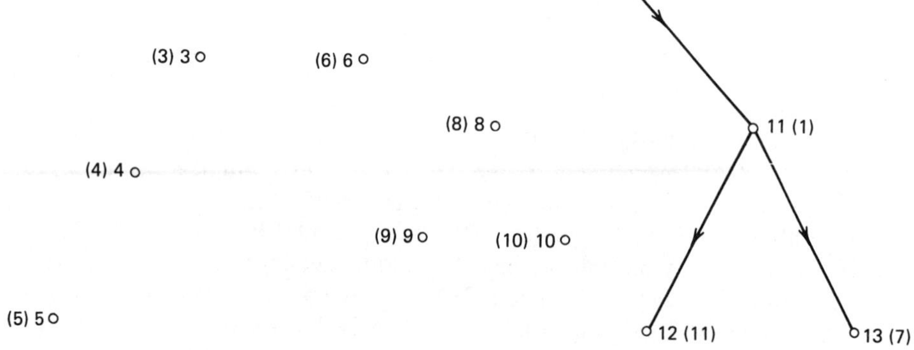

(b)

Figure 14.23.

of the DOM array and the semi-dominators will be as given below:

Vertex	DOM	Semi-Dominator (SEMI)
2	1	1
3	2	2
4	2	2
5	1	1
6	2	2
7	1	1
8	7	7
9	1	1
10	1	1
11	1	1
12	11	11
13	11	7

For every vertex $w \neq 13$, DOM(w)=SEMI(w). So for all vertices w except 13, IDOM(w)=DOM(w). For vertex 13, we compute

$$\begin{aligned}\text{DOM}(13) &= \text{DOM}(\text{DOM}(13))\\ &= \text{DOM}(11)\\ &= 1.\end{aligned}$$

So IDOM(13)=1. This completes step S4 of the algorithm and we get the dominator tree shown in Fig. 14.22*b*.

Clearly, the complexity of the above algorithm depends crucially on the implementation of LINK and EVAL instructions. Tarjan [14.44] discusses two methods which use path compression. One of these is described below.

To represent the forest built by the LINK instructions, the algorithm uses two arrays, ANCESTOR and LABEL. Initially ANCESTOR(v)=0 and LABEL(v)=v for each vertex v. In general, ANCESTOR(v)=0 only if v is a tree root in the forest; otherwise, ANCESTOR(v) is ancestor of v in the forest.

The algorithm maintains the labels so that they satisfy the following property. Let v be any vertex, let r be the root of the tree in the forest containing v, and let $v=v_k, v_{k-1},\ldots, v_0=r$ be such that ANCESTOR(v_i)= v_{i-1} for $1 \leq i \leq k$. Let x be a vertex such that SEMI(x) is minimum among vertices $x \in \{\text{LABEL}(v_i) | 1 \leq i \leq k\}$. Then we have the following property.

x is a vertex such that SEMI(x) is minimum among vertices x satisfying

$$r \xrightarrow{+} x \xrightarrow{*} v.$$

To carry out LINK(v, w), the algorithm assigns ANCESTOR(w)=v. To carry out EVAL(v), the algorithm follows ancestor pointers and determines the sequence $v=v_k, v_{k-1},\ldots, v_0=r$ such that ANCESTOR(v_i)=v_{i-1} for $1 \leq i \leq k$. If $v=r$, then the algorithm sets EVAL(v)=v. Otherwise the algorithm sets ANCESTOR$(v_i)=r$ for $2 \leqslant i \leqslant k$, simultaneously updating labels as follows (to maintain the property mentioned before):

$$\text{If SEMI(LABEL}(v_{i-1})) < \text{SEMI(LABEL}(v_i)),$$
$$\text{then LABEL}(v_i) = \text{LABEL}(v_{i-1})$$

Then the algorithm sets EVAL$(v)=$LABEL(v).

Tarjan [14.44] has shown that the complexity of implementing $(n-1)$ LINKs and $(m+n-1)$ EVALs using the method described above is $O(m\log n)$. If we use the more sophisticated implementation of LINK and EVAL instructions, also described in [14.44], then the algorithm would require $O(m\alpha(m,n))$time, where $\alpha(m,n)$ is the functional inverse of Ackermann's function defined in the previous section.

For other dominator algorithms see Aho and Ullman [14.45] and Purdom and Moore [14.46].

14.7 FURTHER READING

Frank and Frisch [14.47], Knuth [14.48], Aho, Hopcroft, and Ullman [14.40], Christofides [14.49], Lawler [14.50], Reingold, Nievergelt, and Deo [14.51], Minieka [14.52], and Even [14.53] are other good references for graph and combinatorial algorithms. Aho, Hopcroft, and Ullman [14.40] and Horowitz and Sahni [14.41] may be consulted for a discussion on the different methods of computer representation of graphs.

The running time of all the algorithms considered in this book is bounded above by some polynomial in the number of vertices and the number of edges of a graph. Let $\mathcal{P}$ be the class of all problems which can be solved by a polynomial-time algorithm. There are a large number of problems for which no polynomial-time algorithms are known to exist. Many of these can be solved in polynomial time by a nondeterministic algorithm. $\mathcal{NP}$ denotes the class of all such problems. A problem is $\mathcal{NP}$-hard if a deterministic polynomial-time algorithm for its solution can be used to find a deterministic polynomial-time algorithm for every problem in $\mathcal{NP}$. An $\mathcal{NP}$-hard problem in $\mathcal{NP}$ is called $\mathcal{NP}$-complete. Thus if a deterministic polynomial-time algorithm is found for any problem in the $\mathcal{NP}$-complete class, then such an algorithm exists for every problem in $\mathcal{NP}$. In a pioneering paper Cook [14.54] showed that the satisfiability problem is $\mathcal{NP}$-complete. Karp [14.55] has demonstrated that a large number of problems are $\mathcal{NP}$-complete. Garey and Johnson [14.56] and Even [14.53] give an elegant introduction to the theory of $\mathcal{NP}$-completeness and the proof of $\mathcal{NP}$-completeness of several graph problems. See also [14.40], [14.51], and Horowitz and Sahni [14.57].

In the literature many interesting graph algorithms have been reported. Some of these are listed at the end of the next chapter.

14.8 EXERCISES

14.1 A *transitive reduction* of a directed graph $G=(V, E)$ is defined to be any graph $G'=(V', E')$ with as few edges as possible such that the transitive closure of G' is equal to the transitive closure of G. Design an algorithm to find a transitive reduction of a directed graph. How is this algorithm related to the transitive closure algorithm? (Aho, Garey, and Ullman [14.58].)

14.2 Using DFS, design algorithms for the following problems:

(a) Topological sorting of the vertices of a graph.

(b) Finding bridges of a graph.

(c) Finding a spanning forest.

(d) Finding a set of fundamental circuits and a set of fundamental cutsets.

(e) Testing whether a graph is bipartite.

(f) Testing whether a subset of edges is a cut of a graph.

14.3 Consider a graph G and let T be a DFS spanning forest of G. Let $\text{ND}(v)$ be the number of descendants of v (including itself). Prove that a vertex w is a descendant of v if and only if

$$\text{DFN}(v) \leqslant \text{DFN}(w) < \text{DFN}(v) + \text{ND}(v).$$

14.4 Modify the DFS algorithm to include computation of $\text{ND}(v)$.

14.5 Let T be a DFS tree of a connected graph G. Let G_s be a complete subgraph of G. Show that all the vertices of G_s lie on one directed path in T.

14.6 Let T be a DFS tree of a directed graph G. If C is a directed circuit in G and v is the vertex on C with minimum depth-first number, show that v is an ancestor in T of every vertex on C.

14.7 *Breadth-First Search* (*BFS*) explores a connected graph G as follows:

S1. To begin with no vertices of G are labeled.

S2. Select vertex s and label it with 0.

S3. Set $i=0$.

S4. Let S be the set of all unlabeled vertices adjacent to at least one vertex labeled i.

S5. If S is empty STOP. Otherwise label all the vertices in S with $i+1$.

S6. Set $i=i+1$ and go to step S4.

Show that the edges traversed while labeling the vertices form a spanning tree of G.

14.8 Show how the BFS algorithm can be used to compute the distance from a vertex s to all the vertices in a connected graph G. (By distance between vertices u and v we mean the length of a u–v path having the smallest number of edges.)

14.9 Show that the set of back edges of a reducible program graph is unique, that is, all DFS spanning forests have the same set of back edges (Hecht and Ullman [14.35]).

14.10 Use the result of Theorem 6.10 to design an algorithm to find all the spanning trees in a connected graph.

14.11 Let T be a tree on n vertices. Let $\{1,2,\ldots,n\}$ be the vertex set of T. We can associate a unique sequence $(t_1, t_2, \ldots, t_{n-2})$ with T as follows: Let s_1 be the first vertex of degree 1 in T. Then the vertex adjacent to s_1 is taken as t_1. Now remove s_1 from T. If s_2 is the first vertex of degree 1 in $T-s_1$, then the vertex adjacent to s_2 in $T-s_1$ is taken as t_2. Remove s_2 and repeat the operation until t_{n-2} is defined. The sequence $(t_1, t_2, \ldots, t_{n-2})$ is called the *Prüfer sequence* associated with T. Clearly, two different trees have different Prüfer sequences.

Given a sequence $(t_1, t_2, \ldots, t_{n-2})$ with each $t_i \in \{1,2,\ldots,n\}$. Design an algorithm to construct a tree T for which $(t_1, t_2, \ldots, t_{n-2})$ is the Prüfer sequence.

(The number of $(n-2)$-letter sequences that can be formed from the set $\{1,2,\ldots,n\}$ is n^{n-2}. Each one of these is the Prüfer sequence of a spanning tree of K_n. So there is a one-to-one correspondence between the sequences constructed using any $n-2$ letters (not necessarily distinct) of the set $\{1,2,\ldots,n\}$ and the spanning trees of K_n. Thus the number of spanning trees of K_n is n^{n-2}. This proof is due to Prüfer [14.59].)

14.12 Use the result of Theorem 9.10 to design an algorithm to find the chromatic number of a graph.

14.9 REFERENCES

14.1 D. Gries, *Compiler Construction for Digital Computers*, Wiley, New York, 1971.

14.2 S. Warshall, "A Theorem on Boolean Matrices," *J. ACM*, Vol. 9, 11–12 (1962).

14.3 H. S. Warren, "A Modification of Warshall's Algorithm for the Transitive Closure of Binary Relations," *Comm. ACM*, Vol. 18, 218–220 (1975).

14.4 V. L. Arlazarov, E. A. Dinic, M. A. Kronrod, and I. A. Faradžev, "On Economical Construction of the Transitive Closure of a Directed Graph," *Soviet Math. Dokl.*, Vol. 11, 1209–1210 (1970).

14.5 V. Strassen, "Gaussian Elimination is Not Optimal," *Numerische Math.*, Vol. 13, 354–356 (1969).

14.6 M. J. Fischer and A. R. Meyer, "Boolean Matrix Multiplication and Transitive Closure," *Conf. Record, IEEE 12th Annual Symp. on Switching and Automata Theory*, 1971, pp. 129–131.

14.7 M. E. Furman, "Application of a Method of Fast Multiplication of Matrices in the Problem of Finding the Transitive Closure of a Graph," *Soviet Math. Dokl.*, Vol. 11, 1252 (1970).

14.8 I. Munro, "Efficient Determination of the Transitive Closure of a Directed Graph," *Information Proccessing Lett.*, Vol. 1, 56–58 (1971).

14.9 P. E. O'Neil and E. J. O'Neil, "A Fast Expected Time Algorithm for Boolean Matrix Multiplication and Transitive Closure," *Inform. and Control*, Vol. 22, 132–138 (1973).

14.10 J. Eve and R. Kurki-Suonio, "On Computing the Transitive Closure of a Relation," *Acta Informatica*, Vol. 8, 303–314 (1977).

14.11 P. Purdom, "A Transitive Closure Algorithm," *BIT*, Vol. 10, 76–94 (1970).

14.12 C. P. Schnorr, "An Algorithm for Transitive Closure with Linear Expected Time," *SIAM J. Computing*, Vol. 7, 127–133 (1978).

14.13 M. M. Syslo and J. Dzikiewicz, "Computational Experience with Some Transitive Closure Algorithms," *Computing*, Vol. 15, 33–39 (1975).

14.14 A. Pnueli, A. Lempel, and S. Even, "Transitive Orientation of Graphs and Identification of Permutation Graphs," *Canad. J. Math.*, Vol. 23, 160–175 (1971).

14.15 P. C. Gilmore and A. J. Hoffman, "A Characterization of Comparability Graphs and of Interval Graphs," *Canad. J. Math.*, Vol. 16, 539–548 (1964).

14.16 S. Even, A. Pnueli, and A. Lempel, "Permutation Graphs and Transitive Graphs," *J. ACM*, Vol. 19, 400–410 (1972).

14.17 S. Even, *Algorithmic Combinatorics*, Macmillan, New York, 1973.

14.18 C. L. Liu, *Introduction to Combinatorial Mathematics*, McGraw-Hill, New York, 1968.

14.19 R. E. Tarjan, "Depth-First Search and Linear Graph Algorithms," *SIAM J. Computing*, Vol. 1, 146–160 (1972).

14.20 J. Hopcroft and R. Tarjan, "Efficient Algorithms for Graph Manipulation," *Comm. ACM*, Vol. 16, 372–378 (1973).

14.21 A. V. Aho, J. E. Hopcroft, and J. D. Ullman, "On Finding the Least Common Ancestors in Trees," *SIAM J. Computing*, Vol. 5, 115–132 (1976).

14.22 J. Cocke, "Global Common Subexpression Elimination," *SIGPLAN Notices*, Vol. 5, 20–24 (1970).

14.23 J. D. Ullman, "Fast Algorithms for the Elimination of Common Subexpressions," *Acta Informatica*, Vol. 2, 191–213 (1973).

14.24 K. Kennedy, "A Global Flow Analysis Algorithm," *Int. J. Computer Math.*, Vol. 3, 5–16 (1971).

14.25 M. Schaefer, *A Mathematical Theory of Global Program Optimization*, Prentice-Hall, Englewood Cliffs, N.J., 1973.

14.26 A. V. Aho and J. D. Ullman, *The Theory of Parsing, Translation and Compiling, Vol. II—Compiling*, Prentice-Hall, Englewood Cliffs, N.J., 1973.

14.27 F. E. Allen, "Control Flow Analysis," *SIGPLAN Notices*, Vol. 5, 1–19 (1970).

14.28 F. E. Allen, *Program Optimization, Annual Review in Automatic Programming*, Vol. 5, Pergamon, New York, 1969.

14.29 M. S. Hecht, *Flow Analysis of Computer Programs*, Elsevier, New York, 1977.

14.30 M. S. Hecht and J. D. Ullman, "Flow Graph Reducibility," *SIAM J. Computing*, Vol. 1, 188–202 (1972).

14.31 J. Cocke and R. E. Miller, "Some Analysis Techniques for Optimizing Computer Programs," *Proc. 2nd Int. Conf. on System Sciences*, Honolulu, Hawaii, 1969.

14.32 J. E. Hopcroft and J. D. Ullman, "An $n \log n$ Algorithm for Detecting Reducible Graphs," *Proc. 6th Annual Princeton Conf. on Information Sciences and Systems*, Princeton, N.J., 1972, pp. 119–122.

14.33 R. E. Tarjan, "Testing Flow Graph Reducibility," *Proc. 5th Annual ACM Symp. on Theory of Computing*, 1973, pp. 96–107.

14.34 R. E. Tarjan, "Testing Flow Graph Reducibility," *J. Comput. Sys. Sci.*, Vol. 9, 355–365 (1974).

14.35 M. S. Hecht and J. D. Ullman, "Characterizations of Reducible Flow Graphs," *J. ACM*, Vol. 21, 367–375 (1974).

14.36 J. M. Adams, J. M. Phelan, and R. H. Stark, "A Note on the Hecht-Ullman Characterization of Non-Reducible Flow Graphs," *SIAM J. Computing*, Vol. 3, 222–223 (1974).

14.37 M. Fischer, "Efficiency of Equivalence Algorithms," in *Complexity of Computer Computations* (R. E. Miller and J. W. Thatcher, Eds.), Plenum Press, New York, 1972, pp. 153–168.

14.38 J. E. Hopcroft and J. D. Ullman, "Set Merging Algorithms," *SIAM J. Computing*, Vol. 2, 294–303 (1973).

14.39 R. E. Tarjan, "On the Efficiency of a Good but Not Linear Set Union Algorithm," *J. ACM*, Vol. 22, 215–225 (1975).

14.40 A. V. Aho, J. E. Hopcroft, and J. D. Ullman, *The Design and Analysis of Computer Algorithms*, Addison-Wesley, Reading, Mass., 1974.

14.41 E. Horowitz and S. Sahni, *Fundamentals of Data Structures*, Computer Science Press, Potomac, Md., 1976.

14.42 T. Lengauer and R. E. Tarjan, "A Fast Algorithm for Finding Dominators in a Flow Graph," *Trans. on Prog. Lang. and Sys.*, Vol. 1, 121–141 (1979).

14.43 R. E. Tarjan, "Finding Dominators in Directed Graphs," *SIAM J. Computing*, Vol. 3, 62–89 (1974).

14.44 R. E. Tarjan, "Applications of Path Compression on Balanced Trees," *J. ACM*, Vol. 26, 690–715 (1979).

14.45 A. V. Aho and J. D. Ullman, *Principles of Compiler Design*, Addison-Wesley, Reading, Mass., 1977.

14.46 P. W. Purdom and E. F. Moore, "Algorithm 430: Immediate Predominators in a Directed Graph," *Comm. ACM*, Vol. 15, 777–778 (1972).

14.47 H. Frank and I. T. Frisch, *Communication, Transmission and Transportation Networks*, Addison-Wesley, Reading, Mass., 1971.

14.48 D. E. Knuth, *The Art of Computer Programming, Vol. 3: Sorting and Searching*, Addison-Wesley, Reading, Mass., 1973.

14.49 N. Christofides, *Graph Theory: An Algorithmic Approach*, Academic Press, New York, 1975.

14.50 E. L. Lawler, *Combinatorial Optimization: Networks and Matroids*, Holt, Rinehart and Winston, New York, 1976.

14.51 E. M. Reingold, J. Nievergelt, and N. Deo, *Combinatorial Algorithms: Theory and Practice*, Prentice Hall, Englewood Cliffs, N.J., 1977.

14.52 E. Minieka, *Optimization Algorithms for Networks and Graphs*, Marcel Dekker, New York, 1978.

14.53 S. Even, *Graph Algorithms*, Computer Science Press, Potomac, Md., 1979.

14.54 S. A. Cook, "The Complexity of Theorem Proving Procedures," *Proc. 3rd ACM Symp. on Theory of Computing*, 1971, pp. 151–158.

14.55 R. M. Karp, "Reducibility among Combinatorial Problems," in *Complexity of Computer Computations* (R. E. Miller and J. W. Thatcher, Eds.) Plenum Press, New York, 1972, pp. 85–104.

14.56 M. R. Garey and D. S. Johnson, *Computers and Intractability, A Guide to the Theory of NP-Completeness*, Freeman, San Francisco, Ca., 1979.

14.57 E. Horowitz and S. Sahni, *Fundamentals of Computer Algorithms*, Computer Science Press, Potomac, Md., 1978.

14.58 A. V. Aho, M. R. Garey, and J. D. Ullman, "The Transitive Reduction of a Directed Graph," *SIAM J. Computing*, Vol. 1, 131–137 (1972).

14.59 H. Prüfer, "Neuer Beweis eines Satzes über Permutationen," *Arch. Math. Phys.*, Vol. 27, 742–744 (1918).

Chapter 15

Algorithmic Optimization

In this chapter, we study several algorithms which concern optimization problems on graphs. These problems arise in a variety of applications relating to operations research and computer science. We discuss algorithms relating to the following main topics:

1. Shortest paths,
2. Optimum trees,
3. Matchings in a graph,
4. Network flows,
5. Optimum branchings.

15.1 SHORTEST PATHS

Let G be a connected directed graph in which each directed edge is associated with a real positive number called the *length* of the edge. The length of an edge directed from a vertex i to a vertex j is denoted by $w(i, j)$. If there is no edge directed from vertex i to vertex j, then $w(i, j)=\infty$. The length of a directed path in G is the sum of the lengths of the edges in the path. A minimum length directed s–t path is called a *shortest path* from s to t. The length of a shortest directed s–t path is called the *distance* from s to t, and it is denoted as $d(s, t)$. Clearly, $d(i, i)=0$ for all i.

In this section we consider the following two problems:

1. Find the shortest paths from a specified vertex s to all other vertices in G.
2. Find the shortest paths between all the ordered pairs of vertices in G.

These two problems arise in several optimization problems. For example, finding a minimum cost flow in a transport network involves finding a shortest path from the source to the sink in the network [15.1].

15.1.1 Shortest Paths from a Specified Vertex s to all Other Vertices in a Graph

We now develop an algorithm due to Dijkstra [15.2], which finds shortest paths from a specified vertex s to all the other vertices in an n-vertex connected directed graph G. The following ideas form the basis of Dijkstra's algorithm.

Let V denote the vertex set of G, and let S be a subset of V such that $s \in S$. $\bar{S}$ will denote the complement of S in V. Thus $\bar{S} = V - S$.

Among all the directed paths from s to the vertices in $\bar{S}$, let P be a path with minimum length. The length of P is then called the distance $d(s, \bar{S})$ from s to $\bar{S}$. Let P: $s, \ldots, u, v$. Then clearly $v \in \bar{S}$ and $u \in S$. Further, the s–u section of P must be a shortest s–u path with all its vertices in S. Therefore,

$$d(s, \bar{S}) = d(s, u) + w(u, v).$$

From this we can see that the distance $d(s, \bar{S})$ can be computed using the formula

$$d(s, \bar{S}) = \min_{\substack{u \in S \\ v \in \bar{S}}} \{d(s, u) + w(u, v)\}. \tag{15.1}$$

Now, if v is the vertex in $\bar{S}$ such that $d(s, \bar{S}) = d(s, u) + w(u, v)$ for some $u \in S$, then clearly

$$d(s, v) = d(s, \bar{S}). \tag{15.2}$$

Dijkstra's algorithm constructs a sequence $S_0 = \{s\}, S_1, S_2, \ldots$ of subsets of V such that the following conditions are satisfied:

1. If $s = u_0, u_1, u_2, \ldots, u_{n-1}$ are the vertices of V such that $d(s, u_1) \leqslant d(s, u_2) \leqslant d(s, u_3) \leqslant \cdots \leqslant d(s, u_{n-1})$, then $S_i = \{s, u_1, u_2, \ldots, u_i\}$ for $i > 0$.
2. When the set S_i has been determined, the shortest paths from s to $u_1, u_2, \ldots, u_i$ will be known.

If the sets S_i are defined as above, then by (15.2) we have

$$d(s, u_{i+1}) = d(s, \bar{S}_i). \tag{15.3}$$

Thus determining S_{i+1} from S_i involves computing $d(s, \bar{S}_i)$.

The subsets $S_1, S_2, \ldots, S_{n-1}$ can be constructed as follows: By (15.1),

$$d(s, \bar{S}_0) = \min_{\substack{u \in S_0 \\ v \in \bar{S}_0}} \{d(s,u) + w(u,v)\} = \min_{v \in \bar{S}_0} \{w(s,v)\}.$$

Hence by (15.3), u_1 is a vertex with the property

$$d(s, u_1) = \min_{v \in \bar{S}_0} \{w(s,v)\}. \tag{15.4}$$

If P_i denotes a shortest path from s to the vertex u_i, then clearly P_1: s, u_1.
Suppose that the subsets $S_0, S_1, \ldots, S_i$ and the paths $P_1, P_2, \ldots, P_i$ have been determined. Now to determine S_{i+1}, we first compute $d(s, \bar{S}_i)$ using (15.1). By (15.3), u_{i+1} is a vertex in $\bar{S}_i$ with the property

$$d(s, u_{i+1}) = d(s, \bar{S}_i).$$

By (15.1), there is a vertex $j \in S_i$ such that

$$d(s, u_{i+1}) = d(s, u_j) + w(u_j, u_{i+1}).$$

Therefore we can obtain P_{i+1} by adjoining the edge (u_j, u_{i+1}) to the path P_j.
If we are interested only in a shortest path from a specified vertex s to another specified vertex t, then we can terminate the above procedure after we have determined the first set S_i which contains t.
It is clear that in the above procedure we need to compute the minimum in (15.1) at each stage. If this minimum were to be computed from scratch at each stage, then determining S_i from S_{i-1} would require $(i-1) \times (n-i)$ additions and $\{i(n-i)-1\}$ comparisons. The total number of operations for the entire algorithm would then be

$$\sum_{i=1}^{n-1} \{(2i-1)(n-i)-1\}$$

resulting in an overall complexity of $O(n^3)$. However, many of these additions and comparisons would be repeated unnecessarily.
Dijkstra, in his algorithm, avoids such repeated additions and comparisons by storing computational information from one stage to the next. This is achieved by a labeling procedure which, as we shall see, improves the complexity of the algorithm to $O(n^2)$. The following ideas lead to Dijkstra's labeling procedure. Suppose we do the labeling so that for $i = 0, 1, 2, \ldots$ the label $l_i(v)$ for vertex v satisfies the following:

1. $l_0(s) = 0$ and $l_0(v) = \infty$ for all $v \neq s$.

2. For $i \geqslant 1$,

$$l_i(v) = d(s, v), \qquad \text{for all } v \in S_{i-1},$$

$$l_i(v) = \min_{u \in S_{i-1}} \{d(s, u) + w(u, v)\}, \qquad \text{for all } v \in \bar{S}_{i-1}.$$

Clearly, then, u_i is a vertex with the property

$$d(s, u_i) = d(s, \bar{S}_{i-1}) = \min_{v \in \bar{S}_{i-1}} \{l_i(v)\}.$$

Now we can compute $l_{i+1}(v)$ from $l_i(v)$ as follows:

1. For $v \in S_i$, $l_{i+1}(v) = l_i(v) = d(s, v)$.
2. For $v \in \bar{S}_i$,

$$\begin{aligned} l_{i+1}(v) &= \min_{u \in S_i} \{d(s, u) + w(u, v)\} \\ &= \min\{l_i(v), d(s, u_i) + w(u_i, v)\} \\ &= \min\{l_i(v), l_i(u_i) + w(u_i, v)\}. \end{aligned} \qquad (15.5)$$

Then u_{i+1} is to be chosen such that

$$\begin{aligned} d(s, u_{i+1}) &= d(s, \bar{S}_i) \\ &= \min_{v \in \bar{S}_i} \{l_{i+1}(v)\}. \end{aligned} \qquad (15.6)$$

Note that the label of u_i does not change after the set S_i has been determined.

Thus Dijkstra's algorithm begins with the labels $l_0(s) = 0$ and $l_0(v) = \infty$ for all $v \neq s$. As the algorithm progresses, the labels are modified according to (15.5). The labels $l_{n-1}(v)$ would give the distance from s to v.

It is clear that determining u_{i+1} involves computing $l_{i+1}(v)$ for all $v \in \bar{S}_i$ and then finding the minimum of these labels. For $i \geqslant 1$ the former computations (15.5) require $(n-i-1)$ additions and $(n-i-1)$ comparisons, whereas the latter computations (15.6) require $\{(n-i)-2\}$ comparisons. Thus, clearly, the complexity of Dijkstra's algorithm is $O(n^2)$.

A description of Dijkstra's algorithm is presented next. In this description, LABEL is an array in which the current labels of the vertices are stored. A vertex becomes permanently labeled when it is set equal to u_i for some i. We use an array PERM to indicate which of the vertices are permanently labeled. If PERM$(v) = 1$, then v is a permanently labeled vertex. Note that in such a case the label of v is equal to $d(s, v)$. We start with PERM$(s) = 1$ and PERM$(v) = 0$ for all $v \neq s$.

PRED is an array which keeps a record of the vertices from which the vertices get permanently labeled. If a vertex v is permanently labeled, then

$$v, \mathrm{PRED}(v), \mathrm{PRED}(\mathrm{PRED}(v)), \ldots, s$$

are the vertices in a shortest directed s–v path.

Algorithm 15.1 Shortest Paths (Dijkstra)

S0. G is the given directed graph with lengths associated with its edges. Shortest paths from vertex s to all other vertices in G are required.

S1. (Initialize.) Set LABEL(s)=0, PERM(s)=1, and PRED(s)=s. For all $v \neq s$, set LABEL(v)=∞, PERM(v)=0, and PRED(v)=v.

S2. Set $i=0$ and $u=s$. (u is the latest vertex permanently labeled. Now it is s.)

S3. (Compute LABEL(v) and update the entries of the PRED array.) Set $i=i+1$. Do the following for each vertex v which is not yet labeled permanently:

1. Set $M=\min\{\mathrm{LABEL}(v), \mathrm{LABEL}(u)+w(u,v)\}$.
2. If $M<\mathrm{LABEL}(v)$, then set LABEL(v)=M and PRED(v)=u.

S4. (Identify vertex u_i.) Find among all vertices, which are not yet permanently labeled, a vertex w with the smallest label. (In case of a tie the choice can be made arbitrarily.) Set PERM(w)=1 and $u=w$. ($u_i=w$, and it is the latest vertex labeled permanently.)

S5. If $i<n-1$, then go to step S3. Otherwise HALT. (All the shortest paths are found. The vertex labels give the lengths of the shortest paths. Now v, PRED(v), PRED(PRED(v)), …, s are the vertices in a shortest directed s–v path.) ■

Note that in a computer program ∞ is represented by as high a number as necessary. Further, if the final label of a vertex v is equal to ∞, then it means that there is no directed path from s to v.

To illustrate Dijkstra's algorithm, consider the graph G in Fig. 15.1, in which the length of an edge is shown next to the edge. In Fig. 15.2, we have shown the entries of the LABEL and PRED arrays.

For any i, the circled entries in the LABEL array correspond to the permanently labeled vertices, namely, the vertices in S_i. The entry with the mark * is the label of the latest vertex permanently labeled, namely, the vertex u_i. The shortest paths from s and the corresponding distances are obtained from the final values of the entries in the PRED and LABEL arrays.

In our discussions thus far we have assumed that all the lengths are nonnegative. Dijkstra's algorithm is not valid if some of the lengths are

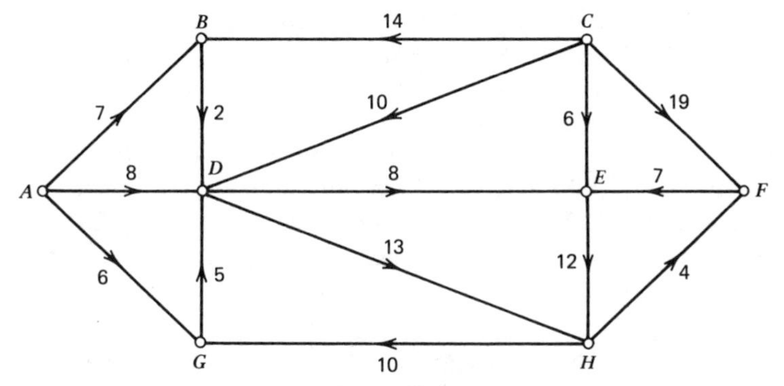

Figure 15.1.

	Vertices							
i	A	B	C	D	E	F	G	H
0	⓪*	∞	∞	∞	∞	∞	∞	∞
1	⓪	7	∞	8	∞	∞	⑥*	∞
2	⓪	⑦*	∞	8	∞	∞	⑥	∞
3	⓪	⑦	∞	⑧*	∞	∞	⑥	∞
4	⓪	⑦	∞	⑧	⑯*	∞	⑥	21
5	⓪	⑦	∞	⑧	⑯	∞	⑥	㉑*
6	⓪	⑦	∞	⑧	⑯	㉕*	⑥	㉑
7	⓪	⑦	(∞)*	⑧	⑯	㉕	⑥	㉑

(a)

PRED(A)=A	PRED(E)=D
PRED(B)=A	PRED(F)=H
PRED(C)=C	PRED(G)=A
PRED(D)=A	PRED(H)=D

From	To	Shortest Path
A	B	A, B
A	C	No path
A	D	A, D
A	E	A, D, E
A	F	A, D, H, F
A	G	A, G
A	H	A, D, H

(b)

Figure 15.2. Illustration of Dijkstra's algorithm. The LABEL array is shown in (a).

negative (why?). However, a modification of this algorithm makes it applicable for general networks which have no directed circuits of negative length. Johnson [15.3] has shown that in the worst case this modified algorithm has a complexity of $O(n2^n)$ and not $O(n^3)$, as claimed in [15.4].

Sometimes we may be interested in getting the second- (and third-, etc.) shortest paths. This and related problems are discussed in Dreyfus [15.5], Hu [15.6], Frank and Frisch [15.7], Christofides [15.8], Lawler [15.9], and Spira and Pan [15.10]. For algorithms designed for sparse graphs see Johnson [15.11] and Wagner [15.12].

15.1.2 Shortest Paths between All Pairs of Vertices

Suppose that we are interested in finding the shortest paths between all the $n(n-1)$ ordered pairs of vertices in an n-vertex directed graph. A straightforward approach to get these paths would be to use Dijkstra's algorithm n times. However, there are algorithms which are computationally more efficient than this. These algorithms are applicable even when the lengths are negative, but there are no negative-length directed circuits. Now we discuss one of these algorithms. This algorithm, due to Floyd [15.13], is based on Warshall's procedure (Algorithm 14.1) for computing transitive closure.

Given an n-vertex directed graph G with lengths associated with its edges. Let the vertices of G be denoted as $1,2,\ldots,n$. Assume that there are no negative-length directed circuits in G. Let $W=[w_{ij}]$ be the $n\times n$ matrix of direct lengths in G, that is, w_{ij} is the length of the directed edge (i, j) in G. We set $w_{ij}=\infty$ if there is no edge (i, j) directed from i to j. We also set $w_{ii}=0$ for all i.

Starting with the matrix $W^{(0)}=W$, Floyd's algorithm constructs a sequence $W^{(1)},W^{(2)},\ldots,W^{(n)}$ of $n\times n$ matrices so that the entry $w_{ij}^{(n)}$ in $W^{(n)}$ would give the distance from i to j in G. The matrix $W^{(k)}=[w_{ij}^{(k)}]$ is constructed from the matrix $W^{(k-1)}=[w_{ij}^{(k-1)}]$ according to the following rule:

$$w_{ij}^{(k)}=\min\left\{w_{ij}^{(k-1)},\quad w_{ik}^{(k-1)}+w_{kj}^{(k-1)}\right\}. \tag{15.7}$$

Let $P_{ij}^{(k)}$ denote a path of minimum length among all the directed i–j paths which use as internal vertices only those from the set $\{1,2,\ldots,k\}$. The following theorem proves the correctness of Floyd's algorithm.

THEOREM 15.1. For $0\leqslant k\leqslant n$, $w_{ij}^{(k)}$ is equal to the length of $P_{ij}^{(k)}$.

Proof

Proof follows along the same lines as that for Theorem 14.1. ■

Usually, in addition to the shortest lengths, we are also interested in obtaining the paths which have these lengths. Recall that in Dijkstra's algorithm we use the PRED array to keep a record of the vertices which occur in the shortest paths. This is achieved in Floyd's algorithm as described below.

As we construct the sequence $W^{(0)}, W^{(1)}, \ldots, W^{(n)}$, we also construct another sequence $Z^{(0)}, Z^{(1)}, \ldots, Z^{(n)}$ of matrices such that the entry $z_{ij}^{(k)}$ of $Z^{(k)}$ gives the vertex which immediately follows vertex i in $P_{ij}^{(k)}$. Clearly, then

$$z_{ij}^{(0)} = \begin{cases} j, & \text{if } w_{ij} \neq \infty. \\ 0, & \text{if } w_{ij} = \infty. \end{cases} \tag{15.8}$$

Given $Z^{(k-1)} = [z_{ij}^{(k-1)}]$, $Z^{(k)} = [z_{ij}^{(k)}]$ is obtained according to the following rule: Let

$$M = \min\left\{ w_{ij}^{(k-1)}, w_{ik}^{(k-1)} + w_{kj}^{(k-1)} \right\}.$$

Then

$$z_{ij}^{(k)} = \begin{cases} z_{ij}^{(k-1)}, & \text{if } M = w_{ij}^{(k-1)}. \\ z_{ik}^{(k-1)}, & \text{if } M < w_{ij}^{(k-1)}. \end{cases} \tag{15.9}$$

This rule is similar to the one given in step S3 of Algorithm 15.1 for updating the PRED array. The justification for this rule is as follows. If

$$M = w_{ij}^{(k-1)},$$

then

$$\text{length of } P_{ij}^{(k)} = \text{length of } P_{ij}^{(k-1)}.$$

Therefore $z_{ij}^{(k)}$ is the same as $z_{ij}^{(k-1)}$.

On the other hand, if $M < w_{ij}^{(k-1)}$, then $P_{ij}^{(k)}$ is the concatenation of the paths $P_{ik}^{(k-1)}$ and $P_{kj}^{(k-1)}$ in that order. So

$$z_{ij}^{(k)} = z_{ik}^{(k-1)}.$$

It should be clear that the shortest i–j path is given by the sequence $i, i_1, i_2, \ldots, i_p, j$ of vertices, where

$$i_1 = z_{ij}^{(n)}, \qquad i_2 = z_{i_1 j}^{(n)}, \qquad i_3 = z_{i_2 j}^{(n)}, \ldots, \qquad \text{and } j = z_{i_p j}^{(n)}. \tag{15.10}$$

Note that in (15.7), $w_{ij}^{(k)} = w_{ij}^{(k-1)}$ if $w_{ik}^{(k-1)}$ or $w_{kj}^{(k-1)}$ is equal to ∞. This simple observation is made use of in the following description of Floyd's algorithm. We have also incorporated in this algorithm a test to detect the presence of a negative-length directed circuit.

Algorithm 15.2 Shortest Paths between All Pairs of Vertices (Floyd)

S1. $W=[w_{ij}]$ is the $n\times n$ matrix of direct lengths in the given directed graph G. Here $w_{ii}=0$ for all $i=1,2,\ldots,n$. $Z=[z_{ij}]$ is an $n\times n$ matrix in which

$$z_{ij}=\begin{cases} j, & \text{if } w_{ij}\neq\infty. \\ 0, & \text{if } w_{ij}=\infty. \end{cases}$$

S2. Set $k=0$.

S3. Set $k=k+1$. For all $i\neq k$ such that $w_{ik}\neq\infty$, and all $j\neq k$ such that $w_{kj}\neq\infty$, do the following:

1. Set $M=\min\{w_{ij}, w_{ik}+w_{kj}\}$.
2. If $M<w_{ij}$, then set $z_{ij}=z_{ik}$ and $w_{ij}=M$.

S4.

1. If any $w_{ii}<0$, then vertex i is in some negative-length directed circuit, and so HALT.
2. If every $w_{ii}\geqslant 0$ and $k=n$, then $[w_{ij}]$ gives the lengths of all the shortest paths, and z_{ij} gives the first vertex after vertex i in a shortest directed i–j path. HALT.
3. If all $w_{ii}\geqslant 0$ but $k<n$, then go to step S3. ■

Floyd's algorithm is the most efficient known algorithm for finding all the shortest paths. It is also valid for finding a shortest path in a network with negative lengths. For another efficient algorithm for finding all the shortest paths see Tabourier [15.14].

Yen [15.15] has given an algorithm for finding all the shortest paths. Subsequently, Williams and White [15.16] pointed out an error in this paper and presented a corrected version.

Pierce [15.17] gives an exhaustive bibliography on algorithms for the shortest path and related problems.

15.2 MINIMUM WEIGHTED PATH LENGTH TREES

An m-ary tree is a directed tree in which the out-degree of each vertex is at most m. Recall that in a directed tree the in-degree of the root is equal to zero, and for all the other vertices the in-degree is equal to 1. Usually an m-ary tree is drawn with the root at the top, all edges pointing downward, and all the vertices which are at the same distance from the root lined up on a horizontal line. A 3-ary tree is shown in Fig. 15.3.

Let T be an m-ary tree, and let $M=\{0,1,2,\ldots,m-1\}$ be an alphabet with m letters. Suppose that we assign to each edge in T a letter from the set M such that no two edges incident out of a vertex are assigned the same letter.

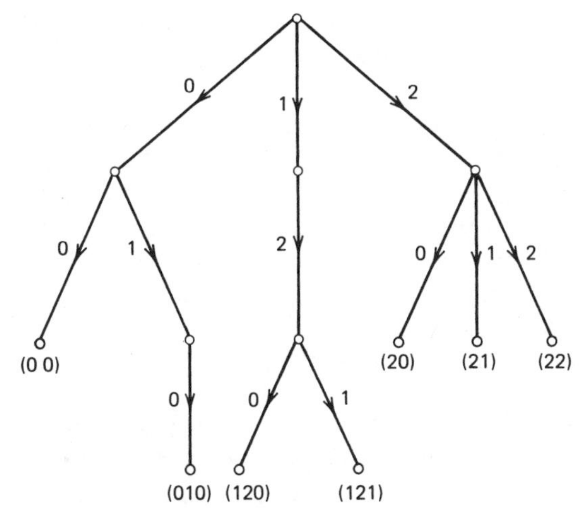

Figure 15.3. A 3-ary tree.

Then we can associate with each vertex of T a word which is obtained by concatenating the letters assigned to the edges which occur as we traverse from the root to the vertex. The words so associated with the leaves of T are called *code words*, and they are said to form a *prefix code*.

For example, in the case of the 3-ary tree shown in Fig. 15.3, the code words are 00, 010, 120, 121, 22, 20, and 21.

An interesting and useful property of the words in a prefix code is that no word is the beginning of another.

Suppose that we have encoded L messages $M_1, M_2, \ldots, M_L$ into the words of a prefix code. If we transmit a sequence of some of these encoded messages over a communication channel, then we would get, at the receiving end of the channel, a sequence of letters formed by concatenating the code words which correspond to the transmitted messages. To recover the messages from this sequence, we need to decompose the sequence into the words of the prefix code. The process of decomposing a sequence into code words is called *decoding*, and it can be easily done using the tree corresponding to the prefix code.

For example, consider the sequence 120202200 which is formed by the concatenation of some of the words of the prefix code corresponding to the tree shown in Fig. 15.3. To decode this sequence we scan from left to right the letters in the sequence. As we scan, we traverse the tree, starting from the root, along the edges which correspond to the scanned letters, until we reach a leaf. Now the code word corresponding to this leaf is the first word in the given sequence. Thus we obtain 120 as the first word in the sequence 120202200. We then repeat the decoding process with the remaining sequence 202200 and identify 20, 22, 00 as the second, third, and fourth code words in 120202200.

It is clear from the decoding process described above that the cost of decoding a code word is proportional to the number of letters in the word. If w_i is the frequency of occurrence of the message M_i, then the expected decode cost is related to $\Sigma_{i=1}^{L} w_i l_i$, where l_i is the length of the path from the root to the leaf corresponding to the message M_i. The sum $\Sigma_{i=1}^{L} w_i l_i$ is called the *weighted path length* of the tree. Thus the expected decode cost can be minimized by choosing the lengths of the code words in such a way that the resulting tree has minimum weighted path length. This leads us to the problem of constructing, for a given set of weights $w_1, w_2, \ldots, w_L$, an m-ary tree with minimum weighted path length.

Before we consider a solution to the above problem, we discuss another situation where this problem occurs.

Suppose that we have L lists $S_1, S_2, \ldots, S_L$. Assume that each list consists of integers arranged in nondecreasing order. Now we would like to merge these lists into a single list whose elements are also arranged in nondecreasing order. To do this we may merge first any two of these lists, say S_1 and S_2, and obtain a new list S'. We may then merge any two lists of the set $\{S', S_3, S_4, \ldots, S_L\}$ and continue merging until a single list is obtained. A binary tree can be used to describe such a merging policy.

For example, consider the tree shown in Fig. 15.4 which describes a way of merging five lists S_1, S_2, S_3, S_4, and S_5. Here, first we merge S_1 and S_2 to yield S_1'. Then S_1' and S_3 are merged to yield S_2'. In the next step S_4 and S_5 are merged and we get S_3'. Finally S_2' and S_3' are merged to yield the required single list.

One way of merging any two lists is as follows. Pick the first element, that is, the smallest element from each of the two given lists. Put the smaller of these two elements as the first element in the required merged list. Remove the element so selected from its list. Repeat this operation on the same two

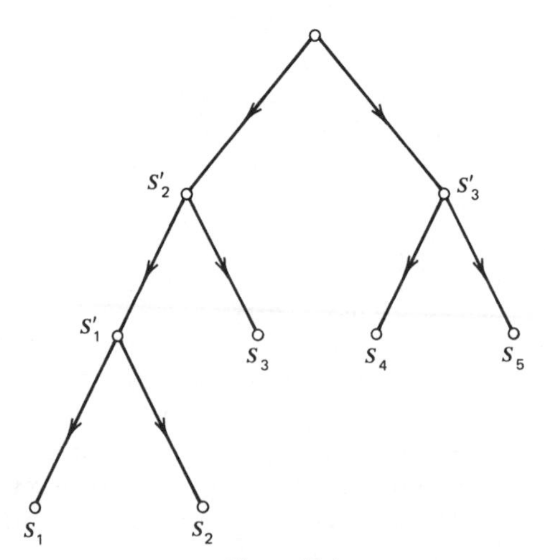

Figure 15.4.

lists, one of which is now shorter than before, until they merge into a single list.

Clearly, the cost of merging any two lists is proportional to the total number of elements in the lists. Therefore, the cost of merging the lists $S_1, S_2, \ldots, S_L$ is equal to $\Sigma_{i=1}^{L}|S_i|l_i$, where l_i is the number of merging processes in which the elements of list S_i are involved.

For example, the cost of merging $S_1, S_2, \ldots, S_5$ as shown in Fig. 15.4 is equal to

$$3|S_1|+3|S_2|+2|S_3|+2|S_4|+2|S_5|.$$

It may be seen from Fig. 15.4 that l_i in this case is, in fact, equal to the length of the path from the root to the leaf corresponding to S_i. Thus the problem of minimizing the cost of merging the lists $S_1, S_2, \ldots, S_L$ is equivalent to the problem of constructing, for the weights $|S_1|$, $|S_2|, \ldots, |S_L|$, a binary tree with minimum weighted path length.

Huffman [15.18] has given an elegant solution to the above problem. The remainder of this section is devoted to the development of Huffman's algorithm. Consider an m-ary tree T with L leaves. Let $l_1, l_2, \ldots, l_L$ denote the lengths of the paths from the root to the leaves of T. The l_i's are called *path lengths* of the corresponding leaves, and the vector $\lambda=(l_1, l_2, \ldots, l_L)$ is called the *path length vector* of T. The *characteristic sum* $S(\lambda, m)$ of λ is defined as follows:

$$S(\lambda, m)=\sum_{i=1}^{L} m^{-l_i}. \tag{15.11}$$

First we establish a necessary and sufficient condition for a vector $\lambda=(l_1, l_2, \ldots, l_L)$ to be the path length vector of an m-ary tree. Of course, we assume that each l_i is a positive integer.

LEMMA 15.1. The characteristic sum of the path length vector $\lambda=(l_1, l_2, \ldots, l_L)$ of an m-ary tree is less than or equal to 1.

Proof

Proof is by induction on the maximum length in λ.

If the maximum length in λ is equal to 1, then all the l_i's will be equal to 1. Clearly, in such a case, the lemma follows immediately. Let the lemma be true for every path length vector in which the maximum length is less than k. Assume that the maximum length in λ is equal to k. Let, in the corresponding m-ary tree T, $v_1, v_2, \ldots, v_{d^+(r)}$ be the sons of the root r. Note that the out-degree $d^+(r)$ of r is less than or equal to m. Let T_i denote the subtree of T with v_i as its root. Clearly, the path lengths of T_i are shorter by 1 than their counterparts in T. Thus if S_i is the characteristic sum of the path length vector of T_i, then

$$S(\lambda, m)=\frac{1}{m}(S_1+S_2+\cdots+S_{d^+(r)}).$$

Since $d^+(r)\leqslant m$ and by the inductive hypothesis $S_i\leqslant 1$ for all i, it follows that $S(\lambda, m)\leqslant 1$. ■

LEMMA 15.2. The vector $\lambda=(l_1, l_2, \ldots, l_L)$, where each l_i is a positive integer, is the path length vector of an m-ary tree if it satisfies the characteristic sum condition

$$\sum_{i=1}^{L} m^{-l_i}\leqslant 1. \tag{15.12}$$

Proof

Arrange the lengths l_i so that $l_1\leqslant l_2\leqslant\cdots\leqslant l_L$. Let μ denote the number of lengths equal to l_L. Then

$$l_{L-\mu}<l_{L-\mu+1}=l_{L-\mu+2}=\cdots=l_L.$$

Now, multiplying (15.12) by m^{l_L}, we get

$$\sum_{i=1}^{L-\mu} m^{l_L-l_i}+\mu\leqslant m^{l_L}. \tag{15.13}$$

Let $\mu=\sigma m+\rho$, where σ and ρ are nonnegative and $0\leqslant\rho\leqslant m-1$, and define π as follows:

$$\pi=\begin{cases} 0, & \text{if } \rho=0. \\ m-\rho, & \text{if } \rho>0. \end{cases}$$

Then we can add π to the left-hand side of (15.13) without violating the inequality. Thus

$$\sum_{i=1}^{L-\mu} m^{l_L-l_i}+\mu+\pi\leqslant m^{l_L}. \tag{15.14}$$

Note that $\mu+\pi=\tau m$, where

$$\tau=\begin{cases} \sigma, & \text{if } \rho=0. \\ \sigma+1, & \text{if } \rho>0. \end{cases} \tag{15.15}$$

Now, dividing (15.14) by m^{l_L}, we get

$$\sum_{i=1}^{L-\mu} m^{-l_i}+\tau m^{-(l_L-1)}\leqslant 1. \tag{15.16}$$

The quantity on the left-hand side of the above inequality is the characteristic sum of a vector of $L-\mu+\tau$ lengths in which the maximum length is shorter by 1 than the maximum length in the given vector λ.

Now we prove the lemma by induction on the size of the maximum length. If $l_L=1$, so that $l_i=1$ for all i, then (15.12) requires that $L\leqslant m$. Clearly the lemma is true in this case because the vector λ is the path length vector of an m-ary tree consisting of L edges with all of them incident on the root.

Assume next that the lemma is true for all vectors in which the maximum length is less than k.

Let the maximum length in λ be equal to k. As we have just described above, we can construct a vector λ' of $L-\mu+\tau$ lengths, with its maximum length equal to $k-1$. In λ' there are at least τ lengths equal to $k-1$, and the remaining $L-\mu$ lengths of λ' are the same as the first $L-\mu$ lengths of λ. Further note that $\tau=\sigma$ or $\sigma+1$.

Since λ' satisfies the characteristic sum condition (15.16), it follows from the inductive hypothesis that there exists an m-ary tree T' with λ' as its path length vector. Consider now any τ leaves of T' with path lengths equal to $k-1$. If we attach m edges to each one of any σ of these vertices and ρ edges to the remaining vertex, if any, then we get a new m-ary tree T which has $\mu=\sigma m+\rho$ leaves with path lengths equal to k. The remaining path lengths of T, are the same as the first $L-\mu$ path lengths of λ. Thus for T, λ is the path length vector. ∎

A path length vector $\lambda=(l_1,l_2,\ldots,l_L)$ is said to be *optimal* for the weight vector $W=(w_1,w_2,\ldots,w_L)$ if it minimizes the sum

$$\sum_{i=1}^{L} w_i l_i.$$

If λ is optimal for W, then $l_i\geqslant l_j$ implies that $w_i\leqslant w_j$. So we may assume that $l_1\leqslant l_2\leqslant\cdots\leqslant l_L$ and $w_1\geqslant w_2\geqslant\cdots\geqslant w_L$. The case $m=1$ being trivial, we assume in the following discussion that $m\geqslant 2$. The next result is the basis of Huffman's algorithm for constructing an optimal path length vector for a given weight vector.

THEOREM 15.2. Let $\lambda=(l_1,l_2,\ldots,l_L)$ be an optimal path length vector for the weight vector $W=(w_1,w_2,\ldots,w_L)$. Assume that $l_1\leqslant l_2\leqslant\cdots\leqslant l_L$ and $w_1\geqslant w_2\geqslant\cdots\geqslant w_L$. Then $l_{L-d+1}=l_{L-d+2}=\cdots=l_L$, where $d=2$ if $m=2$, and in case $m>2$, d is given by

$$d=\begin{cases} m, & \text{if } L\equiv 1(\text{mod}\, m-1).\\ m-1, & \text{if } L\equiv 0(\text{mod}\, m-1).\\ \rho, & \text{if } L\equiv\rho(\text{mod}\, m-1) \text{ and } 2\leqslant\rho\leqslant m-2.\end{cases} \tag{15.17}$$

Proof

Essentially the theorem states that in an optimal path length vector at least d lengths are equal to the maximum length. It is a simple exercise to prove the theorem for the case $m=2$. So we consider only the case $m>2$.

It follows from the characteristic sum condition that

$$m^{l_L}-\sum_{i=1}^{L} m^{l_L-l_i}\geqslant 0. \tag{15.18}$$

We now show that

$$m-2\geqslant m^{l_L}-\sum_{i=1}^{L} m^{l_L-l_i}. \tag{15.19}$$

If (15.19) is not true, then

$$m^{l_L}-\sum_{i=1}^{L} m^{l_L-l_i}\geqslant m-1.$$

Thus

$$\sum_{i=1}^{L-1} m^{l_L-l_i}+m\leqslant m^{l_L}.$$

Dividing this by m^{l_L} we get

$$\sum_{i=1}^{L-1} m^{-l_i}+m^{-(l_L-1)}\leqslant 1.$$

But the above contradicts that λ is an optimal vector. Therefore (15.19) is true.

Since $m\equiv 1(\operatorname{mod} m-1)$, it follows that $m^k\equiv 1(\operatorname{mod} m-1)$ for every nonnegative integer k. Therefore

$$\begin{aligned} m^{l_L}-\sum_{i=1}^{L} m^{l_L-l_i} &\equiv 1-L(\operatorname{mod} m-1)\\ &\equiv 1-\rho(\operatorname{mod} m-1), \end{aligned} \tag{15.20}$$

where $L=\rho(\operatorname{mod} m-1)$. Now

$$1-\rho(\operatorname{mod} m-1)=\begin{cases} 0, & \text{if } \rho=1.\\ 1, & \text{if } \rho=0.\\ m-\rho, & \text{if } 2\leqslant\rho\leqslant m-2. \end{cases} \tag{15.21}$$

By (15.18) and (15.19) $m^{l_L}-\sum_{i=1}^{L} m^{l_L-l_i}$ is nonnegative and less than $m-1$.

Thus we get from (15.20) and (15.21)

$$m^{l_L}-\sum_{i=1}^{L} m^{l_L-l_i}=\begin{cases}0, & \text{if } \rho=1.\\ 1, & \text{if } \rho=0.\\ m-\rho, & \text{if } 2\leqslant\rho\leqslant m-2.\end{cases} \tag{15.22}$$

Assume now that j is the last index such that $l_j<l_L$. Thus

$$l_1\leqslant l_2\leqslant\cdots\leqslant l_j<l_{j+1}=l_{j+2}=\cdots=l_L.$$

We now use (15.22) to prove that $L-j\geqslant d$. This would then establish the theorem. We can rewrite (15.22) as

$$m^{l_L}-\sum_{i=1}^{j} m^{l_L-l_i}=\begin{cases}L-j, & \text{if } \rho=1.\\ L-j+1, & \text{if } \rho=0.\\ L-j+m-\rho, & \text{if } 2\leqslant\rho\leqslant m-2.\end{cases}$$

Since the left-hand side of the above equation is divisible by m, we get the following:

If $\rho=1$, then $L-j=km$ for some positive integer k.
If $\rho=0$, then $L-j+1=km$ for some positive integer k.
If $2\leqslant\rho\leqslant m-2$, then $L-j+m-\rho=km$ for some positive integer k.

Thus by (15.17),

if $\rho=1$,	then $L-j\geqslant m=d$.
if $\rho=0$,	then $L-j\geqslant m-1=d$.
if $2\leqslant\rho\leqslant m-2$,	then $L-j\geqslant\rho=d$. ∎

Now we describe Huffman's algorithm for constructing an optimal path length vector for a given weight vector W_0. We denote the number of weights in a weight vector W_i as L_i. If $L_0\leqslant m$, then the corresponding optimal path length vector will consist of all 1's. So we assume that $L_0>m$.

Starting with W_0, Huffman's algorithm constructs a sequence $W_1, W_2,\ldots, W_s$ of weight vectors such that $L_0>L_1>\cdots>L_s=m$. For each L_i we define d_i as in (15.17), that is,

$$d_i=2, \quad \text{if } m=2.$$

Otherwise,

$$d_i=\begin{cases}m, & \text{if } L_i\equiv 1(\operatorname{mod} m-1).\\ m-1, & \text{if } L_i\equiv 0(\operatorname{mod} m-1).\\ \rho, & \text{if } L_i\equiv\rho(\operatorname{mod} m-1) \text{ and } 2\leqslant\rho\leqslant m-2.\end{cases} \tag{15.23}$$

We use d_i in constructing W_{i+1} from W_i.

The algorithm then starts with

$$\lambda_s = \left(\overleftrightarrow{1,1,\ldots,1}^{m}\right)$$

and constructs a sequence of path length vectors $\lambda_s, \lambda_{s-1}, \ldots, \lambda_0$ such that λ_i is the optimal path length vector for W_i. We assume that the lengths in each λ_i are arranged in the same order as the weights in the corresponding W_i.

Algorithm 15.3 Optimal Path Length Vector (Huffman)

S1. W_0 is the given weight vector. Set $i=0$.

S2. Let $W_i=(w_1, w_2, \ldots, w_{L_i})$, with $w_1 \geqslant w_2 \geqslant \cdots \geqslant w_{L_i}$. Construct W_{i+1} as follows:

1. Compute d_i as in (15.23).
2. Compute $P_i = w_{L_i-d_i+1} + w_{L_i-d_i+2} + \cdots + w_{L_i}$, that is, P_i is the sum of the last d_i weights in W_i.
3. W_{i+1} consists of the weights $w_1, w_2, \ldots, w_{L_i-d_i}$, and P_i arranged in nonincreasing order.

S3. Set $i=i+1$. If $L_i=m$, go to step S4. Otherwise to go step S2.

S4. Set

$$\lambda_i = \left(\overleftrightarrow{1,1,\ \ldots,\ 1}^{m}\right).$$

(Note i is now equal to s.) Set $i=i-1$.

S5. Construct λ_i from λ_{i+1} as follows:

1. Each one of the first $L_i - d_i$ weights in W_i is assigned a length which is equal to its length in W_{i+1}.
2. Each one of the last d_i weights in W_i is assigned a length which is one more than the length of P_i in W_{i+1}.

S6. If $i=0$, then HALT; λ_0 is an optimal path length vector for W_0. Otherwise set $i=i-1$ and go to step S5. ■

Clearly, λ_s is a path length vector. By (15.23) $d_i \leqslant m$ for all i. The construction in the algorithm ensures that λ_i, for all i, is a path length vector.

We have illustrated in Fig. 15.5 Huffman's algorithm with $W_0 = (9,7,6,5,5,3,3,2,1)$ and $m=3$. The trees T_3, T_2, T_1, T_0 corresponding to the path length vectors λ_3, λ_2, λ_1, and λ_0 are shown in Fig. 15.6. (Note that T_i is constructed from T_{i+1} by attaching d_i edges to the leaf in T_{i+1} representing the weight P_i.)

In the following we prove the correctness of Huffman's algorithm.

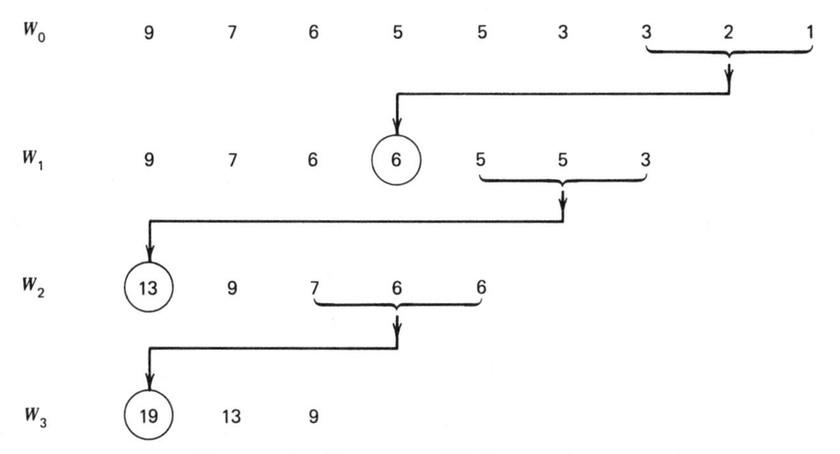

Figure 15.5. Illustration of Huffman's algorithm.

THEOREM 15.3. Let $W_0, W_1, \ldots, W_s$ and $\lambda_0, \lambda_1, \ldots, \lambda_s$ be the weight vectors and the path length vectors as constructed in Huffman's algorithm. For every $0 \leqslant i \leqslant s$, λ_i is optimal for W_i.

Proof

Clearly the path length vector

$$\lambda_s = (\overbrace{1, 1, \ldots, 1}^{m})$$

is optimal for W_s. We now show that if λ_{i+1} is optimal for W_{i+1}, then λ_i is optimal for W_i.

Assume the contrary, namely, λ_{i+1} is optimal for W_{i+1}, but λ_i is not optimal for W_i. Then let λ_i^* be an optimal path length vector for W_i. Let

$$W_i = (p_1, p_2, \cdots, p_{L_i}) \text{ with } p_1 \geqslant p_2 \geqslant \cdots \geqslant p_{L_i},$$

$$\lambda_i = (l_1, l_2, \cdots, l_{L_i}),$$

$$W_{i+1} = (q_1, q_2, \cdots, q_{L_{i+1}}) \text{ with } q_1 \geqslant q_2 \geqslant \cdots \geqslant q_{L_{i+1}},$$

$$\lambda_{i+1} = (t_1, t_2, \cdots, t_{L_{i+1}}),$$

$$\lambda_i^* = (l_1^*, l_2^*, \cdots, l_{L_i}^*).$$

Since λ_i is not optimal for W_i, we have

$$\sum_{k=1}^{L_i} p_k l_k > \sum_{k=1}^{L_i} p_k l_k^*. \tag{15.24}$$

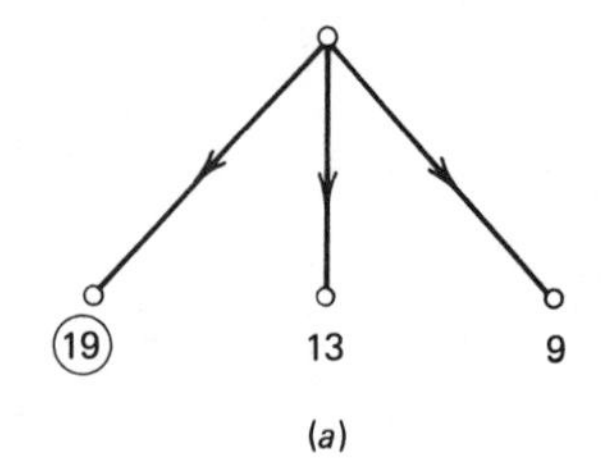

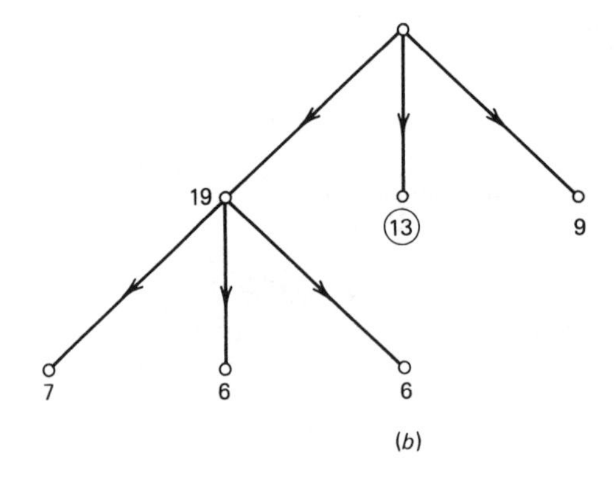

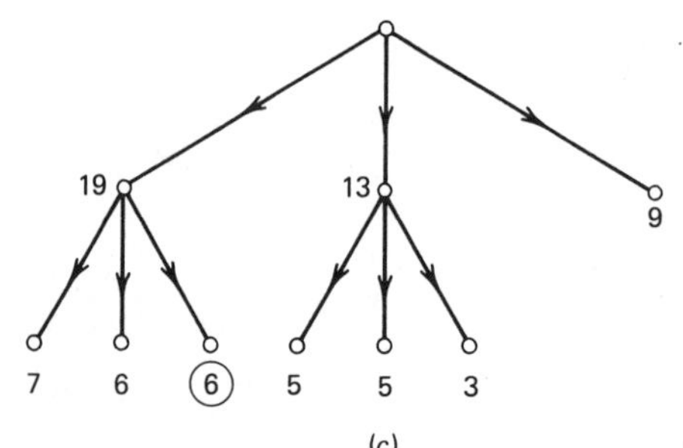

Figure 15.6. (a) Tree T_3. (b) Tree T_2. (c) Tree T_1. (d) Tree T_0.

It follows from step S5 in Huffman's construction that, for some $1 \leqslant u \leqslant L_{i+1}$,

$$q_u = p_{L_i - d_i + 1} + p_{L_i - d_i + 2} + \cdots + p_{L_i}.$$

(Note that q_u is the same as P_i, and the remaining q_i's are $p_1, p_2, \ldots, p_{L_i - d_i}$.) Also $t_u = l_{L_i} - 1$. Thus

$$\sum_{k=1}^{L_{i+1}} q_k t_k = \sum_{k=1}^{L_i} (p_k l_k) - q_u. \tag{15.25}$$

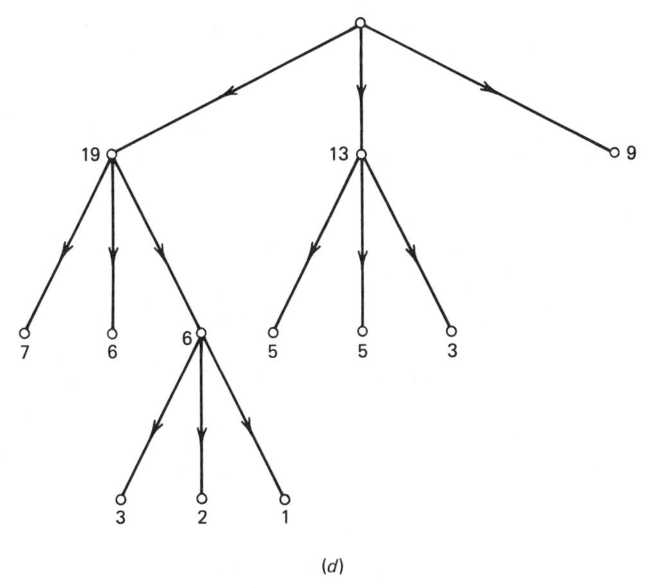

(d)

Figure 15.6. (*continued*)

Since λ_i^* is an optimal vector, we have, by Theorem 15.2, that

$$l^*_{L_i-d_i+1}=l^*_{L_i-d_i+2}=\cdots=l^*_{L_i}.$$

So if we use the same type of transformation as the one that relates λ_i to λ_{i+1}, then we can get from λ_i^* a path length vector $\lambda_{i+1}^*=(t_1^*, t_2^*,\ldots, t^*_{L_{i+1}})$ which satisfies

$$\sum_{k=1}^{L_{i+1}} q_k t_k^* = \sum_{k=1}^{L_i} (p_k l_k^*) - q_u. \tag{15.26}$$

Now we can show, using (15.24), (15.25), and (15.26), that

$$\sum_{k=1}^{L_{i+1}} q_k t_k^* < \sum_{k=1}^{L_{i+1}} q_k t_k,$$

implying that λ_{i+1} is not an optimal vector, a contradiction. ■

15.3 OPTIMUM BINARY SEARCH TREES

An *ordered tree* is a directed tree with an order defined for the sons of every vertex of the tree. Usually we draw an ordered tree with the root at the top and the sons of each vertex arranged in their assigned order from left to right.

In the case of an ordered binary tree, the subtree whose root is the left son of a vertex v is called the *left subtree* of v. Similarly the *right subtree* of v is defined.

Let $A=\{a_1, a_2,\dots, a_n\}$ be a set with its elements ordered as $a_1<a_2<\cdots<a_n$. A *binary search tree* for the set A is an ordered binary tree in which each vertex v is labeled by an element $l(v)$ of A such that:

1. For each vertex u in the left subtree of v, $l(u)<l(v)$.
2. For each vertex u in the right subtree of v, $l(u)>l(v)$.
3. For each element x of A there is exactly one vertex v such that $l(v)=x$.

Suppose A is a subset of a universal set S. Let then $B=\{b_0, b_1,\dots, b_n\}$ be a set such that:

1. b_i, $1\leq i\leq n-1$, represents the set of all elements $x\in S-A$ with $a_i<x<a_{i+1}$.
2. b_0 represents the set of all elements $x\in S-A$ with $x<a_1$.
3. b_n represents the set of all elements $x\in S-A$ with $x>a_n$.

An *extended binary search tree* for A is a binary search tree for A with $n+1$ leaves representing the elements of B. Note that in an extended binary search tree the out-degree of every internal vertex is equal to 2.

For example, the trees shown in Fig. 15.7 are two different extended binary search trees over the set $A=\{a_1, a_2, a_3, a_4\}$. Note that in an extended binary search tree leaves appear from left to right in the order $b_0, b_1,\dots, b_n$.

In the following we refer to an extended binary search tree for a set A simply as a *binary search tree* for A.

Given a subset A of a universal set S and a binary search tree T for A. The set S may be the set of all names over the English alphabet, and the ordering may be lexicographic. Suppose we are interested in determining whether an element $x\in S$ belongs to A. Then a common approach is to compare x with the element which corresponds to the root of T. Four cases now arise and we may proceed as follows:

Case 1 There is no root (the binary search tree T is empty). Hence x is not in A, and the search terminates unsuccessfully.

Case 2 x matches the element at the root. Hence the search terminates successfully.

Case 3 x is less than the element at the root. So the search continues down the left subtree of the root.

Case 4 x is greater than the element at the root. So the search continues down the right subtree of the root.

Clearly, successful searches terminate at the internal vertices and unsuccessful searches terminate at the leaves of T.

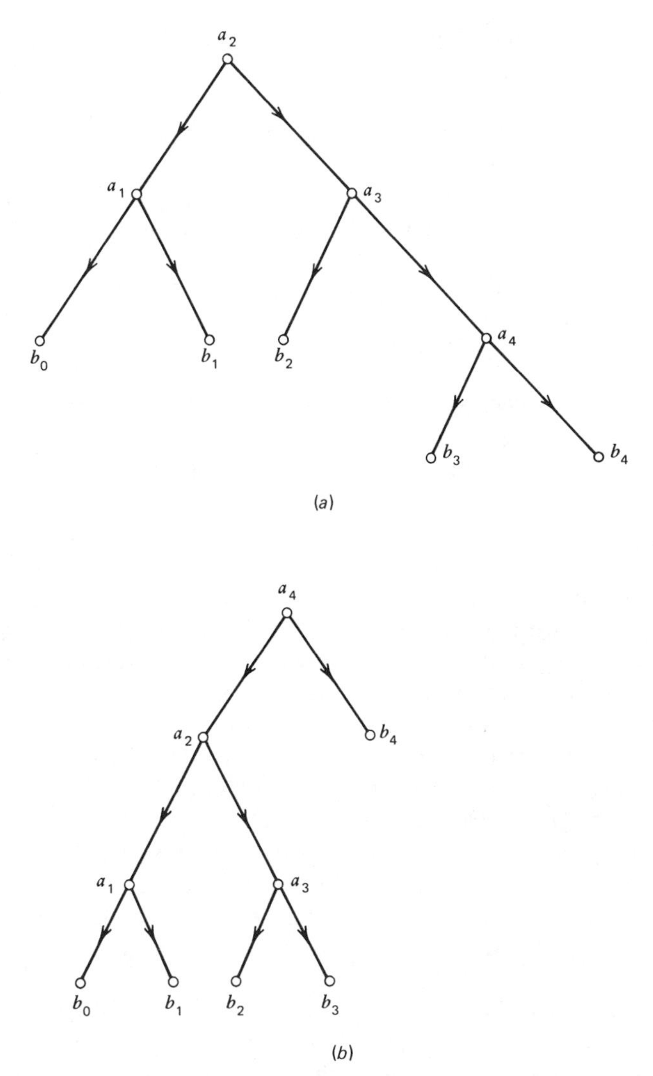

Figure 15.7. Examples of binary search trees.

Suppose we define the *depth of a vertex* v as the length of the path from the root to v. Then we can see that the number of comparisons made before a search terminates successfully at an internal vertex v of T is equal to 1 more than the depth of v. On the other hand if the search terminates unsuccessfully at a leaf, then the number of comparisons made is equal to the depth of the leaf.

Let $p_1, p_2, \ldots, p_n$ denote the access frequencies of the elements $a_1, a_2, \ldots, a_n$, respectively, and let $q_0, q_1, \ldots, q_n$ denote the frequencies with which the searches terminate at the leaves of T representing $b_0, b_1, \ldots, b_n$, respectively. Then the average search time in T (over all searches) is proportional to the cost of the tree T defined as

$$\sum_{i=1}^{n} p_i(\text{depth of } a_i+1)+\sum_{i=0}^{n} q_i(\text{depth of } b_i).$$

For example, if $p_1=0.2, p_2=0.2, p_3=0.1, p_4=0.1, q_0=0.1, q_1=0.1, q_2=0.05, q_3=0.05, q_4=0.1$, then the cost of the tree shown in Fig. 15.7*a* is

$$\{p_1(2)+p_2(1)+p_3(2)+p_4(3)\}+\{q_0(2)+q_1(2)+q_2(2)+q_3(3)+q_4(3)\}=2.05,$$

and the cost of the tree in Fig. 15.7*b* is

$$\{p_1(3)+p_2(2)+p_3(3)+p_4(1)\}+\{q_0(3)+q_1(3)+q_2(3)+q_3(3)+q_4(1)\}=2.4.$$

The above example motivates the following problem. Given nonnegative weights p_i and q_i, construct a minimum cost binary search tree for the set $A=\{a_1, a_2, \ldots, a_n\}$ whose elements are ordered as $a_1<a_2<\cdots<a_n$. We shall often find it convenient to refer to such a tree as an *optimum binary search tree* for the weights $q_0, p_1, q_1, p_2, \ldots, q_{n-1}, p_n, q_n$.

The algorithm we now discuss for constructing an optimum binary search tree is based on the following useful property of such trees.

Let T be an optimum binary search tree for the weights $q_0, p_1, q_1, p_2, \ldots, p_n, q_n$. If a_k is the element at the root of T, then the left subtree of the root of T is an optimum binary search tree for the weights $q_0, p_1, q_1, p_2, \ldots, p_{k-1}, q_{k-1}$, and the right subtree of the root of T is an optimum binary search tree for the weights $q_k, p_{k+1}, \ldots, p_n, q_n$.

For $0\leqslant i<j\leqslant n$, let T_{ij} denote an optimum binary search tree for the set $\{q_i, p_{i+1}, q_{i+1}, \ldots, p_j, q_j\}$. Let r_{ij} denote the root of T_{ij}, and let c_{ij} denote the cost of T_{ij}. The weight w_{ij} of T_{ij} is defined as equal to $q_i+(p_{i+1}+q_{i+1})+\cdots+(p_j+q_j)$. T_{ii} will be a tree consisting of the leaf b_i. So $c_{ii}=0$ and $w_{ii}=q_i$.

Suppose a_k is the root of T_{ij}, $i<j$. Then, as we have just observed, the left subtree of a_k is $T_{i,k-1}$, which is an optimum binary search tree for the set $\{q_i, p_{i+1}, \ldots, p_{k-1}, q_{k-1}\}$, and the right subtree of a_k is T_{kj}, which is an optimum binary search tree for the set $\{q_k, p_{k+1}, \ldots, p_j, q_j\}$ (Fig. 15.8). Since the depth of a vertex in T_{ij} is 1 more than its depth in $T_{i,k-1}$ or T_{kj}, it follows that

$$\begin{aligned} c_{ij} &= c_{i,k-1}+w_{i,k-1}+c_{kj}+w_{kj}+p_k \\ &= w_{ij}+c_{i,k-1}+c_{kj}. \end{aligned} \tag{15.27}$$

It should be now clear that we can find the root r_{ij} of T_{ij} by computing the value of k, $i<k\leqslant j$, which minimizes the sum in (15.27). This forms the basis of Algorithm 15.4, which computes r_{ij} and c_{ij} in the order of increasing values of $j-i$. This algorithm is from Gilbert and Moore [15.19].

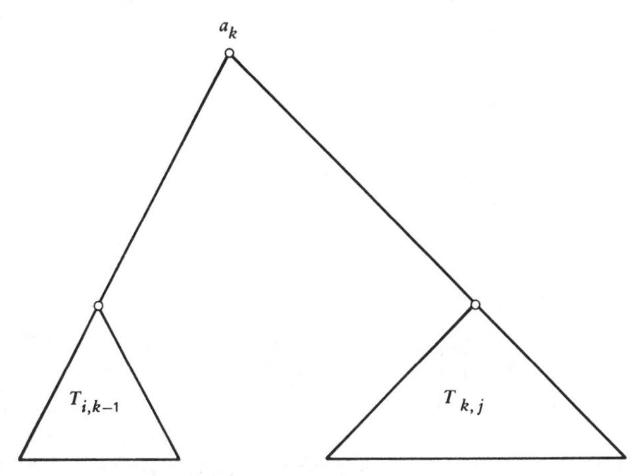

Figure 15.8. Subtree T_{ij} with root a_k.

Algorithm 15.4 Optimum Binary Search Tree (Gilbert and Moore)

S1. Given nonnegative weights $p_1, p_2, \ldots, p_n$ and $q_0, q_1, \ldots, q_n$. For $i = 0, 1, 2, \ldots, n$, set

$$w_{ii} = q_i,$$

$$c_{ii} = 0,$$

$$r_{ii} = b_i.$$

S2. For $l = 1, 2, \ldots, n$ do step S3.

S3. For $i = 0, 1, 2, \ldots, n - l$ do step S4.

S4. Set

$$j = i + l,$$

$$w_{ij} = w_{i,j-1} + p_j + q_j.$$

Let m be a value of k, $i < k \leqslant j$, for which the sum $c_{i,k-1} + c_{kj}$ is minimum. Set

$$c_{ij} = w_{ij} + c_{i,m-1} + c_{mj},$$

$$r_{ij} = a_m.$$

S5. HALT. ■

After computing r_{ij}'s using the above algorithm, we can easily construct T_{0n}

using the following procedure:

1. r_{0n} is the root of T_{0n}. If $r_{0n}=a_k$, then a_k has $r_{0,k-1}$ as its left son and r_{kn} as its right son.
2. In general, consider an internal vertex a_m. If $a_m=r_{ij}$, then a_m has $r_{i,m-1}$ as its left son and r_{mj} as its right son.

We now illustrate Algorithm 15.4 and the above procedure to construct an optimum binary search tree.

Consider the four elements $a_1<a_2<a_3<a_4$ with $p_1=2$, $p_2=1$, $p_3=3$, and $p_4=1$ and $q_0=1$, $q_1=2$, $q_2=3$, $q_3=2$, and $q_4=1$. In Table 15.1 we show the values of w_{ij}, r_{ij}, and c_{ij} computed by Algorithm 15.4. In this table the entries in a row are computed from left to right. Further, computation of the entries in any row begins only after all the entries in the preceding row have been computed. In Fig. 15.9 we show the corresponding optimum binary search tree.

It is easy to show that Algorithm 15.4 has complexity $O(n^3)$ and the procedure to construct the tree from the table of r_{ij}'s is of complexity $O(n)$. Thus an optimum binary search tree can be constructed in $O(n^3)$ time.

Knuth [15.20] has shown that the root of T_{ij} lies between the roots of $r_{i,j-1}$ and $r_{i+1,j}$. In other words, $r_{i,j-1}\leqslant r_{i,j}\leqslant r_{i+1,j}$. So in step S4 we may restrict our search for m to the range between $r_{i,j-1}$ and $r_{i+1,j}$. With this modification the complexity of Algorithm 15.4 becomes $O(n^2)$. Several generalizations of Knuth's algorithm are discussed in Itai [15.21].

Table 15.1

$l=j-i$ ↓ \ $i\rightarrow$	0	1	2	3	4
0	$w_{00}=1$ $c_{00}=0$ $r_{00}=b_0$	$w_{11}=2$ $c_{11}=0$ $r_{11}=b_1$	$w_{22}=3$ $c_{22}=0$ $r_{22}=b_2$	$w_{33}=2$ $c_{33}=0$ $r_{33}=b_3$	$w_{44}=1$ $c_{44}=0$ $r_{44}=b_4$
1	$w_{01}=5$ $c_{01}=5$ $r_{01}=a_1$	$w_{12}=6$ $c_{12}=6$ $r_{12}=a_2$	$w_{23}=8$ $c_{23}=8$ $r_{23}=a_3$	$w_{34}=4$ $c_{34}=4$ $r_{34}=a_4$	
2	$w_{02}=9$ $c_{02}=14$ $r_{02}=a_2$	$w_{13}=11$ $c_{13}=17$ $r_{13}=a_3$	$w_{24}=10$ $c_{24}=14$ $r_{24}=a_3$		
3	$w_{03}=14$ $c_{03}=27$ $r_{03}=a_2$	$w_{14}=13$ $c_{14}=23$ $r_{14}=a_3$			
4	$w_{04}=16$ $c_{04}=34$ $r_{04}=a_3$				

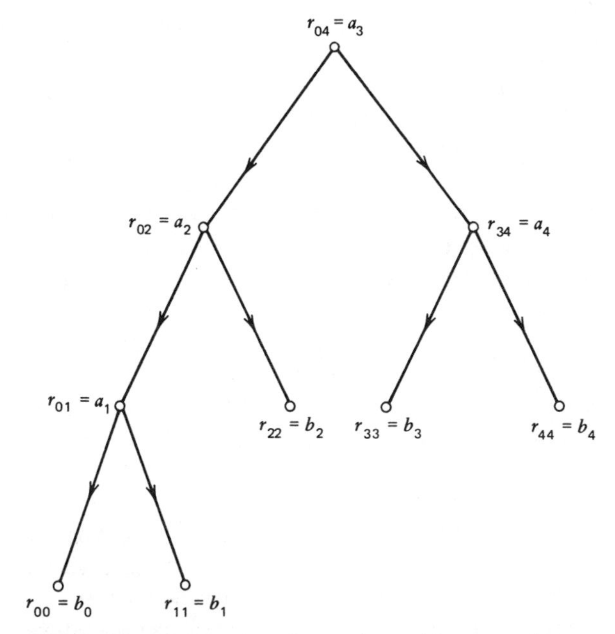

Figure 15.9. Construction of an optimum binary search tree.

Hu and Tucker [15.22] have given a $O(n^2)$ algorithm for constructing optimum binary search trees in the special case in which all p_i's are equal to zero. See also Hu [15.23]. Knuth [15.24] has shown that this algorithm can be implemented in $O(n \log n)$ time. For the same problem another $O(n \log n)$ algorithm closely related to the Hu-Tucker algorithm is given in [15.25]. This algorithm is easier to justify and is based on variational methods.

Extensive research has been carried out on search techniques and several questions concerning optimum search trees. A detailed discussion on these topics and several related references may be found in Knuth [15.24] and Reingold, Nievergelt, and Deo [15.26]. See also a recent paper by Miyakawa, Yuba, Sugito, and Hoshi [15.27].

15.4 MAXIMUM MATCHING IN A GRAPH

We consider in this section the problem of constructing a maximum matching in a graph. First we discuss a basic approach due to Edmonds [15.28] for constructing such a matching. We then describe an algorithm due to Gabow [15.29], which is essentially an efficient implementation of Edmonds' approach.

A more efficient algorithm for the case of bipartite graphs is discussed in Section 15.5. Applications relating to the optimal personnel assignment and timetable scheduling problems are considered in Section 15.6.

15.4.1 Edmonds' Approach

Edmonds's algorithm is based on Berge's theorem (Theorem 8.20) which states that a matching is maximum if and only if there is no augmenting path relative to the matching. So, given a graph and an initial matching M, we may proceed as follows to get a maximum matching.

Find an augmenting path P with respect to M. Get the matching $M \oplus P$ which has one more edge than M. With respect to this new matching, find an augmenting path and proceed as before. Repeat this until we get a matching with respect to which there is no augmenting path. Then by Berge's theorem such a matching is maximum.

Thus the problem essentially reduces to finding an augmenting path relative to a given matching in an efficient way. The most important idea in this context is that of a "blossom" introduced by Edmonds, and this is described below.

To find an augmenting path relative to a matching M, we have to start our search necessarily at an unsaturated vertex, say u. If there exists an augmenting path P from u to u' (note that u' is also an unsaturated vertex), then, in P, u' is adjacent to either u or a saturated vertex v. Such a vertex v will be at an even distance from u in the path P; that is, there exists an alternating path of even length from u to v. This implies that the search for an augmenting path should be done only at a selected group of vertices, namely, those to which there are alternating paths of even length from u.

For example, let $v_1, v_2, \ldots, v_r$ be the vertices adjacent to u (Fig. 15.10). If any one of them is unsaturated, then we have found an augmenting path. Otherwise let $u_1, u_2, \ldots, u_r$ be their respective mates in the matching M. At this stage the selected group consists of the vertices $u, u_1, u_2, \ldots, u_r$. We then pick a vertex, say u_1, from the selected group which is not yet examined. If u_1 has a neighbor which is unsaturated, then we have found an augmenting path.

Otherwise, suppose that u_1 is not adjacent to any vertex in the selected group. If $v_1', v_2', \ldots, v_s'$ are those vertices adjacent to u_1, such that $v_i' \neq v_j$ for all i and j, then their mates $u_1', u_2', \ldots, u_s'$ also join the selected group of vertices.

If we find, while searching a vertex in the selected group, that it is adjacent to some other vertex already in the selected group, then an odd circuit (that is, a circuit of odd length) is created. This circuit, which is a closed alternating path of odd length, is called a *blossom*. For example, in Fig. 15.11, the addition of the edge (u_9, u_7) creates a blossom $(u_2, v_6, u_6, v_9, u_9, u_7, v_7, u_2)$. Before the addition of this edge the selected group consisted of the vertices $u, u_1, u_2, \ldots, u_9$. But once the blossom is created, the vertices v_6, v_7, and v_9 also join the selected group, because we can now find alternating paths of even length from u to these vertices. For example, in Fig. 15.11, $(u, v_2, u_2, v_7, u_7, u_9, v_9)$ is an alternating path of even length to v_9. So once a blossom is created, we find that all the vertices of that blossom join the selected group.

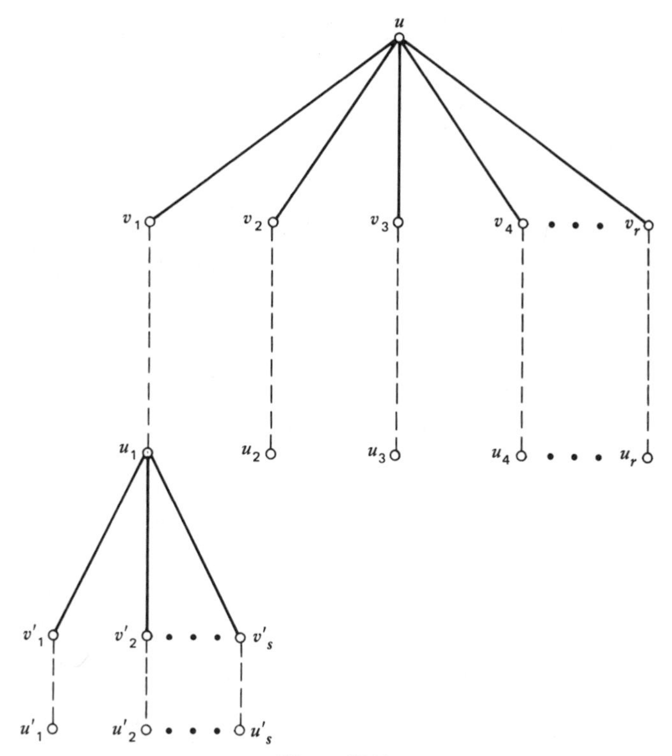

Figure 15.10.

In Edmonds' algorithm once a blossom is created, all the vertices in the blossom are replaced by a single vertex; the new vertex is called a *pseudo-vertex*. All the vertices that were adjacent to one or more vertices of the blossom will now be adjacent to this new vertex. Thus we get a reduced graph. We will continue the search for an augmenting path in this reduced graph. This procedure is repeated whenever we find a blossom.

In this method if we finally end up with some reduced graph without finding an augmenting path, then it implies that there exists no augmenting path relative to the current matching in the original graph as well. Hence the current matching is maximum.

On the other hand, if we find an augmenting path in some reduced graph, then it clearly implies that there exists an augmenting path P in the original graph also. To find the path P, we will have to trace back carefully by expanding the blossoms obtained previously.

The shrinking and expansion of blossoms required in Edmonds' algorithm might lead to a complexity of $O(n^4)$ for the algorithm, where n is the number of vertices in the graph.

Gabow avoids the shrinking and expansion operations by recording the pertinent structure of blossoms using an efficient labeling technique and

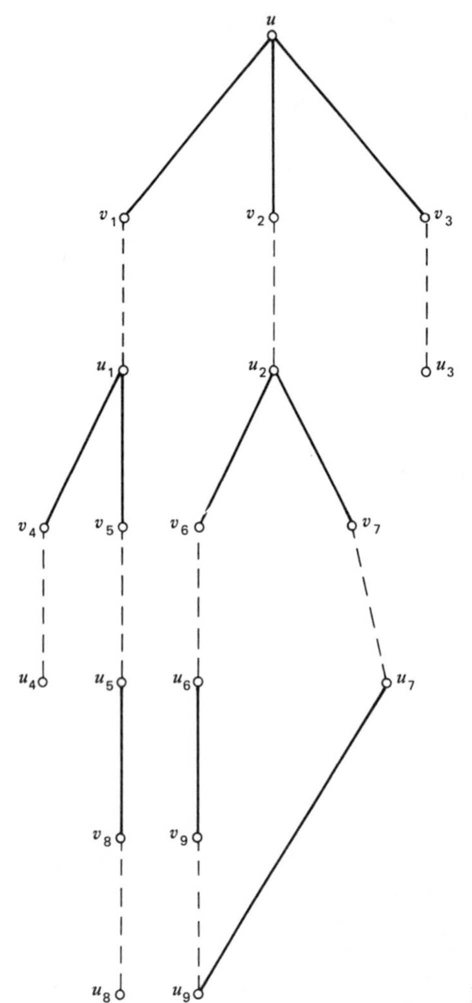

Figure 15.11. Blossom formation.

suitable arrays. This helps to achieve a complexity of $O(n^3)$. The labeling technique used by Gabow is similar to those in the matching algorithms of Balinski [15.30], Witzgall and Zahn [15.31], and Kameda and Munro [15.32].

15.4.2 Gabow's Method

First we discuss the basic strategy and define the different arrays used in Gabow's algorithm.

Let the given graph have n vertices and m edges. The algorithm begins by numbering the vertices and the edges of the graph. The vertices are numbered from 1 to n, and the edges are numbered as $n+2, n+4, \ldots, n+2m$. The number of an edge (x, y) is denoted by $N(x, y)$. A dummy vertex numbered 0 is also used.

END is an array which has entries numbered from $n+1$ to $n+2m$. For each edge, there are two consecutive entries in END containing the numbers of the end vertices of the edge. Thus if edge (v, w) has number k (where $k=n+2i$ for some $1 \leqslant i \leqslant m$), then END$(k-1)=v$ and END$(k)=w$. So given the number of an edge, its end vertices can easily be determined using this array.

Gabow's algorithm constructs a number of matchings starting with an initial matching which may be empty. It terminates with a maximum matching. A matching is stored in an array called MATE. This array has an entry for each vertex. Edge (v, w) is matched if MATE$(v)=w$ and MATE$(w)=v$.

A vertex v is called *outer* with respect to a fixed unsaturated vertex u if and only if there exists an alternating path of even length from u to v. It is clear that this path $P(v)$ when traced from v to u starts with a matched edge. Thus $P(v)=(v, v_1, \ldots, u)$, where (v, v_1) is a matched edge.

If an edge joining an outer vertex v to an unsaturated vertex $u' \neq u$ is scanned, then the algorithm finds the augmenting path as

$$(u')*P(v)=(u', v, v_1, \ldots, u),$$

where $*$ denotes concatenation. If no such edge is ever scanned, then the vertex u is not in an augmenting path.

LABEL is an array which has an entry for every vertex. The LABEL entry of an outer vertex v is used to find the alternating path $P(v)$.

The LABEL entry for an outer vertex is interpreted as a start label or a vertex label or an edge label.

Start Label The start vertex u has a start label. LABEL(u) is set to 0 in this case. Now the alternating path $P(u)=(u)$.

Vertex Label If LABEL$(v)=i$, where $1 \leqslant i \leqslant n$, then v is said to have a vertex label. In this case v is an outer vertex, and LABEL(v) is the number of another outer vertex. Path $P(v)$ is defined as $(v, \text{MATE}(v))*P(\text{LABEL}(v))$.

Edge Label If LABEL$(v)=n+2i$, $1 \leqslant i \leqslant m$, then v is said to have an edge label. Now v is an outer vertex, and LABEL(v) contains the number of an edge joining two outer vertices, say x and y. Thus LABEL$(v)=N(x, y)$. The edge label $N(x, y)$ of the vertex v indicates that there is an alternating path $P(v)$ of even length from v to the start vertex u, which passes through the edge (x, y). The path $P(v)$ can be defined in terms of paths $P(x)$ and $P(y)$. If v is in path $P(x)$, let $P(x, v)$ denote the portion of $P(x)$ from x to v along $P(x)$. Then $P(v)=\text{rev}\, P(x, v)*P(y)$, where the first term denotes the reverse of the path from x to v.

LABEL$(v)<0$ when v is a nonouter vertex. To start with, all the vertices are nonouter and we assign -1 as LABEL value to all the vertices.

The algorithm also uses an array called FIRST. If v is an outer vertex, FIRST(v) is the first nonouter vertex in $P(v)$. If the path $P(v)$ does not contain a nonouter vertex, then FIRST(v) is set to 0. FIRST(v)=0 if v is nonouter.

An array called OUTER is used to store the outer vertices encountered during the search for an augmenting path. The search graph is grown at the outer vertices in order of their appearance during the search. A breadth-first search is done at these outer vertices.

Gabow's algorithm (as presented below) consists of three procedures: PROC-EDMONDS, PROC-LABEL, and PROC-REMATCH.

PROC-EDMONDS is the main procedure. It starts a search for an augmenting path from each unsaturated vertex. It scans the edges of the graph, deciding to assign labels or to augment the matching.

When the presence of an augmenting path is detected (step E3 in Algorithm 15.5), PROC-REMATCH is invoked. This procedure computes a new matching which has one more edge than the current matching.

If a blossom is created (step E4) while scanning the edge (x, y), then PROC-LABEL is invoked. Now x and y are outer vertices. PROC-LABEL performs the following:

1. The value of a variable JOIN is set to the first nonouter vertex which is in both $P(x)$ and $P(y)$.
2. All nonouter vertices preceding JOIN in $P(x)$ or $P(y)$ now become outer vertices. They are assigned the edge label $N(x, y)$. This edge label indicates that to each one of these vertices there is an alternating path of even length from the start vertex which passes through the edge (x, y).
3. Now JOIN is the first nonouter vertex in $P(x)$ as well as in $P(y)$. So the entries of the FIRST array corresponding to all the vertices which precede JOIN in $P(x)$ or $P(y)$ are set to JOIN.

A description of Gabow's algorithm now follows. In each step appropriate comments and explanations are given in parentheses.

Algorithm 15.5 Maximum Matching (Gabow)

PROC-EDMONDS

E0. (Initialize.) G is the given graph. Number the vertices of G from 1 to n and the edges as $n+2, n+4, \ldots, n+2m$. Create a dummy vertex 0. For $0 \leqslant i \leqslant n$, set LABEL($i$)$=-1$, FIRST($i$)=0, and MATE($i$)=0. (To start with, all the vertices are nonouter and unsaturated.) Set $u=0$.

E1. (Find an unsaturated vertex.) Set $u=u+1$. If $u>n$, then HALT; now MATE contains a maximum matching. Otherwise if vertex u is saturated, repeat step E1. If u is unsaturated, add u to the OUTER array. Set LABEL(u)=0. (Assign a start label to u and begin a new search.)

E2. (Choose an edge.) Choose an edge (x, y) (where x is an outer vertex) which has not yet been examined at x. If no such edge exists, go to step E7.

(**Note** Edges (x, y) can be chosen in an arbitrary order. We adopt a breadth-first search: an outer vertex $x = x_1$ is chosen, and edges (x_1, y) are chosen in succeeding executions of step E2. When all such edges have been chosen, the vertex x_2 that was labeled immediately after x_1 is chosen, and the process is repeated for $x = x_2$. This breadth-first search requires maintaining a list of outer vertices $x_1, x_2, \ldots$. The OUTER array is used for this purpose.)

E3. (Presence of an augmenting path is detected.) If y is unmatched and $y \neq u$, carry out PROC-REMATCH and then go to step E7.

E4. (A blossom is created.) If y is outer, then carry out PROC-LABEL and then go to step E2.

E5. (Assign a vertex label.) Set v=MATE(y). If v is outer, go to step E6. If v is nonouter, set LABEL(v)=x, FIRST(v)=y, and add v to the OUTER array. (Now y is encountered for the first time in this search; its mate v is a new outer vertex. This fact is noted in the OUTER array.) Then go to step E6.

E6. (Get next edge.) Go to step E2. (A closed alternating path of even length is obtained; so edge (x, y) adds nothing.)

E7. (Stop the search.) Set LABEL(i)= -1, for $0 \leqslant i \leqslant n$. Then go to step E1. (All the vertices are made nonouter for the next search.)

PROC-LABEL

L0. (Initialize.) Set r=FIRST(x) and s=FIRST(y). If $r = s$, then go to step L6. (There is no nonouter vertex in the blossom.) Otherwise flag the vertices r and s. (Steps L1 and L2 find JOIN by advancing alternately along paths $P(x)$ and $P(y)$. Flags are assigned to non-outer vertices r in these paths. This is done by setting LABEL(r) to a negative edge number; that is, LABEL(r)= $-N(x, y)$. This way, each invocation of PROC-LABEL uses a distinct flag value.)

L1. (Switch paths.) If $s \neq 0$, interchange r and s. (r is a flagged nonouter vertex, alternately in $P(x)$ and $P(y)$.)

L2. (Get the nonouter vertex.) Set r=FIRST(LABEL(MATE(r))). (r is set to the next nonouter vertex in $P(x)$ or $P(y)$.) If r is not flagged, flag r and go to step L1. Otherwise set JOIN=r. (We have found the JOIN.) Go to step L3.

L3. (Label vertices in $P(x)$, $P(y)$; that is, all nonouter vertices between x and JOIN or y and JOIN will be assigned edge labels, namely,

$N(x, y)$.) Set v=FIRST(x) and do step L4; then set v=FIRST(y) and do step L4. Then go to step L5.

L4. (Label a nonouter vertex v.) If $v \neq$JOIN, set LABEL(v)=$N(x, y)$ and FIRST(v)=JOIN, and add v to the OUTER array. Then set v=FIRST(LABEL(MATE(v))). (Get the next nonouter vertex.) Repeat step L4. Otherwise (that is, v=JOIN and hence we have assigned edge labels to all the nonouter vertices in the concerned path) continue as specified in step L3 (that is, return to step L3).

L5. (Update FIRST.) For each outer vertex i, if FIRST(i) is outer, set FIRST(i)=JOIN (that is, JOIN is the new first nonouter vertex in $P(i)$).

L6. (Edge labeling is over.) End the procedure.

PROC-REMATCH

R0. (Obtain the augmenting path.) Compute $P(x)$ as described below:

1. If x has an edge label $N(v, w)$, then compute $P(v)$ and $P(w)$. If x lies in $P(v)$, then

$$P(x) = (\text{rev}\, P(v, x)) * P(w).$$

Otherwise,

$$P(x) = (\text{rev}\, P(w, x)) * P(v).$$

2. If x has a vertex label, then

$$P(x) = (x, \text{MATE}(x)) * P(\text{LABEL}(x)).$$

The augmenting path P_a is then given by

$$P_a = (y) * P(x).$$

R1. (Augment the current matching.) Obtain a new matching by removing from the current matching all the matched edges in P_a and adding to it all the unmatched edges in P_a. (That is, if M is the current matching, then $M \oplus P_a$ is the new matching.) Modify suitably the entries in the MATE array and end the procedure. ■

It should be pointed out that in the above algorithm, a search for an augmenting path is made from a vertex only once. Suppose that the search from an unsaturated vertex u terminates without finding an augmenting path. Let S_u denote this search. The *Hungarian subgraph H* for vertex u is the subgraph which consists of all the edges containing an outer vertex of S_u and all the vertices in these edges. Edmonds [15.28] has shown that we can ignore the Hungarian subgraph H in searches after S_u. This suggests that we can modify Algorithm 15.5 by changing step E2 as follows:

E2′. (Choose an edge.) Choose an edge.... If no such edge exists, go to step E1.

Step E2′ now causes step E7, which unlabels vertices to be skipped after S_u.

This modification speeds up the algorithm if the graph has no perfect matching. However, it does not change the worst-case complexity of $O(n^3)$.

For discussions regarding the complexity and the proof of the correctness of Algorithm 15.5, see [15.29].

Now we illustrate Gabow's algorithm.

For the graph in Fig. 15.12a, with the initial matching shown in dashed lines, Gabow's algorithm proceeds as follows.

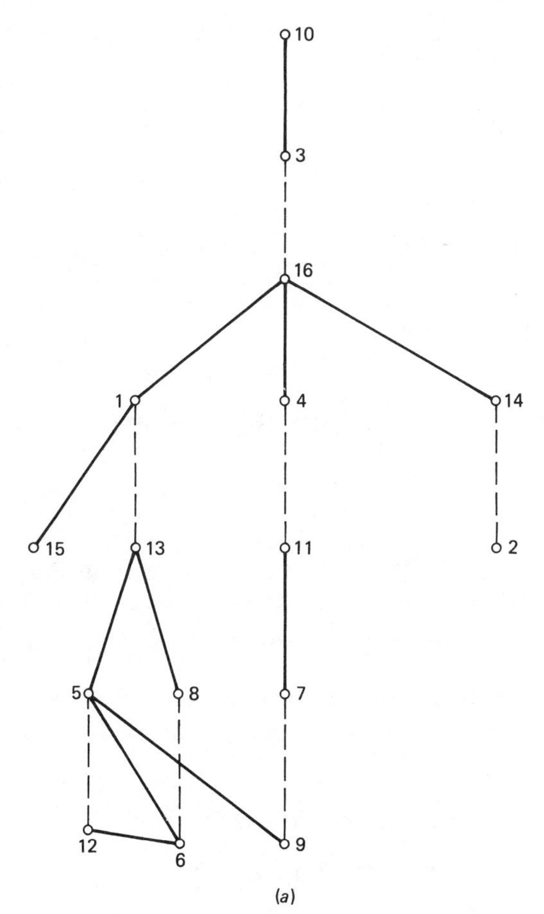

(a)

Figure 15.12.

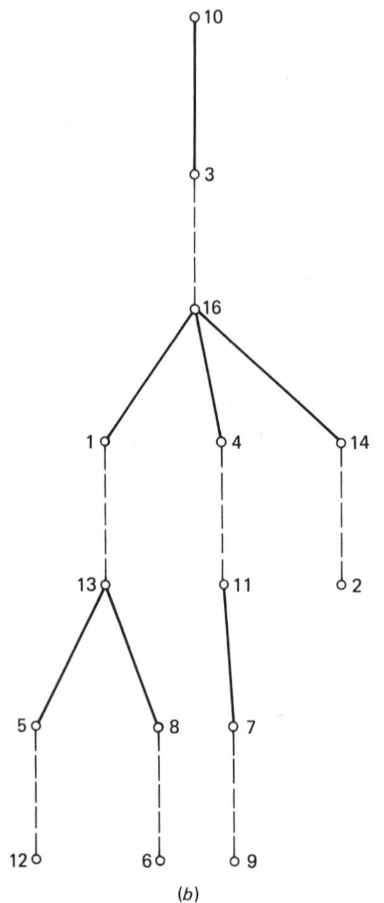

Figure 15.12. (*continued*)

The start vertex is 10. After completing the search at the outer vertices 10, 16, 13, 11, and 2, the search graph will be as shown in Fig. 15.12*b*. At this stage, the entries of the LABEL and FIRST arrays are shown in Table 15.2.

The OUTER array contains the vertices 10, 16, 13, 11, 2, 12, 6, and 9, in that order.

When we search at vertex 12, we examine the edge (12,6) and get the blossom (13,5,12,6,8,13). Now the LABEL and FIRST entries get changed as follows:

$$\text{LABEL}(5) = \text{LABEL}(8) = N(12,6);$$

$$\text{FIRST}(i) = 1,\ i = 5, 12, 6, 8.$$

Vertices 5 and 8 are now placed in the OUTER array.

Table 15.2

Vertex Number	LABEL	FIRST
1	−1	0
2	16	14
3	−1	0
4	−1	0
5	−1	0
6	13	8
7	−1	0
8	−1	0
9	11	7
10	0	0
11	16	4
12	13	5
13	16	1
14	−1	0
15	−1	0
16	10	3

Next, when we search at vertex 6, we examine the edge (6,5), and this creates another blossom (6,5,12,6). But all the vertices of this blossom are already outer, and so nothing gets changed.

Then we search at vertex 9. The edge (9,5) is examined and the blossom (9,7,11,4,16,1,13,8,6,12,5,9) is created. This again changes the LABEL and FIRST entries for some of the vertices. The resulting values are shown in Table 15.3.

Table 15.3

Vertex Number	LABEL	FIRST
1	$N(9,5)$	3
2	16	14
3	−1	0
4	$N(9,5)$	3
5	$N(12,6)$	3
6	13	3
7	$N(9,5)$	3
8	$N(12,6)$	3
9	11	3
10	0	0
11	16	3
12	13	3
13	16	3
14	−1	0
15	−1	0
16	10	3

Now the OUTER array contains the vertices 10, 16, 13, 11, 2, 12, 6, 9, 5, 8, 1, 7, and 4, in that order.

All the vertices in the OUTER array up to 9 have now been searched. Search continues with the remaining vertices. Searching at 5 and 8 does not add any new vertex to the OUTER array. While searching at vertex 1, we examine the edge (1,15) and find that 15 is unsaturated. So an augmenting path is noticed.

The augmenting path is $(15)*P(1)$.

We now use the procedure given in step R0 to compute $P(1)$. Vertex 1 has the edge label $N(9,5)$. So to compute $P(1)$, we need $P(9)$ and $P(5)$. Further, since vertex 5 has the edge label $N(12,6)$, we need $P(12)$ and $P(6)$ to compute $P(5)$.

Vertex 12 has a vertex label. So

$$\begin{aligned} P(12) &= (12, \text{MATE}(12))*P(\text{LABEL}(12)) \\ &= (12,5)*P(13) \\ &= (12,5)*(13,\text{MATE}(13))*P(\text{LABEL}(13)) \\ &= (12,5,13,1)*P(16) \\ &= (12,5,13,1)*(16,\text{MATE}(16))*P(\text{LABEL}(16)) \\ &= (12,5,13,1,16,3,10). \end{aligned}$$

Similarly

$$P(6) = (6,8,13,1,16,3,10)$$

and

$$P(9) = (9,7,11,4,16,3,10).$$

Since vertex 5 lies on $P(12)$,

$$\begin{aligned} P(5) &= (\text{rev}\, P(12,5))*P(6) \\ &= (5,12)*(6,8,13,1,16,3,10) \\ &= (5,12,6,8,13,1,16,3,10). \end{aligned}$$

Now we find that vertex 1 lies on $P(5)$. Therefore

$$\begin{aligned} P(1) &= (\text{rev}\, P(5,1))*P(9) \\ &= (1,13,8,6,12,5)*(9,7,11,4,16,3,10) \\ &= (1,13,8,6,12,5,9,7,11,4,16,3,10). \end{aligned}$$

Thus the augmenting path is

$$(15)*P(1) = (15,1,13,8,6,12,5,9,7,11,4,16,3,10).$$

After the augmentation, we get a new matching consisting of the edges (15,1), (13,8), (6,12), (5,9), (7,11), (4,16), (3,10), and (14,2). Since all the vertices of the graph are saturated in this matching, it is a maximum matching (in fact, a perfect matching).

Edmonds [15.33] and Gabow [15.34] discuss an efficient algorithm for the weighted matching problem.

Lawler [15.9] discusses, in considerable detail, the matching and related problems.

15.5 MAXIMUM MATCHING IN A BIPARTITE GRAPH

The problem of finding a maximum matching in a bipartite graph has a wide variety of applications. For example, it occurs as a subproblem in the solution of the Hitchcock transportation problem [15.1]. Further, scheduling a timetable in certain cases involves partitioning the edge set of a bipartite graph into disjoint matchings of the graph. Finding an element of the partition, in turn, requires finding a maximal matching in a bipartite graph. See Bondy and Murty [15.35].

In view of the variety of its applications, the computational complexity of this problem is of interest. Hopcroft and Karp [15.36] have shown how to construct a maximum matching in a bipartite graph in steps proportional to $n^{5/2}$, where n is the number of vertices in the graph. The philosophy of their approach is based on some interesting contributions (they have made) to the theory of matching. These contributions and their bipartite matching algorithm are discussed in this section.

15.5.1 Philosophy of Hopcroft and Karp's Approach

All maximum matching algorithms developed so far start with a matching (which may not be maximum) and obtain, if it exists, a matching of greater cardinality by locating an augmenting path. The choice of an augmenting path can be made in an arbitrary manner. The complexity of these algorithms is $O(n^3)$. Hopcroft and Karp have shown that if the augmentation is done along a shortest path, then a maximum matching can be obtained in $O(n^{5/2})$ phases, where each phase involves finding a maximal set of vertex-disjoint shortest augmenting paths relative to a matching. We now prove this result.

Let M be a matching. An augmenting path P is called a *shortest path relative to M* if P has the smallest length among all the augmenting paths relative to M.

LEMMA 15.3. Let M and N be two matchings in a graph G. If $|M|=s$ and $|N|=r$ with $r>s$, then $M \oplus N$ contains at least $r-s$ vertex-disjoint augmenting paths relative to M.

Proof

Consider the induced subgraph G' of G on the edge set $M \oplus N$. By Theorem 8.19 each (connected) component of G' is either:

1. A circuit of even length, with edges alternately in $M-N$ and $N-M$; or
2. A path whose edges are alternately in $M-N$ and $N-M$.

Let the components of G' be $C_1, C_2, \ldots, C_k$, where each C_i has vertex set V_i

and edge set E_i. Let

$$\delta(C_i)=|E_i \cap N|-|E_i \cap M|.$$

Then $\delta(C_i)$ is -1 or 0 or 1 for every i; and $\delta(C_i)=1$ if and only if C_i is an augmenting path relative to M.

Now

$$\sum_{i=1}^{k} \delta(C_i)=|N-M|-|M-N|=|N|-|M|=r-s.$$

Hence there are at least $r-s$ components of G', such that $\delta(C_i)=1$. These components are vertex-disjoint, and each is an augmenting path relative to M. ■

LEMMA 15.4. Let M be a matching. Let $|M|=r$ and suppose that the cardinality of a maximum matching is s. Then there exists an augmenting path relative to M of length at most

$$2\left\lfloor \frac{r}{s-r} \right\rfloor+1.$$

Proof

Let S be a maximum matching. Then by the previous lemma, $M \oplus S$ contains at least $s-r$ vertex-disjoint (and hence edge-disjoint) augmenting paths relative to M. Altogether, these paths contain at most r edges from M. So one of these paths will contain at most $\lfloor r/(s-r) \rfloor$ edges from M, and hence at most

$$2\left\lfloor \frac{r}{s-r} \right\rfloor+1$$

edges altogether. ■

LEMMA 15.5. Let M be a matching, P a shortest augmenting path relative to M, and P' an augmenting path relative to $M \oplus P$. Then $|P'| \geqslant |P|+|P \cap P'|$.

Proof

Let $N=M \oplus P \oplus P'$. Then N is a matching, and $|N|=|M|+2$. So $M \oplus N$ contains two vertex-disjoint augmenting paths P_1 and P_2 relative to M.

Since $M \oplus N=P \oplus P'$, $|P \oplus P'| \geqslant |P_1|+|P_2|$. But $|P_1| \geqslant |P|$ and $|P_2| \geqslant |P|$, because P is a shortest augmenting path. So $|P \oplus P'| \geqslant |P_1|+|P_2| \geqslant 2|P|$. Then from the identity $|P \oplus P'|=|P|+|P'|-|P \cap P'|$, we get $|P'| \geqslant |P|+|P \cap P'|$. ■

Suppose that we compute, starting with a matching $M_0=\varnothing$, a sequence of matchings $M_1, M_2,\ldots, M_i,\ldots$, where $M_{i+1}=M_i \oplus P_i$ and P_i is a shortest augmenting path relative to M_i. Then from Lemma 15.5, $|P_{i+1}| \geqslant |P_i| + |P_i \cap P_{i+1}|$. Hence we have the following.

LEMMA 15.6. $|P_i| \leqslant |P_{i+1}|$. ■

THEOREM 15.4. For all i and j such that $|P_i|=|P_j|$, P_i and P_j are vertex-disjoint.

Proof

Proof is by contradiction.

Assume that $|P_i|=|P_j|$, $i<j$, and P_i and P_j are not vertex-disjoint. Then there exist k and l such that $i \leqslant k < l \leqslant j$, P_k and P_l are not vertex-disjoint, and for each r, $k<r<l$, P_r is vertex-disjoint from P_k and P_l. Then P_l is an augmenting path relative to $M_k \oplus P_k$, so $|P_l| \geqslant |P_k| + |P_k \cap P_l|$. But $|P_l|=|P_k|$. So $|P_k \cap P_l|=0$. Thus P_k and P_l have no edges in common. But if P_k and P_l had a vertex v in common, then they would have in common that edge incident on v which is in $M_k \oplus P_k$. Hence P_k and P_l are vertex-disjoint, and a contradiction is obtained. ■

The main result of this section now follows.

THEOREM 15.5. Let s be the cardinality of a maximum matching. The number of distinct integers in the sequence $|P_0|, |P_1|,\ldots,|P_i|,\ldots$ is less than or equal to $2\lfloor \sqrt{s} \rfloor + 2$.

Proof

Let $r=\lfloor s-\sqrt{s} \rfloor$. Then $|M_r|=r$, and by Lemma 15.4,

$$|P_r| \leqslant 2\frac{\lfloor s-\sqrt{s} \rfloor}{\left(s-\lfloor s-\sqrt{s} \rfloor\right)} + 1 \leqslant 2\lfloor \sqrt{s} \rfloor + 1.$$

Thus for each $i<r$, $|P_i|$ is one of the $\lfloor \sqrt{s} \rfloor + 1$ positive odd integers less than or equal to $2\lfloor \sqrt{s} \rfloor + 1$. Also $|P_{r+1}|,\ldots,|P_s|$ contribute at most $s-r=\lceil \sqrt{s} \rceil$ distinct integers, and so the total number of distinct integers in the sequence $|P_0|, |P_1|,\ldots$ is less than or equal to $\lfloor \sqrt{s} \rfloor + 1 + \lceil \sqrt{s} \rceil \leqslant 2\lfloor \sqrt{s} \rfloor + 2$. ■

In view of Lemma 15.6 and Theorems 15.4 and 15.5, we may regard the computation of the sequence $M_0, M_1, M_2,\ldots$ as consisting of at most $2\lfloor \sqrt{s} \rfloor + 2$ phases, such that the augmenting paths found in each phase are vertex-disjoint and of the same length. Since all the augmenting paths in a phase are vertex-disjoint, they are also augmenting paths relative to the matching with which the phase is begun. This leads Hopcroft and Karp to suggest the following alternate way of describing the computation of a maximum matching.

Step 0. Start with a null matching M, that is, $M=\varnothing$.

Step 1. Let $l(M)$ be the length of a shortest augmenting path relative to M. Find a maximal set of paths $\{Q_1, Q_2, \ldots, Q_t\}$ with the following properties:

1. for each i, Q_i is an augmenting path relative to M, and $|Q_i| = l(M)$.
2. The Q_i are vertex-disjoint.

HALT if no such path exists.

Step 2. Set $M = M \oplus Q_1 \oplus Q_2 \oplus \cdots \oplus Q_t$; go to step 1.

It is clear from our previous discussion that steps 1 and 2 of the above computation will be executed at most $2\lfloor \sqrt{s} \rfloor + 2$ times, that is, $O(n^{1/2})$ times. Further, the complexity of the computation depends crucially on the complexity of implementing step 1. In a general graph, implementing this step is quite involved, since it requires generation of all the augmenting paths relative to a given matching and then selecting from them a maximal set of shortest paths which are vertex-disjoint.

However, in the special case of bipartite graphs, an $O(n^2)$ implementation of step 1 is possible so that the complexity of the computation for such special graphs is $O(n^{5/2})$. In the next subsection we discuss this result which is also due to Hopcroft and Karp.

15.5.2 Hopcroft and Karp's Approach for Maximum Bipartite Matching

Let G be a connected bipartite graph with bipartition (X, Y). We refer to the vertices in X as boys and those in Y as girls.

Hopcroft and Karp's implementation of the step 1 mentioned earlier is in two stages. If a matching M is not maximum, then in the first stage, a directed graph G^* is constructed such that the shortest augmenting paths relative to M are in one-to-one correspondence with the directed paths of G^* which begin at a free girl and end at a free boy. In the second stage a maximal set of such directed paths in G^* with the property that they are vertex-disjoint is constructed. These paths give the required shortest augmenting paths in G relative to M.

The graph G^* is constructed as follows.

First we assign directions to the edges of G in such a way that augmenting paths relative to M become directed paths. This is done by directing each unmatched edge so that it is directed from a girl to a boy, and each matched edge is directed from a boy to a girl. Let $\overline{G}$ denote the resulting directed graph.

Let L_0 be the set of free boys so that all the edges incident on L_0 will be unmatched, that is, directed from girls to boys.

Now consider the set L_1 of girls such that there exists a directed edge from each of these girls to at least one of the boys in L_0.

If the set L_1 contains at least one free girl, then we have found an augmenting path showing that the length of a shortest augmenting path relative to M is 1. Let L_1' be the set of free girls in L_1, and let E_0' be the set of directed edges connecting L_1' to L_0. Then the required graph G^* has the vertex set $L_0 \cup L_1'$ and the edge set E_0'.

Otherwise (that is, if L_1 does not contain a free girl) let E_0 be the set of edges connecting L_1 to L_0, and let L_2 be the set of mates of the girls in L_1. Let E_1 be the corresponding matched edges. It is clear that L_2 is a set of boys, and the edges in E_1 are directed from L_2 to L_1.

In general, suppose that we have constructed the vertex sets $L_0, L_1, \ldots, L_{2k}$ and the edge sets $E_0, E_1, \ldots, E_{2k-1}$. Now consider the set of girls L_{2k+1} which do not occur in $L_0, L_1, \ldots, L_{2k}$ and such that each of these girls is adjacent to at least one of the boys in L_{2k}.

If L_{2k+1} contains at least one free girl, then we have found an augmenting path, showing that the length of a shortest augmenting path relative to M is $2k+1$. Let L_{2k+1}' be the set of free girls in L_{2k+1}, and let E_{2k}' be the set of directed edges connecting L_{2k+1}' to L_{2k}.

Otherwise (that is, L_{2k+1} does not contain a free girl) let E_{2k} be the set of edges connecting L_{2k+1} to L_{2k}, and let L_{2k+2} be the set of mates of the girls in L_{2k+1}. Let E_{2k+1} be the corresponding matched edges.

It is clear that the procedure described above will end in one of the following two ways.

1. For some k, L_{2k+1} contains at least one free girl. In this case, we stop the procedure here and the required graph G^* has the vertex set $(L_0 \cup L_1 \cup \cdots \cup L_{2k} \cup L_{2k+1}')$ and the edge set $(E_0 \cup E_1 \cup \cdots \cup E_{2k-1} \cup E_{2k}')$.
2. The set L_{2k+1} is empty for some k, that is, we can get no new girls which are adjacent to the boys of L_{2k}. This shows that the given graph does not contain any augmenting path relative to M. Hence the matching M is maximum.

The following properties of G^* are immediate:

1. For odd i, L_i consists of girls. Otherwise L_i consists of boys.
2. Each edge in G^* is directed from a vertex in L_{i+1} to a vertex in L_i for some i.
3. G^* has no directed circuits.
4. The shortest augmenting paths in G relative to M are in one-to-one

correspondence with the directed paths in G^* which begin at a free girl and end at a free boy.

Further, it is clear that the complexity of constructing G^* is $O(n^2)$. In view of property 4, generating a maximal set of vertex-disjoint shortest augmenting paths in G relative to M is equivalent to generating a maximal set of vertex-disjoint directed paths in G^* which begin at a free girl and end at a free boy. Such directed paths in G^* can be generated in $O(n^2)$ steps by a straightforward depth-first search in G^*, as we shall see in Algorithm 15.6.

We now present Hopcroft and Karp's algorithm for maximum bipartite matching. We assume, without loss of generality, that the given graph is connected. The algorithm consists of two procedures: PROC-HOPKARP and PROC-AUGMENT.

PROC-HOPKARP is the main procedure. Given a matching M, it tests whether M is maximum. If not, it constructs the graph G^* and invokes PROC-AUGMENT.

PROC-AUGMENT generates a maximal set of vertex-disjoint directed augmenting paths in G^* and does the necessary augmentations. The augmenting paths constructed in each execution of this procedure correspond to those in a single phase of the sequence $P_0, P_1, \ldots$.

In PROC-AUGMENT, we use an array PATH in which the vertices of a path are stored as the path is generated. POINT is a variable which points to the top element in PATH.

Algorithm 15.6 Maximum Bipartite Matching (Hopcroft and Karp)

PROC-HOPKARP

H1. G is the given connected bipartite graph. Let M_0 be the null matching, that is, $M_0 = \varnothing$. Set $i = 0$.

H2. Construct from G the directed graph $\bar{G}$ by directing each unmatched edge (relative to M_i) of G from a girl to a boy, and each matched edge (relative to M_i) from a boy to a girl.

H3. Let L_0 be the set of free boys relative to M_i. If L_0 is not empty, go to step H4. Otherwise HALT. (The matching M_i is maximum.)

H4. Construct as described below the sequence $L_0, L_1, \ldots$ of subsets of the vertex set V of $\bar{G}$ and the sequence $E_0, E_1, \ldots$ of subsets of the edge set E of $\bar{G}$.

1. $E_i = \{(u, v) | (u, v) \in \bar{E}, v \in L_i, u \notin L_0 \cup L_1 \cup \cdots \cup L_i\}, i = 0, 1, 2, \ldots$.
2. $L_{i+1} = \{u | \text{for some } v, (u, v) \in E_i\}, i = 0, 1, 2, \ldots$.
3. $m^* = \min\{i \mid L_i \cap \{\text{free girls}\} \neq \varnothing\}$.

H5. If m^* is defined, go to step H6; otherwise HALT. (There is no augmenting path relative to M_i and hence M_i is a maximum matching.)

H6. Construct the subgraph $G^*=(V^*, E^*)$ of $\overline{G}$, where

$$V^*=L_0\cup L_1\cup \cdots \cup L_{m^*-1}\cup\{(L_{m^*}\cap\{\text{free girls}\})\}$$

and

$$E^*=E_0\cup E_1\cup \cdots \cup E_{m^*-2}\cup\{(u,v)|v\in L_{m^*-1} \text{ and } u\in\{\text{free girls}\}\}.$$

H7. Carry out the procedure PROC-AUGMENT and then go to step H2.

PROC-AUGMENT

A1. Set POINT=0. (Clear the array PATH.) Select a free girl y in G^*, which is not yet labeled "examined," and go to step A2. If no such free girl exists, go to step A8.

A2. Label y "examined." Set POINT=POINT+1 and PATH(POINT)= y. (y is now placed on top of the array PATH.)

A3. Select an edge (y, x) which is not yet labeled "examined" and go to step A5. If no such edge exists (we can get no augmenting path which uses vertex y), do the following:

1. If POINT=1, then go to step A1.
2. If POINT>1, then go to step A4.

A4. Set POINT=POINT−2 and y=PATH(POINT). (Remove the top two elements of the array PATH.) Go to step A3.

A5. Label (y, x) "examined." If x is already labeled "examined" (x already occurs in some shortest augmenting path or there is no such path which uses x), then go to step A3. Otherwise label x "examined" and set POINT=POINT+1 and PATH(POINT)=x. (x is now placed on top of the array PATH.)

A6. If x is free relative to M_i, then go to step A7. Otherwise get y=mate of x and go to step A2.

A7. An augmenting path P relative to M_i is now found. The vertices of P are PATH(1), PATH(2),..., PATH(POINT). Set $M_{i+1}=M_i\oplus P$. (M_{i+1} is the new matching after augmentation.) Set $i=i+1$ and go to step A1.

A8. (All shortest vertex-disjoint augmenting paths in G^* have been found and the corresponding augmentations done.) End the procedure. ■

It can be shown that the complexity of executing PROC-AUGMENT is $O(n^2)$. Since in any phase construction of G^* and execution of PROC-AUGMENT are done only once, and by Theorem 15.5, there are at most $O(n^{1/2})$ phases, it follows that the complexity of Algorithm 15.6 is $O(n^{5/2})$.

As an example, consider the graph G shown in Fig. 15.13a, where the dashed lines indicate the edges of a matching M. In this graph the girls are numbered from 1 to 12 and the boys from −1 to −10.

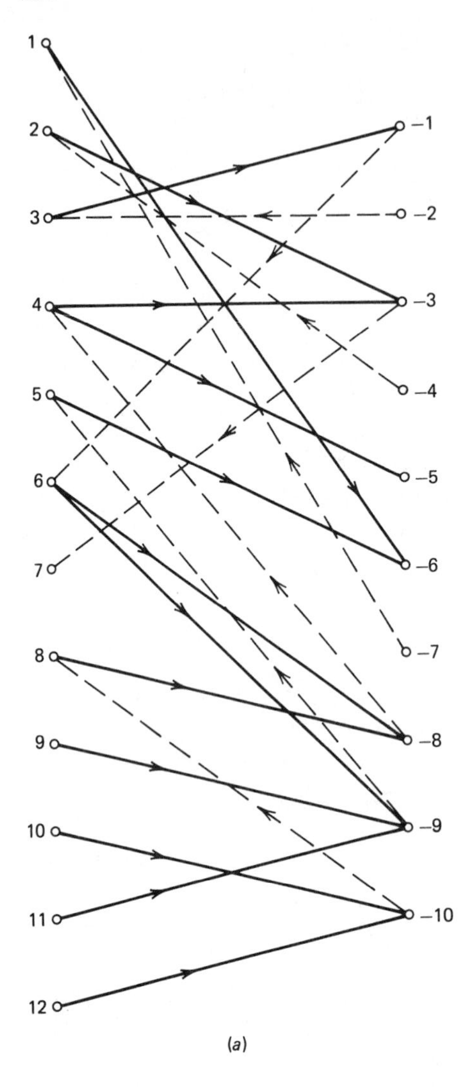

Figure 15.13.

First we direct the edges of G so that the unmatched edges are directed from a girl to a boy and the matched ones are directed from a boy to a girl. The resulting directions of edges are as shown in Fig. 15.13*a*.

The sets L_0, L_1, L_2, and L_3 are then obtained as follows:

$$L_0=\{-5,-6\},$$

$$L_1=\{1,4,5\},$$

$$L_2=\{-7,-8,-9\},$$

$$L_3=\{6,8,9,11\}.$$

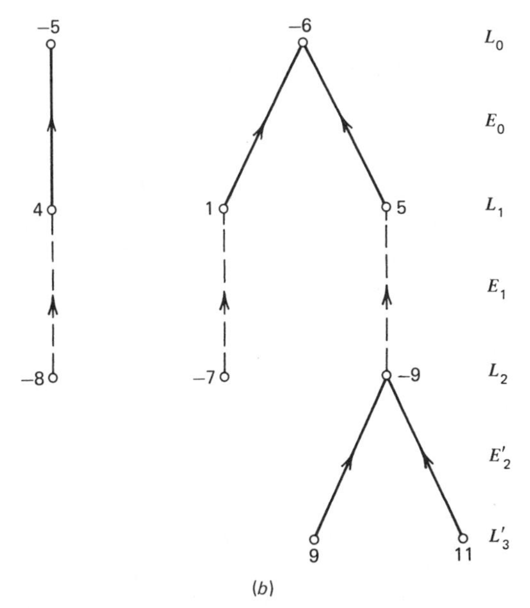

(b)

Figure 15.13. (*continued*)

Now $m^* = 3$. Since vertices 6 and 8 of L_3 are not free girls, they are not in the graph G^*. The graph G^* relative to M is shown in Fig. 15.13*b*.

Note that there are only two augmenting paths in G^*. However, they are not vertex-disjoint. So PROC-AUGMENT would generate only one augmenting path for the phase which begins with the matching M. If $\{-6, 5, -9, 9\}$ is the augmenting path generated, then after the augmentation we get a new matching in which the new matched edges are $(-6, 5)$ and $(-9, 9)$. The edge $(5, -9)$, which was a matched edge in M, now becomes unmatched.

For further reading we recommend Even and Tarjan [15.37] where it is shown, as a special case of a more general result, that a maximum matching in a bipartite graph can be constructed in $O(n^{5/2})$ steps.

Even and Kariv [15.38] have developed a $O(n^{5/2})$ algorithm for maximum matching in general graphs. This algorithm uses some techniques based on breadth-first search and depth-first search. Some of the ideas presented earlier by Gabow [15.29] and Hopcroft and Karp [15.36] are also used in developing this algorithm.

15.6 PERFECT MATCHING, OPTIMAL ASSIGNMENT, AND TIMETABLE SCHEDULING

The optimal assignment and the timetable scheduling problems, the study of which involves the theory of matching, are discussed in this section. Obtaining an optimal assignment requires as a first step the construction of a perfect matching in an appropriate bipartite graph. With this in view, we first discuss an algorithm for constructing a perfect matching in a bipartite graph.

15.6.1 Perfect Matching

Consider the following personnel assignment problem in which n available workers are qualified for one or more of n available jobs, and we are interested to know whether we can assign jobs to all the workers, one job per worker, for which they are qualified. If we represent the workers by one set $X=\{x_1, x_2,\ldots, x_n\}$ of vertices and the jobs by the other set $Y=\{y_1, y_2,\ldots, y_n\}$ of vertices of a bipartite graph G, in which x_i is joined to y_j if and only if the worker x_i is qualified for the job y_j, then it is clear that the personnel assignment problem is to find whether the graph G has a perfect matching or not.

One method of finding a solution for this problem would be to apply Algorithm 15.6 and find a maximum matching. If this matching consists of n edges, then it shows that the graph has a perfect matching, and the maximum matching obtained is nothing but a perfect matching.

The main drawback of the above method is that if the graph does not have a perfect matching, then we will know this only at the end of the procedure. Now we discuss an algorithm which either finds a perfect matching of G or stops when it finds a subset S of X such that $|\Gamma(S)|<|S|$, where $\Gamma(S)$ is the set of vertices adjacent to those in S. Clearly, by Hall's theorem (Theorem 8.13) there exists no perfect matching in the latter case.

The basic idea behind the algorithm is very simple. As usual, we start with an initial matching M. If M saturates all the vertices in X, then it is the one that we are looking for. Otherwise, as in the general case, we choose an unsaturated vertex u in X and systematically search for an augmenting path P starting from u. While looking for such a path P, we keep a count of the number of vertices selected from set X, the number of their neighbors, and the number of vertices selected from set Y.

The bipartite nature of the graph assures us that we can get no odd circuit during our search, and hence blossoms are not created. As we have seen in Section 15.4.1, a closed alternating path of even length is not of help in augmenting the given matching M. Hence the search graph which we develop is always a tree. This tree is called a *Hungarian tree*. At any stage, if we find an augmenting path, we perform the augmentation and get the new matching which saturates one more vertex in X and proceed as before. If such a path does not exist, then we would have obtained a set $S\subseteq X$, violating the necessary and sufficient condition for the existence of a perfect matching.

Let M be a matching in G, and let u be an unsaturated vertex in X. A tree H in G is called an *M-alternating tree* rooted at u if:

1. u belongs to the vertex set of H; and

2. For every vertex v of H, the unique path from u to v in H is an M-alternating path (that is, an alternating path relative to M).

Let us denote by S the subset of vertices of X and by T the subset of vertices of Y which occur in H.

The alternating tree is grown as follows. Initially, H consists of only the vertex u. It is then grown in such a way that at any stage, there are two possibilities.

1. All vertices of H except u are saturated (for example, see Fig. 15.14a).
2. H contains an unsaturated vertex different from u (for example, see Fig. 15.14b), in which case we have an augmenting path and hence we get a new matching.

In the first case, either $\Gamma(S)=T$ or $T\subset\Gamma(S)$.

1a. $\Gamma(S)=T$. Since $|S|=|T|+1$ in the tree H, we get in this case $|\Gamma(S)|=|S|-1$, and so the set S does not satisfy the necessary and sufficient condition required by Hall's theorem. Hence we conclude that there exists no perfect matching in G.

1b. $T\subset\Gamma(S)$. So there exists a vertex y in Y which does not occur in T, but which occurs in $\Gamma(S)$. Let this vertex y be adjacent to vertex x in S. If y is saturated with the vertex z as its mate, then we grow H by adding the

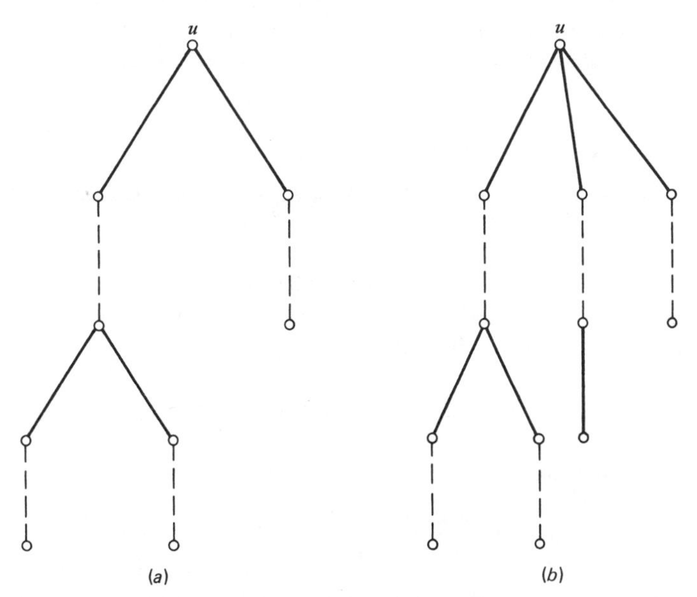

Figure 15.14. Examples of alternating trees.

vertices y and z and the edges (x, y) and (y, z). We are then back to the first case. If y is unsaturated, we grow H by adding the vertex y and the edge (x, y), resulting in the second case. The path from u to y in H is an augmenting path relative to M.

The method described above is presented in the following algorithm.

Algorithm 15.7 Perfect Matching

S1. Let G be a bipartite graph with bipartition (X, Y) and $|X|=|Y|$. Let M_0 be the null matching, that is, $M_0=\varnothing$. Set $i=0$.

S2. If all the vertices in X are saturated in the matching M_i, then HALT. (M_i is a perfect matching in G.) Otherwise pick an unsaturated vertex u in X and set $S=\{u\}$ and $T=\varnothing$.

S3. If $\Gamma(S)=T$, then HALT. (Now $|\Gamma(S)|<|S|$ and hence there is no perfect matching in G.) Otherwise select a vertex y from $\Gamma(S)-T$.

S4. If y is not saturated in M_i, go to step S5. Otherwise set $z=$mate of y, $S=S\cup\{z\}$ and $T=T\cup\{y\}$, and then go to step S3.

S5. (An augmenting path P is found.) Set $M_{i+1}=M_i \oplus P$ and $i=i+1$. Go to step S2. ■

As an example, consider the bipartite graph G shown in Fig. 15.15*a*. In this graph, the edges of an initial matching M are shown in dashed lines. The vertex x_1 is not saturated in M. The M-alternating tree rooted at x_1 is now developed. We terminate the growth of this tree as shown in Fig. 15.15*b* when we locate the augmenting path x_1, y_1, x_2, y_3. We then augment M and obtain the new matching shown in Fig. 15.15*c*.

Vertex x_4 is not saturated in this new matching. So we proceed to develop the alternating tree rooted at x_4, with respect to the new matching. This tree terminates as shown in Fig. 15.15*d*. Further growth of this tree is not possible since at this stage $\Gamma(S)=T$, where $S=\{x_1, x_2, x_3, x_4\}$ and $T=\{y_1, y_2, y_3\}$. Hence the graph in Fig. 15.15*a* has no perfect matching.

Sometimes we may be interested in finding perfect matchings having a specific property. Itai, Rodeh, and Tanimoto [15.39] discuss an algorithm applicable for a class of such problems.

15.6.2 Optimal Assignment

Consider an assignment problem in which each worker is qualified for all the jobs. Here it is obvious that every worker can be assigned a job (of course, we assume, as before, that there are n workers and n jobs). In fact, any maximum matching performs this, and we have got $n!$ such matchings. A problem of interest in this case is to take into account the effectiveness of the workers in their various jobs, and then to make that assignment which maximizes the

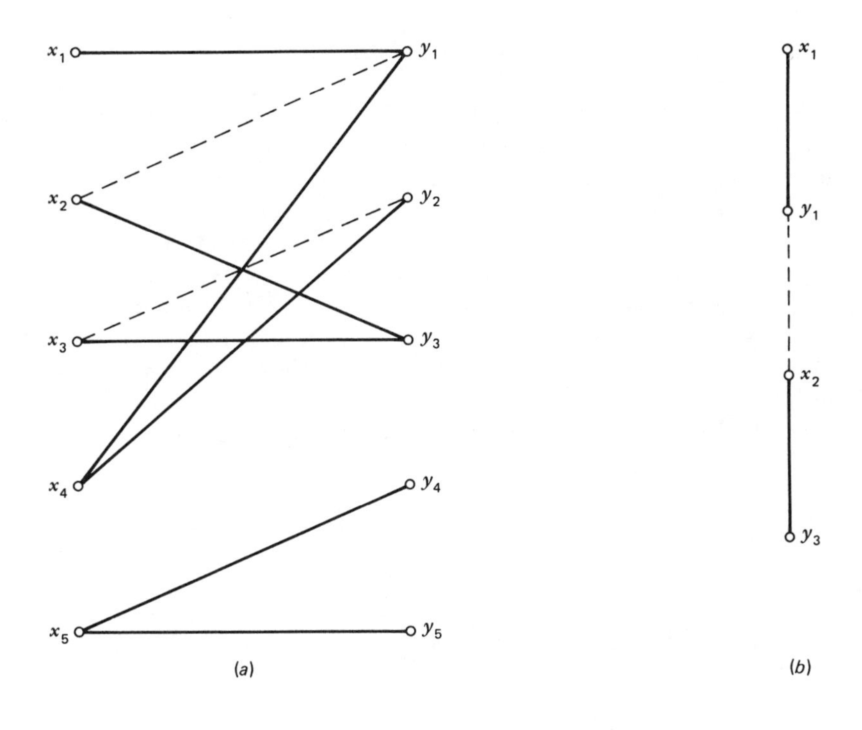
x_1
x_2
x_3
x_4
x_5
y_1
y_2
y_3
y_4
y_5
(a)
x_1
y_1
x_2
y_3
(b)

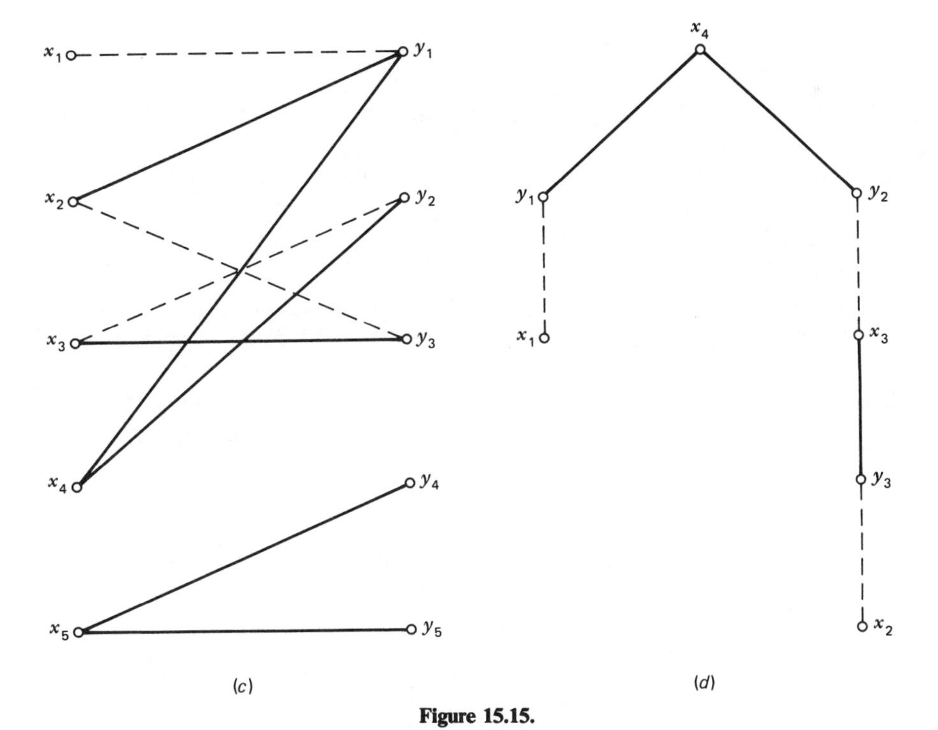
x_1
x_2
x_3
x_4
x_5
y_1
y_2
y_3
y_4
y_5
(c)
x_4
y_1
y_2
x_1
x_3
y_3
x_2
(d)

Figure 15.15.

total effectiveness of the workers. The problem of finding such an assignment is known as the *optimal assignment problem*.

The bipartite graph for this problem is a complete one; that is, if $X=\{x_1, x_2,\ldots, x_n\}$ represents the workers and $Y=\{y_1, y_2,\ldots, y_n\}$ represents the jobs, then for all i and j, x_i is adjacent to y_j. Also, we assign a weight $w_{ij}=w(x_i, y_j)$ to every edge (x_i, y_j), which represents the effectiveness of worker x_i in job y_j (measured in some units). Then the optimal assignment problem corresponds to finding a maximum weight perfect matching in this weighted graph. Such a matching is referred to as an *optimal matching*.

We now discuss a $O(n^4)$ algorithm due to Kuhn and Munkres [15.40], [15.41] for the optimal assignment problem. We follow the treatment given in Bondy and Murty [15.35].

A *feasible vertex labeling* is a real-valued function f on the set $X \cup Y$ such that

$$f(x)+f(y) \geqslant w(x, y), \qquad \text{for all } x \in X \text{ and } y \in Y.$$

$f(x)$ is then called the *label* of the vertex x.

For example, the following labeling is a feasible vertex labeling:

$$\begin{aligned} f(x)&=\max\{w(x, y)\}, &&\text{if } x \in X,\\ f(y)&=0, &&\text{if } y \in Y. \end{aligned}$$

From this it should be clear that there always exists a feasible vertex labeling irrespective of what the weights are.

For a given feasible vertex labeling f, let E_f denote the set of all those edges (x, y) of G such that $f(x)+f(y)=w(x, y)$. The spanning subgraph of G with the edge set E_f is called the *equality subgraph* corresponding to f. We denote this subgraph by G_f.

The following theorem relating equality subgraphs and optimal matchings forms the basis of the Kuhn-Munkres algorithm.

THEOREM 15.6. Let f be a feasible vertex labeling of a graph $G=(V, E)$. If G_f contains a perfect matching M^*, then M^* is an optimal matching in G.

Proof

Suppose that G_f contains a perfect matching M^*. Since G_f is a spanning subgraph of G, M^* is also a perfect matching in G. Let $w(M^*)$ denote the weight of M^*, that is,

$$w(M^*)=\sum_{e \in M^*} w(e).$$

Since each edge $e \in M^*$ belongs to the equality subgraph and the vertices of the edges of M^* cover each vertex of G exactly once, we get

$$w(M^*) = \sum_{e \in M^*} w(e)$$
$$= \sum_{v \in V} f(v). \qquad (15.28)$$

On the other hand, if M is any perfect matching in G, then

$$w(M) = \sum_{e \in M} w(e)$$
$$\leqslant \sum_{v \in V} f(v). \qquad (15.29)$$

Now combining (15.28) and (15.29), we see that

$$w(M^*) \geqslant w(M).$$

Thus M^* is an optimal matching in G. ■

In the Kuhn-Munkres algorithm, we first start with an arbitrary feasible vertex labeling f and find the corresponding G_f. We will choose an initial matching M in G_f and apply Algorithm 15.7. If a perfect matching is obtained in G_f, then, by Theorem 15.6, this matching is optimal. Otherwise Algorithm 15.7 terminates with a matching M' that is not perfect, giving an M'-alternating tree H that contains no M'-augmenting path and which cannot be grown further in G_f. We then modify f to a feasible vertex labeling f' with the property that both M' and H are contained in $G_{f'}$, and H can be extended in $G_{f'}$. We make such a modification in the feasible vertex labeling whenever necessary, until a perfect matching is found in some equality subgraph. Details of the Kuhn-Munkres algorithm are presented below:

Algorithm 15.8 Optimal Assignment (Kuhn and Munkres)

S1. G is the given complete bipartite graph with bipartition (X, Y) and $|X| = |Y|$. $W = [w_{ij}]$ is the given weight matrix. Set $i = 0$.

S2. Start with an arbitrary feasible vertex labeling f in G. Find the equality subgraph G_f and then select an initial matching M_i in G_f.

S3. If all the vertices in X are saturated in M_i, then M_i is a perfect matching, and hence by Theorem 15.6, it is an optimal matching. So HALT. Otherwise let u be an unsaturated vertex in X. Set $S = \{u\}$ and $T = \varnothing$.

S4. Let $\Gamma_f(S)$ be the set of vertices which are adjacent in G_f to the vertices in S. If $\Gamma_f(S) \supset T$, then go to step S5. Otherwise (that is, if $\Gamma_f(S) = T$)

compute

$$d_f = \min_{\substack{x \in S \\ y \notin T}} \{ f(x) + f(y) - w(x, y) \} \tag{15.30}$$

and get a new feasible vertex labeling f' given by

$$f'(v) = \begin{cases} f(v) - d_f, & \text{if } v \in S. \\ f(v) + d_f, & \text{if } v \in T. \\ f(v), & \text{otherwise.} \end{cases} \tag{15.31}$$

(Note that $d_f > 0$ and $\Gamma_f(S) = T$.)
Replace f by f' and G_f by $G_{f'}$.

S5. Select a vertex y from $\Gamma_f(S) - T$. If y is not saturated in M_i, go to step S6. Otherwise set $z =$ mate of y in M_i, $S = S \cup \{z\}$ and $T = T \cup \{y\}$, and then go to step S4.

S6. (An augmenting path P is found.) Set $M_{i+1} = M_i \oplus P$ and $i = i + 1$. Go to step S3. ■

To illustrate the Kuhn-Munkres algorithm consider a complete bipartite graph G having the following weight matrix $W = [w_{ij}]$:

$$W = \begin{bmatrix} 4 & 4 & 1 & 3 \\ 3 & 2 & 2 & 1 \\ 5 & 4 & 4 & 3 \\ 1 & 1 & 2 & 2 \end{bmatrix}.$$

An initial feasible vertex labeling f of G may be chosen as follows:

$$f(x_1) = 4, \quad f(x_2) = 3, \quad f(x_3) = 5, \quad f(x_4) = 2;$$

$$f(y_1) = f(y_2) = f(y_3) = f(y_4) = 0.$$

The equality subgraph G_f is shown in Fig. 15.16*a*. Applying Algorithm 15.6, we find that G_f has no perfect matching because for the set $S = \{x_1, x_2, x_3\}$, $T = \Gamma(S) = \{y_1, y_2\}$.

Using (15.30) we compute

$$d_f = 1.$$

The following new labeling f' is then obtained using (15.31):

$$f'(x_1) = 3, \quad f'(x_2) = 2, \quad f'(x_3) = 4, \quad f'(x_4) = 2;$$

$$f'(y_1) = 1, \quad f'(y_2) = 1, \quad f'(y_3) = 0, \quad f'(y_4) = 0.$$

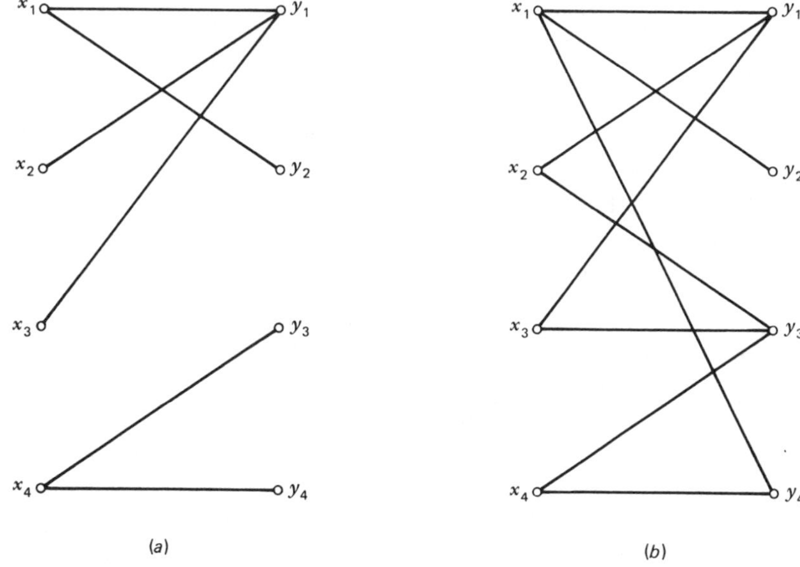

Figure 15.16.

The equality subgraph $G_{f'}$ is shown in Fig. 15.16*b*. Using Algorithm 15.6 on $G_{f'}$, we obtain the perfect matching M consisting of the edges (x_1, y_2), (x_2, y_1), (x_3, y_3), and (x_4, y_4). This matching is an optimal matching.

In a recent paper Megido and Tamir [15.42] have discussed a $O(n \log n)$ algorithm for a class of weighted matching problems which arise in certain applications relating to scheduling and optimal assignment.

15.6.3 Timetable Scheduling

In a school there are p teachers $x_1, x_2, \ldots, x_p$ and q classes $y_1, y_2, \ldots, y_q$. Given that teacher x_i is required to teach class y_j for p_{ij} periods, we would like to schedule a timetable having the minimum possible number of periods. This is a special case of what is known as the *timetable scheduling* problem.

Suppose that we construct a bipartite graph $G=(X, Y)$ in which the vertices in X represent the teachers and those in Y represent classes, and vertex $x_i \in X$ is connected to vertex $y_j \in Y$ by p_{ij} parallel edges. Since in any one period each teacher can teach at most one class and each class can be taught by at most one teacher, it follows that a teaching schedule for one period corresponds to a matching in G, and conversely each matching corresponds to a possible assignment of teachers to classes for one period. Thus the timetable scheduling problem mentioned above is to partition the edges of G into as few matchings as possible.

By Corollary 8.22.1, the minimum number of matchings in any partition of the edge set of a bipartite graph G is equal to the maximum degree in G. The proof of this theorem also suggests the following procedure for determining a partition having the smallest number of matchings:

Step 1. Let G be the given bipartite graph. Set $i=0$ and $G_0=G$.

Step 2. Construct a matching M_i of G_i which saturates all the maximum degree vertices in G_i.

Step 3. Remove M_i from G_i. Let G_{i+1} denote the resulting graph. If G_{i+1} has no edges, then $M_0, M_1, \ldots, M_i$ is a required partition of the edge set of G. Otherwise set $i=i+1$ and go to step 1.

Clearly, the complexity of the above procedure depends on the complexity of implementing step 2 which requires finding a matching which saturates all the maximum degree vertices in a bipartite graph $G=(X,Y)$. Such a matching may be found as follows. (See proof of Theorem 8.21.)

Let X_a denote the set of maximum degree vertices in X, and let Y_b denote the set of maximum degree vertices in Y. Let G_a denote the subgraph of G formed by the edges incident on the vertices in X_a. Similarly G_b is the subgraph on the edges incident on the vertices in Y_b.

By Theorem 8.22 there is a matching M_a which saturates all the vertices in X_a. The matching M_a is a maximum matching in G_a. Similarly in G_b there is a maximum matching M_b which saturates all the vertices in Y_b. Following the procedure used in the proof of Theorem 8.21, we can find from M_a and M_b a matching M which saturates the vertices in X_a and Y_b. M is then a required matching saturating all the maximum degree vertices in G.

The complexity of finding M_a and M_b is the same as the complexity of finding a maximum matching, that is, $O(n^{5/2})$, where n is the number of vertices in G. It is easy to show that the complexity of constructing M from M_a and M_b is $O(n^2)$. Thus the overall complexity of implementing step 2 is $O(n^{5/2})$.

Since step 2 will be repeated Δ times, where Δ is the maximum degree in G, and $\Delta \leqslant n$, it follows that the complexity of constructing the required timetable is $O(n^{7/2})$.

A discussion of the general form of the timetable scheduling problem and references related to this may be found in Even, Itai, and Shamir [15.43].

15.7 FLOWS IN A TRANSPORT NETWORK

A *transport network* N is a connected directed graph which has no self-loops and which satisfies the following conditions:

1. There is only one vertex with zero in-degree; this vertex is called the *source* and is denoted by s.
2. There is only one vertex with zero out-degree; this vertex is called the *sink* and is denoted by t.

3. Each directed edge $e=(i, j)$ in N is associated with a nonnegative real number called the *capacity* of the edge; it is denoted by $c(e)$ or $c(i, j)$. If there is no edge e directed from i to j, then we define $c(e)=0$.

A transport network represents a model for the transportation of a commodity from its production center to its market through communication routes. The *capacity* of an edge may then be considered as representing the maximum rate at which a commodity can be transported along the edge.

A *flow* f in a transport network N is an assignment of a nonnegative real number $f(e)=f(i, j)$ to each edge $e=(i, j)$ such that the following conditions are satisfied:

1. $f(i, j) \leqslant c(i, j)$ for every edge (i, j) in N. (15.32)
2. $\Sigma_{\text{all } j} f(i, j)=\Sigma_{\text{all } j} f(j, i)$ for all $i \neq s, t$. (15.33)

The value $f(e)$ of the flow in edge e may be considered as the rate at which material is transported along e under the flow f. Condition (15.32), called the *capacity constraint*, requires that the rate of flow along an edge cannot exceed the capacity of the edge. Condition (15.33), called the *conservation condition*, requires that for each vertex i, except the source and the sink, the rate at which material is transported into i is equal to the rate at which it is transported out of i.

As an example, a transport network N with a flow f is shown in Fig. 15.17. In this figure, next to each edge e we have shown the capacity $c(e)$ and the flow $f(e)$ in that order.

The value of a flow f, denoted by val(f), is defined as

$$\operatorname{val}(f)=\sum_{\text{all } j} f(s, j). \tag{15.34}$$

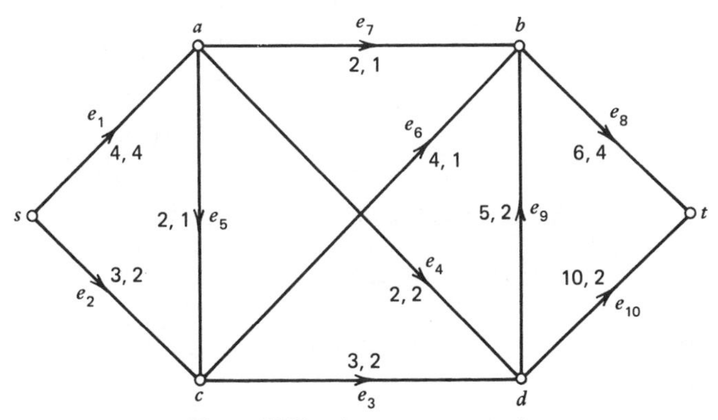

Figure 15.17. A transport network.

Soon we prove using the conservation condition that

$$\text{val}(f) = \sum_{\text{all } j} f(s, j) = \sum_{\text{all } j} f(j, t). \tag{15.35}$$

This would only confirm the intuitively obvious fact that, because of the conservation condition, the total amount of material transported out of the source is equal to the total amount transported into the sink.

A flow f^* in a transport network N is said to be *maximum* if there is no flow f in N such that $\text{val}(f) > \text{val}(f^*)$.

In the following subsection we establish the maximum flow minimum cut theorem of Ford and Fulkerson. We then discuss Ford-Fulkerson's labeling algorithm for determining a maximum flow in a transport network. In the concluding subsection we prove Menger's theorems using the maximum flow minimum cut theorem.

15.7.1 Maximum Flow Minimum Cut Theorem

A cut $\langle S, \bar{S} \rangle$ in a transport network N is said to separate the source s and the sink t if $s \in S$ and $t \in \bar{S}$. Such a cut will be referred to as an s–t cut. The *capacity* $c(K) = c(S, \bar{S})$ of a cut $K = \langle S, \bar{S} \rangle$ is defined as

$$c(K) = \sum_{\substack{i \in S \\ j \in \bar{S}}} c(i, j). \tag{15.36}$$

Note that the capacities of the edges which are directed from $\bar{S}$ to S do not contribute to the capacity of the cut $K = \langle S, \bar{S} \rangle$. We denote by $f(S, \bar{S})$ the sum of the flows in the edges directed from S to $\bar{S}$. The quantity $f(\bar{S}, S)$ is defined in a similar way.

For example, consider the cut $K = \langle S, \bar{S} \rangle$ in the transport network shown in Fig. 15.17, where

$$S = \{a, b, c, s\} \qquad \text{and} \qquad \bar{S} = \{d, t\}.$$

The edges directed from S to $\bar{S}$ are e_3, e_4, and e_8. Hence

$$c(S, \bar{S}) = c(e_3) + c(e_4) + c(e_8) = 11,$$

$$f(S, \bar{S}) = f(e_3) + f(e_4) + f(e_8) = 8,$$

$$f(\bar{S}, S) = f(e_9) = 2.$$

THEOREM 15.7. For any flow f and any s–t cut $\langle S, \bar{S} \rangle$ in a transport network N,

$$\text{val}(f) = f(S, \bar{S}) - f(\bar{S}, S). \tag{15.37}$$

Proof

From the definitions of a flow and the value of a flow, we have

$$\sum_{\text{all } j} f(i,j) - \sum_{\text{all } j} f(j,i) = \begin{cases} \text{val}(f), & \text{if } i=s. \\ 0, & \text{if } i\in S-\{s\}. \end{cases}$$

Summing these equations over all the vertices in S, we get

$$\sum_{i\in S}\sum_{\text{all } j} f(i,j) - \sum_{i\in S}\sum_{\text{all } j} f(j,i) = \text{val}(f). \tag{15.38}$$

In the above equation, both $f(i,j)$ and $-f(i,j)$, for $i\in S$ and $j\in S$, appear exactly once on the left-hand side and get canceled out. So (15.38) simplifies to

$$\sum_{i\in S}\sum_{j\in \bar{S}} f(i,j) - \sum_{i\in S}\sum_{j\in \bar{S}} f(j,i) = \text{val}(f).$$

Thus

$$f(S,\bar{S}) - f(\bar{S},S) = \text{val}(f). \quad \blacksquare$$

Note that (15.34) is a special case of (15.37).

Corollary 15.7.1 For any flow f and any s–t cut $K=\langle S,\bar{S}\rangle$ in a transport network N,

$$\text{val}(f) \leq c(S,\bar{S}). \tag{15.39}$$

Proof

Since every $f(i,j)$ is nonnegative, we get from (15.37)

$$\begin{aligned} \text{val}(f) &= f(S,\bar{S}) - f(\bar{S},S) \\ &\leq f(S,\bar{S}) \\ &\leq c(S,\bar{S}). \quad \blacksquare \end{aligned}$$

An edge (i,j) is said to be *f-saturated* if $f(i,j)=c(i,j)$ and *f-unsaturated* otherwise; it is *f-positive* if $f(i,j)>0$, and *f-zero* if $f(i,j)=0$.

Note that equality in (15.39) holds if and only if $f(\bar{S},S)=0$ and $f(S,\bar{S})=c(S,\bar{S})$. In other words, $\text{val}(f)=c(S,\bar{S})$ if and only if all the edges directed from S to $\bar{S}$ are f-saturated and those directed from $\bar{S}$ to S are f-zero.

An s–t cut K in a transport network N is a *minimum cut* if there is no s–t cut K' in N such that $c(K')<c(K)$.

Corollary 15.7.2 Let f be a flow and K be an s–t cut such that $\text{val}(f)=c(K)$. Then f is a maximum flow and K is a minimum s–t cut.

Proof

Let f^* be a maximum flow and K^* be a minimum s–t cut. By Corollary 15.7.1,

$$\text{val}(f^*) \leqslant c(K^*).$$

So we get

$$\text{val}(f) \leqslant \text{val}(f^*) \leqslant c(K^*) \leqslant c(K).$$

Since, by hypothesis, $\text{val}(f)=c(K)$, it follows that $\text{val}(f)=\text{val}(f^*)=c(K^*)=c(K)$. Thus f is a maximum flow and K is a minimum s–t cut. ■

We now proceed to show that the value of a maximum flow is in fact equal to the capacity of a minimum cut.

Consider a transport network N with a flow f. Let

$$P: \quad s=u_0 \overset{e_1}{\text{---}} u_1 \overset{e_2}{\text{---}} u_2 \cdots u_{i-1} \overset{e_i}{\text{---}} u_i \cdots u_{k-1} \overset{e_k}{\text{---}} u_k=v$$

be a path in N from the source to some vertex v. Note that P need not be a directed path.

Edge e_i of P is called a *forward edge* of P if it is oriented from u_{i-1} to u_i. Otherwise it will be called a *reverse edge* of P. For each edge e_i in P, let

$$\epsilon_i(P)=\begin{cases} c(e_i)-f(e_i), & \text{if } e_i \text{ is a forward edge.} \\ f(e_i), & \text{if } e_i \text{ is a reverse edge.} \end{cases} \tag{15.40}$$

With the path P we associate a nonnegative number $\epsilon(P)$ defined by

$$\epsilon(P)=\min_i \{\epsilon_i(P)\}. \tag{15.41}$$

Note that $\epsilon(P) \geqslant 0$.

We call a path *f-unsaturated* if all the forward edges of the path are f-unsaturated and all the reverse edges of the path are f-positive.

An s–t path P is called an *f-augmenting path* if P is f-unsaturated. From (15.40) and (15.41) it follows that for an s–t path P, $\epsilon(P)>0$ if and only if P is f-augmenting. Given an s–t path P in the network N, we can define a new flow $\hat{f}$ as follows:

$$\hat{f}(e)=\begin{cases} f(e)+\epsilon(P), & \text{if } e \text{ is a forward edge of } P. \\ f(e)-\epsilon(P), & \text{if } e \text{ is a reverse edge of } P. \\ f(e), & \text{otherwise.} \end{cases}$$

We can easily see that

$$\text{val}(\hat{f}) = \text{val}(f) + \epsilon(P).$$

Thus $\text{val}(\hat{f}) > \text{val}(f)$ if and only if P is f-augmenting. In other words, a flow f is not maximum if there exists an f-augmenting path.

As an example, consider the network N of Fig. 15.17. Let flow f be as shown in this figure. In the path P consisting of e_2, e_5, e_7, and e_8, the edges e_2, e_8, and e_7 are forward edges, and e_5 is the only backward edge. With respect to the flow f, $\epsilon_2(P)=1$, $\epsilon_5(P)=1$, $\epsilon_7(P)=1$, and $\epsilon_8(P)=2$. Hence $\epsilon(P)=\min\{\epsilon_2(P), \epsilon_5(P), \epsilon_7(P), \epsilon_8(P)\}=1$.

Since $\epsilon(P)>0$, P is f-augmenting. The revised flow $\hat{f}$ based on the path P is as shown in Fig. 15.18. Note that $\hat{f}$ is obtained by increasing the flows in all the forward edges of P by $\epsilon(P)$ and decreasing the flows in the backward edges of P by $\epsilon(P)$. The flows in the remaining edges remain unaltered.

THEOREM 15.8. A flow f in a transport network N is maximum if and only if there is no f-augmenting path.

Proof

Necessity If there is an f-augmenting path P in N, then clearly f is not a maximum flow because the revised flow $\hat{f}$ based on P has a larger value than f.

Sufficiency Suppose that N contains no f-augmenting path. Let S denote the set of all vertices in N which can be reached from the source by f-unsaturated paths. Clearly, $s \in S$. Further, $t \in \bar{S}$ because there is no f-augmenting path in N.

Now we show that $\text{val}(f) = c(S, \bar{S})$, thereby proving (by Corollary 15.7.2) that f is a maximum flow and $\langle S, \bar{S} \rangle$ is a minimum cut.

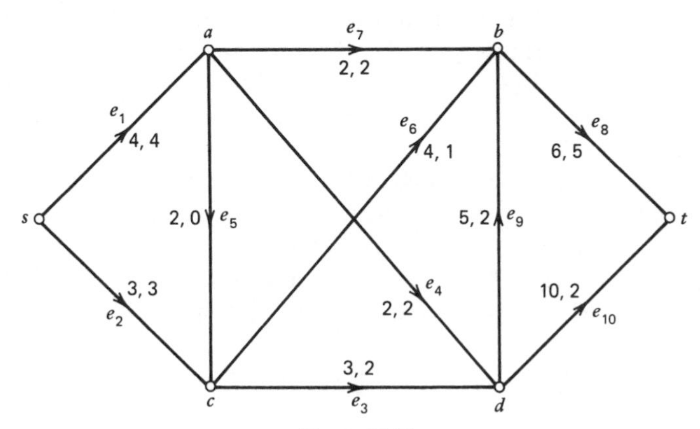

Figure 15.18.

Consider a directed edge (v, w) with $v \in S$ and $w \in \bar{S}$. Since $v \in S$, there is an f-unsaturated s–v path Q. The edge (v, w) must be f-saturated, for otherwise Q can be extended to yield an f-unsaturated s–w path, and this is not possible since $w \in \bar{S}$. Similarly we can show that every edge (v, w) directed from $\bar{S}$ to S should be f-zero. Thus $f(S, \bar{S}) = c(S, \bar{S})$ and $f(\bar{S}, S) = 0$. Hence

$$\text{val}(f) = f(S, \bar{S}) - f(\bar{S}, S) = c(S, \bar{S}).$$

It now follows from Corollary 15.7.2 that f is a maximum flow and $\langle S, \bar{S} \rangle$ is a minimum cut. ∎

In the course of the proof of the above theorem we have also established the following well-known result due to Ford and Fulkerson [15.44] and Elias, Feinstein and Shannon [15.45].

THEOREM 15.9 (MAX-FLOW MIN-CUT THEOREM). In a transport network the value of a maximum flow is equal to the capacity of a minimum cut. ∎

The max-flow min-cut theorem can be used to prove several results of combinatorial significance. We discuss one of these results in Section 15.7.4. For others, Ford and Fulkerson [15.1] and Berge [15.46] may be consulted.

15.7.2 Ford-Fulkerson's Labeling Algorithm

We now discuss an algorithm due to Ford and Fulkerson [15.44] for determining a maximum flow in a transport network. This algorithm, based on Theorem 15.8, consists of two phases.

Given a flow f. In the first phase we use a labeling procedure to check whether an f-augmenting path exists. If such a path does not exist, then by Theorem 15.8 the present flow f is maximum. Otherwise we proceed to the second phase in which, using the labels generated in the first phase, we determine an f-augmenting path P and obtain the revised flow $\hat{f}$ based on P. We then repeat phase 1 with the new flow $\hat{f}$. Note that $\text{val}(\hat{f}) > \text{val}(f)$.

In the first phase a label of the form (d_v, Δ_v) is assigned to a vertex v. The first symbol d_v in the label indicates the vertex from which v receives its label. It would also indicate the direction of labeling—forward or backward. Soon it will become clear that when a vertex v gets labeled, it means that there exists an f-unsaturated s–v path P and that for this path P, $\epsilon(P) = \Delta_v$.

The first phase begins by labeling the source s as $(-, \infty)$; here the value of d_s is irrelevant. The labeling of the vertices then proceeds according to the following rules:

Suppose a vertex u is labeled and a vertex v is unlabeled. Let e be an edge connecting u and v.

Forward Labeling If $e=(u, v)$, then forward labeling of v from u along e is possible if $c(e)>f(e)$. If such a labeling is done, then v gets the label (u^+, Δ_v), where

$$\Delta_v=\min\{\Delta_u, c(e)-f(e)\}.$$

Backward Labeling If $e=(v, u)$, then backward labeling of v from u along e is possible if $f(e)>0$. If such a labeling is done, then v gets the label (u^-, Δ_v), where

$$\Delta_v=\min\{\Delta_u, f(e)\}.$$

In the first phase a vertex gets labeled at most once. This phase terminates when either (1) the vertex t receives a label, or (2) with t unlabeled, no more vertices can be labeled.

If t receives a label in the first phase, then it follows from the rules for labeling that there exists an f-augmenting path P and that $\epsilon(P)=\Delta_t$. In the second phase the path P is traced back using the d_v symbols, and the revised flow $\hat{f}$ based on P is also determined. Phase 1 is then repeated using the new flow $\hat{f}$. If phase 1 terminates without labeling t, then it means that there is no f-augmenting path, and hence the present flow f is maximum.

A description of the Ford-Fulkerson's algorithm is presented below.

Algorithm 15.9 Maximum Flow in a Transport Network (Ford and Fulkerson)

S1. Select any flow f in the given transport network. We may select $f(e)=0$ for every edge e in N.

S2. (Phase 1 begins.) Label s as $(-, \infty)$.

S3. If there exists an unlabeled vertex which can be labeled through either forward labeling or backward labeling, then select one such vertex v, label it, and then go to step S4. Otherwise go to step S7.

S4. If $v=t$, then go to step S5. (Phase 1 ends.) Otherwise go to step S3.

S5. (Phase 2 begins.) Let the label of v be (d_v, Δ_v). Now do the following:

1. If $d_v=u^+$, then set $f(u, v)=f(u, v)+\Delta_t$.
2. If $d_v=u^-$, then set $f(v, u)=f(v, u)-\Delta_t$.

S6. If $u=s$, erase all the labels (Phase 2 ends) and go to step S2. Otherwise set $v=u$ and go to step S5.

S7. (The present flow f is maximum.) HALT. ■

We now illustrate the above algorithm.

Consider the transport network N shown in Fig. 15.19. In this figure, next to each edge e we have shown $c(e)$ and $f(e)$ in this order. We take as an initial flow $f(e)=0$ for all the edges in N. Starting with the label $(-, \infty)$ for

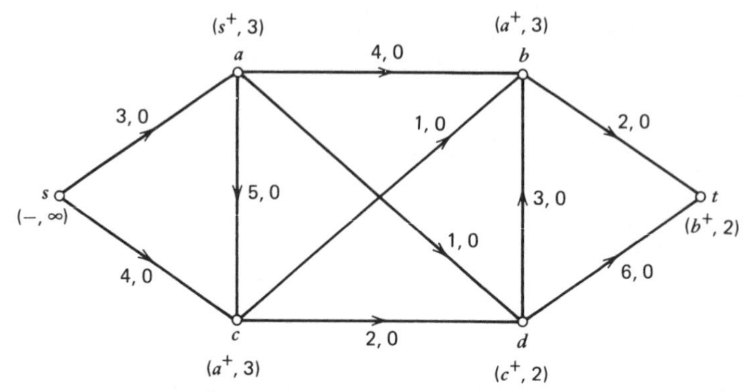

Figure 15.19. Network for illustrating Ford-Fulkerson's labeling algorithm.

the source s, we label (step S3 of Algorithm 15.9) the vertices a, b, c, d, and t in this order. The resulting labels are

$$\begin{aligned} a&: \quad (s^+,3), \\ b&: \quad (a^+,3), \\ c&: \quad (a^+,3), \\ d&: \quad (c^+,2), \\ t&: \quad (b^+,2). \end{aligned}$$

Phase 1 now terminates because vertex t has just been labeled. In phase 2 we determine an augmenting path P and the revised flow f_1 as follows (steps S5 and S6).

The first symbol in the label of t is b^+. This means that in the path P vertex b precedes t. The first symbol in the label of b indicates that a precedes b in P. Similarly we see that s precedes a in P. Thus

$$P: \quad s \rightarrow a \rightarrow b \rightarrow t$$

All the edges in P are forward edges. So to obtain the revised flow f_1 we increase the flows in all the edges of P by $\Delta_t = 2$. The flow f_1 has a value equal to 2, and it is shown in Fig. 15.20a.

We now erase the labels of all the vertices. Starting with f_1 we then obtain a new set of labels as shown in Fig. 15.20a. An augmenting path with respect to f_1 consists of the vertices s, c, d, t in this order. All the edges in this path P are forward edges. The flows in these edges are increased by 2. The resulting flow f_2 is shown in Fig. 15.20b.

A new set of labels based on f_2 is shown in Fig. 15.20b. An augmenting path with respect to f_2 consists of the vertices s, c, b, a, d, t. The edge connecting a and b is a backward edge in this path. All the other edges of this path

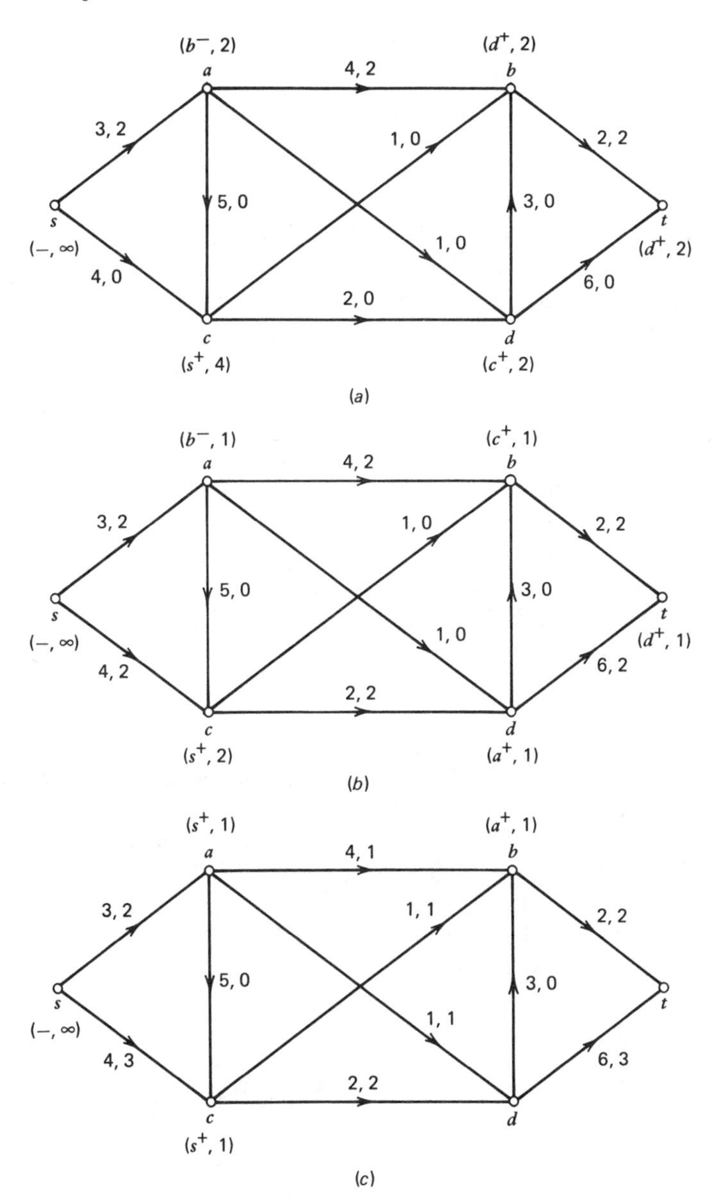

Figure 15.20. Illustration of Ford-Fulkerson's labeling algorithm.

are forward edges. We now increase the flows in the forward edges by 1 and decrease the flow in the backward edge by 1. The resulting flow f_3 is shown in Fig. 15.20*c*.

Starting with f_3 we then proceed to label the vertices. This process terminates as in Fig. 15.20*c* without labeling vertex t. Thus there is no augmenting path relative to the flow f_3. So f_3 is a maximum flow.

Let S denote the set of labeled vertices in Fig. 15.20*c*. Thus $S=\{s, a, b, c\}$. Then it is clear from the proof of Theorem 15.8 that the cut $\langle S, \bar{S}\rangle$ is a minimum cut and val(f_3)=capacity of $\langle S, \bar{S}\rangle$.

15.7.3 Edmonds and Karp's Modification of the Labeling Algorithm

In Ford-Fulkerson's algorithm which we described in the previous subsection, vertices can be labeled in any order. In other words, selection of an augmenting path (when it exists) can be made in any arbitrary manner. We now illustrate, with an example, a problem to which this arbitrariness might lead.

Consider the transport network N shown in Fig. 15.21. Suppose we start the labeling algorithm with the zero flow and alternately use the paths P_1: s, a, b, t and P_2: s, b, a, t as augmenting paths. At each step the flow value increases by exactly 1, and the maximum flow of $2M$ is achieved after $2M$ augmentation steps. Thus the number of computational steps used in this case is not bounded by a function of the number of vertices and the number of edges in N. This number is in fact a function of the capacity M which can be arbitrarily large.

Furthermore, Ford and Fulkerson [15.1] showed that their algorithm may fail if the capacities are irrational numbers. They gave an example where the flow converges in infinitely many steps to one fourth of the value of a maximum flow.

To avoid this problem, Edmonds and Karp [15.4] have suggested a refinement to the labeling algorithm: at each step, augment the flow along a shortest path. Here by a shortest path we refer to a path having the smallest number of edges. It can be easily seen that a shortest augmenting path will be selected if in the labeling procedure we scan on a "first-labeled first-scanned" basis; that is, if vertex v has been labeled before vertex u, then scan v before u. Here to scan a labeled vertex v means to label (whenever possible) all the unlabeled vertices adjacent to v.

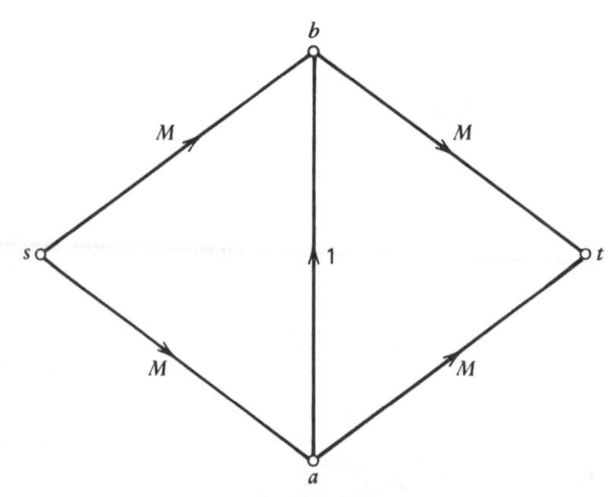

Figure 15.21.

Edmonds and Karp have also shown that this refinement will guarantee that the number of computational steps required to implement the labeling algorithm will be independent of the capacities. We now proceed to develop this result.

Consider a flow f in a transport network N. Let

$$P: \quad s=u_0 \xrightarrow{e_1} u_1 \xrightarrow{e_2} u_2 \cdots u_{i-1} \xrightarrow{e_i} u_i \cdots u_{k-1} \xrightarrow{e_k} u_k=t$$

be an augmenting path. Recall that

$$\epsilon_i(P)=\begin{cases} c(e_i)-f(e_i), & \text{if } e_i \text{ is a forward edge.} \\ f(e_i), & \text{if } e_i \text{ is a reverse edge.} \end{cases}$$

Further

$$\epsilon(P)=\min_i \{\epsilon_i(P)\}.$$

Thus $\epsilon(P)=\epsilon_i(P)$, for some i. The corresponding e_i is then called a *bottleneck*.

Assume that the labeling algorithm starts with an initial flow f_0 and constructs a sequence of flows $f_1, f_2, f_3, \ldots$.

Note that when an edge e is a bottleneck in the forward direction in an augmenting path, then during the augmentation it gets saturated. Also, if e is a bottleneck in the backward direction, then during the augmentation the flow through e is reduced to zero. This observation leads to the following result.

LEMMA 15.7. If $k<p$ and e is a bottleneck in the forward (backward) direction both in the augmenting path which changes f_k to f_{k+1} and in the augmenting path which changes f_p to f_{p+1}, then there exists an l, $k<l<p$, such that e is used as a backward (forward) edge in the augmenting path which changes f_l to f_{l+1}. ■

Let $\lambda^i(u,v)$ denote the length of a shortest f_i-unsaturated path from u to v. Here, of course, the path need not be directed. Further note that an edge e is used as a forward edge in the path only if it is not saturated, that is, $f_i(e)<c(e)$, and e is used as a backward edge only if $f_i(e)>0$.

LEMMA 15.8. For every vertex v and every $k=0,1,2\ldots,$

$$\lambda^k(s,v)\leqslant\lambda^{k+1}(s,v), \tag{15.42}$$

$$\lambda^k(v,t)\leqslant\lambda^{k+1}(v,t). \tag{15.43}$$

Proof

We shall prove (15.42) and the proof of (15.43) will follow in a similar manner.

Suppose there is no f_{k+1}-unsaturated path from s to v. Then $\lambda^{k+1}(s,v)$ is assumed to be infinite and (15.42) follows trivially. So assume that

$$P: \quad s=u_0 \overset{e_1}{—} u_1 \overset{e_2}{—} u_2 \overset{e_3}{—} u_3 \cdots u_{p-1} \overset{e_p}{—} u_p=v$$

is a shortest f_{k+1}-unsaturated path from s to v.

Suppose that e_i is used as a forward edge in P. Then $f_{k+1}(e_i)<c(e_i)$. Therefore, (1) either $f_k(e_i)<c(e_i)$, or (2) $f_k(e_i)=c(e_i)$, and so e_i has been used as a backward edge in the augmenting path which changes f_k to f_{k+1}.

In the former case,

$$\lambda^k(s,u_i)\leqslant\lambda^k(s,u_{i-1})+1, \tag{15.44}$$

because a shortest f_k-unsaturated path from s to u_{i-1} followed by e_i in the forward direction is an f_k-unsaturated path from s to u_i. Clearly, in the latter case

$$\lambda^k(s,u_{i-1})=\lambda^k(s,u_i)+1. \tag{15.45}$$

In either case (15.44) holds.

Similarly we can show that (15.44) holds when e_i is used as a backward edge in the path P.

Now summing (15.44) for $i=1,2,\ldots,p$ and noting that $\lambda^k(s,u_0)=0$, we get

$$\lambda^k(s,u_p)\leqslant p=\lambda^{k+1}(s,v).$$

Thus (15.42) follows. ■

LEMMA 15.9. If the "first-labeled first-scanned" principle is used and if $k<l$ and e is used as a forward (backward) edge in the augmenting path which changes f_k to f_{k+1} and as a backward (forward) edge in the augmenting path which changes f_l to f_{l+1}, then

$$\lambda^l(s,t)\geqslant\lambda^k(s,t)+2.$$

Proof

Assume that e is directed from u to v. Then

$$\lambda^k(s,v)=\lambda^k(s,u)+1, \tag{15.46}$$

because e is used as a forward edge while augmenting f_k. Further,

$$\lambda^l(s,t)=\lambda^l(s,v)+1+\lambda^l(u,t), \tag{15.47}$$

because e is used as a backward edge while augmenting f_l.

But

$$\lambda^l(s,v)\geqslant\lambda^k(s,v)$$

and

$$\lambda^l(u,t)\geqslant\lambda^k(u,t). \tag{15.48}$$

Now we get from (15.46), (15.47), and (15.48)

$$\lambda^l(s,t)\geqslant\lambda^k(s,u)+\lambda^k(u,t)+2$$
$$=\lambda^k(s,t)+2. \quad \blacksquare$$

THEOREM 15.10 (EDMONDS AND KARP). If, in Ford-Fulkerson's labeling algorithm, each flow augmentation is done along a shortest augmenting path, then a maximum flow can be obtained after no more than $m(n+2)/2$ augmentations, where m is the number of edges and n is the number of vertices in the transport network.

Proof

Consider any edge e directed from u to v. Consider the sequence of flows $f_{k_1}, f_{k_2}, \ldots,$ where $k_1<k_2<\ldots,$ such that e is used as a forward edge while augmenting f_{k_i} and is a bottleneck. By Lemma 15.7 there exists a sequence $l_1, l_2, \ldots,$ such that

$$k_1<l_1<k_2<l_2<k_3<\cdots,$$

and e is used as a backward edge while augmenting f_{l_i}.

By Lemma 15.8,

$$\lambda^{k_i}(s,t)+2\leqslant\lambda^{l_i}(s,t)$$

and

$$\lambda^{l_i}(s,t)+2\leqslant\lambda^{k_{i+1}}(s,t).$$

So

$$\lambda^{k_1}(s,t)+4(j-1)\leqslant\lambda^{k_j}(s,t).$$

Since

$$\lambda^{k_j}(s,t)\leqslant n-1$$

and

$$\lambda^{k_1}(s,t) \geqslant 1,$$

we get

$$1+4(j-1) \leqslant n-1$$

or

$$j \leqslant \frac{n+2}{4}.$$

Thus e can be used as a bottleneck in the forward direction at most $(n+2)/4$ times. Similarly it can be used as a bottleneck in the backward direction at most $(n+2)/4$ times. Thus each edge can serve as a bottleneck at most $(n+2)/2$ times. Therefore the total number of augmentations is at most $m(n+2)/2$. ■

In the proof of the above theorem no assumption has been made about the nature of the capacities, except that they are nonnegative. Thus it is clear from this theorem that if the "first-labeled first-scanned" principle is used, then Algorithm 15.9 will terminate after a finite number of augmentations for any real nonnegative capacities. Since an augmenting path can be found in $O(m)$ steps, it follows from Theorem 15.10 that Algorithm 15.9 is of complexity $O(m^2n)$.

Zadeh [15.47] has characterized a class of networks for which $O(n^3)$ augmentations are necessary when each augmentation is done along a shortest augmenting path. Thus the upper bound in Theorem 15.10 cannot be improved upon except for a linear scale factor.

Dinic [15.48] has shown how to improve the efficiency of Algorithm 15.9 by finding in one application of the labeling procedure all shortest augmentations. The idea here is similar to what Hopcroft and Karp [15.36] have used in constructing a maximum matching in a bipartite graph. Dinic's algorithm is of complexity $O(mn^2)$.

Recently Karzanov [15.49] has given a $O(n^3)$ algorithm. Subsequently, Malhotra, Pramodh Kumar, and Maheswari [15.50] have given a simpler $O(n^3)$ algorithm. This algorithm and Dinic's algorithm are both described in Even [15.51].

Even and Tarjan [15.37] have studied the complexity of Dinic's algorithm for special classes of transport networks. As we mentioned earlier in Section 15.5, they have used this result to show that a maximum matching in a bipartite graph can be constructed in $O(n^{5/2})$ time.

15.7.4 Menger's Theorems Revisited

In this section we prove Menger's theorems for both directed and undirected graphs (Theorems 15.12 through 15.15) using the max-flow min-cut theorem. We may recall that we had stated earlier Menger's theorems, without proofs, for undirected graphs (see Theorems 8.9 and 8.12). In the following discussions, for the sake of generality, we shall permit the source to have nonzero in-degree and the sink to have nonzero out-degree.

Our proof of Menger's theorems is based on the following result.

THEOREM 15.11. Let N be a transport network with source s and sink t and in which each edge has unit capacity. Then

1. The value of a maximum flow in N is equal to the maximum number r of edge-disjoint directed s–t paths in N.
2. The capacity of a minimum cut in N is equal to the minimum number q of edges whose removal destroys all directed s–t paths in N.

Proof

1. Let f^* be a maximum flow in N and let N^* be the directed graph obtained by removing from N all its f^*-zero edges. Since the capacity of every edge is unity, it is clear that $f^*(e)=1$ for every edge e in N^*. Thus

   a. $d^+_{N^*}(s)-d^-_{N^*}(s)=\text{val}(f^*)=d^-_{N^*}(t)-d^+_{N^*}(t)$.

   b. $d^+_{N^*}(v)=d^-_{N^*}(v)$ for all $v\neq s, t$.

   Here $d^+_{N^*}(x)$ and $d^-_{N^*}(x)$ denote, respectively, the out-degree and the in-degree in the graph N^* of the vertex x.
   Therefore (Exercise 5.8) there are val(f^*) edge-disjoint directed s–t paths in N^* and hence also in N. Thus

   $$\text{val}(f^*)\leqslant r. \tag{15.49}$$

   Now let $\{P_1, P_2, \ldots, P_r\}$ be a collection of r edge-disjoint directed s–t paths in N. Define a flow f such that

   $$f(e)=\begin{cases} 1, & \text{if } e \text{ is in some } P_i. \\ 0, & \text{otherwise.} \end{cases}$$

   Clearly,

   $$\text{val}(f)=r.$$

   Since f^* is a maximum flow, we have

   $$\text{val}(f^*)\geqslant r. \tag{15.50}$$

   Combining (15.49) and (15.50), we get

   $$\text{val}(f^*)=r.$$

2. Let $K^* = \langle S, \bar{S} \rangle$ be a minimum s–t cut in N. Suppose we remove from N the set of edges $(S, \bar{S})$. Then in the resulting directed graph there will be no directed s–t paths. So

$$\text{cap}(K^*) = |(S, \bar{S})| \geq q. \tag{15.51}$$

Now let Z denote a set of q edges whose removal destroys all directed s–t paths in N, and let S denote the set of all vertices reachable from s by a directed path containing no edge from Z. Clearly, $K = \langle S, \bar{S} \rangle$ is an s–t cut in N. Further, $(S, \bar{S}) \subseteq Z$. So,

$$\text{cap}(K^*) \leq \text{cap}(K) = |(S, \bar{S})| \leq |Z| = q. \tag{15.52}$$

Combining (15.51) and (15.52), we get

$$\text{cap}(K^*) = q. \quad \blacksquare$$

THEOREM 15.12. Let s and t be two vertices in a directed graph G. Then the maximum number of edge-disjoint directed s–t paths in G is equal to the minimum number of edges whose removal destroys all directed s–t paths in G.

Proof

From G construct a transport network N with s as the source and t as the sink by assigning unit capacity to each edge of G. The theorem then follows from Theorem 15.11 and the max-flow min-cut theorem. ■

For an undirected graph G, let $D(G)$ denote the directed graph obtained by replacing each edge e of G by a pair of oppositely oriented edges having the same end vertices as e. It can be easily shown that

1. There exists a one-to-one correspondence between the paths in G and the directed paths in $D(G)$; and
2. For any two vertices s and t, the minimum number of edges whose removal destroys all s–t paths in G is equal to the minimum number of edges whose removal destroys all directed s–t paths in $D(G)$.

The undirected version of Theorem 15.12 now follows from the above observations.

THEOREM 15.13. Let s and t be two vertices of an undirected graph G. The maximum number of edge-disjoint s–t paths in G is equal to the minimum number of edges whose removal destroys all s–t paths in G. ■

Vertex analogs of Theorems 15.12 and 15.13 are proved next.

Let s and t be any two nonadjacent vertices in a directed graph $G=(V, E)$. From G construct a directed graph G' as follows:

1. Split vertex $v \in V-\{s,t\}$ into two new vertices v' and v'' and connect them by a directed edge (v', v'').
2. Replace each edge of G having $v \in V-\{s,t\}$ as terminal vertex by a new edge having v' as terminal vertex.
3. Replace each edge of G having $v \in V-\{s,t\}$ as initial vertex by a new edge having v'' as initial vertex.

A graph G and the corresponding graph G' are shown in Fig. 15.22. It is not difficult to prove the following:

1. Each directed s–t path in G' corresponds to a directed s–t path in G which is obtained by contracting all edges of the type (v', v''); conversely, each directed s–t path in G corresponds to a directed s–t path in G' obtained by splitting all the vertices other than s and t of the path.

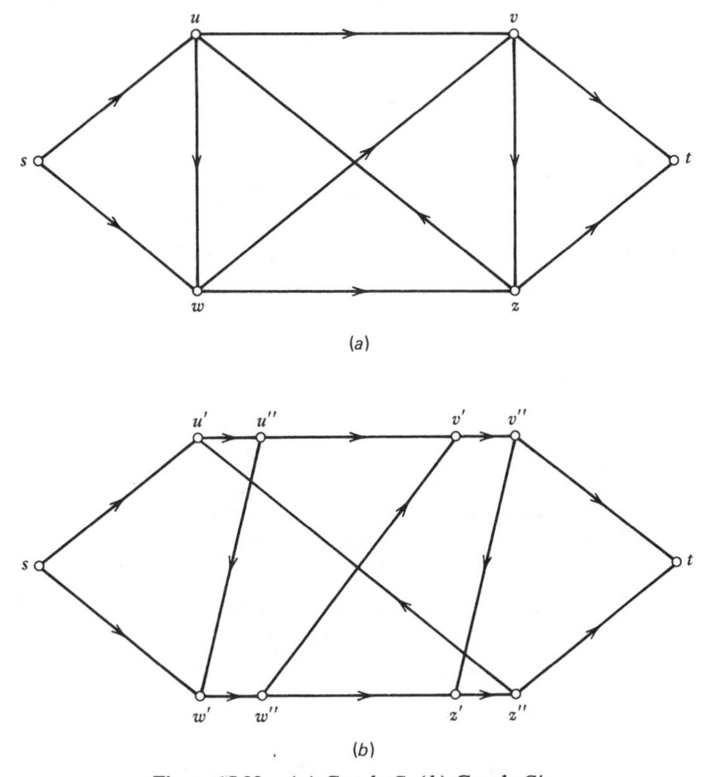

Figure 15.22. (a) Graph G. (b) Graph G'.

2. Two directed s–t paths in G' are edge-disjoint if and only if the corresponding paths in G are vertex-disjoint.
3. The maximum number of edge-disjoint directed s–t paths in G' is equal to the maximum number of vertex-disjoint directed s–t paths in G.
4. The minimum number of edges whose removal destroys all directed s–t paths in G' is equal to the minimum number of vertices whose removal destroys all directed s–t paths in G.

From these observations we get the following vertex analog of Theorem 15.12.

THEOREM 15.14. Let s and t be two nonadjacent vertices in a directed graph G. Then the maximum number of vertex-disjoint directed s–t paths in G is equal to the minimum number of vertices whose removal destroys all directed s–t paths in G. ■

The vertex analog of Theorem 15.13 now follows.

THEOREM 15.15. Let s and t be two nonadjacent vertices in an undirected graph G. Then the maximum number of vertex-disjoint s–t paths in G is equal to the minimum number of vertices whose removal destroys all s–t paths in G.

Proof

Apply Theorem 15.14 to the graph $D(G)$. ■

Standard reference on network flow theory is Ford and Fulkerson [15.1], where the minimum cost flow problem is also discussed. See also Berge [15.46], Berge and Ghouila-Houri [15.52], Christofides [15.8], Frank and Frisch [15.7], Hu [15.53], and Lawler [15.9].

15.8 OPTIMUM BRANCHINGS

Consider a weighted directed graph $G=(V, E)$. Let $w(e)$ be the weight of edge e. The weight of a subgraph of G is defined to be equal to the sum of the weights of all the edges in the subgraph.

A subgraph G_s of G is a *branching* in G if G_s has no directed circuits and the in-degree of each vertex of G_s is at most 1. Clearly each component of G_s is a directed tree. A branching of maximum weight is called an *optimum branching*.

In this section we discuss an algorithm due to Edmonds [15.54] for computing an optimum branching of G. Our discussion here is based on Karp [15.55].

An edge $e=(i, j)$ directed from vertex i to vertex j is *critical* if

1. $w(e)>0$; and
2. $w(e) \geqslant w(e')$ for every edge $e'=(k, j)$ incident into j.

A spanning subgraph H of G is a *critical subgraph* of G if

1. Every edge of H is critical; and
2. The in-degree of every vertex of H is at most 1.

A directed graph G and a critical subgraph H of G are shown in Fig. 15.23.
It is easy to see that

1. Each component of a critical subgraph contains at most one circuit, and such a circuit will be a directed circuit; and
2. A critical subgraph with no circuits is an optimum branching of G.

Consider a branching B. Let $e=(i, j)$ be an edge not in B, and let e' be the edge of B incident into vertex j. Then e is *eligible* relative to B if

$$B'=(B \cup e)-e'$$

is a branching.

For example, the edges $\{e_1, e_2, e_3, e_4, e_7, e_8\}$ form a branching B of the graph in Fig. 15.23. The edge e_6, not in B, is eligible relative to B since

$$(B \cup e_6)-e_7$$

is a branching of this graph.

The following two lemmas are easy to prove, and they lead to Theorem 15.16 which forms the basis of Karp' proof of the correctness of Edmonds' algorithm. In the following, the edge set of a subgraph H will also be denoted by H.

LEMMA 15.10. Let B be a branching, and let $e=(i, j)$ be an edge not in B. Then e is eligible relative to B if and only if in B there is no directed path from j to i. ■

LEMMA 15.11. Let B be a branching and let C be a directed circuit such that no edge of $C-B$ is eligible relative to B. Then $|C-B|=1$. ■

THEOREM 15.16. Let H be a critical subgraph. Then there exists an optimum branching B such that, for every directed circuit C in H, $|C-B|=1$.

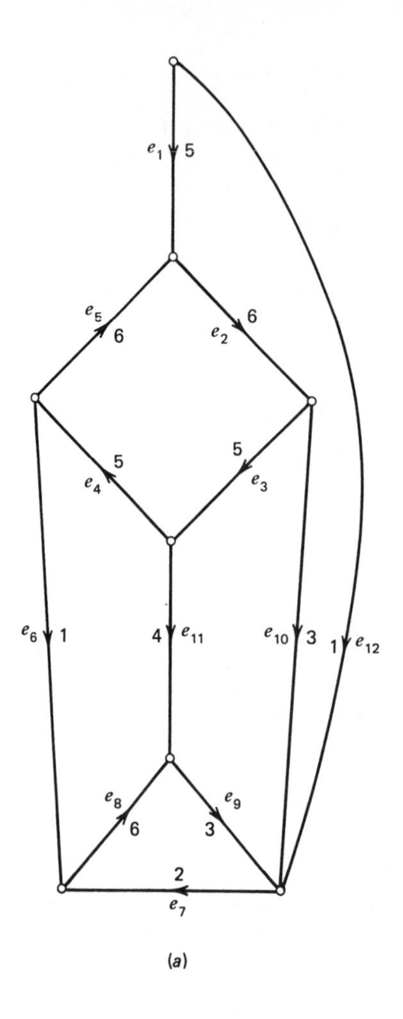

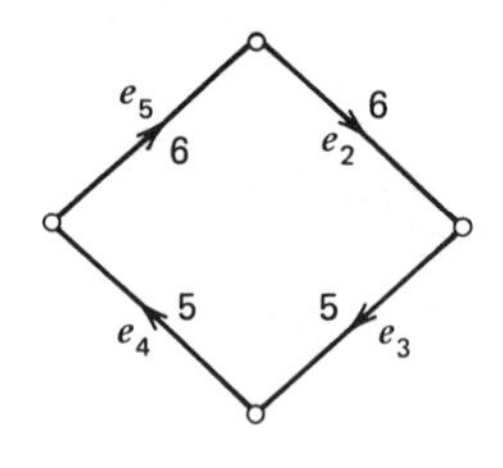

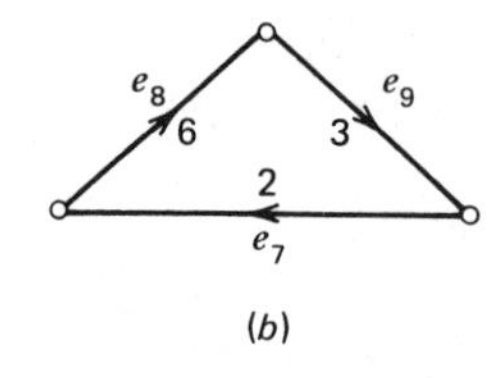

Figure 15.23. (*a*) A directed graph *G*. (*b*) A critical subgraph of *G*.

Proof

Let B be an optimum branching which, among all optimum branchings, contains a maximum number of edges of the critical subgraph.

Consider any edge $e \in H-B$. Let e be incident into vertex j, and let e' be the edge of B incident into j. If e were eligible, then

$$(B \cup e)-e'$$

would also be an optimum branching, containing a larger number of edges of H than B does; a contradiction. Thus no edge of $H-B$ is eligible relative to B. So, by Lemma 15.11, for each circuit C in H, $|C-B|=1$. ■

Let $C_1, C_2, \ldots, C_k$ be the directed circuits in H. Note that no two circuits of H can have a common edge. In other words, these circuits are edge-disjoint. For each C_i, let e_i^0 be an edge of minimum weight in C_i.

Corollary 15.16.1 There exists an optimum branching B such that

1. $|C_i-B|=1$, $i=1,2,\ldots,k$; and
2. If no edge of $B-C_i$ is incident into a vertex in C_i, $i=1,2,\ldots,k$, then

$$C_i-B=e_i^0. \tag{15.53}$$

Proof

Let, among all optimum branchings satisfying (1), B be a branching containing a minimum number of edges from the set $\{e_1^0, e_2^0, \ldots, e_k^0\}$. We now show that B satisfies (2).

If not, suppose, for some i, $e_i^0 \in B$, but no edge of $B-C_i$ is incident into a vertex in C_i. Let $e=C_i-B$. Then $(B-e_i^0)\cup e$ is clearly an optimum branching which satisfies (1), but has fewer edges than B from the set $\{e_1^0, e_2^0, \ldots, e_k^0\}$. This is a contradiction. ■

The above result is very crucial in the development of Edmonds' algorithm. It suggests that we can restrict our search for optimum branchings to those which satisfy (15.53).

Now we construct, from the given graph G, a simpler graph G'. We also show how to construct from an optimum branching of G' an optimum branching of G which satisfies (15.53).

As before, let H be the critical subgraph of G and let $C_1, C_2, \ldots, C_k$ be the directed circuits in H. The graph G' is constructed by contracting all the edges in each C_i, $i=1,2,\ldots,k$. In G', vertices of each circuit C_i are represented by a single vertex a_i, called a *pseudo-vertex*. The weights of the edges of G' are the same as those of G, except for the weights of the edges incident into the pseudo-vertices. These weights are modified as follows.

Let $e=(i, j)$ be an edge of G such that j is a vertex of some circuit C_r and i is not in C_r. Then in G', e is incident into the pseudo-vertex a_r. Define $\tilde{e}$ as the unique edge in C_r which is incident into vertex j. Then in G' the weight of e, denoted by $w'(e)$, is given by

$$w'(e)=w(e)-w(\tilde{e})+w(e_r^0). \tag{15.54}$$

For example, consider the edge e_1 incident into the directed circuit $\{e_2, e_3, e_4, e_5\}$ of the critical subgraph of the graph G of Fig. 15.23. Then $\tilde{e}_1 = e_5$, and the weight of e_1 in G' is given by

$$\begin{aligned} w'(e_1) &= w(e_1)-w(e_5)+w(e_4) \\ &= 5-6+5 \\ &= 4. \end{aligned}$$

Note that e_4 is a minimum weight edge in the circuit $\{e_2, e_3, e_4, e_5\}$.

Let E and E', respectively, denote the edge sets of G and G'. We now show how to construct from a branching B' of G' a branching B of G which satisfies (15.53) and vice versa.

For any branching B of G which satisfies (15.53) it is easy to see that

$$B'=B \cap E' \tag{15.55}$$

is a branching of G'. Further B' as defined above is unique for a given B.

Next consider a branching B' of G'. For each C_i, let us define C_i' as follows:

1. If the in-degree in B' of a pseudo-vertex a_i is zero, then

$$C_i' = C_i - e_i^0.$$

2. If the in-degree in B' of a_i is nonzero, and e is the edge of B' incident into a_i, then

$$C_i' = C_i - \tilde{e}.$$

Then it is easy to see that

$$B = B' \bigcup_{i=1}^{k} C_i' \tag{15.56}$$

is a branching of G which satisfies (15.53). Further B as defined above is unique for a given B'.

Thus we conclude that there is a one-to-one correspondence between the set of branchings of G which satisfy (15.53) and the set of branchings of G'.

Furthermore, the weights of the corresponding branchings B and B' satisfy

$$w(B)-w(B')=\sum_{i=1}^{k} w(C_i)-\sum_{i=1}^{k} w(e_i^0). \tag{15.57}$$

This property of B and B' implies that if B is an optimum branching of G which satisfies (15.53), then B' is an optimum branching of G' and vice versa. Thus we have proved the following theorem.

THEOREM 15.17. There exists a one-to-one correspondence between the set of all optimum branchings in G which satisfy (15.53) and the set of all optimum branchings in G'. ■

Edmonds' algorithm for constructing an optimum branching is based on the above theorem and is as follows:

Algorithm 15.10 Optimum Branching (Edmonds)

S1. From the given graph $G=G_0$ construct a sequence of graphs $G_0, G_1, G_2, \ldots, G_k$, where

1. G_k is the first graph in the sequence whose critical subgraph is acyclic; and
2. G_i, $1 \leqslant i \leqslant k$, is obtained from G_{i-1} by contracting the circuits in the critical subgraph H_{i-1} of G_{i-1} and altering the weights as in (15.54).

S2. Since H_k is acyclic, it is an optimum branching in G_k. Let $B_k = H_k$. Construct the sequence $B_{k-1}, B_{k-2}, \ldots, B_0$, where

1. B_i, $0 \leqslant i \leqslant k-1$, is an optimum branching of G_i; and
2. B_i, for $i \geqslant 0$, is constructed by expanding, as in (15.56), pseudo-vertices in B_{i+1}. ■

As an example let G_0 be the graph in Fig. 15.23*a*, and let H_0 be the graph in Fig. 15.23*b*. H_0 is the critical subgraph of G_0. After contracting the edges of the circuits in H_0 and modifying the weights, we obtain the graph G_1 shown in Fig. 15.24*a*. The critical subgraph H_1 of G_1 is shown in Fig. 15.24*b*. H_1 is acyclic. So it is an optimum branching of G_1. An optimum branching of G_0 is obtained from H_1 by expanding the pseudo-vertices a_1 and a_2 (which correspond to the two directed circuits in H_0), and it is shown in Fig. 15.24*c*.

Tarjan [15.56] gives a $O(m \log n)$ implementation of Edmonds' algorithm, where m is the number of edges and n is the number of vertices. See also Chu and Liu [15.57] and Bock [15.58] who have independently discovered Edmonds' algorithm.

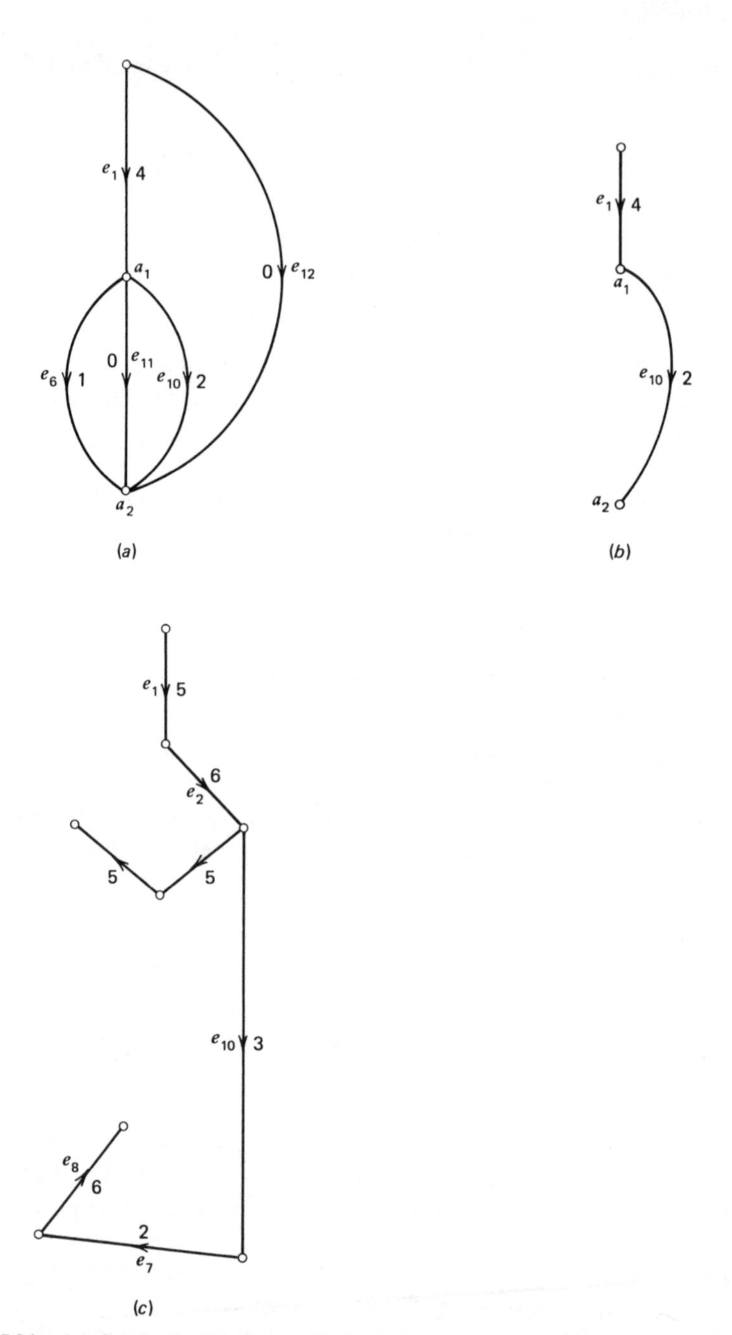

Figure 15.24. (*a*) Graph G_1. (*b*) H_1, a critical subgraph of G_1. (*c*) An optimum branching of graph G of Fig. 15.23*a*.

15.9 FURTHER READING

Many interesting algorithms which concern graph problems have been reported in the literature. Some of these are listed below with appropriate references which lead to related papers.

Isomorphism Corneil and Gotlieb [15.59], Weinberg [15.60], Hopcroft and Tarjan [15.61], and Hopcroft and Wong [15.62].

Planarity Hopcroft and Tarjan [15.63], Deo [15.64], Reingold, Nievergelt, and Deo [15.26], and Even [15.51].

Connectivity Hopcroft and Tarjan [15.65], Kleitman [15.66], Even and Tarjan [15.37], and Even [15.67], [15.51].

Chromatic Number Corneil and Graham [15.68] and McDiarmid [15.69].

Graph Coloring Welsh and Powell [15.70], Wood [15.71], Matula, Marble, and Isaacson [15.72], Christofides [15.8], and Brélaz [15.73].

Independent Sets and Stability Number Paull and Unger [15.74], Bron and Kerbosch [15.75], Akkoyunlû [15.76], Tsukiyama, Ide, Ariyoshi, and Shirakawa [15.77], Tarjan and Trojanowski [15.78], and Chvátal [15.79].

Minimum Spanning Trees Kruskal [15.80], Prim [15.81], Kerschenbaum and Van Slyke [15.82], Yao [15.83], Cheriton and Tarjan [15.84], Spira and Pan [15.10], Gabow [15.85]. See also Section 10.9 in Part I of this book.

Augmentation Problems Eswaran and Tarjan [15.86] and Rosenthal and Goldner [15.87].

Parallel Computations Reghbati and Corneil [15.88].

Branchings Edmonds [15.89], Fulkerson and Harding [15.90], Tarjan [15.91], and Lovász [15.92].

Network Optimization Golden and Magnanti [15.93].

Scheduling Fujii, Kasami, and Ninomiya [15.94], Coffman and Graham [15.95], Coffman and Denning [15.96], Sethi [15.97], [15.98], Ullman [15.99], and Lam and Sethi [15.100].

Graph-Theoretic Study of Linear Equations Rose [15.101], Ohtsuki [15.102], Rose, Tarjan, and Lueker [15.103], and Tarjan [15.104].

Approximation Algorithms Horowitz and Sahni [15.105] and Garey and Johnson [15.106].

15.10 EXERCISES

15.1 Prove the validity of the following modification of Dijkstra's algorithm for finding the shortest paths between a specified vertex s and all the other vertices in a weighted directed graph G which has no directed circuits of negative lengths. In the following, E denotes the edge set of G, and $w(u,v)$ denotes the length of the directed edge (u,v).

S1. Set $\sigma(s)=0$, and set $\sigma(u)=\infty$ for $u\neq s$.

S2. Set $S=\{s\}$.

S3. If $S=\varnothing$, HALT; otherwise select u^* such that $u^*\in S$ and $\sigma(u^*)=\min_{u\in S}\{\sigma(u)\}$.

S4. For each v such that $(u^*,v)\in E$, set $\sigma(v)=\min\{\sigma(v),\sigma(u^*)+w(u^*,v)\}$. If this process decreases $\sigma(v)$, set $S=S\cup\{v\}$.

S5. Set $S=S-\{u^*\}$, and go to step S3.

(Edmonds and Karp [15.4].)

15.2 Prove the validity of the following algorithm due to Dantzig [15.107] for finding the shortest paths between all pairs of vertices in a weighted graph which has no directed circuits of negative length. Here $W^k(i,j)$ is the distance from i to j where $1\leqslant i,j\leqslant k$ and no vertices higher than k are used on the path. Here $w(i,j)=W^1(i,j)$ is the length of the edge (i,j) and $w(i,j)=\infty$ if no such edge exists. Also $w(i,i)=0$ for all i.

S1. Set $k=2$.

S2. For $1\leqslant i<k$ do

$$W^k(i,k)=\min_{1\leqslant j<k}\{w(i,k),W^{k-1}(i,j)+w(j,k)\}$$

$$W^k(k,i)=\min_{1\leqslant j<k}\{w(k,i),w(k,j)+W^{k-1}(j,i)\}.$$

S3. For $1\leqslant i,j<k$ do

$$W^k(i,j)=\min\{W^{k-1}(i,j),W^k(i,k)+W^k(k,j)\}.$$

S4. If $k=n$, STOP; otherwise set $k=k+1$ and go to step S2.

Also show how negative-length circuits can be detected while carrying out the above algorithm.

15.3. Using topological sorting, design an algorithm to find the shortest path from a vertex s to all the other vertices in an acyclic weighted directed graph.

15.4 Suppose the shortest path from i to j is not unique. Which path is chosen by Floyd's algorithm (Algorithm 15.2)?

15.5 Prove that if a path length vector $(l_1, l_2, \ldots, l_n)$ satisfies the characteristic sum condition and if $l_1 \leqslant l_2 \leqslant \cdots \leqslant l_n$ and $l_j < l_{j+1} = l_{j+2} = \cdots = l_n$, then the vector $(l_1, l_2, \ldots, l_j, l_{j+1} - 1, l_{j+d+1}, l_{j+d+2}, \ldots, l_n)$ also satisfies the characteristic sum condition and the path lengths are still in a nondecreasing order. (Here d is the same as in (15.17).)

15.6 Using the result of Exercise 15.5 show how to construct an m-ary tree in which the order of the leaves from left to right is the same as in the path length vector.

15.7 Find a minimum cost binary search tree for the weights $(1,1,2,3,3,4,4)$.

15.8 Prove that replacing step E2 in Algorithm 15.5 by step

"**E2′**. (Choose an edge.) Choose an edge .... If no such edge exists, go to step E1."

does not affect the validity of the algorithm for finding a maximum matching. (Edmonds [15.28].)

15.9 Show that if there exists no directed s–t path in a transport network N, then the value of a maximum flow and the capacity of a minimum cut are both zero.

15.10 If $\langle S, \bar{S} \rangle$ and $\langle T, \bar{T} \rangle$ are minimum cuts in a transport network N, show that $\langle S \cup T, \overline{S \cup T} \rangle$ and $\langle S \cap T, \overline{S \cap T} \rangle$ are also minimum cuts in N.

15.11 Show that in any transport network with integer capacities, there is a maximum flow f such that $f(e)$ is an integer for every edge e in N.

15.12 Consider a transport network in which each vertex $v \neq s, t$ is associated with a nonnegative integer $m(v)$. Show how a maximum flow in which the flow into each vertex $v \neq s, t$ is not greater than $m(v)$ can be found by applying the labeling algorithm to a modified network. (See the construction used in establishing Theorem 15.14.)

15.13 Consider a transport network N in which a lower bound is also specified for flow in each edge.

(a) Find a necessary and sufficient condition for the existence of a flow in N.

(b) Modify the labeling algorithm to find a maximum flow in N. (See Even [15.108].)

15.14 Prove that in a transport network with a lower bound for flow in each edge, there exists a flow if and only if every edge e is in a directed circuit or is in a directed path from s to t or is in a directed path from t to s.

15.15 Describe a method for locating in a transport network N an edge e which has the property that increasing its capacity increases the maximum flow in N.

15.16 An optimum branching may not be an optimum spanning directed tree. Show how Edmonds' algorithm can be modified to find an optimum spanning directed tree.

15.17 (a) Prove the validity of the following algorithm due to Prim [15.81] for finding a minimum weight spanning tree in a connected weighted graph $G=(V, E)$:
Select any vertex v in G. Among all the edges incident on v, select an edge e_1 with minimum weight. Contract e_1 and let G^1 be the resulting graph. Repeat this process on G^1 till e_{n-1} is defined. The edges $e_1, e_2, \ldots, e_{n-1}$ form a minimum weight spanning tree of G.

(b) Design a $O(n^2)$ implementation of Prim's algorithm.

15.18 Prove Hall's theorem (Theorem 8.13) using the max-flow min-cut theorem (Theorem 15.9) and vice versa.

15.19 Let $G=(X, Y)$ be a bipartite graph in which there is a complete matching of X into Y. Then it is known that there exists a vertex $v \in X$ such that for every edge e incident on v there exists a complete matching containing e. (See Exercise 8.16.) The straightforward algorithm to locate such a vertex v will require $O(m\,n^{5/2})$ time. Try to construct an algorithm with a better complexity.

15.11 REFERENCES

15.1 L. R. Ford and D. R. Fulkerson, *Flows in Networks*, Princeton Univ. Press, Princeton, N.J., 1962.

15.2 E. W. Dijkstra, "A Note on Two Problems in Connexion with Graphs," *Numerische Math.*, Vol. 1, 269–271 (1959).

15.3 D. B. Johnson, "A Note on Dijkstra's Shortest Path Algorithm," *J. ACM*, Vol. 20, 385–388 (1973).

15.4 J. Edmonds and R. M. Karp, "Theoretical Improvements in Algorithmic Efficiency for Network Flow Problems," *J. ACM*, Vol. 19, 248–264 (1972).

15.5 S. E. Dreyfus, "An Appraisal of Some Shortest-Path Algorithms," *Operations Research*, Vol. 17, 395–412 (1969).

15.6 T. C. Hu, "A Decomposition Algorithm for Shortest Paths in a Network," *Operations Research*, Vol. 16, 91–102 (1968).

15.7 H. Frank and I. T. Frisch, *Communication, Transmission and Transportation Networks*, Addison-Wesley, Reading, Mass., 1971.

15.8 N. Christofides, *Graph Theory: An Algorithmic Approach*, Academic Press, New York, 1975.

15.9 E. L. Lawler, *Combinatorial Optimization: Networks and Matroids*, Holt, Rinehart and Winston, New York, 1976.

15.10 P. M. Spira and A. Pan, "On Finding and Updating Spanning Trees and Shortest Paths," *SIAM J. Comput.*, Vol. 4, 375–380 (1975).

15.11 D. B. Johnson, "Efficient Algorithms for Shortest Paths in Sparse Networks," *J. ACM*, Vol. 24, 1–13 (1977).

15.12 R. A. Wagner, "A Shortest Path Algorithm for Edge-Sparse Graphs," *J. ACM*, Vol. 23, 50–57 (1976).

15.13 R. W. Floyd, "Algorithm 97: Shortest Path," *Comm. ACM*, Vol. 5, 345 (1962).

15.14 Y. Tabourier, "All Shortest Distances in a Graph: An Improvement to Dantzig's Inductive Algorithm," *Discrete Math.*, Vol. 4, 83–87 (1973).

15.15 J. Y. Yen, "Finding the Lengths of All Shortest Paths in N-Node, Non-Negative Distance Complete Networks Using $N^3/2$ Additions and N^3 Comparisons," *J. ACM*, Vol. 19, 423–424 (1972).

15.16 T. A. Williams and G. P. White, "A Note on Yen's Algorithm for Finding the Length of All Shortest Paths in N-Node Non-Negative Distance Networks," *J. ACM*, Vol. 20, 389–390 (1973).

15.17 A. R. Pierce, "Bibliography on Algorithms for Shortest Path, Shortest Spanning Tree and Related Circuit Routing Problems (1956–1974)," *Networks*, Vol. 5, 129–149 (1975).

15.18 D. A. Huffman, "A Method for the Construction of Minimum Redundancy Codes," *Proc. IRE*, Vol. 40, 1098–1101 (1952).

15.19 E. N. Gilbert and E. F. Moore, "Variable-Length Binary Encodings," *Bell Sys. Tech. J.*, Vol. 38, 933–968 (1959).

15.20 D. E. Knuth, "Optimum Binary Search Trees," *Acta Informatica*, Vol. 1, 14–25 (1971).

15.21 A. Itai, "Optimal Alphabetic Trees," *SIAM J. Comput.*, Vol. 5, 9–18 (1976).

15.22 T. C. Hu and A. C. Tucker, "Optimal Computer Search Trees and Variable-Length Alphabetical Codes," *SIAM J. Appl. Math.*, Vol. 21, 514–532 (1971).

15.23 T. C. Hu, "A New Proof of the T-C Algorithm," *SIAM J. Appl. Math.*, Vol. 25, 83–94 (1973).

15.24 D. E. Knuth, *The Art of Computer Programming, Vol. 3: Sorting and Searching*, Addison-Wesley, Reading, Mass., 1973.

15.25 A. M. Garsia and M. L. Wachs, "A New Algorithm for Minimum Cost Binary Trees," *SIAM J. Comput.*, Vol. 6, 622–642 (1977).

15.26 E. M. Reingold, J. Nievergelt, and N. Deo, *Combinatorial Algorithms: Theory and Practice*, Prentice-Hall, Englewood Cliffs, N.J., 1977.

15.27 M. Miyakawa, T. Yuba, Y. Sugito, and M. Hoshi, "Optimum Sequence Trees," *SIAM J. Comput.*, Vol. 6, 201–234 (1977).

15.28 J. Edmonds, "Paths, Trees and Flowers," *Canad. J. Math.*, Vol. 17, 449–467 (1965).

15.29 H. N. Gabow, "An Efficient Implementation of Edmonds' Algorithm for Maximum Matching on Graphs," *J. ACM*, Vol. 23, 221–234 (1976).

15.30 M. L. Balinski, "Labelling to Obtain a Maximum Matching," in *Combinatorial Mathematics and Its Applications*, (R. C. Bose and T. A. Dowling, Eds.), Univ. North Carolina Press, Chappel Hill, N.C., 1967, pp. 585–602.

15.31 D. Witzgall and C. T. Zahn, Jr., "Modification of Edmonds' Algorithm for Maximum Matching of Graphs," *J. Res. Nat. Bur. Std.*, Vol. 69B, 91–98 (1965).

15.32 T. Kameda and I. Munro, "A $O(|V|\cdot|E|)$ Algorithm for Maximum Matching of Graphs," *Computing*, Vol. 12, 91–98 (1974).

15.33 J. Edmonds, "Maximum Matching and a Polyhedron with 0, 1 Vertices," *J. Res. Nat. Bur. Std.*, Vol. 69B, 125–130 (1965).

15.34 H. Gabow, "An Efficient Implementation of Edmonds' Maximum Matching Algorithm," Tech. Rep. 31, *Stanford Univ. Comp. Science Dept.*, 1972.

15.35 J. A. Bondy and U. S. R. Murty, *Graph Theory with Applications*, Macmillan, London, 1976.

15.36 J. E. Hopcroft and R. M. Karp, "An $n^{5/2}$ Algorithm for Maximum Matching in Bipartite Graphs," *SIAM J. Comput.*, Vol. 2, 225–231 (1973).

15.37 S. Even and R.E. Tarjan, "Network Flow and Testing Graph Connectivity," *SIAM J. Comput.*, Vol. 4, 507–518 (1975).

15.38 S. Even and O. Kariv, "An $O(n^{5/2})$ Algorithm for Maximum Matching in General Graphs," *Proc. 16th Annual Symp. on Foundations of Comp. Science*, IEEE, 1975, pp. 100–112.

15.39 A. Itai, M. Rodeh, and S. L. Tanimoto, "Some Matching Problems for Bipartite Graphs," *J. ACM*, Vol. 25, 517–525 (1978).

15.40 H. W. Kuhn, "The Hungarian Method for the Assignment Problem," *Naval Res. Logist. Quart.*, Vol. 2, 83–97 (1955).

15.41 J. Munkres, "Algorithms for the Assignment and Transportation Problems," *J. SIAM*, Vol. 5, 32–38 (1957).

15.42 N. Megido and A. Tamir, "An $O(N \log N)$ Algorithm for a Class of Matching Problems," *SIAM J. Comput.*, Vol. 7, 154–157 (1978).

15.43 S. Even, A. Itai, and A. Shamir, "On the Complexity of Time-Table and Multicommodity Flow Problems," *SIAM J. Comput.*, Vol. 5, 691–703 (1976).

15.44 L. R. Ford and D. R. Fulkerson, "Maximal Flow through a Network," *Canad. J. Math.*, Vol. 8, 399–404 (1956).

15.45 P. Elias, A. Feinstein, and C. E. Shannon, "A Note on the Maximum Flow through a Network," *IRE Trans. Information Theory*, Vol. IT-2, 117–119 (1956).

15.46 C. Berge, *Graphs and Hypergraphs*, North-Holland, Amsterdam, 1973.

15.47 N. Zadeh, "Theoretical Efficiency of the Edmonds-Karp Algorithm for Computing Maximal Flows," *J. ACM*, Vol. 19, 184–192 (1972).

15.48 E. A. Dinic, "Algorithm for the Solution of a Problem of Maximum Flow in a Network with Power Estimation," *Soviet Math. Dokl.*, Vol. 11, 1277–1280 (1970).

15.49 A. V. Karzanov, "Determining the Maximal Flow in a Network by the Method of Preflows," *Soviet Math. Dokl.*, Vol. 15, 434–437 (1974).

15.50 V. M. Malhotra, M. Pramodh Kumar, and S. N. Maheswari, "An $O(V^3)$ Algorithm for Maximum Flows in Networks," *Information Processing Lett.*, Vol. 7, 277–278 (1978).

15.51 S. Even, *Graph Algorithms*, Computer Science Press, Potomac, Md., 1979.

15.52 C. Berge and A. Ghouila-Houri, *Programming, Games and Transportation Networks*, Wiley, New York, 1962.

15.53 T. C. Hu, *Integer Programming and Network Flows*, Addison-Wesley, Reading, Mass., 1969.

15.54 J. Edmonds, "Optimum Branchings," *J. Res. Nat. Bur. Std.*, Vol. 71B, 233–240 (1967).

15.55 R. M. Karp, "A Simple Derivation of Edmonds' Algorithm for Optimum Branchings," *Networks*, Vol. 1, 265–272 (1972).

15.56 R. E. Tarjan. "Finding Optimum Branchings," *Networks*, Vol. 7, 25–35 (1977).

15.57 Y. Chu and T. Liu, "On the Shortest Arborescence of a Directed Graph," *Scientia Sinica [Peking]*, Vol. 4, 1396–1400, (1965); *Math. Rev.*, Vol. 33, #1245 (D. W. Walkup).

15.58 F. C. Bock, "An Algorithm to Construct a Minimum Directed Spanning Tree in a Directed Network," in *Developments in Operations Research*, (B. Avi-Itzak, Ed.), Gordon and Breach, New York, 1971, pp. 29–44.

15.59 D. G. Corneil and C. C. Gotlieb, "An Efficient Algorithm for Graph Isomorphism," *J. ACM*, Vol. 17, 51–64 (1970).

15.60 L. Weinberg, "A Simple and Efficient Algorithm for Determining Isomorphism of Planar Triply Connected Graphs," *IEEE Trans. Circuit Theory*, Vol. CT-13, 142–148 (1966).

15.61 J. E. Hopcroft and R. E. Tarjan, "A $V \log V$ Algorithm for Isomorphism of Triconnected Planar Graphs," *J. Comput. Syst. Sci.*, Vol. 7, 323–331 (1973).

15.62 J. E. Hopcroft and J. K. Wong, "Linear Time Algorithm for Isomorphism on Planar Graphs," *Proc. 6th Annual ACM Symp. on Theory of Computing*, 1974, pp. 172–184.

15.63 J. E. Hopcroft and R. E. Tarjan, "Efficient Planarity Testing," *J. ACM*, Vol. 21, 549–568 (1974).

15.64 N. Deo, "Note on Hopcroft and Tarjan's Planarity Algorithm," *J. ACM*, Vol. 23, 74–75 (1976).

15.65 J. E. Hopcroft and R. E. Tarjan, "Dividing a Graph into Triconnected Components," *SIAM J. Comput.*, Vol. 2, 135–158 (1973).

15.66 D. J. Kleitman, "Methods for Investigating the Connectivity of Large Graphs," *IEEE Trans. Circuit Theory*, Vol. CT-16, 232–233 (1969).

15.67 S. Even, "An Algorithm for Determining whether the Connectivity of a Graph is at Least k," *SIAM J. Comput.*, Vol. 4, 393–396 (1975).

15.68 D. G. Corneil and B. Graham, "An Algorithm for Determining the Chromatic Number of a Graph," *SIAM J. Comput.*, Vol. 2, 311–318 (1973).

15.69 C. McDiarmid, "Determining the Chromatic Number of a Graph," *SIAM J. Comput.*, Vol. 8, 1–14 (1979).

15.70 D. J. A. Welsh and M. B. Powell, "An Upper Bound for the Chromatic Number of a Graph and Its Applications to Timetabling Problems," *The Computer J.*, Vol. 10, 85–86 (1967).

15.71 D. C. Wood, "A Technique for Colouring a Graph Applicable to Large Scale Timetabling Problems," *The Computer J.*, Vol. 10, 317–319 (1969).

15.72 D. Matula, G. Marble, and J. Isaacson, "Graph Colouring Algorithms," in *Graph Theory and Computing*, (R. C. Read, Ed.), Academic Press, New York, 1972, pp. 109–122.

15.73 D. Brélaz, "New Methods to Color the Vertices of a Graph," *Comm. ACM*, Vol. 22, 251–256 (1979).

15.74 M. C. Paull and S. H. Unger, "Minimizing the Number of States in Incompletely Specified Sequential Switching Functions," *IRE Trans. Elect. Comput.*, Vol. EC-8, 356–357 (1959).

15.75 C. Bron and J. Kerbosch, "Finding All Cliques of an Undirected Graph—Algorithm 457," *Comm. ACM*, Vol. 16, 575–577 (1973).

15.76 E. A. Akkoyunlû, "The Enumeration of Maximal Cliques of Large Graphs," *SIAM J. Comput.*, Vol. 2, 1–6 (1973).

15.77 S. Tsukiyama, M. Ide, H. Ariyoshi, and I. Shirakawa, "A New Algorithm for Generating All the Maximal Independent Sets," *SIAM J. Comput.*, Vol. 6, 505–517 (1977).

15.78 R. E. Tarjan and A. E. Trojanowski, "Finding a Maximum Independent Set," *SIAM J. Comput.*, Vol. 6, 537–546 (1977).

15.79 V. Chvátal, "Determining the Stability Number of a Graph," *SIAM J. Comput.*, Vol. 6, 643–662 (1977).

15.80 J. B. Kruskal, Jr., "On the Shortest Spanning Subtree of a Graph and the Travelling Salesman Problem," *Proc. Am. Math. Soc.*, Vol. 7, 48–50 (1956).

15.81 R. C. Prim, "Shortest Connection Networks and Some Generalizations," *Bell Sys. Tech. J.*, Vol. 36, 1389–1401 (1957).

15.82 A. Kerschenbaum and R. Van Slyke, "Computing Minimum Spanning Trees Efficiently," *Proc. 25th Ann. Conf. of the ACM*, 1972, pp. 518–527.

15.83 A. C. Yao, "An $O(|E|\log\log|V|)$ Algorithm for Finding Minimum Spanning Trees," *Information Processing Lett.*, Vol. 4, 21–23 (1975).

15.84 D. Cheriton and R. E. Tarjan, "Finding Minimum Spanning Trees," *SIAM J. Comput.*, Vol. 5, 724–742 (1976).

15.85 H. N. Gabow, "Two Algorithms for Generating Spanning Trees in Order," *SIAM J. Comput.*, Vol. 6, 139–150 (1977).

15.86 K. P. Eswaran and R. E. Tarjan, "Augmentation Problems," *SIAM J. Comput.*, Vol. 5, 653–665 (1976).

15.87 A. Rosenthal and A. Goldner, "Smallest Augmentations to Biconnect a Graph," *SIAM J. Comput.*, Vol. 6, 55–66 (1977).

15.88 E. Reghbati and D. G. Corneil, "Parallel Computations in Graph Theory," *SIAM J. Comput.*, Vol. 7, 230–237 (1978).

15.89 J. Edmonds, "Edge-Disjoint Branchings," in *Combinatorial Algorithms*, (R. Rustin, Ed.), Algorithmics Press, New York, 1973, pp. 91–96.

15.90 D. R. Fulkerson and G. C. Harding, "On Edge-Disjoint Branchings," *Networks*, Vol. 6, 97–104 (1976).

15.91 R. E. Tarjan, "Edge-Disjoint Spanning Trees and Depth-First Search," *Acta Informatica*, Vol. 6, 171–185 (1976).

15.92 L. Lovász, "On Two Minimax Theorems in Graph," *J. Combinatorial Theory B*, Vol. 21, 96–103 (1976).

15.93 B. L. Golden and T. L. Magnanti, "Deterministic Network Optimization: A Bibliography," *Networks*, Vol. 7, 149–183 (1977).

15.94 M. Fujii, T. Kasami, and K. Ninomiya, "Optimal Sequencing of Two Equivalent Processors," *SIAM J. Appl. Math.*, Vol. 17, 784–789 (1969); Erratum, Vol. 20, 141 (1971).

15.95 E. G. Coffman, Jr., and R. L. Graham, "Optimal Scheduling for Two-Processor Systems," *Acta Informatica*, Vol. 1, 200–213 (1972).

15.96 E. G. Coffman, Jr., and P. J. Denning, *Operating System Theory*, Prentice Hall, Englewood Cliffs, N.J., 1973.

15.97 R. Sethi, "Scheduling Graphs on Two Processors," *SIAM J. Comput.*, Vol. 5, 73–82 (1976).

15.98 R. Sethi, "Algorithms for Minimal Length Schedules," in *Computer and Job Scheduling Theory* (E. G. Coffman, Jr., Ed.), Wiley, New York, 1976, pp. 51–99.

15.99 J. D. Ullman, "Complexity of Sequencing Problems," in *Computer and Job Scheduling Theory*, (E. G. Coffman, Jr., Ed.), Wiley, New York, 1976, pp. 139–164.

15.100 S. Lam and R. Sethi, "Worst Case Analysis of Two Scheduling Algorithms," *SIAM J. Comput.*, Vol. 6, 518–536 (1977).

15.101 D. J. Rose, "A Graph-Theoretic Study of Numerical Solutions of Sparse Positive Definite Systems of Linear Equations," in *Graph Theory and Computing* (R. C. Read, Ed.), Academic Press, New York, 1972, pp. 183–217.

15.102 T. Ohtsuki, "A Fast Algorithm for Finding an Optimal Ordering for Vertex Elimination on a Graph," *SIAM J. Comput.*, Vol. 5, 133–145 (1976).

15.103 D. J. Rose, R. E. Tarjan, and G. S. Lueker, "Algorithmic Aspects of Vertex Elimination on Graphs," *SIAM J. Comput.*, Vol. 5, 266–283 (1976).

15.104 R. E. Tarjan, "Graph Theory and Gauss Elimination," in *Sparse Matrix Computations*, (J. R. Bunch and D. J. Rose, Eds.), Academic Press, New York, 1976.

15.105 E. Horowitz and S. Sahni, *Fundamentals of Computer Algorithms*, Computer Science Press, Potomac, Md.,1978.

15.106 M. R. Garey and D. S. Johnson, "Approximation Algorithms for Combinatorial Problems: An Annotated Bibliography," in *Algorithms and Complexity: New Directions and Recent Results* (J. Traub, Ed.) Academic Press, New York, 1976.

15.107 G. B. Dantzig, "All Shortest Routes in a Graph," in *Theory of Graphs*, Gordon and Breach, New York, 1967, pp. 91–92.

15.108 S. Even, *Algorithmic Combinatorics*, Macmillan, New York, 1973.

Author Index

Subject Index